The Practical Bioinformatician

editor

Limsoon Wong

Institute for Infocomm Research, Singapore

World Scientific

NEW JERSEY · LONDON · SINGAPORE · BEIJING · SHANGHAI · HONG KONG · TAIPEI · CHENNAI

THE PRACTICAL BIOINFORMATICIAN

Edited by

Limsoon Wong
Institute for Infocomm Research
Singapore

Published by

World Scientific Publishing Co. Pte. Ltd.
5 Toh Tuck Link, Singapore 596224
USA office: 27 Warren Street, Suite 401-402, Hackensack, NJ 07601
UK office: 57 Shelton Street, Covent Garden, London WC2H 9HE

British Library Cataloguing-in-Publication Data
A catalogue record for this book is available from the British Library.

THE PRACTICAL BIOINFORMATICIAN

ISBN-13 978-981-238-846-9
ISBN-10 981-238-846-X
ISBN-13 978-981-256-665-2 (pbk)
ISBN-10 981-256-665-1 (pbk)

DEDICATION

This book is dedicated to the memory of Dr. G. Christian Overton, a pioneer in the field of bioinformatics and a researcher with unparalleled vision for the use of computational approaches to solve biological problems.

Dr. Overton received his Bachelor of Science degree in mathematics and physics from the University of New Mexico in 1971 and his Ph.D. in biophysics from the Johns Hopkins University in 1978. He then did a postdoc with Dr. Davor Solter at the Wistar Institute, during which time he became increasingly drawn to computers, and in particular to the promise of artificial intelligence. He therefore returned to school part-time, and took a Master's degree in Computer and Information Science at the University of Pennsylvania. After receiving this degree in 1986, he briefly left academia to work for Burroughs (later to become Unisys) and returned to the University of Pennsylvania in 1991 to lead the informatics component of the Center for Chromosome 22, a collaboration between researchers at the University of Pennsylvania, Children's Hospital of Philadelphia, and other local institutions. During this time, he developed strong ties with researchers in computer science, statistics and biology. These collaborations eventually led in 1997 to the establishment of the Center for Bioinformatics at the University of Pennsylvania, of which Dr. Overton was the founding Director. The Center for Bioinformatics is an interdisciplinary venture between the Schools of Medicine, Arts and Sciences, and Engineering and Applied Science.

Dr. Overton was an Associate Professor in the Department of Genetics, and held a secondary appointment in the Department of Computer and Information Science. Dr. Overton's unique training in biophysics, developmental biology and computer science, and his deep understanding of biomedical and computational problems, enabled him to establish many exciting research projects, effectively bridging the gap between collaborators in numerous fields. Dr. Overton's own research interests focused on problems associated with database integration, genome annotation, gene prediction, and the recognition of regulatory elements within the sea of nucleotides that comprise the human genome. Through his use of advanced computational techniques to help in the construction of genomic databases, Dr.

Overton also gained considerable recognition in computer science. In addition to his research, Dr. Overton was an Editor for the *Journal of Computational Biology*, *Bioinformatics*, and *Gene/Gene-COMBIS*, as well as a Member of the Board of Directors for the International Society for Computational Biology.

Peter Buneman
University of Ediburgh

Susan B. Davidson
University of Pennsylvania

Limsoon Wong
Institute for Infocomm Research

26 November 2003

PREFACE

Over the past decade, computer scientists have increasingly been enlisted as "bioinformaticians" to assist molecular biologists in their research. This book is conceived as a practical introduction to bioinformatics for these computer scientists. While it is not possible to be exhaustive in coverage, the chapters are in-depth discussions by expert bioinformaticians on both general techniques and specific approaches to a range of selected bioinformatics problems. Let us provide a brief overview of these chapters here.

A practical bioinformatician must learn to speak the language of biology. We thus begin with Chapter 1 by Ng and Chapter 2 by Schönbach that form an overview of modern molecular biology and of the planning and execution of bioinformatics experiments. At the same time, a practical bioinformatician must also be conversant in a broad spectrum of topics in computer science—data mining, machine learning, mathematical modeling, sequence alignment, data integration, database management, workflow development, and so on. These diversed topics are surveyed in three separate chapters, *viz.* Chapter 3 by Li *et al.* which provides an in-depth review of data mining techniques that are amongst the key computing technologies for the analysis of biological data; Chapter 10 by Brown *et al.* which discusses the advances through the past thirty years in both global and local alignment, and present methods for general purpose homology that are widely adopted; and Chapter 17 by Wong which reviews some of the requirements and technologies relevant to data integration and warehousing.

DNA sequences contain a number of functional and regulatory sites, such as the site where the transcription of a gene begins, the site where the translation of a gene begins, the site where an exon of a gene ends and an intron begins, and so on. The next several chapters of the book deal with the computational recognition of these sites, *viz.* Chapter 4 by Li *et al.* which is an in-depth survey spanning two decades of research of methods for computational recognition of translation initiation sites from mRNA, cDNA, and DNA sequences; and Chapters 5–7 by Bajić *et al.* which discuss general frameworks, conceptual issues, and performance tuning related to the use of statistical and neural network modeling for computational recognition of promoters and regulatory sites. The recognition of translation initiation sites is among the simplest problems in the recognition of functional sites from DNA sequences. On the other hand, among the toughest problems in the recognition of functional sites from DNA sequences is the determination of locations of promoters and related regulatory elements and functional sites. Thus we hope these four chapters together can bring out clearly the whole range of approaches to this group of problems.

We next move to analysis of RNA sequences. We consider the problem of predicting the secondary structure of RNAs, which is relevant to applications such as function classification, evolution study, and pseudogene detection. Chapter 8 by Sung provides a detailed review of computational methods for predicting secondary structure from RNA sequences. Unlike the prediction methods introduced in earlier chapters for recognition of functional sites from DNA sequences, which are mostly data mining and machine learning methods, the methods described in this chapter come from the realm of mathematical modeling.

After looking at RNA sequences, we move on to protein sequences, and look at aspects relating to protein function prediction. For example, each compartment in a cell has a unique set of functions, and it is thus reasonable to assume that the compartment or membrane in which a protein resides is a determinant of its function. So we have Chapter 9 by Horton *et al.* to discuss various aspects of protein subcellular localization in the context of bioinformatics and review the twenty years of progress in predicting protein subcellular localization. As another example, the homology relationship between a protein and another protein is also suggestive of the function of that protein. So we have Chapter 12 by Kaplan *et al.* to describe two bioinformatics tools, ProtoNet and PANDORA. ProtoNet uses an approach of protein sequence hierarchical clustering to detect remote protein relatives. PANDORA uses a graph-based method to interpret complex protein groups through their annotations. As a third example, motifs are commonly used to classify protein sequences and to provide functional clues on binding sites, catalytic sites, and active sites, or structure/functions relations. So we have Chapter 16 by Schönbach and Matsuda to present a case study of a workflow for mining new motifs from the FANTOM1 mouse cDNA clone collection by a linkage-clustering method, with an all-to-all sequence comparison, followed by visual inspection, sequence, topological, and literature analysis of the motif candidates.

Next we sample the fascinating topic of phylogenetics—the study of the origin, development, and death of a taxon—based on sequence and other information. Chapter 11 by Meng is an introduction to phylogenetics using a case study on Saururaceae. In contrast to earlier chapters, which emphasize the computational aspect, this chapter is written from the perspective of a plant molecular biologist, and emphasizes instead the care that must be exercised in the use of computational tools and the analysis that must be performed on the results produced by computational tools.

The genomics and proteomics efforts have helped identify many new genes and proteins in living organisms. However, simply knowing the existence of genes and proteins does not tell us much about the biological processes in which they participate. Many major biological processes are controlled by protein interaction networks and gene regulation networks. Thus we have Chapter 13 by Tan and Ng to give an overview of the various current methods for discovering protein-protein interactions experimentally and computationally.

The development of microarray technology in the last decade has made possible the simultaneous monitoring of the expression of thousands of genes. This development offers great opportunities in advancing the diagnosis of dieases, the treatment of diseases, and the understanding of gene functions. Chapter 14 by Li and Wong is an in-depth survey of several approaches to some of the gene expression analysis challenges that accompany these opportunities. On the other hand, Chapter 15 by Lin *et al.* presents a method for selecting probes in the design of a microarray to profile genome-wide gene expression of a given genome.

Biological data is being created at ever-increasing rates as different high-throughput technologies are implemented for a wide variety of discovery platforms. It is crucial for researchers to be able to not only access this information but also to integrate it well and synthesize new holistic ideas about various topics. So it is appropriate that we devote the remaining chapters of this book to the issues of integrating databases, cleansing databases, and large-scale experimental and computational analysis workflows as follows. Chapter 18 by Kolatkar and Lin demonstrates the construction of a purpose-built integrated database PPDB using the powerful general data integration engine Kleisli. Chapter 19 by Wu and Barker presents the classification-driven rule-based approach in PIR database to the functional annotation of proteins. Wu and Barker also provide two case studies: the first looks at error propagation to secondary databases using the example of IMP Dehydrogenase; the second looks at transitive identification error using the example of His-I bifunctional proteins. Finally, Chapter 20 by Scheetz and Casavant describes the sophisticated informatics tools and workflow underlying the large-scale effort in EST-based gene discovery in Rat, Human, Mouse, and other species being conducted at the University of Iowa.

Limsoon Wong
11 December 2003

CONTENTS

CHAPTER 1

MOLECULAR BIOLOGY FOR
THE PRACTICAL BIOINFORMATICIAN

See-Kiong Ng

Institute for Infocomm Research
skng@i2r.a-star.edu.sg

Bioinformatics is a marriage of computer science with molecular biology. A practical bioinformatician must learn to speak the language of biology to enable fruitful cross-fertilization. However, the complexity of biological knowledge is daunting and the technical vocabulary that describes it is ever-expanding. In order to sift out the core information necessary for formulating a solution, it can be difficult for a non-biologically trained bioinformatician not to be lost in the labyrinths of confounding details. The purpose of this introductory chapter is therefore to provide an overview of the major foundations of modern molecular biology, so that a non-biologically trained bioinformatician can begin to appreciate the various intriguing problems and solutions described in the subsequent chapters of this book.

ORGANIZATION.

Section 1. We begin with a brief history of the major events in modern molecular biology, motivating the marriage of biology and computer science into bioinformatics.

Section 2. Then we describe the various biological parts that make up our body.

Section 3. Next we describe the various biological processes that occurs in our body.

Section 4. Finally, we describe the various biotechnological tools that have been developed by scientists for further examination of our molecular selves and our biological machineries.

1. Introduction

In the 1930s—years before the first commercial computer was created—a brilliant mathematician named Alan Turing conceived a theoretical computing machine that essentially encapsulated the essence of digital computing. Turing's sublime creation was as simple as it was elegant. A "Turing machine" consists of three key components—a long re-writable tape divided into single-digit cells each inscribed with either a 0 or a 1, a read/write head scanning the tape cells, and a table of simple instructions directing it, such as "if in State 1 and scanning a 0: print 1, move right, and go into State 3". This deceptively simple concoction of Turing has since been proven to be able to compute anything that a modern digital computer can compute.

About a decade and a half later in the 1950s, James Watson and Francis Crick deduced the structure of the DNA.[875] Their revelation also unveiled the uncanny parallel between Turing's machines and Nature's own biological machinery of life. With few exceptions, each of our cells in our body is a biochemical Turing machine. Residing in the brain of each cell is a master table of genetic instructions encoded in three billion DNA letters of A's, C's, G's, and T's written on about six feet of tightly-packed DNA, just like Turing's tape. This master table of genetic instructions—also known as the "genome"—contains all the instructions for everything our cells do from conception until death. These genetic instructions on the DNA are scanned by the ribosome molecules in the cells. Just like Turing's read/write heads, the ribosome molecules methodically decipher the encoded instructions on the tape to create the various proteins necessary to sustain life.

While the computer's alphabet contains only 1 and 0, the genetic instructions on the DNA are also encoded with a very simple set of alphabet containing only four letters—A, C, G, and T. Reading the genetic sequence on the DNA is like reading the binary code of a computer program, where all we see are seemingly random sequences of 1's and 0's. Yet when put together, these seemingly meaningless sequences encode the instructions to perform such complex feats as compute complicated arithmetic on the computer, or perform sophisticated functions of life in the cells. The DNA is indeed our book of life; the secrets of how life functions are inscribed amongst its many pages. This book of life was previously only readable by the ribosome molecules in our cells. Thanks to the tireless efforts of international scientists in the Human Genome Project, the book of life is now completely transcribed from the DNA into digitally-readable files.[467, 859] Each and everyone of us can now browse our book of life just like the ribosome molecules in the cells. Figure 1 outlines the various significant events leading up to the so-called "genome era" of mapping and sequencing the human genome.

While the book of life reads like gibberish to most of us, therein lies the many secrets of life. Unraveling these secrets requires the decryption of the 3 billion letter-long sequence of A's, C's, G's, T's. This has led to the necessary marriage of biology and computer science into a new discipline known as "bioinformatics". However, the complexity of biological knowledge is daunting and the technical vocabulary that describes it is ever-increasing. It can be difficult for a non-biology trained bioinformatician to learn to speak the language of biology. The purpose of

Year	Event
1865	Gregor Mendel discovered genes
1869	DNA was discovered
1944	Avery and McCarty demonstrated that DNA is the major carrier of genetic information
1953	James Watson and Francis Crick deduced the three-dimensional structure of DNA[875]
1960	Elucidation of the genetic code, mapping DNA to peptides (proteins)
1970's	Development of DNA sequencing techniques
1985	Development of polymerase chain reaction (PCR) for amplifying DNA[585, 586]
1980-1990	Complete sequencing of the genomes of various organisms
1989	Launch of the Human Genome Project
2001	The first working draft of the human genome was published[467, 859]
2003	The reference sequence of the human genome was completed

Fig. 1.   Time line for major events in modern molecular biology leading to the genome era.

this chapter is to give an introduction to the major foundations of modern molecular biology. We hope to provide sufficient biological information so that a practical bioinformatician can appreciate various problems and solutions described in subsequent chapters of this book and—ultimately—the chapters in the book of life.

The rest of this chapter is divided into three main sections. In Section 2: *Our Molecular Selves*, we describe various biological parts that make up our body. In Section 3: *Our Biological Machineries*, we describe various biological processes that occurs in our body. Finally, in Section 4: *Tools of the Trade*, we describe various biotechnological tools that have been developed by scientists for examination of our molecular selves and our biological machineries.

2. Our Molecular Selves

No matter how complex we are, life begins with a single cell—the "zygote"—a cell resulting from the fusion of a sperm from a male and an ovum from a female. This cell must contain all the programmatic instructions for it to develop into a complex multi-cellular organism in due course. In this section, we describe how nature organizes such genetic information at the molecular level. We also study how this complex information is permuted for diversity and passed on from one generation to another.

2.1. *Cells, DNAs, and Chromosomes*

Despite the outward diversity that we observe, all living organisms are strikingly similar at the cellular and molecular levels. We are all built from basic units called "cells". Each cell is a complex automaton capable of generating new cells which are self-sustaining and self-replicating. Figure 2 shows the structure of the cell together with its genetic components.

The "brain" of each cell is a set of DNA molecules that encode the requisite genetic information for life, usually found in the nucleus of the cell. The name DNA is an acronym for "deoxyribonucleic acid". The DNA molecules of all organisms—animals and plants alike—are chemically and physically the same. A DNA molecule is made up of four types of base molecules or "nucleotides". Each nucleotide comprises a phosphate group, a sugar (deoxyribose), and a base— either an adenine (A), a cytosine (C), a guanine (G), or a thymine (T). As such, we refer to the nucleotides by their bases, and we represent the content of a DNA molecule as a sequence of A's, C's, G's, and T's.

Each DNA in the cell is a highly compressed macromolecule. It comprises two intertwining chains of millions of nucleotides in a regular structure commonly known as the "double helix", first described by Watson and Crick in 1953.[875] The two strands of the double helix are held together by hydrogen bonds between specific pairings of the nucleotides: an A on one strand is always paired to a T on the other strand, and a G paired to a C. This is why the term "base pairs" is often also used to refer to the nucleotides on the double helix.

As a result of the specific pairings of the nucleotides, knowing the sequence of one DNA strand implies the sequence of the other. In other words, each strand contains all the information necessary to reconstruct the complementary strand. In fact, this is the basis for DNA replication in the cell as well as in the laboratories, where single strands of complementary DNA are repeatedly used as templates for DNA replication; see polymerase chain reaction or PCR in Section 4.1.2 for example. In practice, we therefore refer to the genetic recipe on the DNA as a single string containing letters from A, C, G, and T. This means that not only is the sequence of the DNA important, but the direction is also important. With a strand of DNA, one end is called the 5'-end and the other end the 3'-end. A DNA sequence—for example, 5'-GATCATTGGC-3'—is always written in a left-to-right fashion with the "upstream" or 5'-end to the left and the 3'-end to the right.

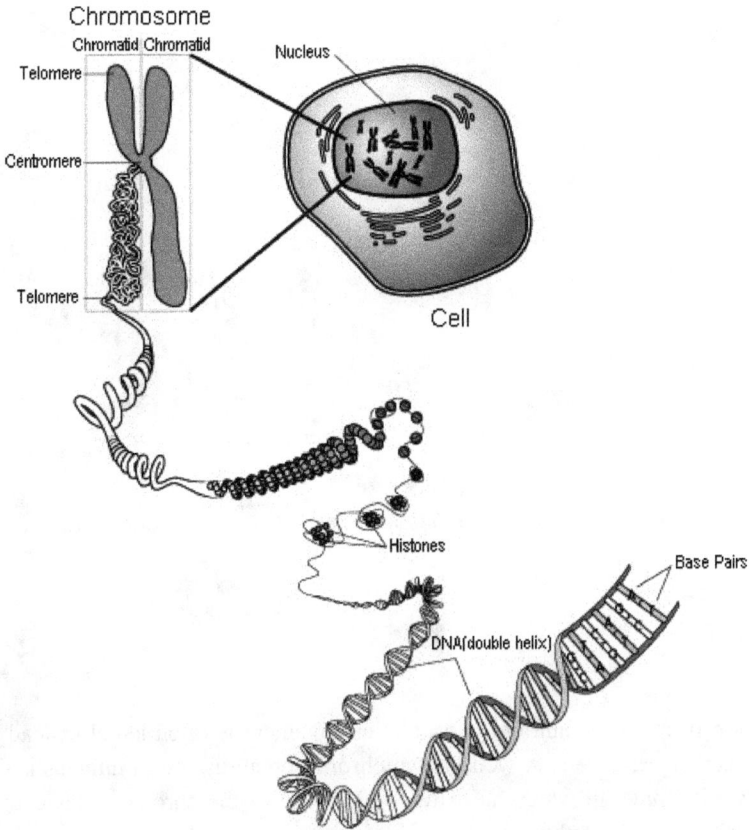

Fig. 2. The cell, chromosome, and DNA. (*Image credit: National Human Genome Research Institute.*)

This is also how it is scanned by the ribosomes in our cells; see Section 3.1 for more details. The corresponding complementary or "antisense" strand—namely, 3'-CTAGTAACCG-5' for the example DNA sequence—can then serve as the template for replicating the original DNA sequence using the nucleotide pairing rules.

Typically, the DNA in a cell is arranged in not one but several physically separate molecules called the "chromosomes". This particular arrangement forms—in effect—a distributed DNA database for the genetic information in the cells. While different species may have different number of chromosomes, the specific arrangement amongst all members in the same species is always consistent. Any

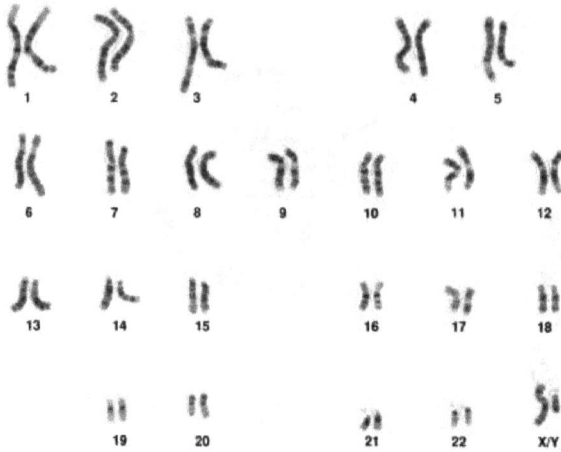

Fig. 3.   23 pairs of human chromosomes. (*Image credit: National Human Genome Research Institute.*)

aberration from the default chromosomal arrangement is often lethal or lead to serious genetic disorders. A well-known chromosomal disease in humans is the Down's Syndrome, in which an extra copy of one of the chromosomes causes mental retardation and physical deformation.

The chromosomes are usually organized in homologous (matching) pairs—each chromosome pair containing one chromosome from each parent. In humans, there are 23 pairs of homologous chromosomes ranging in length from about 50 million to 250 million base pairs. The human chromosomes are numbered from 1 to 22, with X/Y being the sex chromosomes. Figure 3 shows a photograph of how the 23 pairs of human chromosomes look like under a microscope. Collectively, the genetic information in the chromosomes are called the "genome". As each cell divides, the entire genome in the DNA is copied exactly into the new cells. This mechanism of copying the entire genetic blueprint in each and every cell is rather remarkable considering that the human body contains approximately 100 trillion cells. In theory, any of these 100 trillion cell on our body possesses the full complement of the genetic instructions for building a complex living organism like ourselves from it—if we can decipher the many secrets within the pages of our book of life in the DNA.

2.2. *Genes and Genetic Variations*

In 1865, an Augustinian monk in Moravia named Gregor Mendel published a then-obscure paper, "Versuche über Pflanzen-Hybriden" (Experiments in Plant Hybridization), describing his experiments on peas in the monastery garden. His paper was rediscovered *post mortem* 35 years later, and Mendel is now accredited as the Father of Genetics. In his legendary experiments, Mendel mated pea plants with different pairs of traits—round vs. wrinkled seeds, tall or dwarf plants, white or purple flowers, *etc.*—and observed the characteristics of the resulting offsprings. His results defied the then popular thinking which theorized that a tall plant and a short plant would have medium offspring. Based on his experiments, Mendel was able to theorize that the offspring must have received two particles— now called genes—one from each parent. One gene was dominant, the other recessive. His novel concept of genes explained why instead of medium plants, a tall and a short plant would produce some tall plants and some short plants.

Today, we know that Mendel's genes are indeed the basic physical and functional units of heredity. The human genome, for example, is estimated to comprise more than 30,000 genes. Biochemically, genes are specific sequences of bases that encode the recipes on how to make different proteins, which in turn determine the expression of physical traits such as hair color or increased susceptibility to heart diseases. We describe how this works in Section 3.

Genes are linearly ordered along each chromosome like beads on a necklace. However, there are large amounts of non-gene DNA sequences interspersed between the genes. In fact, it is estimated that genes comprise only about 2% of the human genome. The remainder 98% consists of non-coding regions, whose specific biological functions are still unknown. As mentioned earlier, genes are like recipes; cells follow these recipes to make the right proteins for each specific trait or "phenotype"—a protein for red hair cells, for example. The actual versions—or "alleles"—of the genetic recipes that each of us have may differ from one another. *E.g.*, some of us may have inherited the recipe for making red hair cells, while others may have the blonde hair cell recipe at the gene that determines the hair color phenotype. The actual versions of genetic recipes that each of us have is called the "genotype". With some exceptions of genes on the sex chromosomes, each person's genotype for a gene comprises two alleles inherited from each of the parents. Cells can follow the recipe of either allele—a dominant allele always overpowers a recessive gene to express its trait; a recessive gene remains unseen unless in the presence of another recessive gene for that trait. Many of us may actually carry a disease gene allele although we are perfectly healthy—the disease gene allele is probably a recessive one. In this way, even fatal disease genes can "survive" in

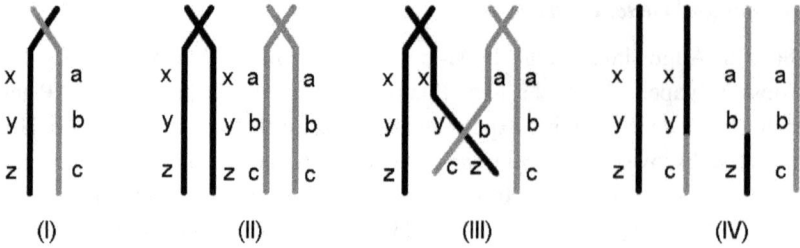

Fig. 4.   Crossing-over during meiosis.

the gene pool unchecked as it is passed on from generations to generations until it is eventually paired up with another copy of the disease allele. Furthermore, common genetic diseases as well as many other traits are often "polygenic", determined by multiple genes acting together. Together with the dominant-recessive control mechanism, many important gene alleles may remain dormant until all the key factors come together through serendipitous genetic shufflings.

A key mechanism of genetic shufflings occur during reproduction, when genetic information are transmitted from parents to their offsprings. Instead of simply passing on one of the chromosomes in the chromosome pairs in each parent to the offspring intact, the parental chromosome pairs "cross over" during a special cell division process called "meiosis". Meiosis is the cell division process for generating the sexual cells which contains only half the genetic material of the normal cells in the parents. Before the cells split into halves to form egg or sperm cells, crossing over occurs to allow interchanging of homologous DNA segments in the chromosomal pairs in each parent.

Figure 4 shows a simplified example on how such DNA recombination during meiosis can give rise to further genetic variations in the offsprings. For illustration, we focus on only three gene loci on a chromosome pair in one of the parents here—say, the biological father. Let us supposed that the father has the genotypes (x, a), (y, b), and (z, c) for the three genes, and the alleles are arranged on the two chromosomes as shown in Part I of the figure. The order of the alleles on a chromosome is also known as the "haplotypes"—the two haplotypes on the father here are (x, y, z) and (a, b, c). As we shall see shortly, meiotic cross-over events ensures diversity in the haplotypes (and hence genotypes) being passed on to the offsprings from each parent.

To prepare for DNA recombination by crossing over, the chromosome pairs double during the first phase of meiosis, resulting in two copies of each chromosome as shown in Part II of the figure. The exact copies are paired together, and the two pairs of homologous chromosomes line up side by side. Then, crossing-over takes place between the chromosomes, allowing recombination of the DNA. Part III of the figure depicts a cross-over between two of the four homologous chromosomes. As a result, four homologous chromosomes are generated, each with a different allelic set or haplotype for the three genes, as shown in Part IV. Each of these four chromosomes then goes into a different sperm cell, and an off-spring may inherit any one of these four chromosome from the father. A similar process occurs in the mother in the formation of egg cells, so that the offspring inherits a randomly juxtaposed genotype from each of the parents. This explains why children tend to look like their parents, but not exactly and certainly not an absolutely equal 50/50 mix. They usually inherit some distinct characteristics from one parent, some from the other and some from their grandparents and great grandparents. In this way, different alleles are continuously being shuffled as they are transmitted from one generation to another, thereby ensuring further haplo-typic and genotypic diversity in the population.

2.2.1. *Mutations and Genetic Diseases*

While meiotic recombination ensures that we do not always inherit the same geno-types from our parents, the types of alleles in the population's gene pool still re-main the same despite the genetic shufflings. New genetic diversity is introduced into the gene pool via such genetic mutations as edit changes in the DNA "letters" of a gene or an alteration in the chromosomes. We call a particular DNA sequence variation a "polymorphism" if it is common, occurring in more than 1% of a popu-lation. By nature of their common occurrence, DNA polymorphisms are typically neutral—that is, they are neither harmful nor beneficial and therefore do not affect the balance of the population too much. The most common type of genetic vari-ation is called a Single Nucleotide Polymorphism (SNP). SNPs are the smallest possible change in DNA, involving only a single base change in a DNA sequence. They are found throughout the human genome with a very high frequency of about 1 in 1,000 bases. This means that there could be millions of SNPs in each human genome. The abundance, stability, and relatively even distribution of SNPs and the ease with which they can be measured make them particularly useful as "genetic markers" or genomic reference points among people to track the flow of genetic information in families or population, and even for predicting an individual's ge-netic risk of developing a certain disease or predicting how an individual responds to a medicine, if a genetic marker is directly linked to a phenotype of interest.

If a particular genetic variation occurs in less than 1% of the population, we call it a "mutation" instead of a polymorphism. Mutations can arise spontaneously during normal cell functions, such as when a cell divides, or in response to environmental factors such as toxins, radiation, hormones, and even diet. On average, our DNA undergoes about 30 mutations during a human lifetime. Fortunately, as a large part of our DNA are non-coding, most mutations tend to happen there and therefore do not cause problems. In addition, nature has also provided us with a system of finely tuned enzymes that find and fix most DNA errors. However, those unrepaired mutations that change a gene's coding recipe may cause disease, and genetic mutations in the sperm or egg cells can cause diseases that pass on to the next generation.

The Human Genome Project has provided us with a readable copy of our book of life written in genetic code. The next step is to begin to understand what is written in it and what it means in terms of human health and disease. Scientists have already used genomic data to pinpoint many genes that are associated with human genetic diseases. For example, the disease genes for cystic fibrosis,[429] breast cancer[315, 899] and Parkinson disease[672] have been identified. It is no longer unrealistic to hope that in the not-too-distant future, such advances in molecular biology can transform the practice of medicine to one that is more effective, personalized, and even preventive.

3. Our Biological Machineries

Although genes get a lot of attention, it is actually the proteins that perform most life functions; see Figure 5 for some examples. The DNA only provides the blueprint for making the proteins—the actual workhorses of our biological machineries. It is through the proteins that our genes influence almost everything about us, including how tall we are, how we process foods, and how we respond to infections and medicines.

Proteins are large, complex molecules made up of long chains of smaller subunits called "amino acids". There are twenty different kinds of amino acids found in proteins. While each amino acid has different chemical properties, their basic structure is fairly similar, as shown in Figure 6. All amino acids have an amino group at one end and a carboxyl group at the other end—where each amino acid differs is in the so-called "R" group which gives an amino acid its specific properties. R can be as simple as a single hydrogen atom—as in the amino acid Glycine—or as complex side chains such as CH_3-S-$(CH_2)_2$ in Methionine.

Like the nucleotides on the DNA, the amino acids are also arranged side-by-side in a protein molecules like beads on long necklaces. A protein can contain from 50 to 5,000 amino acids hooked by peptide bonds from end-to-end. However, these amino acid "necklaces" do not remain straight and orderly in the cell—they

Protein	Function
Hemoglobin	Carry oxygen in our blood to every part of our body.
Enzymes in saliva, stomach, and small intestines	Help digest food in our body.
Muscle proteins like actin and myosin	Enable all muscular movements from blinking to running.
Ion channel proteins	Control signaling in the brain by allowing small molecules into and out of nerve cells.
Receptor proteins	Hang around on the cell surface or vicinity to help transmit signal to proteins on the inside of the cells.
Antibodies	Defend our body against foreign invaders such as bacteria and viruses.

Fig. 5.   Some interesting proteins in our body.

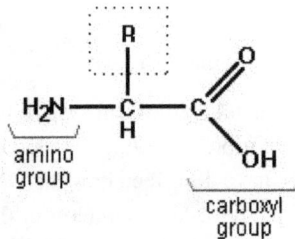

Fig. 6.   Structure of an amino acid.

twist and buckle, fold in upon themselves, form knobs of amino acids, and so on. Unlike the DNA where the linear sequence of the nucleotides pretty much determines the function—with proteins, it is mostly their three-dimensional structures that dictate how they function in the cell. Interaction between proteins is a three-dimensional affair—a protein interacts with other proteins via "lock-and-key" arrangements. Misfolding of a protein can thus lead to diseases. *E.g.*, the disease cystic fibrosis is known to be caused by the misfolding of a protein called CFTR (cystic fibrosis transmembrane conductance regulator). The misfolding—in this case, due to the deletion of a single amino acid in CFTR—disrupts the molecule's function in allowing chloride ions to pass through the outer membranes of cells. This functional disruption causes thick mucus to build up in the lungs and digestive organs, and it often results in the death of patients at an early age.

The shapes of proteins are therefore a critical determinant for the proper functioning of biological faculties in our body. Without knowing the three-dimensional structures of the proteins, we cannot fully understand or make any predictions about the phenotype of the organism. The "Protein Folding Problem" remains one of the most fundamental unsolved problems in bioinformatics. While some sections on the proteins fold into regular recognizable shapes such as spiral-shaped "alpha helices" or pleated "beta sheets", scientists are still unable to reliably predict the final 3-D structure from the amino acid sequence of a protein.

Clearly, solving the folding problem has many rewarding implications. For example, if we are able to predict how a protein folds in the cell, we can theoretically design exact drugs on a computer to, say, inhibit its function without a great deal of costly experimentation. Much concerted efforts by the bioinformatics community has been invested in solving this important problem. Over the years, the structural bioinformaticians have made significant progress with this problem, as demonstrated by the results reported in the regular community-wide experiment on the Critical Assessment of Techniques for Protein Structure Prediction (CASP).[580]

3.1. *From Gene to Protein: The Central Dogma*

A healthy body is a complex dynamic biological system that depends on the continuous interplay of thousands of proteins, acting together in just the right amounts and in just the right places. We have described how the DNA contains the genetic recipes for making proteins. We have also pointed out that the proteins are the actual workhorses that perform most life functions. Now, how does a sequence of DNA bases turn into a chain of amino acids and form a protein in the cell?

For a cell to make a protein, the information from a gene recipe is first copied—base by base—from a strand of DNA in the cell's nucleus into a strand of messenger RNA (mRNA). Chemically, the RNA—or ribonucleic acid—and the DNA are very similar. RNA molecules are also made up of four different nucleotides {A, C, G, U}—the nucleotide U (uracil) in RNA replaces the T (thymine) in DNA. Like thymine, the uracil also base-pairs to adenine. After copying the genetic recipes on the DNA in the nucleus, the mRNA molecules then travel out into the cytoplasm, and becomes accessible to cell organelles there called ribosomes. Here, each ribosome molecule reads the specific genetic code on an mRNA, and translates the genetic code into the corresponding amino acid sequence based on a genetic coding scheme; see Figure 9 in Section 3.2. With the help of transfer RNA (tRNA) molecules that transport different amino acids in the cell to the ribosome molecule as needed, the prescribed protein molecule is

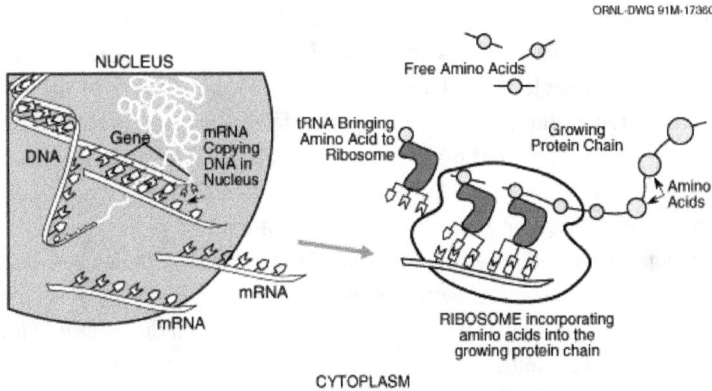

ORNL-DWG 91M-17360

Fig. 7. From gene to protein. (*Image credit: U.S. Department of Energy Human Genome Program.*)

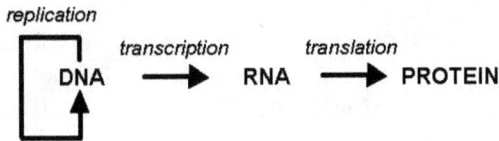

Fig. 8. The Central Dogma of Molecular Biology.

assembled—amino acid by amino acid—as instructed by the genetic recipe. Figure 7 illustrates how information stored in DNA is ultimately transferred to protein in the cell.

Figure 8 provides a schematic view of the relationship between the DNA, RNA, and protein in terms of three major processes:

(1) Replication—the process by which the information in the DNA molecule in one cell is passed on to new cells as the cell divides and the organism grows. The double-stranded complementary structure of the DNA molecule, together with the nucleotide pairing rules described earlier in Section 2.1 provide the framework for single DNA chains to serve as templates for creating complementary DNA molecules. Entire genetic blueprint can thus be passed on from cell to cell through DNA replication. In this way, virtually all the cells in our body have the full set of recipes for making all the proteins necessary to sustain life's many different functions.

(2) Transcription—the process by which the relevant information encoded in DNA is transferred into the copies of messenger RNA molecules during

the synthesis of the messenger RNA molecules. Just as in the DNA replication process, the DNA chains also serve as templates for synthesizing complementary RNA molecules. Transcription allows the amount of the corresponding proteins synthesized by the protein factories—the ribosomes in the cytoplasm—to be regulated by the rate at which the respective mRNAs are synthesized in the nucleus. Microarray experiments—see Section 4.2.2—measure gene expression with respect to the amount of corresponding mRNAs present in the cell, and indirectly infer the amount of the corresponding proteins—the gene products—present in the cell.

(3) Translation—the process by which genetic information on the mRNA is transferred into actual proteins. Protein synthesis is carried out by the ribosomes, and it involves translating the genetic code transcribed on the mRNA into a corresponding amino-acid string which can then fold into the functional protein. The genetic code for the translation process has been "cracked" by scientists in the 1960s—we describe it in more details in the next section.

This multi-step process of transferring genetic information from DNA to RNA to protein is known as the "Central Dogma of Molecular Biology"—often also considered as the backbone of molecular biology. This multi-step scheme devised by nature serves several important purposes. The installation of the transcription process protects the central "brain" of the entire system—the DNA—from the caustic chemistry of the cytoplasm where protein synthesis occurs. At the same time, the transcription scheme also allows easy amplification of gene information—many copies of an RNA can be made from one copy of DNA. The amount of proteins can then be controlled by the rate at which the respective mRNAs are synthesized in the nucleus, as mentioned earlier. In fact, the multi-step DNA-protein pathway in the dogma allows multiple opportunities for controlling in different circumstances. For example, in eukaryotic cells, after copying the genetic recipe from the DNA, the mRNA is spliced before it is processed by a ribosome. Alternative splicing sites on the mRNA allows more than one type of proteins to be synthesized from a single gene.

3.2. The Genetic Code

But how does a mere four letter alphabet in the DNA or RNA code for all possible combinations of 20 amino acids to make large number of different protein molecules? Clearly, to code for twenty different amino acids using only letters from {A, C, G, U}, a scheme of constructing multi-letter "words" from the 4-letter alphabet to represent the amino acids is necessary.

It turns out that the genetic code is indeed a triplet coding scheme—a run of three nucleotides, called a "codon", encodes one amino acid. A two-letter scheme is clearly inadequate as it can only encode $4^2 = 16$ different amino acids. Since $4^3 = 64 > 20$, a triplet coding scheme is sufficient to encode the 20 different amino acids. Figure 9 shows all possible mRNA triples and the amino acids that they specify—as revealed by industrious scientists such as Marshall Nirenberg and others who "cracked" the code in the 1960s.[177, 480, 616] The triplet scheme can theoretically code for up to 64 distinct amino acids. As there are only 20 distinct amino acids to be coded, the extraneous coding power of the triplet coding scheme is exploited by nature to encode many of the 20 amino acids with more than one mRNA triplet, thus introducing additional robustness to the coding scheme.

Since the genetic code is triplet-based, for each genetic sequence there are three possible ways that it can be read in a single direction. Each of these possibility is called a "reading frame", and a proper reading frame between a start codon and an in-frame stop codon is called an "open reading frame" or an ORF. Typically, the genetic messages work in such a way that there is one reading frame that makes sense, and two reading frames that are nonsense. We can illustrate this point using an analogy in English. Consider the following set of letters:

```
shesawthefatmanatethehotdog
```

The correct reading frame yields a sentence that makes sense:

```
she saw the fat man ate the hot dog
```

The other two alternative reading frames produce nonsensical sentences:

```
s hes awt hef atm ana tet heh otd og
sh esa wth efa tma nat eth eho tdo g
```

It is thus possible to computationally identify all the ORFs of a given genetic sequence given the genetic translation code shown in Figure 9.

3.3. *Gene Regulation and Expression*

Each of us originates from a single cell—the "zygote"—a single fertilized cell that contains the complete programmatic instructions for it to develop into a complex multi-cellular organism. We now know that these genetic instructions are encoded in the DNA found in nucleus of the cell, and we know how cells execute the genetic instructions under the Central Dogma. But how do cells in our body which is made up of trillions of cells with many different varieties, shapes, and sizes, selectively executes the genetic instructions so that a bone cell makes bone proteins and a liver cell makes liver proteins, and not vice versa?

S.-K. Ng

1st	2nd position				3rd
position	U	C	A	G	position
	Phe	Ser	Tyr	Cys	U
U	Phe	Ser	Tyr	Cys	C
	Leu	Ser	*stop*	*stop*	A
	Leu	Ser	*stop*	Trp	G
	Leu	Pro	His	Arg	U
C	Leu	Pro	His	Arg	C
	Leu	Pro	Gln	Arg	A
	Leu	Pro	Gln	Arg	G
	Ile	Thr	Asn	Ser	U
A	Ile	Thr	Asn	Ser	C
	Ile	Thr	Lys	Arg	A
	Met	Thr	Lys	Arg	G
	Val	Ala	Asp	Gly	U
G	Val	Ala	Asp	Gly	C
	Val	Ala	Glu	Gly	A
	Val	Ala	Glu	Gly	G

Fig. 9. The Genetic Code. AUG is also known as the "start codon" that signals the initiation of translation. However, translation does not always start at the first AUG in an mRNA—predicting the actual translation initiation start sites is an interesting bioinformatics problem.[658, 869, 928] The three combinations that do not specify any amino acids—UAA, UAG, and UGA—are "stop codons" that code for chain termination.

The distinct variety of cells in our body indicates that different subsets of genes must be "switched on" in different cells to selectively generate different kinds of proteins necessary for the cells' particular biological functions. In order for a single zygote to develop into a complex multi-cellular organism comprising trillions of cells of different varieties, the workings of the genetic machinery in our cells must be a highly regulated process—the types of genes switched on at any particular time must be controlled precisely, and the amount of each protein expressed by each active gene regulated according to different cellular events.

There are regulatory proteins in the cells that recognize and bind to specific sequences in the DNA such as the "promoters"—sequences upstream of a coding gene that contains the information to turn the gene on or off—to influence the transcription of a gene. Some genes are normally inactive—their transcription is blocked by the action of repressor proteins. On the occasion when the production of the protein is needed, the corresponding gene is induced by the arrival of an inducer molecule that binds to the repressor protein, rendering it inactive and re-enabling transcription of the gene. Other genes are normally active and must be constantly transcribed. For these genes, their repressor proteins are produced in the inactive form by default. When the production of a protein needs to be reduced or stopped, transcription of these genes is blocked with the arrival and binding of a suitable corepressor molecule that forms a complex with the repressor molecule to act as a functional repressor that blocks the gene's transcription. More detail of transcription is discussed in greater details in Chapter 5.

Other than controlling expression of genes at transcription initiation, regulation can also occur at the other points in the gene-to-protein pathway as governed by the Central Dogma depicted in Figure 8. Regulation can occur post-transcription in the "alternative splicing" process mentioned earlier, allowing biologically different proteins to be generated from the same gene under different circumstances. Regulation can also take place at translational initiation by modulating the activities of translational initiation factors such as the phosphorylation of initiation factors and their regulated association with other proteins. Regulation can also be implemented post-translation—modification such as glycosylation and acetylation can be employed to switch certain proteins on or off as necessary. More detail of translation is discussed in Chapter 4.

In short, the biological mechanisms that regulate our gene expression constitute a highly parallel and finely tuned system with elaborate control structure that involves extensive multi-factorial feedback and other circumstantial signals. At best, the sequencing of the genomes by the Human Genome Project and other sequencing projects can only give us the recipes for making the different proteins in our cells. But to begin to answer the who, the what, the when, the how, and the why of the various biological processes in our body, we need much more information than just the genetic recipes. In the rest of this chapter, we describe several key "tools of the trade" that biologists have acquired thus far in helping them decipher our book of life. In addition to these conventional developments in experimental tools, the genome era has also seen the creation of "bioinformatics"—a new tool of the trade from the marriage of computer science with biology. The emergence of bioinformatics has brought about a novel *in silico* dimension to modern molecular biology research. We focus on describing the conventional experimental tools here, and leave it to the other authors of this book to describe the wide array of bioinformatics applications in the subsequent chapters.

4. Tools of the Trade

To manage the many challenges in genomics research, molecular biologists have developed an interesting arsenal of ingenious technological tools to facilitate their study of the DNA. We can categorize the basic operations for manipulating the DNA for scientific study as follows:

(1) Cutting—the DNA is a fairly long molecule; thus to perform experiments on DNA in the laboratory, scientists must be able to cut the DNA into smaller strands for easy manipulation. The biologically-derived "restriction enzymes" provide the experimental apparatus for cutting DNA at specific regions.

(2) Copying—the cut DNA must be replicated or "amplified" to a sufficient quantity so that experimental signals be detected. The "polymerase chain reaction"—or PCR—is an ingenious solution devised by biologists for rapidly generating millions of copies from a DNA fragment.

(3) Separating—oftentimes, a mixture of different DNA fragments is resulted; they must be separated for individual identification, and this can be done by a process called "electrophoresis".

(4) Matching—one way of rapidly identifying a strand of DNA is matching by "hybridization". Hybridization uses the complementary nature of DNA strands due to specific pairings of the nucleotides to match identical—or rather complementary—strands of DNA. This is also the basis for DNA microarrays which allows scientists to monitor whole-genome genetic expression in the cell in parallel with quantitative DNA-DNA matching.

In the rest of this chapter, we describe the various staple biotechnological tools of the trade for the above basic operations in further details. We then conclude the chapter by describing how these basic operations are put together to enable two key applications in the genome era—the sequencing of the genome and the concurrent monitoring of gene expression by microarray.

4.1. Basic Operations

4.1.1. Cutting DNA

To study the DNA in the laboratory, we must be able to cut the long DNA at specific pre-determined points to produce defined and identifiable DNA fragments for further manipulation such as the joining of DNA fragments from different origins in recombinant DNA research. A pair of "molecular scissors" that cuts discriminately only at specific sites in their sequence is needed.

Fortunately, nature has provided us with a class of such molecular scissors called "restriction enzymes". Many bacteria make these enzymatic molecular scissors to protect themselves from foreign DNA brought into their cells by viruses. These restriction enzymes degrade foreign DNAs by cutting the area that contain specific sequences of nucleotides.

A restriction enzyme functions by scanning the length of a DNA molecule for a particular 4 to 6-nucleotide long pattern—the enzyme's so-called "recognition site". Once it encounters its specific recognition sequence, the enzyme binds to the DNA molecule and makes one cut at the designated cleavage site in each of the DNA strand of the double helix, breaking the DNA molecule into fragments. The enzyme is said to "digest" the DNA, and the process of cutting the DNA is therefore called a restriction digest or a digestion.

Since restriction enzymes are isolated from various strains of bacteria, they have such names as *Hind*II and *Eco*RI, where the first part of the name refers to the strain of bacteria which is the source of the enzyme—*e.g.*, *Haemophilus influenzae* Rd and *Escherichia coli* RY 13—and the second part of the name is a Roman numeral typically indicating the order of discovery. The first sequence-specific restriction enzyme called "*Hind*II" was isolated around 1970 by a group of researchers working with *Haemophilus influenzae* bacteria at Johns Hopkins University.[779] *Hind*II always cuts DNA molecules at a specific point within the following 6-base pair sequence pattern:

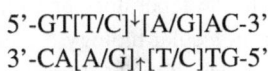

$$5'\text{-GT[T/C]}^{\downarrow}\text{[A/G]AC-3'}$$
$$3'\text{-CA[A/G]}_{\uparrow}\text{[T/C]TG-5'}$$

Restriction enzymes are useful experimental tools for molecular biologists for manipulating DNA because they allow the molecular biologists to cut DNA at specific pre-determined locations. Different restriction enzymes recognize different sequence patterns. Ever since the discovery of "*Hind*II", molecular biologists have already isolated many restriction enzymes from different strains of bacteria in their arsenal of specialized molecular scissors, enabling them to cut DNA at hundreds of distinct DNA cleavage sites in the laboratory.

4.1.2. *Copying DNA*

Another essential operation in the laboratory is the replication of DNA. For experimental analysis, the DNA must often be replicated or cloned many times to provide sufficient material for experimental manipulation. DNA can be replicated either *in vivo* or *in vitro*—for clarity, we use the term "cloning" for replicating DNA in host cells, and "amplification" for the test-tube equivalent.

Cloning is done by a cut-and-paste process to incorporate a target DNA fragment into the DNA of a "vector" such as a bacteriophage—a virus that infects bacteria—for *in vivo* replication. Scientists carry out the cut-and-paste process on a DNA molecule in the following steps. The source and vector DNA are first isolated and then restriction enzymes are used to cut the two DNAs. This creates ends in both DNAs that allows the source DNA to connect with the vector. The source DNA are then bonded to the vector using a DNA ligase enzyme that repairs the cuts and creates a single length of DNA. The DNA is then transferred into a host cell—a bacterium or another organism where the recombinant DNA is replicated and expressed. As the cells and vectors are small and it is relatively easy to grow a lot of them, copies of a specific part of a DNA or RNA sequence can be selectively produced in an unlimited amount. Genome-wide DNA libraries can be constructed in this way for screening studies.

In comparison, amplification is done by the polymerase chain reaction or PCR—an *in vitro* technique that allows one to clone a specific stretch of DNA in the test tube, without the necessity of cloning and sub-cloning in bacteria. The basic ingredients for a PCR process are: a DNA template, two primers, and DNA polymerase—an enzyme that replicates DNA in the cell. The template is the DNA to be replicated; in theory, a single DNA molecule suffice for generating millions of copies. The primers are short chains of nucleotides that correspond to the nucleotide sequences on either side of the DNA strand of interest. These flanking sequences can be constructed in the laboratory or simply purchased from traditional suppliers of reagents for molecular biology.

The PCR process was invented by Kary Mullis[585] in the mid 1980's and it has since become a standard procedure in molecular biology laboratories today. In fact, the process has been fully-automated by PCR machines, often also known as thermal cyclers. Replication of DNA by PCR is an iterative process. By iterating the replication cycle, millions of copies of the specified DNA region can be rapidly generated at an exponential rate by a combinatorial compounding process. Each PCR cycle is a 3-step procedure, as depicted in Figure 10:

(1) Denaturation—the first step of a PCR cycle is to create single-stranded DNA templates for replication. In its normal state, the DNA is a two-stranded molecule held together by hydrogen bonds down its center. Boiling a solution of DNA adds energy and breaks these bonds, making the DNA single-stranded. This is known as "denaturing"—or metaphorically, "melting"—the

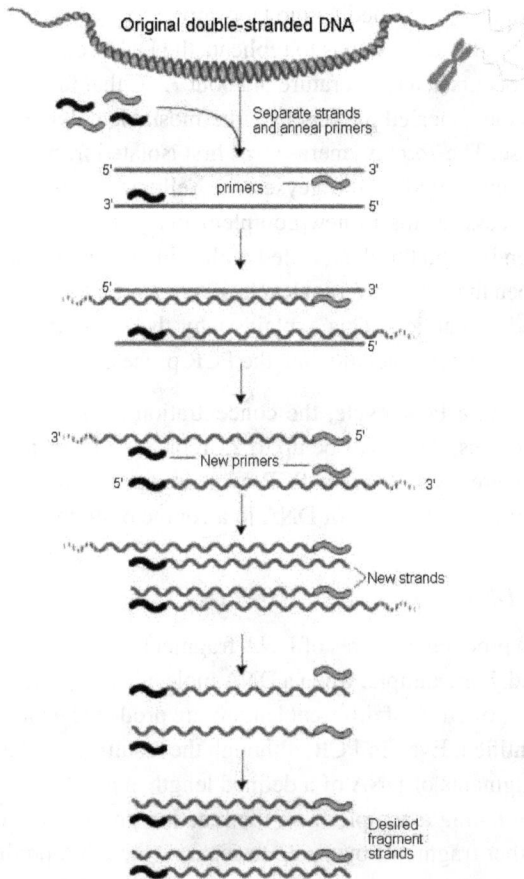

Original double-stranded DNA

Separate strands
and anneal primers

5' ——————————— 3'
primers
3' ——————————— 5'

5' 3'

5'

3' 5'

New primers

5' 3'

New strands

Desired
fragment
strands

Fig. 10. Polymerase Chain Reaction. (*Image credit: National Human Genome Research Institute.*)

DNA, and it usually takes place around 94°C. This is, however, a generally unstable state for DNA, and it will spontaneously re-form a double-helix if permitted to cool slowly.

(2) Primer annealing—the second step is to specify a region on the DNA to be replicated by creating a sort of "content-addressable pointer" to the start and the end of the target region. To do so, we label the starting point for DNA synthesis with a synthetic oligonucleotide primer that anneals—at a lower temperature, say about 30–65°C to the single-stranded DNA templates at that

point. By supplying a pair of flanking primers, only the DNA region between the marked points is amplified by the DNA polymerase in due process.

(3) DNA synthesis—the third step is to replicate the DNA region specified by the primers. This occurs at a temperature of about 72°C that facilitates elongation of DNA from the annealed primers by a thermostable polymerase such as the *Taq* polymerase. The *Taq* polymerase was first isolated from *Thermus aquaticus*—a bacterium found in the geysers in Yellowstone National Park. The elongation process results in new complementary strands on the templates, which can then be denatured, annealed with primers, and replicated again by cycling between the three different levels of temperature for DNA denaturing, annealing, and elongation. This explains why the term "thermal cyclers" is also used for machines that automate the PCR process.

By going through a PCR cycle, the concentration of the target sequence is doubled. After n cycles, there will be up to 2^n replicated DNA molecules in the resulting PCR mixture. Typically, the PCR cycle is repeated for about 30–60 cycles, generating millions of copies of DNA in a couple of hours.

4.1.3. *Separating DNA*

Experiments often produce mixtures of DNA fragments which must then be separated and identified. For example, when a DNA molecule is digested by restriction enzymes, fragments of DNA of different lengths are produced which must then be separated and identified. Even in PCR, although the resulting product is expected to contain only fragments of DNA of a defined length, a good experimentalist always verifies it by taking a sample from the reaction product and check for the presence of any other fragments by trying to separate the DNA produced.

The common way to separate macromolecules such as DNA and proteins in the laboratory is by size or their electric charge. When exposed to an electric field, a mixture of macromolecules travels through a medium—such as an agarose or acrylamide gel—at different rates depending on their physical properties. By labeling the macromolecules with a radioactive or fluorescent molecule, the separated macromolecules can then be seen as a series of bands spread from one end of the gel to the other. If we know what molecules are expected to be in the mixture, we may then deduce the identities of the molecules in individual bands from their relative size difference. We can run molecular weight standards or DNAs with known sizes on the gel together with the unknown DNA molecules if it is necessary to determine the actual sizes of these DNA molecules; see Figure 11. We can then use the calibration curve generated from the gel positions of the molecular size standards to interpolate the actual sizes of the unknown DNA molecules.

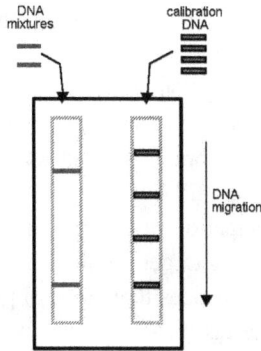

Fig. 11.   Gel Electrophoresis.

This laboratory technique of separating and identifying biomolecules is called "gel electrophoresis". Gel electrophoresis can also be used for the isolation and purification of macromolecules. Once the molecules in the mixture are separated into discrete bands on the gels, they can be individually retrieved from the gel for further processing by a "blotting" procedure that transfers the separated macro-molecules from the gel matrix onto an inert support membrane. If DNA is the type of macromolecules being separated, the process is called "Southern blotting". It is called a "Northern blot" and a "Western blot" respectively if mRNA and proteins are the macromolecules being separated.

4.1.4. *Matching DNA*

A direct way to identify DNA is by sequence similarity. This can be deter-mined experimentally using a technique called "DNA hybridization". Hybridiza-tion is based on the property that complementary nucleic acid strands bind quite specifically to each other—the underlining principle being the base-pairing of nu-cleotides, *viz.* A-T and G-C for DNA, and A-U and G-C for RNA. The more similar two single-stranded DNA molecules are in terms of their sequences, the stronger they bind or hybridize to each other.

In DNA hybridization experiments, a small nucleic acid sequence—either a DNA or RNA molecule—can be used as a "probe" to detect complementary se-quences within a mixture of different nucleic acid sequences. For example, we can chemically synthesize an oligonucleotide of a sequence complementary to the gene of interest. Just like in gel electrophoresis, we can label the probe with a dye or other marker for detection.

Typically, different DNA or RNA fragments—for example, those obtained by blotting from gel electrophoresis—are spotted and immobilized on a membrane and then allowed to react with a labeled probe. The membrane is then washed extensively to remove non-specifically bound probes, and the spots where the probe have remained bound are visualized. In this way, the sequence identity of the DNA or RNA in those spots on the membrane is revealed. It is important to note that some degree of hybridization can still occur between DNA with inexact match with the probe. For accurate detection, it is important to design the probe such that it is sufficiently specific for its target sequence so that it does not hybridize with any other DNA or RNA sequences that are also present. Fortunately, probe specificity can be achieved quite easily—even with a short probe length of say 20 nucleotides, the chance that two randomly occurring 20-nucleotide sequence are matching is very low, namely 1 in 4^{20} or approximately 1 in 10^{12}.

4.2. *Putting It Together*

We conclude this section with two applications in the genome era—the sequencing of the genome, and the concurrent monitoring of gene expression. Genome sequencing allows us to read our book of life letter by letter. Gene expression profiling allows us to take snapshots of the genetic expression of our cells in different conditions and localization to reveal how they responds to different needs.

4.2.1. *Genome Sequencing*

The Human Genome Project was an ambitious project started in 1990 that aims to read—letter by letter—the 3 billion units of human DNA in 15 years. Thanks to the concerted efforts of international scientists and the timely emergence of many innovative technologies, the project was completed ahead of its schedule— a working draft was published in 2001, followed with a more accurate reference sequence in 2003. We describe here, how the various basic tools of the trade that we have described in the previous section are put together to achieve the Herculean task of reading the formidable human genome.

As there are no technology that allows us to effectively and reliably read the A's, C's, G's, and T's from one end of the genome to the other end in a single pass, we must first cut the DNA molecule into smaller fragments. For this, we employ the restriction enzymes to cut the DNA into small manageable fragments as described earlier. Then, in a process ingeniously modified from the DNA replication process in PCR, we read out sequence on a DNA fragment letter by letter.

To do so, the four major ingredients for a PCR reaction are put together into a sequencing mixture: the source DNA template, its corresponding primers, the DNA polymerase, and sufficient quantity of free nucleotides for replication. The only difference from standard PCR is that for sequencing reactions, two different classes of the free nucleotides are added. In addition to the normal free nucleotides {A, C, G, T}, we also add a few modified ones attached with an extra fluorescent dye {A*, C*, G*, T*}. These colored nucleotides have the special property that when they are attached to a growing strand during the elongation process, it stops the new DNA strand from growing any further. Even though we have used the same "*" symbol to denote the attachment of a dye to a nucleotide, in a sequencing reaction, a different colored dye is attached to each of the four kinds of bases.

The reaction is then iterated over many cycles to replicate many complementary copies of the source DNA template just like in the PCR process. However, unlike regular PCR, the presence of the colored nucleotides introduces many occurrences of incomplete replication of the source DNA at each cycle. This results in many new DNA templates in the mixture—each starting at the same primer site but ending at various points downstream with a colored nucleotide. In due course, the completed sequencing reaction ends up containing an array of colored DNA fragments, each a one-sided complementary substring of the original sequence, differing by a base at a time. For example, a sequencing reaction for a DNA template with the source sequence ATGAGCC is going to end up with the following products in its sequencing mixture:

[primer]-T*
[primer]-T-A*
[primer]-T-A-C*
[primer]-T-A-C-T*
[primer]-T-A-C-T-C*
[primer]-T-A-C-T-C-G*
[primer]-T-A-C-T-C-G-G*

Note that the sequence of the longest colored DNA fragment generated, TACTCGG, is the complementary sequence of the source sequence.

The DNA molecules produced in the sequencing reaction must then be separated and identified using an electrophoretic process. The products of sequencing reactions are fed into an automated sequencing machine. There, a laser excites the fluorescent dyes attached to the end of each colored DNA fragment, and a camera detects the color of the light emitted by the excited dyes. In this way, as the DNA molecules passes down the gel in increasing number of nucleotides, the sequencing machine detects the last nucleotide of each fragment, and the sequence can be read out letter by letter by a computer. Figure 12 shows an example of how the differently colored bands for each DNA fragments can be read out into a sequence.

Each single sequencing reaction allows us to read out the sequence of a few

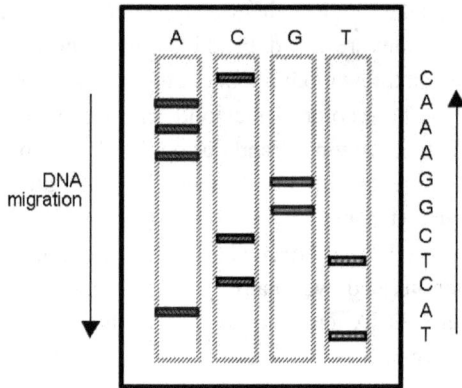

Fig. 12.   Reading out the sequence.

hundred letters of DNA at a time. This amounts to a small fraction of a page in our Book of Life. By exploiting the overlaps in the sequences, the various fragments can then stitched together by bioinformaticians into an intact whole on the computer using sequence assembly programs. By the cross-disciplinary efforts of the biologists and the bioinformaticians, scientists have managed to assemble our Book of Life ahead of the planned schedule. We can now begin to read it in its entirety, letter by letter, to uncover the secrets of how life works.

4.2.2. Gene Expression Profiling

The sequencing of genomes by the Human Genome Project and other sequencing projects can only give us the recipes for making the different proteins in our cells. Our body is a complex and dynamic biological system that depends on the interplay of thousands of proteins in the cells, acting together in just the right amounts and in just the right places. To begin to uncover the underlying networks and intricacies of the various biomolecules, we need to be able to peep into the cell to discover what genes are turned on—or expressed—at different times and places. This can be done by gene expression profiling, using the so-called microarray which is based on the laboratory technique of DNA matching by hybridization described previously.

There are two key recent technological advancements that make gene expression profiling a possibility for scientists. First, the advent of high-throughput DNA sequencing—as described in the previous section—has resulted in the availability of many genome sequences. As a result, the DNA sequences of the genes on the genome are now known to scientists for experimentation by hybridization studies. Second, the emergence of DNA microarray technology has made it possible for scientists to systematically and efficiently profile the types and amounts of mRNA produced by a cell to reveal which genes are expressed in different conditions.

A microarray is a solid support such as a glass microscope slide, a nylon membrane, or a silicon chip, onto which DNA molecules are immobilized at fixed spots. The DNA samples may be applied to the array surface by pins, nibs or ink-jet technology,[632] or they may be directly synthesized onto the array by *in situ* photolithographic synthesis of oligonucleotides.[142,518] With the help of robotics, tens of thousands of spots—each containing a large number of identical DNA molecules—can be placed on a small array of dimension, say, less than an inch. In a whole-genome scan by gene expression, for instance, each of these spots would contain a unique DNA fragment that identify a gene in the genome.

Microarrays make use of our ability to match DNA by hybridization as described in Section 4.1.4. Through the preferential binding of complementary single-stranded nucleic acid sequences, we can rapidly determine the identity of an mRNA and the degree it is expressed in the cell by hybridization probing. Let us suppose that we are interested in comparing the gene expression levels of four different genes, A, B, C, and D, in a particular cell type in two different states, namely: a healthy and diseased state. First, a microarray slide is prepared with unique DNA fragments from genes A, B, C, and D immobilized onto known locations on the array as probing "targets". Then, the mRNAs from both the healthy and diseased cells are extracted. The mRNAs are then used as templates to create corresponding cDNAs that are labeled with different fluorescent tags to mark their cellular origins—for example, a green dye for the healthy cells and a red dye for the diseased cells.

The two color-labeled samples from the healthy and diseased cells are then mixed and incubated with the microarray. The labeled molecules—or "probes"—then bind to the spots on the array corresponding to the genes expressed in each cell. The excess samples are washed away after the hybridization reaction. Note that in the terminology of Affymetrix, a "probe" refers to the unlabeled oligonucleotides synthesized on their microarray—known as GeneChip—rather than the solution of labeled DNA to be hybridized with a microarray as described here.

The microarray is then placed in a "scanner" to read out the array. Just as in the case of sequencing machines, the fluorescent tags on the molecules are excited by laser. A camera, together with a microscope, captures the array with a digital image which can then be analyzed by image quantitation software. The software identifies each spot on the array, measures its intensity and compares to the background. For each target gene, if the RNA from the healthy sample is in abundance, the spot is green. If the RNA from the diseased sample is in relative abundance, it is red. If there are no differential expression between the two cellular states, the spot is yellow. And if the gene is never expressed in the cell regardless of whether it is healthy or not, then its corresponding spot does fluoresce and it appears black. In this way, we estimate the relative expression levels of the genes in both samples using the fluorescence intensities and colors for each spot.

Microarray technology is a significant advance because it enables scientists to progress from studying the expression of one gene in several days to hundreds of thousands of gene expressions in a single day, or even just several hours. To cope with the huge amount of data generated by microarray studies, increasingly sophisticated bioinformatics methods and tools are required to derive new knowledge from the expression data. We discuss the analysis of microarray data with bioinformatics methods in Chapter 14. In general, it can be said that the new tools and technologies for biological research in the genome era and post-genome era have led to huge amounts of data being generated at a rate faster than we can interpret it. Thus far, informatics has played a critical role in the representation, organization, distribution and maintenance of these data in the digital form, thereby helping to make the unprecedented growth of biological data generation a manageable process. The new challenge, however, is for informatics to evolve into an integral element in the science of biology—helping to discover useful knowledge from the data generated so that ultimately, the immense potential of such ingenious inventions as PCR and microarrays can be unleashed for understanding, treating, and preventing diseases that affect the quality of human life each day.

5. Conclusion

As bioinformatics becomes an indispensable element for modern genomics, a practical bioinformatician must learn to speak the language of biology and be appreciative of how the data that we analyze are derived. This introductory chapter provides a brief overview of the major foundations of modern molecular biology and its various tools and applications. Its objective is to prime a non-biologically trained bioinformatician with sufficient basic knowledge to embark—together with the biologists—on the exciting and rewarding quest to uncover the many intriguing secrets in our book of life.

As practical bioinformaticians, we must keep in mind that biology—just like information technology—is now a rapidly evolving field. Our self-education in biology should be a firm commitment of a continuous process—it should not stop at this chapter. The future discovery of noteworthy methodologies and pivotal findings in molecular biology definitely requires the involved participation of the bioinformaticians together with the biologists—ultimately, it is those who are truly trained in the cross-disciplinary spirit who are able to make the most remarkable contributions.

CHAPTER 2

STRATEGY AND PLANNING OF BIOINFORMATICS EXPERIMENTS

Christian Schönbach

RIKEN Genomic Sciences Center
schoen@gsc.riken.jp

Genomics has revolutionized biology by the generation of huge data sets. However, it has also created a data analysis and interpretation bottleneck. Fancy programming techniques and newest technologies are irrelevant to dealing with this bottleneck if we do not have a proper understanding of the specific questions that a biologist is seeking answers to from these data sets.

This chapter briefly discusses the planning and execution of bioinformatics experiments.

ORGANIZATION.

Section 1. We briefly explain that a novice Bioinformatician needs to learn to identify problems of interest, and then develop a solution strategy before going into details.

Section 2. Then we provide a short example to illustrate the reasoning and overall strategy for identifying alternative splice variants from cDNA libraries for the construction of a variant expression array.

Section 3. Finally, we touch on the steps that are often necessary in the planning and execution of such a bioinformatics experiment.

1. Multi-Dimensional Bioinformatics

The discovery of the DNA-double helix 50 years ago[875] triggered a new type of biology that enabled biologists to link biological phenomena to the function of genes. The main strategies are cloning, sequencing, and the analysis of specific functions on sequence, transcript, protein, cellular, and organism level. The roadmap from gene to function has changed little except the pace. High-throughput methods facilitate the sequencing and expression of whole genomes and systematic testing of protein interactions within weeks, instead of the years that these projects used to take. While Genomics has revolutionized biology by the generation of huge data sets, it has also created a data analysis and interpretation bottleneck.

To understand the underlying biology, huge amounts of heterogeneous data on different functional levels must be acquired from their sources, integrated, queried, analyzed, and modeled. The types of data include sequence data, gene expression data, luciferase assay data for protein-protein interaction data, phenotype data from knock-out mice, and so on. The different functional levels include transcription, translation, protein-protein interactions, signaling, and so on. The "omes" in genomics, transcriptomics, proteomics, metabolomics, and physiomics have generated multi-dimensional, time-dependent, and noisy data with sometimes analog non-linear interactions. To complicate matters further, the causes and effects of individual components such as transcription, translation, and protein interactions are often unknown. Therefore, bioinformatics needs to keep up with the data growth on multiple levels of biological hierarchy and the increasing overlap of biological and biomedical subdisciplines to obtain a more holistic view of biological phenomena.

In the past five or six years, bioinformatics methods of individual components—such as sequence, microarray, and structure analyses—have improved and become scalable. However, the combination of various data types and cross component analysis towards exploring the path from a gene to a disease are still in their infancy. Bioinformatics books that explain the usage of analysis tools are not the only solution for complex problem solving. A novice Bioinformatician also needs to learn to identify problems of interest in the context above, and then develop a solution strategy before going into the details. The difficulties in successfully identifying, defining, and drafting a solution to bioinformatics problems have also an environment and education component that requires attention.

Bioinformaticans should not have the illusion to efficiently solve all questions that biologists or medical researchers ask. On the other hand, the experimental scientists should not expect Bioinformaticans to be service providers and technical problem solvers. I believe that this traditional view of bioinformatics has done a lot of damage a decade ago and has not been completely eradicated. Part of the problem arose from policy makers that did not perceive bioinformatics early enough as an independent subject that merges biology, computer science, mathematics, and physics. Therefore, Bioinformaticians that can apply mathematics, physics, and computer science to biological problems are still in the minority.

2. Reasoning and Strategy

In this book, we introduce a series of biologically relevant case studies and techniques as a hands-on reference for planning and executing bioinformatics experiments. Here, I provide a brief example to illustrate the reasoning and overall strategy for identifying alternative splice variants from cDNA libraries to construct a variant expression array and protein-protein interaction experiments.

Recent EST-based analysis by Modrek *et al.*[573] and Kan *et al.*[408] estimates that 42–55% percent of the human genes are alternatively spliced. The importance of detecting alternative splice is underlined by the frequently found single base substitutions in human mRNA splice junctions and association with genetic diseases. If we want to make a conservative minimum estimate of alternative splicing of human transcripts we need to map them to genomic sequences. Sequences that cluster and show variations with a canonical splice site motif are candidates for alternative splicing.

Scoring or threshold settings can facilitate the decision on statistically significant results that are biologically relevant. Nevertheless statistically significant results are sometimes biological irrelevant. For example, the unspliced intron of a cDNA clone will produce statistically significant results to genomics DNA sequence. The result is not of biological interest unless the cDNA sequence is an alternative splice candidate or if we want to estimate which introns are the last to be spliced out. If the sequence is an alternative splice variant, it will have statistically significant hits to ESTs and genomic sequences. To increase confidence, the ESTs should be derived from various independent library sources.

When preparing a large-scale bioinformatics experiment to detect potential transcriptional variants for designing a variant microarray or protein-protein interaction screen, it is crucial to keep the number of false positives as low as possible. Error sources to be eliminated are unmapped pieces of genomic sequences and splice candidates that are hit by only one EST.

After minimizing error sources, we need to prioritize the targets by intellectual property status and relevance to disease geno- and pheno-types. A search of the transcripts against a patent database will yield potential new candidates for patenting. Potential disease genotypes can be established by using disease genes of the OMIM database.[318] Potential disease phenotype candidates, such as organ specificity, can be extracted from the tissue and/or organ sources of matched ESTs and expression information of the source clones. Disease MeSH (Medical Subject Headings) of MEDLINE abstracts extracted with the knowledge discovery support system FACTS[592] are applicable for gross classification of mRNA transcript inferred human disease associations and decision support of biomedical experts to assist the targeting process in validating potential human disease genes

3. Planning

The example in Section 2 contains several implicit steps that are often not described in publications but are applicable to any bioinformatics experiment:

- Identify the exact nature of the problem by interviewing domain experts
- Understand the data sources
- Determine
    - problem solving strategy
    - data sources
    - methodology
    - cost and time constraints
- Minimize errors and their propagation
- Evaluate overall strategy with domain experts
- Amend strategy and methodology if necessary
- Determine how to prioritize results
- Perform a test on a characterized data set
- Evaluate test results together with a domain expert
- Amend experimental set-up if required
- Start real experiment

The understanding of data sources is crucial in deriving an appropriate strategy, and must precede the algorithm selection step. Therefore, a significant amount of time should be invested in understanding and questioning the data to identifying potential problems in the data. Understanding the data requires biological knowledge that a bioinformatican should ideally know or obtain by interviewing a domain expert. Fancy programming techniques and newest technologies are irrelevant for the problem identification and data understanding steps of a bioinformatics experiment. The opposite approach of choosing the methodology and algorithm before understanding the data may induce, in the worst case, data fitting. Another common but problematic attitude caused by casual data preparation is to ignore data that do not fit the expected outcome. If the general guidelines and the biological context are taken into account, bioinformatics experiments will produce more biologically relevant results that are of interest for the biologist and medical scientist.

CHAPTER 3

DATA MINING TECHNIQUES FOR
THE PRACTICAL BIOINFORMATICIAN

Jinyan Li

Institute for Infocomm Research
jinyan@i2r.a-star.edu.sg

Huiqing Liu

Institute for Infocomm Research
huiqing@i2r.a-star.edu.sg

Anthony Tung

National University of Singapore
atung@comp.nus.edu.sg

Limsoon Wong

Institute for Infocomm Research
limsoon@i2r.a-star.edu.sg

The use of computing to organize, manage, and analyse biological and clinical data has become an important element of biology and medical research. In the 1990s, large quantities of DNA sequence data were generated by the Human Genome Project and other genome sequencing efforts. Computing projects in algorithms, software, and databases were crucial in the automated assembly and analysis of the genomic data. In the 2000s, we enter the so-called post-genome era. Many genomes have already been completely sequenced. Genome research migrates from raw data generation into scientific knowledge discovery. Computing shifts correspondingly from managing and integrating sequence databases to discovering knowledge from such biological data.

Data mining tools and techniques are amongst the key computing technologies that are playing an important role in the ongoing post-genome extension of human's understanding of biological life. This chapter is an in-depth review of data mining techniques, including feature selection methods, classification methods, clustering methods, and association rules discovery methods. We also briefly mention examples of their use in biomedicine.

ORGANIZATION.

Section 1.   Feature selection is concerned with the issue of distinguishing signal from noise
in data analysis. We begin with an exposition on the curse of dimensionality to motivate the importance of feature selection. Then we present a number of feature selection
methods, including signal-to-noise, t-test, Fisher criterion, entropy, χ^2, information
gain, information gain ratio, Wilcoxon rank sum test, CFS, and PCA.

Section 2.   Classification is the process of finding models, based on selected features from
training data, that separate two or more data classes. We review in this section a number
of classification methods, including decision tree induction, Bayesian inference, hidden Markov models, artificial neural networks, support vector machines, and emerging
patterns.

Section 3.   The problem of association rule mining from a dataset is that of how to generate
association rules that are interesting. We describe the two best-known methods for
mining association rules, *viz.* the Apriori algorithm and the Max-Miner algorithm.

Section 4.   Clustering is the process of grouping a set of objects into clusters so that objects within a cluster have high similarity to each other and high dissimilarity with
objects in other clusters. We first discuss factors that can affect clustering. Then we
describe clustering methods based on the partitioning concept, *viz.* k-means, EM, and
k-mediods. Following that, we describe clustering methods based on the hierarchical
decomposition concept, *viz.* AGNES and DIANA.

1. Feature Selection Methods

Feature selection is concerned with the issue of distinguishing signal from noise
in data analysis. This section begins with subsection on the curse of dimensionality to motivate the importance of feature selection. It is then followed by several
subsections, each presenting a different feature selection method.

1.1. *Curse of Dimensionality*

Let us assume that we would like to analyse data points in a d-dimensional unit
hypercube. How much of each dimension of the unit hypercube is needed if we
would like to sample a proportion p of the total volume of possible data points in
this space? To calculate this, let the amount that we would need of each dimension be x. Then $p = x^d$. Thus $x = \sqrt[d]{p}$. Therefore, there is exponential growth in
the proportion of each dimension that is needed to sample a fixed proportion of a
space, with respect to the number of dimension that the space has. This observation is called the curse of dimensionality.

To appreciate the significance of this curse, let us look at Figure 1. In Part I
of this figure, we see that if we utilize 50% of each of the two dimensions, we

(I) 50% of each dimension is sufficient to cover 25% of a 2-dimensional space

(II) 50% of each dimension is only sufficient to cover 12.5% of a 3-dimensional space

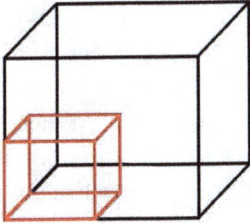

(III) A proportion $p^{1/d}$ of each dimension is needed to cover a proportion p of a d-dimensional space. The graph below plots $p^{1/d}$ vs. d for $p = 1\%$ and $p = 10\%$.

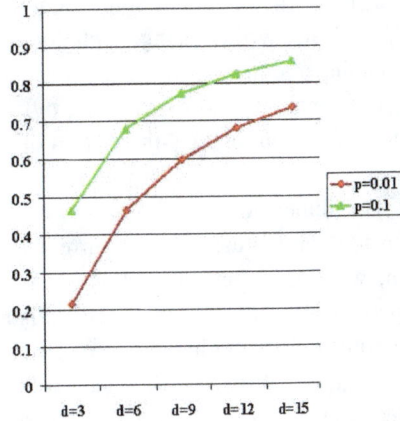

Fig. 1.   Illustrations of the curse of dimensionality.

can cover 25% of the 2-dimensional space. In Part II of this figure, we see that if we utilize 50% of each of the three dimensions, we can cover only 12.5% of the 3-dimensional space. Conversely, we can also see from the graph in Part III of this figure that we would need 46% of each dimension to cover 10% of the volume of a 3-dimensional space, and nearly 86% of each dimension to cover 10% of the volume of a 15-dimensional space.

This curse of dimensionality is an important source of difficulty in effective data mining analysis. In order for a data mining tool to properly analyse a space and to reliably make predictions in that space, the tool needs to sample a certain proportion of the data points in that space. As a consequence of the curse, the tool needs an exponential increase in the number of training samples with respect to an increase in the dimensionality of the samples in order to uncover and learn the relationship of the various dimensions to the nature of the samples.[328]

It is therefore an important issue to determine if a dimension is relevant to the kind of data mining analysis that we would like to perform on a space. If we could determine beforehand that certain dimensions are irrelevant, then we can omit them in our data mining analysis and thus mitigate the curse of dimensionality.

In the data mining tradition, the term "feature" and the term "attribute" are often preferred over the term "dimension." So we use the three terms interchangeably here. Also, a feature f and its value v in a sample are some times referred to as an "item" by the data mining community. A set of items is then called an "itemset," typically written using a notation like $\{f_1 = v_1, ..., f_n = v_n\}$ for an itemset containing features $f_1, ..., f_n$ and associated values $v_1, ..., v_n$. An itemset can also be represented as a vector $\langle v_1, ..., v_n \rangle$, with the understanding that the value of feature f_i is kept in the i-th position of the vector. Such a vector is usually called a feature vector.

The dimensions or features that are relevant are called signals. The dimensions or features that are irrelevant are called noise. In rest of this section, we present several techniques for distinguishing signals from noise, *viz.* signal-to-noise measure,[297] t-test statistical measure,[133] entropy measure,[242] χ^2 measure,[514] information gain measure,[692] information gain ratio,[693] Fisher criterion score,[251] Wilcoxon rank sum test,[742] as well as a correlation-based feature selection method known as CFS[316] and the "unsupervised" approach of principal component analysis.[399]

1.2. *Signal-to-Noise Measure*

Suppose we have two classes $\mathcal{A}$ and $\mathcal{B}$ of sample data points in the space that we would like to analyse. Then a feature is relevant if it contributes to separating samples in $\mathcal{A}$ from those in $\mathcal{B}$. Conversely, a feature is irrelevant if it does not contribute much to separating samples in $\mathcal{A}$ from those in $\mathcal{B}$.

Let us look at Figure 2 which shows the distributions of the values of 3 features in samples of $\mathcal{A}$ and $\mathcal{B}$ as bars in the three charts. Let us compare Chart I and Chart III. The feature represented by Chart I is not as useful as the feature represented by Chart III for distinguishing $\mathcal{A}$ and $\mathcal{B}$. To see this, we appeal to the following intuitive concept for distinguishing a relevant feature from an irrelevant one: If the values of a feature in samples in $\mathcal{A}$ are significantly different from the values of the same feature in samples in $\mathcal{B}$, then the feature is likely to be more relevant than a feature that has similar values in $\mathcal{A}$ and $\mathcal{B}$. More specifically, in order for a feature f to be relevant, its mean value $\mu_f^{\mathcal{A}}$ in $\mathcal{A}$ should be significantly different from its mean value $\mu_f^{\mathcal{B}}$ in $\mathcal{B}$.

Similarly, the feature represented by Chart II of Figure 2 is not as useful as the feature represented by Chart III for distinguishing $\mathcal{A}$ and $\mathcal{B}$. To see this, we

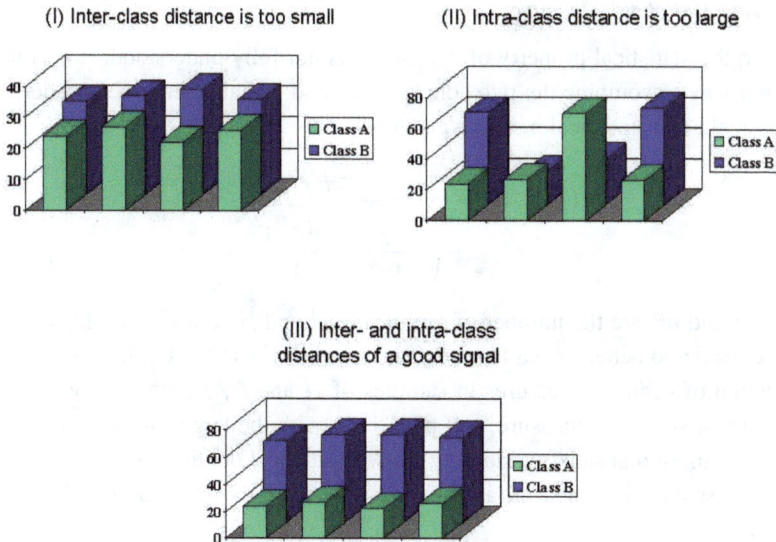

Fig. 2. Characteristics of a feature that is a poor signal (I and II), and of a feature that is a good signal (III).

appeal to a second intuitive concept that if the values of a feature f varies greatly within the same class of samples, then the feature f cannot be a reliable one. More specifically, in order for a feature f to be relevant, the standard deviation σ_f^A and variance $(\sigma_f^A)^2$ of f in $\mathcal{A}$ and the standard deviation σ_f^B and variance $(\sigma_f^B)^2$ of f in $\mathcal{B}$ should be small.

One obvious way to combine these two concepts is the signal-to-noise measure proposed in the first paper[297] that applied gene expression profiling for disease diagnosis:

$$s(f, \mathcal{A}, \mathcal{B}) = \frac{|\mu_f^A - \mu_f^B|}{\sigma_f^A + \sigma_f^B}$$

Then a feature f can be considered better than a feature f' if $s(f, \mathcal{A}, \mathcal{B}) > s(f', \mathcal{A}, \mathcal{B})$. Thus given a collection of candidate features in samples of $\mathcal{A}$ and $\mathcal{B}$, we simply sort them by their signal-to-noise measure and pick those with the largest signal-to-noise measure.

1.3. *T-Test Statistical Measure*

However, the statistical property of $s(f, \mathcal{A}, \mathcal{B})$ is not fully understood. A second and older way to combine the two concepts of good signals mentioned earlier is the classical t-test statistical measure given below:

$$t(f, \mathcal{A}, \mathcal{B}) = \frac{|\mu_f^A - \mu_f^B|}{\sqrt{\dfrac{(\sigma_f^A)^2}{n^A} + \dfrac{(\sigma_f^B)^2}{n^B}}}$$

where n^A and n^B are the number of samples in $\mathcal{A}$ and $\mathcal{B}$ seen. Then a feature f can be considered better than a feature f' if $t(f, \mathcal{A}, \mathcal{B}) > t(f', \mathcal{A}, \mathcal{B})$. Thus given a collection of candidate features in samples of $\mathcal{A}$ and $\mathcal{B}$, we simply sort them by their t-test statistical measure, and pick those with the largest t-test statistical measure, or those that satisfy a suitably chosen statistical confidence level.

The t-test statistical measure is known[58, 133] to require a Student distribution with

$$\frac{\left(\dfrac{(\sigma_f^A)^2}{n^A} + \dfrac{(\sigma_f^B)^2}{n^B}\right)^2}{\dfrac{\left(\dfrac{(\sigma_f^A)^2}{n^A}\right)^2}{n^A - 1} + \dfrac{\left(\dfrac{(\sigma_f^B)^2}{n^B}\right)^2}{n^B - 1}}$$

degrees of freedom. Hence if the values of the feature f is known not to obey this distribution, the t-test statistical measure should not be used for this feature.

1.4. *Fisher Criterion Score*

Closely related to the t-test statistical measure is the Fisher criterion score[251] defined as

$$fisher(f, \mathcal{A}, \mathcal{B}) = \frac{(\mu_f^A - \mu_f^B)^2}{(\sigma_f^A)^2 + (\sigma_f^B)^2}$$

A feature f can be considered better than a feature f' if $fisher(f, \mathcal{A}, \mathcal{B}) > fisher(f', \mathcal{A}, \mathcal{B})$. The Fisher criterion score is known to require a nearly normal distribution,[251] and hence it should not be used if the distribution of the values of a feature is not nearly normal.

Each of $s(f, \mathcal{A}, \mathcal{B})$, $t(f, \mathcal{A}, \mathcal{B})$, and $fisher(f, \mathcal{A}, \mathcal{B})$ iis easy to compute and thus straightforward to use. However, there are three considerations that may make them ineffective. The first consideration is that they require the values of f to follow specific statistical distributions.[133, 251] It is often the case that we do not know what sort of statistical distributions the candidate features have. The second consideration is that the population sizes $n^{\mathcal{A}}$ and $n^{\mathcal{B}}$ should be sufficiently large. Small population sizes can lead to significant underestimates of the standard deviations and variances.

The third consideration is more subtle and we explain using the following example. Let f_1 and f_2 be two features. Suppose f_1 has values ranging from 0 to 99 in class $\mathcal{A}$ with $\mu_{f_1}^{\mathcal{A}} = 75$ and has values ranging from 100 to 199 in class $\mathcal{B}$ with $\mu_{f_1}^{\mathcal{B}} = 125$. Suppose f_2 has values ranging from 25 to 125 in class $\mathcal{A}$ with $\mu_{f_2}^{\mathcal{A}} = 50$ and has values ranging from 100 to 175 in class $\mathcal{B}$ with $\mu_{f_2}^{\mathcal{B}} = 150$. We see that $\mu_{f_2}^{\mathcal{B}} - \mu_{f_2}^{\mathcal{A}} = 100 > 50 = \mu_{f_1}^{\mathcal{B}} - \mu_{f_1}^{\mathcal{A}}$. Suppose the variances of f_1 and f_2 in $\mathcal{A}$ and $\mathcal{B}$ are comparable. Then according to the signal-to-noise and t-statistics measures, f_2 is better than f_1. However, we note that the values of f_1 are distributed so that all those in $\mathcal{A}$ are below 100 and all those in $\mathcal{B}$ are at least 100. In contract, the values of f_2 in $\mathcal{A}$ and $\mathcal{B}$ overlap in the range 100 to 125. Then clearly f_1 should be preferred. The effect is caused by the fact that $t(f, \mathcal{A}, \mathcal{B})$, $s(f, \mathcal{A}, \mathcal{B})$, $fisher(f, \mathcal{A}, \mathcal{B})$ are sensitive to all changes in the values of f, including those changes that may not be important.

So, one can consider alternative statistical measures that are less sensitive to certain types of unimportant changes in the value of f. What types of changes in values of f are not important? One obvious type is those that do not shift the values of f from the range of $\mathcal{A}$ into the range of $\mathcal{B}$. An alternative that takes this into consideration is the entropy measure.[242]

1.5. *Entropy Measure*

Let $P(f, \mathcal{C}, S)$ be the proportion of samples whose feature f has value in the range S and are in class $\mathcal{C}$. The class entropy of a range S with respect to feature f and a collection of classes $\mathcal{U}$ is defined as

$$Ent(f, \mathcal{U}, S) = -\sum_{\mathcal{C} \in \mathcal{U}} P(f, \mathcal{C}, S) \times \log_2(P(f, \mathcal{C}, S))$$

It follows from this definition that the purer the range S is, so that samples in that range are more dominated by one particular class, the closer $Ent(f, \mathcal{U}, S)$ is to the ideal entropy value of 0.

Let T partitions the values of f into two ranges S_1 (of values less than T) and S_2 (of values at least T). We sometimes refer to T as the cutting point of the

(I) A feature with high entropy

(II) A feature with low entropy

(III) A feature with zero entropy

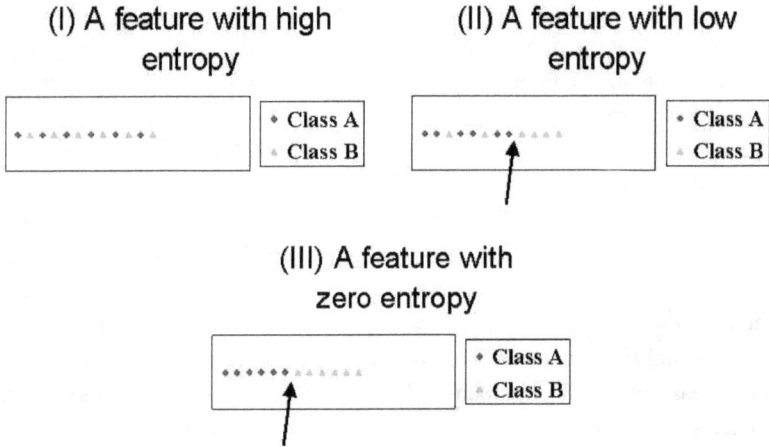

Fig. 3.   Each of the three charts represents a feature. We placed the values that a feature takes on in samples of $\mathcal{A}$ and $\mathcal{B}$ on the horizontal, sorted according to magnitude. Chart I is characteristic of a feature that is a poor signal. Chart II is characteristic of a potentially good signal. Chart III is characteristic of the strongest signal.

values of f. The entropy measure[242] $e(f,\mathcal{U})$ of a feature f is then defined as

$$e(f,\mathcal{U}) = \min\{E(f,\mathcal{U},\{S_1,S_2\}) \mid (S_1,S_2) \text{ is a partitioning of the}$$
$$\text{values of } f \text{ in } \bigcup \mathcal{U} \text{ by some point } T\}$$

Here, $E(f,\mathcal{U},\{S_1,S_2\})$ is the class information entropy of the partition (S_1,S_2). The definition is given below, where $n(f,\mathcal{U},s)$ means the number of samples in the classes in $\mathcal{U}$ whose feature f has value in the range s,

$$E(f,\mathcal{U},S) = \sum_{s \in S} \frac{n(f,\mathcal{U},s)}{n(f,\mathcal{U},\bigcup S)} \times Ent(f,\mathcal{U},s)$$

The entropy measure $e(f,\mathcal{U})$ is illustrated in Figure 3. Chart I of this figure is characteristic of a feature that is a poor signal, because it is not possible to partition the range of the values into an interval containing mostly samples of $\mathcal{A}$ and an interval containing mostly samples of $\mathcal{B}$. In general, the entropy measure for such a feature is a large value. Chart II is characteristic of a potentially good signal, because by partitioning the range of the values at the point indicated by the arrow, one of the interval contains purely samples in $\mathcal{B}$ while the other interval is dominated by samples in $\mathcal{A}$. In general, the entropy measure for such a feature is a small value. Chart III is characteristic of the strongest signal, because by partitioning the range of the values at the point indicated by the arrow, both intervals become pure. The entropy measure for this feature is 0. So under the entropy measure a feature f is more useful than a feature f' if $e(f,\mathcal{U}) < e(f',\mathcal{U})$.

A refinement of the entropy measure is to recursively partition the ranges S_1 and S_2 until some stopping criteria is reached.[242] A stopping criteria is needed because otherwise we can always achieve perfect entropy by partitioning the range into many small intervals, each containing exactly one sample. A commonly used stopping criteria is the so-called minimal description length principle.[242, 888] According to this principle, recursive partitioning within a range S stops iff S is partitioned into ranges S_1 and S_2 such that $n(f,\mathcal{U},S) \times Gain(f,\mathcal{U},S_1,S_2) < \log_2(n(f,\mathcal{U},S)-1)+\log_2(3^k-2)-k\times Ent(f,\mathcal{U},S)+k_1\times Ent(f,\mathcal{U},S_1)+k_2\times Ent(f,\mathcal{U},S_2)$, where $Gain(f,\mathcal{U},S_1,S_2) = Ent(f,\mathcal{U},S) - E(f,\mathcal{U},\{S_1,S_2\})$; and k, k_1, and k_2 are respectively the number of classes that have samples with feature f having values in the range S, S_1, and S_2.

Another refinement is the χ^2 measure.[514] Here, instead of the entropy measure $e(f,\{\mathcal{A},\mathcal{B}\})$ itself, we use the χ^2 correlation of the partitions S_1 and S_2 induced by the entropy measure to the classes $\mathcal{A}$ and $\mathcal{B}$. Some other refinements include the information gain measure and the information gain ratio that are used respectively in ID3[692] and C4.5[693] to induce decision trees. These refinements are described next.

1.6. χ^2 Measure

The χ^2 measure is a feature selection technique that evaluates features individually by measuring the chi-squared statistics with respect to the class. For a numeric feature, we should first "discretize" the range of its values into a number of intervals. For this purpose, it is common to simply use those intervals induced by the computation of the entropy measure presented in Subsection 1.5.

Then the χ^2 measure is given by the follow equation:

$$\chi^2(f,\mathcal{U}) = \sum_{i=1}^{m}\sum_{j=1}^{k} \frac{(A_{ij}(f,\mathcal{U}) - E_{ij}(f,\mathcal{U}))^2}{E_{ij}(f,\mathcal{U})}$$

Here, m is the number of intervals induced in computing $e(f,\mathcal{U})$ or its recursive refinement; k is the number of classes in $\mathcal{U}$; $A_{ij}(f,\mathcal{U})$ is the number of samples in the i-th interval that are of the j-th class; and $E_{ij}(f,\mathcal{U})$ is the expected frequency of $A_{ij}(f,\mathcal{U})$, and hence

$$E_{ij}(f,\mathcal{U}) = R_i(f,\mathcal{U}) \times \frac{C_j(f,\mathcal{U})}{N(\mathcal{U})}$$

where $R_i(f,\mathcal{U})$ is the number of samples in the i-th interval, $C_j(f,\mathcal{U})$ is the number of samples in the j-th class; and using the notation $n^{\mathcal{C}}$ to denote the number of samples seen from class $\mathcal{C}$, we also define $N(\mathcal{U}) = \sum_{\mathcal{C} \in \mathcal{U}} n^{\mathcal{C}}$, as the total number of samples.

We consider a feature f to be more relevant than a feature f' in separating the classes in $\mathcal{U}$ if $\chi^2(f,\mathcal{U}) > \chi^2(f',\mathcal{U})$. Obviously, the $\chi^2(f,\mathcal{U})$ value takes on the worst value of 0 if the feature f has only one interval. The degrees of freedom of the χ^2 statistical measure is $(m-1) \times (k-1)$.[742]

1.7. Information Gain

Recall from Subsection 1.5 that $e(f,\mathcal{U}) = \min\{E(f,\mathcal{U},\{S_1,S_2\}) \mid (S_1,S_2)$ is a partitioning of the values of f in $\bigcup \mathcal{U}$ by some point $T\}$. We can interpret this number as the amount of information needed to identify the class of an element of $\mathcal{U}$.

Let $\mathcal{U} = \{\mathcal{A}, \mathcal{B}\}$, where $\mathcal{A}$ and $\mathcal{B}$ are the two classes of samples. Let f be a feature and S be the range of values that f can take in the samples. Let S be partitioned into two subranges S_1 and S_2. Then the difference between the information needed to identify the class of a sample in $\mathcal{U}$ before and after the value of the feature f is revealed is $Gain(f,\mathcal{U},S_1,S_2) = Ent(f,\mathcal{U},S_1 \cup S_2) - E(f,\mathcal{U},\{S_1,S_2\})$.

Then the information gain is the amount of information that is gained by looking at the value of the feature f, and is defined as

$$g(f,\mathcal{U}) = \max\{Gain(f,\mathcal{U},S_1,S_2) \mid (S_1,S_2) \text{ is a partitioning of the}$$
$$\text{values of } f \text{ in } \bigcup \mathcal{U} \text{ by some point } T\}$$

This information gain measure $g(f,\mathcal{U})$ can also be used for selecting features that are relevant because we can consider a feature f to be more relevant than a feature f' if $g(f,\mathcal{U}) > g(f',\mathcal{U})$. In fact, the ID3 decision tree induction algorithm uses it as the measure for picking discriminatory features for tree nodes.[692]

1.8. Information Gain Ratio

However, $g(f,\mathcal{U})$ tends to favour features that have a large number of values. The information gain ratio is a refinement to compensate for this disadvantage. Let

$$GainRatio(f,\mathcal{U},S_1,S_2) = \frac{Gain(f,\mathcal{U},S_1,S_2)}{SplitInfo(f,\mathcal{U},S_1,S_2)}$$

where $SplitInfo(f,\mathcal{U},S_1,S_2) = Ent(f,\{\mathcal{U}_f^{S_1},\mathcal{U}_f^{S_2}\},S_1 \cup S_2)$, and $\mathcal{U}_f^S = \bigcup_{C\in\mathcal{U}}\{d \in C \mid$ the feature f in sample d has value in range $S\}$. Then the information gain ratio is defined as

$$gr(f,\mathcal{U}) = \max\{GainRatio(f,\mathcal{U},S_1,S_2) \mid (S_1,S_2) \text{ is a partitioning}$$
$$\text{of the values of } f \text{ in } \bigcup\mathcal{U} \text{ by some point } T\}$$

The information gain ratio $gr(f,\mathcal{U})$ can of course be used for selecting features that are relevant because we can consider a feature f to be more relevant than a feature f' if $gr(f,\mathcal{U}) > gr(f',\mathcal{U})$. In fact, the C4.5 decision tree induction algorithm uses it as the measure for picking discriminatory features for tree nodes.[693]

1.9. *Wilcoxon Rank Sum Test*

Another approach to deal with the problems of test statistics that require the feature values to conform to specific distributions is to use non-parametric tests. One of the best known non-parametric tests is the Wilcoxon rank sum test, or the equivalent Mann-Whitney test.

The Wilcoxon rank sum test is a non-parametric test for the equality of two populations' mean or median. It does not require the two populations to conform to a specific distribution, other than that the two distributions have the same general shape.[742] It is thus an alternative to the t-test statistical measure and the Fisher criterion score, both of which require the population distributions to be nearly normal. However, if the assumption on near normality is correct, then the Wilcoxon rank sum test may not be as discriminating as the t-test statistical measure or the Fisher criterion score.

The Wilcoxon rank sum test statistical measure $w(f,\mathcal{U})$ of a feature f in a collection $\mathcal{U} = \{A,B\}$ of classes A and B is obtained using the following procedure:

(1) Sort the values $v_1, v_2, ..., v_{n^A+n^B}$ of f across all the samples in $\mathcal{U}$ in ascending order, so that $v_1 \leq v_2 \leq ... \leq v_{n^A+n^B-1} \leq v_{n^A+n^B}$.
(2) Assign rank $r(v_i)$ to each value v_i above so that ties are resolved by averaging. That is,

$$rank(v_i) = \begin{cases} i & \text{if } v_{i-1} \neq v_i \neq v_{i+1} \\ \dfrac{\sum_{k=0}^n j + k}{n+1} & \text{if } v_{j-1} < v_j = ... = v_i = ... = v_{j+n} < v_{j+n+1} \end{cases}$$

(3) Then

$$w(f, \mathcal{U}) = \sum_{v \in \operatorname{argmin}_{c \in \mathcal{U}} n^c} r(v)$$

That is, the Wilcoxon rank sum test statistical measure is the sum of the ranks for the class that has a smaller number of samples. If the number of samples is same in each class, the choice of which class to use is arbitrary.

To use the Wilcoxon rank sum test to decide if a feature f is relevant, we set up the null hypothesis that: the values of f are not much different in $\mathcal{A}$ and $\mathcal{B}$. Then $w(f, \{\mathcal{A}, \mathcal{B}\})$ is used to to accept or reject the hypothesis. To decide whether to accept or reject the null hypothesis, we have to compare $w(f, \{\mathcal{U}\})$ with the upper and lower critical values derived for a significant level α. For the cardinalities of $n^{\mathcal{A}}$ and $n^{\mathcal{B}}$ that are small, e.g. < 10, the critical values have been tabulated, and can be found in e.g. Sandy.[742]

If either $n^{\mathcal{A}}$ or $n^{\mathcal{B}}$ is larger than what is supplied in the table, the following normal approximation can be used. The expected value of $w(f, \{\mathcal{A}, \mathcal{B}\})$ is:

$$\mu = \frac{n^{\mathcal{A}} \times (n^{\mathcal{A}} + n^{\mathcal{B}} + 1)}{2}$$

assuming class $\mathcal{A}$ has fewer samples than class $\mathcal{B}$. The standard deviation of $w(f, \{\mathcal{A}, \mathcal{B}\})$ is known to be:

$$\sigma = \sqrt{\frac{n^{\mathcal{A}} \times n^{\mathcal{B}} \times (n^{\mathcal{A}} + n^{\mathcal{B}} + 1)}{12}}$$

The formula for calculating the upper and lower critical values is $\mu \pm z_\alpha \times \sigma$, where z_α is the z score for significant level α. If $w(f, \{\mathcal{A}, \mathcal{B}\})$ falls in the range given by the upper and lower critical values, then we accept the null hypothesis. Otherwise, we reject the hypothesis, as this indicates that the values of feature f are significantly different between samples in $\mathcal{A}$ and $\mathcal{B}$. Thus, those features whose Fisher criterion rejects the hypothesis are selected as important features.

1.10. Correlation-Based Feature Selection

All of the preceeding measures provide a rank ordering of the features in terms of their individual relevance to separating $\mathcal{A}$ and $\mathcal{B}$. One would rank the features using one of these measures and select the top n features. However, one must appreciate that there may be a variety of independent "reasons" why a sample is in $\mathcal{A}$ or is in $\mathcal{B}$. For example, there can be a number of different contexts and mechanisms that enable protein translation. If a primary context or mechanism involves n signals, the procedure above may select only these n features and may ignore signals in other secondary contexts and mechanisms. Consequently, concentrating on such top n features may cause us to lose sight of the secondary contexts and mechanisms underlying protein translation.

This issue calls for another concept in feature selection: Select a group of features that are correlated with separating $\mathcal{A}$ and $\mathcal{B}$ but are not correlated with each other. The cardinality in a such a group may suggest the number of independent factors that cause the separation of $\mathcal{A}$ and $\mathcal{B}$. A well-known technique that implements this feature selection strategy is the Correlation-based Feature Selection (CFS) method.[316]

Rather than scoring and ranking individual features, the CFS method scores and ranks the worth of subsets of features. As the feature subset space is usually huge, CFS uses a best-first-search heuristic. This heuristic algorithm embodies our concept above that takes into account the usefulness of individual features for predicting the class along with the level of intercorrelation among them.

CFS first calculates a matrix of feature-class and feature-feature correlations from the training data. Then a score of a subset of features is assigned using the following heuristics:

$$Merit_S = \frac{k\overline{r_{cf}}}{\sqrt{k + k \times (k-1) \times \overline{r_{ff}}}}$$

where $Merit_S$ is the heuristic merit of a feature subset S containing k features, $\overline{r_{cf}}$ is the average feature-class correlation, and $\overline{r_{ff}}$ is the average feature-feature intercorrelation.

Symmetrical uncertainties are used in CFS to estimate the degree of association between discrete features or between features and classes.[316] The formula below measures the intercorrelation between two features or the correlation between a feature X and a class Y which is in the range $[0, 1]$:

$$r_{xy} = 2 \times \frac{H(X) + H(Y) - H(X,Y)}{H(X) + H(Y)}$$

where $H(X) + H(Y) - H(X,Y)$ is the information gain between features and classes, $H(X)$ is the entropy of the feature. CFS starts from the empty set of features and uses the best-first-search heuristic with a stopping criterion of 5 consecutive fully expanded non-improving subsets. The subset with the highest merit found during the search is selected.

1.11. *Principal Component Analysis*

The feature selection techniques that we have described so far are "supervised" in the sense that the class labels of the samples are used. It is also possible to perform feature selection in an "unsupervised" way. For example, the principal component analysis (PCA) approach[399] widely used in signal processing can be used in such a manner.

PCA is a linear transformation method that diagonalizes the covariance matrix of the input data via a variance maximization process. PCA selects features by transforming a high-dimensional original feature space into a smaller number of uncorrelated features called principal components. The first principal component accounts for as much of the variability in the data as possible, and each succeeding component accounts for as much of the remaining variability as possible. Feature selection through PCA can be performed by the following steps:

(1) Calculate the covariance matrix C of data X, where X is a matrix with n rows and m columns. Here, n is the number of samples and m is the number of features. Each column data of X may have to be normalized. Each element $C[i, j]$ of the matrix C is the linear correlation coefficient between the elements of columns i and j of X and is calculated as:

$$C[i, j] = \frac{\sum_{k=1}^{n}(X[k, i] - \mu_{X[-,i]}) \times (X[k, j] - \mu_{X[-,j]})}{\sqrt{\sum_{k=1}^{n}(X[k, i] - \mu_{X[-,i]})^2 \times \sum_{k=1}^{n}(X[k, j] - \mu_{X[-,j]})^2}}$$

where $X[k, i]$ is the kth element in the ith column of X, and $\mu_{X[-,i]}$ is the mean of ith column of X.

(2) Extract eigenvalues λ_i, for $i = 1, 2, ..., m$, using the equation below, where I is an identity matrix:

$$|C - \lambda_i \times I| = 0$$

(3) Compute eigenvectors e_i, for $i = 1, 2, ..., m$, using the equation

$$(C - \lambda_i \times I) \cdot e_i = 0$$

(4) These are the so-called "principal components" of X. Rank the eigenvectors according to the amount of variation in the original data that they account for, which is given by

$$variance_i = \frac{\lambda_i}{\sum_{k=1}^{m} \lambda_k}$$

(5) Select as "feature generators" those eigenvectors that account for most of the variation in the data. In this step, enough eigenvectors that account for some percentage—e.g., 95%—of the variance in the original data are chosen, while the rest are discarded.

Let $g_1, ..., g_{d'}$ be the selected feature generators, with $d' \leq d$. Given a data point X_i in the original d-dimensional space, it is mapped into a data point $Y_i = \langle X_i \cdot g_1, ..., X_i \cdot g_{d'} \rangle$ in the transformed d'-dimensional space. Analysis can then be carried out on the lower dimensional Y_i instead of the high-dimensional X_i.

Indeed, it can be proved that the representation given by PCA is an optimal linear dimension reduction technique in the mean-square sense.[399] It is worth noting that PCA is a unsupervised method since it makes no use of the class attribute. When dealing with k-valued discrete features, one can convert them to k binary features. Each of these new features has a "1" for every occurence of the corresponding kth value of the original discrete feature, and a "0" for all other values. Then the new binary features are treated as numeric features, to which above PCA steps can be applied.

1.12. *Use of Feature Selection in Bioinformatics*

The preceding subsections have provided a fairly comprehensive review of feature selection techniques. These techniques have been used as a key step in the handling of biomedical data of high dimension. For example, their use is prevalent in the analysis of microarray gene expression data.[33, 297, 493, 649] They have also been used in the prediction of molecular bioactivity in drug design.[881] More recently, they have even been used in the analysis of the context of protein translation initiation sites.[494, 515, 928]

Note that in choosing a statistical confidence level for t-test, Wilcoxon rank sum test, and other statistical measures presented in this section, one should be aware of the so-called "multiple comparisons problem," which is especially serious in a microarray setting.[634] In this setting, one typically has several thousand genes or features to choose from and hence performs the t-test, Wilcoxon rank sum test, *etc.* several thousand times. Suppose we perform the tests at the standard $p = 5\%$ confidence level. If there is truly no difference in the values of a feature between the two classes for any feature, one can still expect to observe $n \times 0.05$ significant features, where n is the total number of features under consideration.

A standard conservative solution to this problem is the Bonferroni correction.[711] This correction adjusts the cut-off for significance by dividing the desired confidence level by n; for example, for the 5% confidence level, the adjusted cut-off is $0.05/n$. Of course, this technique is applicable only to feature selection measures with known statistical distributions. For measures with unknown statistical distribution, permutation-based methods are typically used to estimate p values where needed.[297, 634]

2. Classification Methods

As mentioned earlier, classification is the process of finding models that separate two or more data classes. Putting it more prosaicly, we are given some samples of class A and some samples of class B, can we use them as a basis to decide if a new unknown sample is in A or is in B? If among all the features of the samples, there is one whose entropy measure is zero, then we can derive the obvious decision rule based on the cutting point T induced on the range of this feature.

Unfortunately, such zero-entropy features are rarely found in the more difficult classification problems. Hence, we often need to use multiple features in an integrated fashion to make prediction in these more difficult classification problems.

In this section, we describe a number of methods that have been used in the biomedical domain for this purpose, including decision tree induction methods, k nearest neighbours (k-NN), Bayesian methods, hidden Markov models (HMM), artificial neural networks (ANN), support vector machines (SVM), and PCL.

2.1. *Decision Trees*

The most popular group of classification techniques is the idea of decision tree induction. These techniques have an important advantage over machine learning methods such as k-NN, ANN, and SVM, in a qualitative dimension: rules produced by decision tree induction are easy to understand and hence can help greatly in appreciating the underlying reasons that separate the classes A and B.

C4.5 is one of the most widely used decision tree based classifier.[693] C4.5 constructs a decision tree in a recursive process. The process starts by determining the feature that is most discriminatory with regard to the entire training data. Then it uses this feature to split the data into non-overlapping groups. Each group contains multi-class or single class samples, as categorized by this feature. Next, a significant feature of each of the groups is used to recursively partition them until a stopping criteria is satisfied. C4.5 uses the information gain ratio presented in Subsection 1.8 to determine which feature is most discriminatory at each step of its decision tree induction process.

Other algorithms for decision tree induction include CART,[101] ID3,[692] SLIQ,[558] FACT,[520] QUEST,[519] PUBLIC,[699] CHAID,[405] ID5,[852] SPRINT,[769] and BOAT.[281] This group of algorithms are most successful for analysis of clinical data and for diagnosis from clinical data. Some examples are diagnosis of central nervous system involvement in hematooncologic patients,[523] prediction of post-traumatic acute lung injury,[696] identification of acute cardiac ischemia,[763] prediction of neurobehavioral outcome in head-injury survivors,[822] and diagnosis of myoinvasion.[521] More recently, they have even been used to reconstruct molecular networks from gene expression data.[783]

2.2. Bayesian Methods

Another important group of techniques[56, 214, 339, 393, 398, 478, 571, 731] are based on the Bayes theorem. The theorem states that

$$P(h|d) = \frac{P(d|h) \times P(h)}{P(d)}$$

where $P(h)$ is the prior probability that a hypothesis h holds, $P(d|h)$ is the probability of observing data d given some world that h holds, and $P(h|d)$ is the posterior probability that h holds given the observed data d.

Let H be all the possible classes. Then given a test instance with feature vector $\{f_1 = v_1, ..., f_n = v_n\}$, the most probable classification is given by

$$\text{argmax}_{h_j \in H} P(h_j | f_1 = v_1, \ldots, f_n = v_n)$$

Using the Bayes theorem, this is rewritten to

$$\text{argmax}_{h_j \in H} \frac{P(f_1 = v_1, \ldots, f_n = v_n | h_j) \times P(h_j)}{P(f_1 = v_1, \ldots, f_n = v_n)}$$

Since the denominator is independent of h_j, this can be simplified to

$$\text{argmax}_{h_j \in H} P(f_1 = v_1, \ldots, f_n = v_n | h_j) \times P(h_j)$$

However, estimating $P(f_1 = v_1, \ldots, f_n = v_n | h_j)$ accurately may not be feasible unless the training set is sufficiently large.

One of the popular ways to deal with the situation above is that adopted by the Naive Bayes classification method (NB).[214, 393] NB assumes that the effect of a feature value on a given class is independent of the values of other features. This assumption is called class conditional independence. It is made to simplify computation and it is in this sense that NB is considered to be "naive." Under this class conditional independence assumption,

$$\text{argmax}_{h_j \in H} P(f_1 = v_1, \ldots, f_n = v_n | h_j) \times P(h_j)$$

$$= \text{argmax}_{h_j \in H} \prod_i P(f_i = v_i | h_j) \times P(h_j)$$

where $P(h_j)$ and $P(f_i = v_i | h_j)$ can often be estimated reliably from typical training sets.

Some example applications of Bayesian classifiers in the biomedical context are mapping of locus controlling a genetic trait,[288] screening for macromolecular crystallization,[348] classification of cNMP-binding proteins,[554] prediction of carboplatin exposure,[375] prediction of prostate cancer recurrence,[198] prognosis of femoral neck fracture recovery,[461] prediction of protein secondary structure,[34, 420, 808] and reconstruction of molecular networks.[263]

2.3. *Hidden Markov Models*

Related to Bayesian classifiers are hidden Markov models (HMM).[56, 215, 219, 455] A HMM is a stochastic generative model for sequences defined by a finite set S of states, a finite alphabet A of symbols, a transition probability matrix T, and an emission probability matrix E. The system moves from state to state according to T while emitting symbols according to E. In an n-th order HMM, the matrices T and E depend on all n previous states.

HMMs have been applied to a variety of problems in sequence analysis, including protein family classification and prediction,[55, 66, 456] tRNA detection in genomic sequences,[525] methylation guide snoRNA screening,[526] gene finding and gene structure prediction in DNA sequences,[54, 94, 95, 455, 741] protein secondary structure modeling,[256] and promoter recognition.[657, 915]

2.4. *Artificial Neural Networks*

Artificial neural networks (ANN)[56, 141, 729] are another important approach to classification that have high tolerance to noisy data. ANN are networks of highly interconnected "neural computing elements" that have the ability to respond to input stimuli and to learn to adapt to the environment.

Although the architecture of different ANN can differ in several characteristic ways, a typical ANN computing element is a comparator that produces an output when the cumulative effect of the input stimuli exceeds a threshold value. Part I of Figure 4 depicts a single computing element. Each input x_i has an associated weight w_i, which acts to either increase or decrease the input signal to the computing element. The computing element behaves as a monotonic function f producing an output $y = f(net)$, where net is the cummulative input stimuli to the neuron. The number net is usually defined as the weighted sum of the inputs:

$$net = \sum_i x_i \times w_i$$

and the function f is usually defined as a sigmoid:

$$f(net) = \frac{1}{1 + e^{-net}}$$

Such computing elements can be connected in multiple layers into the so-called artificial neural network. Part II of Figure 4 depicts a fully-connected feed-forward artificial neural network with two layers of computing elements. The out-

(I) An artificial neural network computing element

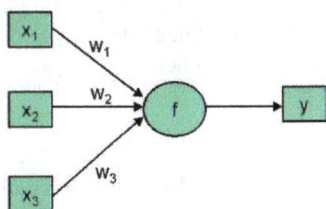

(II) A fully-connected feed-forward artificial neural network with two layers of computing elements.

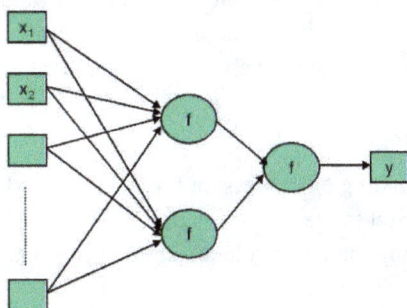

Fig. 4. (I) An artificial neural network computing element. (II) A fully-connected feed-forward artificial neural network with two layers of such computing elements.

put from the two hidden computing elements in the first layer—the so-called hidden layer—are fed as inputs into the computing element at the second layer—the so-called output layer.

The network is used for classification decision as follows. The inputs x_i are fed into the network. Each computing element at the first layer produces its corresponding output, which is fed as input to the computing elements at the next layer. This process continues until an output is produced at the computing element at the final layer.

What makes the ANN work is in how the weights on the links between the inputs and computing elements are chosen. These weights are typically learned from training data using the error back-propagation method.[729]

Let us to introduce some notations before we describe the error back-propagation method. Let v_{ij} denote the weight on the link between x_i and the j-th computing element of the first layer—*i.e.*, the hidden layer—in the artificial neural network. Let w_j denote the weight on the link between the j-th computing element of the first layer and the computing element of the last layer—*i.e.*, the output layer. Let z_j denote the output produced by the j-th computing element of the first layer. Then the output y produced by the artificial neural network for a given training sample is given by

$$y = f\left(\sum_j w_j \times f\left(\sum_i x_i \times v_{ij}\right)\right)$$

For a given training sample, this y may differ from the targeted output t for that particular training sample by an error amount Δ. The crux of ANN training is in how to propagate this error backwards and to use it to adapt the ANN. This is accomplished by adjusting the weights in proportion to the negative of the error gradient. For mathematical convenient, the squared error E can be defined as

$$E = \frac{(t-y)^2}{2}$$

In finding an expression for the weight adjustment, we must differentiate E with respect to weights v_{ij} and w_j to obtain the error gradients for these weights. Applying the chain rule a couple of times and recalling the definitions of y, z_j, E, and f; we derive

$$\frac{\delta E}{\delta w_j} = \frac{\delta E}{\delta \sum_j w_j \times f\left(\sum_i x_i \times v_{ij}\right)} \times \frac{\delta \sum_j w_j \times f\left(\sum_i x_i \times v_{ij}\right)}{\delta w_j}$$

$$= \frac{\delta E}{\delta \sum_j w_j \times f\left(\sum_i x_i \times v_{ij}\right)} \times f\left(\sum_i x_i \times v_{ij}\right)$$

$$= \frac{\delta E}{\delta y} \times \frac{\delta y}{\delta \sum_j w_j \times f\left(\sum_i x_i \times v_{ij}\right)} \times f\left(\sum_i x_i \times v_{ij}\right)$$

$$= (t-y) \times f'\left(\sum_j w_j \times f\left(\sum_i x_i \times v_{ij}\right)\right) \times f\left(\sum_i x_i \times v_{ij}\right)$$

$$= (t-y) \times f'\left(\sum_j w_j \times z_j\right) \times z_j$$

$$= \Delta \times y \times (1-y) \times z_j$$

The last step follows because f is a sigmoid and thus

$$f'(x) = f(x) \times (1 - f(x))$$

Then the adjustment Δ_{w_j} to w_j is defined as below, where η is a fixed learning rate.

$$\Delta_{w_j} = -\eta \times \frac{\delta E}{\delta w_j} = -\eta \times \Delta \times y \times (1-y) \times z_j$$

However, for the weights v_{ij}, we do not have a targeted output to compute errors. So we have to use the errors Δ_{w_j} as surrogates and apply a similar derivation to obtain

$$\Delta_{v_{ij}} = -\eta \times \Delta_{w_j} \times z_j \times (1 - z_j) \times x_i$$

The above derivation provides the adjustments on the weights for one training sample. An "epoch" in the training process is a complete iteration through all training samples. At the end of an epoch, we compute the total error of the epoch as the sum of the squares of the Δ of each sample. If this total error is sufficiently small, the training process terminates. If the number of epoch exceeds some predefined limit, the training process also terminates.

A number of variations are possible. The most obvious are in the configuration of the ANN in terms of the number of layers and in terms of how the computing elements are connected. Another variation is the definition of the function f. For example, the "tansig" function $tanh(net) = (e^{net} - e^{-net})/(e^{net} + e^{-net})$ is another popular choice for f. There are also variations in the method used for propagating errors backwards to adapt the ANN, such as some forms of regularization to prevent overfitting of the ANN to the training data.[82]

Successful applications of artificial neural networks in the biomedical context include protein secondary structure prediction,[687, 713, 725] signal peptide prediction,[164, 231, 611] gene finding and gene structure prediction,[782, 848] T-cell epitope prediction,[362] RNA secondary structure prediction,[800] toxicity prediction,[120] disease diagnosis and outcome prediction,[761, 847, 865] gene expression analysis,[430] as well as protein translation initiation site recognition.[335, 658]

2.5. Support Vector Machines

Let $\phi : D \to D'$ be a function that embeds a training sample in the input space D to a higher dimensional embedding space D'. A kernel $k(x, y) = (\phi(x) \cdot \phi(y))$ is an inner product in D'. It can often be computed efficiently. By formulating a learning algorithm in a way that only requires knowledge of the inner products between points in the embedding space, one can use kernels without explicitly performing the embedding ϕ.

A support vector machine (SVM) is one such learning algorithm. It first embeds its data into a suitable space and then learns a discriminant function to separate the data with a hyperplane that has maximum margin from a small number of critical boundary samples from each class.[124, 855] A SVM's discriminant function $G(T)$ for a test sample T is a linear combination of kernels computed at the training data points and is constructed as

$$G(T) = sign\left(\sum_i \alpha[i] \times Y[i] \times k(T, X[i]) + b\right)$$

where $X[\_]$ are the training data points, $Y[\_]$ are the class labels (which are assumed to have been mapped to 1 or -1) of these data points, $k(\cdot, \cdot)$ is the kernel function, and b and $\alpha[\_]$ are parameters to be learned from the training data.

As it turns out, the training of a SVM is a quadratic programming problem on maximizing the Lagrangian dual objective function

$$\max_\alpha W(\alpha) = \sum_{i=1}^{l} \alpha[i] - \frac{1}{2} \times \sum_{i=1}^{l} \sum_{j=1}^{l} \alpha[i] \times \alpha[j] \times Y[i] \times Y[j] \times k(X[i], X[j])$$

subject to the constraint that $\forall i. 0 \leq \alpha[i] \leq C$ and $\sum_{i=1}^{l} \alpha[i] \times Y[i] = 0$. Here, C is a bound on errors, $X[i]$ and $X[j]$ are the ith and jth training samples, $Y[i]$ and $Y[j]$ are the corresponding class labels (which are mapped to 1 and -1), and $k(\cdot, \cdot)$ is the kernel function. The kernel function normally used is a polynomial kernel

$$k(X, X') = (X \cdot X')^d = \sum_{i_1} \ldots \sum_{i_d} X[i_1] \times \ldots \times X[i_d] \times X'[i_1] \times \ldots \times X'[i_d]$$

In the popular WEKA data mining software package,[888] the quadratic programming problem above is solved by sequential minimal optimization.[668] Once the optimal solution $\alpha[\_]$ is obtained from solving the quadratic programming problem above, we can substitute it into the SVM decision function $G(T) = sign(\sum_i \alpha[i] \times Y[i] \times k(T, X[i]) + b)$ given earlier.

It remains now to find the threshold b. We can estimate b using the fact that the optimal solution $\alpha[\_]$ must satisfy the so-called Karush-Kuhn-Tucker conditions.[328, 668, 757] In particular, for each $\alpha[j] > 0$, $Y[j] \times G(X[j]) = 1$. Hence $Y[j] = sign(\sum_i \alpha[i] \times Y[i] \times k(X[j], X[i]) + b)$. So we estimate b by averaging $Y[i] - \sum_i \alpha[i] \times Y[i] \times k(X[j], X[i])$ for all $\alpha[j] > 0$.

A SVM is largely characterized by the choice of its kernel function. Thus SVMs connect the problem they are designed for to a large body of existing research on kernel-based methods.[124, 700, 855] Besides the polynomial kernel function $k(x, y) = (x \cdot y)^d$, other examples of kernel function include the Gaussian radial basis kernel function,[63, 855] sigmoid kernel function,[757] B_n-spline kernel function,[757] locality-improved kernel function,[940] and so on.

Some recent applications of SVM in the biomedical context include protein homology detection,[388] microarray gene expression data classification,[111] breast cancer diagnosis,[264, 539] protein translation initiation site recognition,[940, 941] as well as prediction of molecular bioactivity in drug design.[881]

2.6. *Prediction by Collective Likelihood of Emerging Patterns*

Prediction by collective likelihood of emerging patterns (PCL)[492, 497] is a classification method that we have been developing during the last couple of years. This method focuses on (a) fast techniques for identifying patterns whose frequencies in two classes differ by a large ratio,[207] which are the so-called emerging patterns; and on (b) combining these patterns to make decision. Note that a pattern is still emerging if its frequencies are as low as 1% in one class and 0.1% in another class, because the ratio indicates a 10 times difference. However, for the purpose of PCL, we use only emerging patterns that are most general and have infinite ratio. That is, we use only emerging patterns that occur in one class of data but not the other class and that are as short as possible.

Basically, the PCL classifier has two phases. Given two training datasets D^A (instances of class A) and D^B (instances of class B) and a test sample T, PCL first discovers two groups of most general emerging patterns from D^A and D^B. Denote the most general emerging patterns of D^A as, $EP_1^A, EP_2^A, \cdots, EP_i^A$, in descending order of frequency. Denote the most general emerging patterns of D^B as $EP_1^B, EP_2^B, \cdots, EP_j^B$, in descending order of frequency.

Suppose the test sample T contains these most general emerging patterns of D^A: $EP_{i_1}^A, EP_{i_2}^A, \cdots, EP_{i_x}^A$, $i_1 < i_2 < \cdots < i_x \leq i$, and these most general emerging patterns of D^B: $EP_{j_1}^B, EP_{j_2}^B, \cdots, EP_{j_y}^B$, $j_1 < j_2 < \cdots < j_y \leq j$. The next step is to calculate two scores for predicting the class label of T. Suppose we use k ($k \ll i$ and $k \ll j$) top-ranked most general emerging patterns of D^A and D^B. Then we define the score of T in the D^A class as

$$score(T, D^A) = \sum_{m=1}^{k} \frac{frequency(EP_{i_m}^A)}{frequency(EP_m^A)},$$

and the score in the D^B class is similarly defined in terms of $EP_{j_m}^B$ and EP_m^B. If $score(T, D^A) > score(T, D^B)$, then T is predicted as the class of D^A. Otherwise it is predicted as the class of D^B. We use the size of D^A and D^B to break tie.

Recall that PCL uses emerging patterns that have infinite ratio and are most general. That is, given two datasets D^A and D^B, an emerging pattern—to be used by PCL—is a pattern such that its frequency in D^A (or D^B) is non-zero but in the other dataset is zero (*i.e.*, infinite ratio between its support in D^A and D^B), and none of its proper subpattern is an emerging pattern (*i.e.*, most general). We can immediately see that if D^A and D^B has n Boolean attributes, then there are 2^n

possible patterns. Hence a naive method to extract emerging patterns requires 2^n scans of the dataset. A more efficient method for extracting emerging patterns is therefore crucial to the operation of PCL.

We briefly discuss here an approach for developing more practical algorithms for finding such emerging patterns. Let us first recall the theoretical observation of Dong and Li[207] that the emerging patterns from D^A to D^B—all of them together, not just most general the ones—form a convex space. Now, it is known that a convex space C can be represented by a pair of borders $\langle L, R \rangle$, so that (a) L is anti-chain, (b) R is anti-chain, (c) each $X \in L$ is more general than some $Z \in R$, (d) each $Z \in R$ is more specific than some $X \in L$, and (e) $C = \{Y \mid \exists X \in L, \exists Z \in R, X \subseteq Y \subseteq Z\}$. Actually, L are the most general patterns in C, and R the most specific patterns in C. We write $[L, R]$ for C.

Dong and Li[207] show that: if D^A and D^B have no duplicate and no intersection and are of the same dimension, then the set of emerging patterns to D^A from D^B is given by $[\{\{\}\}, D^A] - [\{\{\}\}, D^B]$. Having thus reduced emerging patterns to this "border formulation," we derive a more efficient approach to discover them.

The main trick is as follows. Let $D^A = \{A_1, \ldots, A_n\}$ and $D^B = \{B_1, \ldots, B_m\}$. Then

$$
\begin{aligned}
&[\{\{\}\}, D^A] - [\{\{\}\}, D^B] \\
&= [\{\{\}\}, \{A_1, \ldots, A_n\}] - [\{\{\}\}, \{B_1, \ldots, B_m\}] \\
&= [L, \{A_1, \ldots, A_n\}]
\end{aligned}
$$

where

$$
\begin{aligned}
L &= \min(\bigcup_i^n \{\{s_1, \ldots, s_m\} \mid s_j \in A_i - B_j, 1 \le j \le m\}) \\
&= \bigcup_i^n \min\{\{s_1, \ldots, s_m\} \mid s_j \in A_i - B_j, 1 \le j \le m\}
\end{aligned}
$$

which we know how to optimize very well[207, 208, 935] using novel border-based algorithms and constraint-based algorithms based on set enumeration tree.[732]

A number of other classification methods based on the idea of emerging patterns exist. They include CAEP,[209] which uses all of the most-general emerging patterns with ratio exceeding a user-specified threshold; DeEPs,[491] which uses emerging patterns in an instance-based manner; as well as techniques that score emerging patterns in *ad hoc* ways.[498]

Any way, although the PCL classifier is a very recent development, it has already proved to be a good tool for analysing biomedical data. Examples include gene expression data, proteomic data, and translation initiation sites.[492−494, 497, 499, 918]

3. Association Rules Mining Algorithms

A major data mining task is to discover dependencies among data. Association rules[10] are proposed to achieve this goal. In the context of supermarket business data, association rules are often used to describe dependencies such as the presence of some items in a transaction implies, with a certain probability, the presence of other items in the same transaction. An example[10] of such an association rule is the statement that 90% of transactions involving the purchase of bread and butter also involve the purchase of milk. In this section, we first provide some preliminary definitions. Then we outline two algorithms for mining association rules, *viz.* Apriori[12] and Max-Miner.[70]

3.1. *Association Rules*

Recall from Subsection 1.1 that a data point in a feature space can be thought of as an itemset $\{f_1 = v_1, ..., f_n = v_n\}$. It is a convention of the data mining community to write an itemset as $\{f'_1, ..., f'_m\}$ if each v_i is either 1 or 0 and $\{f'_1, ..., f'_m\} = \{f_i| v_i = 1, 1 \leq i \leq n\}$. Under this convention, the itemset is also called a "transaction." Observe that transactions contain only those items whose feature values are 1 rather than those items whose feature values are 0. In this section, we use the term "transaction" instead of "data point." Incidentally, an itemset consisting of k items is called a k-itemset; the number k is called the "length" or "cardinality" of the itemset.

We say that a transaction T contains an itemset X if $X \subseteq T$. We also say that an itemset X occurs in a transaction T if $X \subseteq T$. Let a database $\mathcal{D}$ of transactions and an itemset X be given. Then the support of X in $\mathcal{D}$, denoted $supp^{\mathcal{D}}(X)$, is the percentage of transactions in $\mathcal{D}$ that contain X. The count of X in $\mathcal{D}$, denoted $count^{\mathcal{D}}(X)$, is the number of transactions in $\mathcal{D}$ that contain X. Observe that

$$supp^{\mathcal{D}}(X) = \frac{count^{\mathcal{D}}(X)}{n^{\mathcal{D}}}$$

where $n^{\mathcal{D}}$ is the number of transactions in $\mathcal{D}$. Then given a threshold δ between 0 and 1, an itemset X is said to be a large itemset or a frequent itemset if $supp^{\mathcal{D}}(X) \geq \delta$.

An association rule is an implication of the form $X \rightarrow^{\mathcal{D}} Y$, where X and Y are two itemsets in a dataset $\mathcal{D}$ and $X \cap Y = \{\}$. We often omit the superscript $\mathcal{D}$ if the dataset is understood. The itemset X is called the antecedent of the rule. The itemset Y is called the consequent of the rule. We define the support of the rule as the percentage of the transactions in $\mathcal{D}$ that contain $X \cup Y$. We also define the confidence of the rule as the percentage of the transactions in $\mathcal{D}$ containing X

that also contain Y. Note that the support of the rule $X \to^{\mathcal{D}} Y$ is $supp^{\mathcal{D}}(X \cup Y)$ and the confidence is $count^{\mathcal{D}}(X \cup Y)/count^{\mathcal{D}}(X)$.

The problem of mining association rules is that of how to generate all association rules that have support and confidence greater than or equal to a user-specified minimum support (*minsup*) and a minimum confidence (*minconf*).[10] This problem is solved by decomposing into two sub-problems:

(1) Generate all large itemsets that satisfy the support threshold *minsup*.
(2) For a given large itemset $X = \{f_1, f_2, ..., f_k\}$, generate all association rules and output only those rules that satisfy the confidence threshold *minconf*.

The second sub-problem is simple and straightforward to solve. The first sub-problem is the key efficiency issue of mining association rules, *viz.* to discover all large itemsets whose supports exceeds a given threshold.

A naive approach to solving this problem is to generate all possible itemsets in the dataset $\mathcal{D}$ and then check whether their supports meet the threshold. Obviously, this task rapidly becomes impossible as the number of features increases. This happens because the number of the candidate itemsets increases exponentially when the number of features increases. Fortunately, this problem has been extensively studied in both the database and data mining communities.[10–12, 70, 105, 106, 155, 319, 320, 436, 481, 540, 646, 748, 794, 832, 933] There are two now well-known efficient algorithms, the Apriori algorithm[12] and the Max-Miner algorithm,[70] that partially overcome the difficulty in the naive algorithm. We describe these two algorithms in the next two subsections.

3.2. The Apriori Algorithm

Given a database $\mathcal{D}$ and a support threshold δ, the collection of the large itemsets in $\mathcal{D}$ is the union of all large 1-itemsets, all large 2-itemsets, ..., and all large m-itemsets, where m is the longest length of the large itemsets in $\mathcal{D}$. The Apriori algorithm sequentially outputs the large k-itemsets with k from 1 to m. So, the Apriori algorithm is also called a level-wise algorithm.

The basic idea used in the Apriori algorithm is that any subset of a large itemset is large. This property is called the *a priori* property. It is used in the algorithm to narrow search spaces, generating a strict number of large itemset candidates which is much smaller than the number of the candidates generated by the naive method. In other words, if all the large k-itemsets are known, the Apriori algorithm generates only those $(k + 1)$-itemsets as candidates whose immediate proper subsets (having k items) must all be large. So, the Apriori algorithm does not generate any of those $(k+1)$-itemsets as a candidate for which there exists its an immediate proper subset which is not large.

Briefly, the algorithm comprises the three main steps below, with the iteration parameter k initialized to 1. These three steps are repeated until no new large itemsets are found:

(1) Generate a collection of large itemset candidates, each having k items,
(2) Scan the database $\mathcal{D}$ and compute the supports of the candidates.
(3) $k := k + 1$.

The efficiency Step 1 is key to the efficiency of the whole algorithm, as generating more unnecessary candidates can increase its complexity. Agrawal and Srikant[12] propose the APRIORI-GEN function that makes use of the *a priori* property for candidate generation. We use an example[12] to illustrate the main ideas in the APRIORI-GEN function. Let the collection $\mathcal{D}_3$ of all the large 3-itemsets in some database $\mathcal{D}$ be $\{\{1,2,3\}, \{1,2,4\}, \{1,3,4\}, \{1,3,5\}, \{2,3,4\}\}$. After a joint sub-step in the APRIORI-GEN function, the candidate collection $\mathcal{D}_4$ is $\{\{1,2,3,4\}, \{1,3,4,5\}\}$. After a pruning sub-step, the itemset $\{1,3,4,5\}$ is removed from $\mathcal{D}_4$ because the itemset $\{1,4,5\}$ is not in $\mathcal{D}_3$. As a result, only the itemset $\{1,2,3,4\}$ is a candidate and needs support calculation in Step 2.

The Apriori algorithm achieves good performance by reducing the size of candidate itemsets. However, in some situations where there exist long frequent itemsets (*e.g.*, an itemset with the cardinality of 30), or where a low support threshold is required, the Apriori algorithm still suffers from heavy computational costs. Additional details of the Apriori algorithm and discussions on its effficiency and scalability can be found in Agrawal and Srikant,[12] Bayardo,[70] and Han *et al.*[320]

3.3. *The Max-Miner Algorithm*

The Apriori algorithm explicitly outputs all large itemsets in a level-wise manner. It sequentially finds the large k-itemsets, $1 \le k \le m$, where m is the longest length of the large itemsets. In contrast to the Apriori algorithm, the Max-Miner algorithm[70] only outputs a subset of the collection of all large itemsets with respect to a given support threshold in a database. Specifically, Max-Miner outputs only those large itemsets whose proper supersets are not large. These large itemsets are called *maximal large itemsets*. They can be imagined as a frontier boundary that separates the large itemsets from the non-large itemsets. Since any large itemset is a subset of a maximal large itemset, Max-Miner's output implicitly and concisely represents all large itemsets.

Max-Miner has been shown to perform two or more orders of magnitude better than Apriori on some data sets.[70] This is particularly true when the support threshold is set low, and the data sets are large and high-dimensional. There are two basic ideas behind Max-Miner:

(1) Superset-frequency pruning, which works as follows. Suppose an itemset is known to be large or frequent. Then its subsets must be large or frequent also. Hence, there is no need to generate its subsets for support calculation.

(2) Subset-infrequency pruning, which works as follows. Suppose an itemset is known as non-large or infrequent. Then its supersets must be non-large. Hence, there is no need to generate its supersets for support calculation.

Only the latter pruning technique is used in the Apriori algorithm or its variants.

Max-Miner contains some provision for looking ahead in order to quickly identify long frequent itemsets and short infrequent itemsets to conduct superset-frequency pruning and subset-infrequency pruning. As a result, Max-Miner is highly efficient in comparison to the Apriori algorithm.

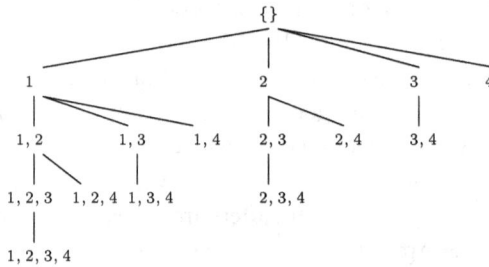

Fig. 5.   A complete set-enumeration tree over four items.

The two pruning strategies of Max-Miner are implemented using the framework of set-enumeration trees[732] by incorporating some heuristic. A simple set-enumeration is shown in Figure 5. See Bayardo[70] for more details of Max-Miner, and Bayardo et al.[69] for its further refinement. Some methods that are closely related to Max-Miner are Pincer-Search by Lin and Kedem,[508] and MaxEclat and MaxClique by Zaki et al.[925]

3.4. *Use of Association Rules in Bioinformatics*

A number of studies[176, 643, 747] have been conducted to make use of the concept of association rules for mining bio-medical data to find interesting rules. We discuss two of them. One discovers association rules from protein-protein interaction data.[643] The other finds expression associations from gene expression profiling data.[176] Both of the studies have revealed interesting patterns or rules. Some of these rules have been confirmed by work reported in biomedical literature, and some are considered to be new hypothesis worthy of wet-experimental validation.

Protein-protein interactions are fundamental biochemical reactions in organisms. It is important to understand general rules about proteins interacting with each other, such as "the protein having a feature A interacts with the protein having the feature B", "this domain interacts with that domain," and so on. Association rules can capture such knowledge patterns. Oyama *et al.*[643] propose a method that is aimed to reveal what types of proteins tend to interact with each other. For example, a popular interaction rule that "a SH3 domain binds to a proline-rich region" is detected by this method.

Oyama *et al.*[643] use 4307 pairs of yeast interaction proteins from four source databases. Functional, primary structural, and other characteristics are used as features to describe the proteins. The total of 5241 features are categorized into seven types, *viz.* YPD categories, EC numbers, SWISS-PROT/PIR keywords, PROSITE motifs, Bias of the amino acids, Segment clusters, and Amino acid patterns. Under $minsup = 9\%$ and $minconf = 75\%$, Oyama *et al.*[643] have discovered a total of 6367 rules. Instead of using only support and confidence, they have also proposed a new scoring measure to indicate the validity of the rules. The number of positive interactions and the number of proteins concerned with the positive interactions are placed at high priority considerations in this scoring measure.

Gene expression profiling data have been widely used to understand functions of genes, biological networks, and cellular states. Creighton and Hanash[176] proposed a method to analyse gene expression data using association rules, aimed at revealing biologically relevant associations between different genes or between environmental effects and gene expression. In this analysis, there are 6316 transcripts which are used as features. These transcripts correspond to 300 diverse mutations and chemical treatments in yeast. The values of these features are their expression levels under different experiments. The expression range of each feature (transcript) is discretized into three categorical values: up, down, and neither-up-nor-down. An expression value greater than 0.2 for the log base 10 of the fold change is categorized as up; a value less than -0.2 as down; all other values as neither-up-nor-down.

Creighton and Hanash[176] use the Apriori algorithm in the mining of the association rules. The minimum support for mining those frequent itemsets is set as 10%. The minimum confidence for those association rules is set as 80%. To avoid the vast majority of redundant rules and also to alleviate the complexity of the mining problem, they suggest an additional criteria besides a minimum support to narrow the seach space of candidate frequent itemsets. The additional criteria is that the method generates only those association rules with the left-hand-side itemset containing a single item, ignoring frequent itemsets that cannot form such a rule. Mining expression data using association rules is claimed to be more useful than clustering methods to uncover gene networks. Association rules discovered from gene expression data can also be used to help relate the expression of genes to their cellular environment.

4. Clustering Methods

Clustering is the process of grouping a set of objects into classes or clusters so that objects within a cluster have high similarity in comparison to one another, but are dissimilar to objects in another clusters. Unlike classifier induction discussed in Section 2, clustering does not rely on predefined classes and labeled training examples to guide the machine learning process. Hence, it is also called unsupervised learning. Clustering algorithms can be separated into four general categories, *viz.* partitioning methods, hierarchical methods, density-based methods and grid-based methods. Of the four, partitioning and hierarchical methods are applicable to both spatial and categorical data while density- and grid-based methods are applicable only to spatial data. As such, this section focuses on the first categories of two methods.

4.1. *Factors Affecting Clustering*

To apply clustering in any domain successfully, the following factors must typically be considered:

(1) type of data,
(2) similarity function, and
(3) goodness function.

It is easy to see that the type of data to be clustered affects the clustering algorithms being applied. For example, clustering a set of DNA sequences is different from clustering a set of gene expression profiles. A useful way for categorizing the data types is to separate them into spatial vs. categorical data. Spatial data are data objects which are mapped into a coordinated space, *e.g.*, the 3-dimensional location of protein molecules. Categorical data are data objects which cannot be represented as objects in a spatial coordinate system, *e.g.*, a set of medical documents. Generally, it is easier to develop efficient clustering algorithms for spatial data through the use of spatial inferences. The clustering of categorical data is more difficult as explicit computation must usually be made for any pair of data objects in order to judge their similarity based on some similarity function.

Since the aim of clustering is to group similar objects together, it is a natural question to ask when are two objects deemed to be similar. The answer comes in the form of a similarity function that is defined by the user based on the application on hand. For spatial data, the similarity function is usually computed from the coordinates of the data objects, such as the Euclidean distance. It is therefore generally easy to estimate the similarity between two groups of objects by picking a representative for each group—such as their mean or their bounding boxes which can be easily computed—and making spatial inference based on the distance between the two representatives. The similarity of two data objects can thus be approximated without knowing their exact coordinate.

Unlike spatial data, the similarity between categorical data is usually more difficult to approximate. The inherent difficulty come from the fact that a good representative for a group of categorical data is usually harder to determine. For example, if we use edit distance as a similarity measurement, finding a good representive for a group of sequences involves the use of multiple alignment, an operation which requires expensive dynamic programming. The other alternative is to pick one of the sequences from the group that is the most similar to all the other sequences, resulting in the need for an all-pairs comparison.

Without any good way to represent a set of data objects, efficient clustering of categorical data is difficult and approximating the similarity between two groups of object becomes computationally intensive too. One usual technique to solve this problem is to derive some transformation that maps categorical data into spatial coordinates such that objects that are similar are mapped into a spatially close region. The objects can then be clustered using the spatial distance as a similarity function instead.

Finally, a goodness function is usually derived to measure the quality of the clustering. This is again application dependent. In general, a goodness function is an aggregate function that is used to gauge the overall similarity between data objects in the clusters. For example, in k-means clustering,[537] the squared distance to the cluster center for each point is summed up as a measure of goodness. In other cases, a test of good fit is done according to some statistical model—*e.g.*, Gaussian—in order to judge the quality of the clusters.

As mentioned earlier, partitioning and hierarchical methods of clustering are applicable to both spatial and categorical data. We discuss them in the next two subsections.

4.2. Partitioning Methods

Partitioning algorithms had long been popular clustering algorithms before the emergence of data mining. Given a set D of n objects in a d-dimensional space and an input parameter k, a partitioning algorithm organizes the objects into k clusters such that the total deviation of each object from its cluster center or from a cluster distribution is minimized. The deviation of a point can be computed differently in different algorithms and is what we earlier called the goodness function.

Three well-known partitioning algorithms are: k-means,[537] Expectation Maximization (EM),[98, 197, 921] and k-medoids.[421] The three algorithms have different ways of representing their clusters. The k-means algorithm uses the centroid (or the mean) of the objects in the cluster as the cluster center. The k-medoids algorithm uses the most centrally located object in the cluster. Unlike k-means and k-medoids, EM clustering uses a distribution consisting of a mean and a $d \times d$ covariance matrix to represent each cluster. Instead of assigning each object to a dedicated cluster, EM clustering assigns each object to a cluster according to a probability of membership that is computed from the distribution of each cluster. In this way, each object has a certain probability of belonging to each cluster, making EM clustering a fuzzy partitioning technique.

Despite their differences in the representation of the clusters, the three partitioning algorithms share the same general approach when computing their solutions. To see this similarity, we first observe that the three algorithms are effectively trying to find the k centers or distributions that will optimize the goodness function. Once the optimal k centers or distributions are found, the membership of the n objects within the k clusters are automatically determined. However, the problem of finding the globally optimal k centers or k distributions is known to be NP-hard.[273] Hence, the three algorithms adopt an iterative relocation technique that finds k locally optimal centers. This technique is shown in Figure 6. The three algorithms differ essentially in their goodness function and in the way they handle Steps 2 and 3 of the iterative relocation technique given in the figure. The general weaknesses of partitioning-based algorithms include a requirement to specify the parameter k and their inability to find arbitrarily shaped clusters.

4.3. Hierarchical Methods

A hierarchical method creates a hierarchical decomposition of the given set of data objects forming a dendrogram—a tree that splits the database recursively into smaller subsets. The dendrogram can be formed in two ways: "bottom-up" or "top-down." The "bottom-up" approach, also called the "agglomerative" approach, starts with each object forming a separate group. It successively merges

---

Input: The number of clusters, k, and a database containing n objects.

Output: A set of k clusters that minimizes a criterion function E.

Method:

(1) Arbitrarily choose k centers/distributions as the initial solution;

(2) (re)compute membership the objects according to present solution.

(3) Update some/all cluster centers/distributions according to new memberships of the objects;

(4) Goto Step 2 if the value of E changes; otherwise, end.

Fig. 6. An outline of the iterative relocation technique used generally by partitioning methods of clustering.

the objects or groups according to some measures like the distance between the two centers of two groups and this is done until all of the groups are merged into one (the topmost level of the hierarchy), or until a termination condition holds. The top-down, also called the "divisive" approach, starts with all the objects in the same cluster. In each successive iteration, a cluster is split into smaller clusters according to some measures until eventually each object is in one cluster, or until a termination condition holds.

AGNES and DIANA are two hierarchical clustering algorithms.[421] AGNES (AGglomerative NESting) is a bottom-up algorithm which starts by placing each object in its own cluster and then merging these atomic clusters into larger and larger clusters, until all of the objects are in a single cluster or until a certain termination condition is satisfied. DIANA (DIvisia ANAlysis), on the other hand, adopts a top-down strategy that does the reverse of AGNES by starting with all objects in one cluster. It subdivides the cluster into smaller and smaller pieces, until each object forms a cluster on its own or until it satisfies certain termination conditions, such as a desired number of clusters is obtained or the distance between the two closest clusters is above a certain threshold distance.

In either AGNES or DIANA, one can specify the desired number of clusters as a termination condition. The widely used measures for distance between clusters are as follows, where m_i is the mean for cluster C_i, n_i is the number of objects in C_i, and $|p - p'|$ is the distance between two objects or points p and p'.

$$d_{min}(C_i, C_j) = min_{p \in C_i, p' \in C_j} |p - p'|$$

$$d_{mean}(C_i, C_j) = |m_i - m_j|$$

$$d_{avg}(C_i, C_j) = \frac{\Sigma_{p \in C_i} \Sigma_{p' \in C_j} |p - p'|}{n_i \times n_j}$$

$$d_{max}(C_i, C_j) = max_{p \in Ci, p' \in C_j} |p - p'|$$

The two algorithms, AGNES and DIANA, often encounter difficulties regarding the selection of merge or split points. Such a decision is critical because once a group of objects are merged or split, the process at the next step will operate on the newly generated clusters. They can neither undo what was done previously, nor perform object swapping between clusters. Thus merge or split decisions, if not well chosen at some step, may lead to low quality clusters. Moreover, the method does not scale well since the decision of merge or split needs to examine and evaluate a good many number of objects or clusters.

To enhance the effectiveness of hierarchical clustering, recent methods have adopted one of the following two approaches. The first approach represented by algorithms like CURE[303] and CHAMELEON[418] utilizes a more complex principle when splitting or merging the clusters. Although the splitting and merging of clusters are still irreversible in this approach, less errors are made because a better method is used for merging or splitting. The second approach represented by algorithms like BIRCH[934] is to obtain an initial result by using a hierarchical agglomerative algorithm and then refining the result using iterative relocation.

4.4. *Use of Clustering in Bioinformatics*

Recent uses of clustering techniques in bioinformatics are mostly in aspects of gene expression analysis. They are amongst the earliest techniques used for disease subtype diagnosis with microarrays, though classification-based approaches have now largely replaced clustering-based approaches to this problem. Neverthess, they are still very useful for disease subtype discovery with microarrays.[20, 21, 73, 83, 816] They are also amongst the earliest techniques used for time series cell cycle studies with microarrays.[225, 277]

In a recent study,[187] six clustering methods—hierarchical clustering using the umweighted pair group method with arithmetic mean,[225] k-means,[324] DIANA,[421] a fuzzy logic-based methods called Fanny,[421] A model-based clustering methods,[59] hierarchical clustering with partial least squares[186]—are compared in the context of transcriptional analysis of budding yeast sporulation as well as two sets of simulated data. The study concludes that DIANA is the most consistent performer with respect to the known profiles of the three test models.

5. Remarks

In this chapter, we have surveyed a large number of techniques for the data mining tasks of feature selection, classification, association rules extraction, and clustering. We have also briefly mentioned some examples of their applications in the analysis of biological and clinical data. There are other varieties of data mining tasks that we have not discussed. We close this chapter by briefly mentioning some of them below.

Outlier analysis deals with objects that do not comply with the general behavior of a model of the data. Most datamining applications discard outliers. However, in some applications the rare events can be more interesting than the more regularly occurring ones. For example, the detection of new particles in nuclear accelerator experiments.

Trend analysis describes and models regularities or trends for objects whose behavior changes over time. The distinctive features of such an analysis include time-series data analysis, periodicity pattern matching, and clustering of time-related data. The inference of gene relationships from large-scale temporal gene expression patterns[205] is an example of this topic.

Feature generation is another important data mining task, especially for the analysis of biological data such as DNA and protein sequences, and clinical data such as images. The techniques described on the main text of this chapter all assume that our data possess explicit features. However, for data such as DNA sequences, protein sequences, and images, the explicit features present in the data are usually inadequate for analysis. For an image file, the features that are explicitly present are the colour and intensity values at each pixel. For a DNA or protein sequence, the features that are explicitly present are the nucleotide or amino acid symbols at each position of the sequence. Such features are at too low a level for good understanding. Feature generation is the task of generating higher-level features that are more suitable for analysis and for understanding the data. However, some knowledge of the problem domain is typically needed to identify and generate high-level features in a robust manner. A recent bioinformatics example where feature generation plays a key role is the recognition of translation initiation sites.[515, 928]

Finally, inference of missing data is a data mining task that is occasionally needed in the analysis of biological and clinical data. There are many reasons for such a need. For example, gene expression microarray experiments often generate datasets with many missing expression values due to limitations in current microarray technologies.[837] As another example, patient medical records usually have many missing values due to missed visits and other reasons. How the missing values can be estimated depends on the situation. Sometimes, a missing value in a sample can be estimated based on a corresponding value in other samples that are most similar to this sample. Sometimes, a missing value in a sample can be inferred from secondary data associated with the sample.

CHAPTER 4

TECHNIQUES FOR RECOGNITION OF
TRANSLATION INITIATION SITES

Jinyan Li

Institute for Infocomm Research
jinyan@i2r.a-star.edu.sg

Huiqing Liu

Institute for Infocomm Research
huiqing@i2r.a-star.edu.sg

Limsoon Wong

Institute for Infocomm Research
limsoon@i2r.a-star.edu.sg

Roland H.C. Yap

National University of Singapore
ryap@comp.nus.edu.sg

Correct prediction of the translation initiation site is an important issue in genomic research. In this chapter, an in-depth survey of half a dozen methods for computational recognization of translation initiation sites from mRNA, cDNA, and genomic DNA sequences are given. These methods span two decades of research on this topic, from the perceptron of Stormo *et al.* in 1982[805] to the systematic method of explicit feature generation and selection of Wong *et al.* in 2002.[928]

ORGANIZATION.

Section 1. We begin with an introduction to the biological background of protein translation initiation. We also explain some of the difficulties involved in recognizing translation initiation sites from mRNA, cDNA, and genomic DNA sequences.

Section 2. We describe the dataset of Pedersen and Nielsen.[658] This is the most popular dataset used for investigating translation initiation sites.

Sections 3–9. After that, we give an in-depth survey of a number of translation initiation site recognition methods. Specifically, we present the method of Stormo *et al.*[805] that is based on perceptrons, the NetStart system of Pedersen and Nielsen[658] that is based on artificial neural networks, the method of Zien *et al.*[940, 941] that is based on kernel engineering of support vector machines, the approach of Wong *et al.*[494, 515, 928] that is based on explicit feature generation and selection, the method of Salamov *et al.*[737] that is based on linear discriminant function, and the method of Hatzigeorgiou[335] that is based on a complicated architecting of two artificial neural networks and the ribosome scanning rule.

Section 10. Finally, the performance of these methods are summarized—most of them achieve close to or above 90% accuracy. A qualitative discussion of these methods are also given.

1. Translation Initiation Sites

Proteins are synthesized from mRNAs by a process called translation. The process can be divided into three distinct stages: initiation, elongation of the polypeptide chain, and termination.[107] The region at which the process initiates is called the Translation Initiation Site (TIS). The coding sequence is flanked by non-coding regions which are the 5' and 3' untranslated regions respectively. The translation initiation site prediction problem is to correctly identify TIS in a mRNA, cDNA, or genomic sequence. This forms an important step in genomic analysis to determine protein coding from nucleotide sequences.

In eukaryotes, the scanning model postulates that the ribosome attaches first to the 5' end of the mRNA and scans along the 5'-to-3' direction until it encounters the first AUG.[451] While this simple rule of first AUG holds in many cases, there are also exceptions. Some mechanisms proposed to explain the exceptions are: leaky scanning where the first AUG is bypassed for reasons such as poor context; reinitiation where a short upstream open reading frame causes a second initiation to occur; and also other alternative proposed mechanisms.[451, 620] Translation can also occur with non-AUG codons, but this is rare in eukaryotes[451] and is not considered here.

The problem of recognizing TIS is compounded in real-life sequence analysis by the difficulty of obtaining full-length and error-free mRNA sequences. Pedersen and Nielsen found that almost 40% of the mRNAs extracted from Gen-Bank contain upstream AUGs.[658] The problem becomes more complex when using unannotated genome data or analyzing expressed sequence tags (ESTs), which usually contain more errors, and are not guaranteed to give the correct 5' end.[87, 88] Thus, the prediction of the correct TIS is a non-trivial task since the biological mechanisms are not fully understood and sequences may have errors and may not be complete.

```
299 HSU27655.1 CAT U27655 Homo sapiens
CGTGTGTGCAGCAGCCTGCAGCTGCCCCAAGCCATGGCTGAACACTGACTCCCAGCTGTG 80
CCCAGGGCTTCAAAGACTTCTCAGCTTCGAGCATGGCTTTTGGCTGTCAGGGCAGCTGTA 160
GGAGGCAGATGAGAAGAGGGAGATGGCCTTGGAGGAAGGGAAGGGGCCTGGTGCCGAGGA 240
CCTCTCCTGGCCAGGAGCTTCCTCCAGGACAAGACCTTCCACCCAACAAGGACTCCCCT
.......................................................... 80
.............................iEEEEEEEEEEEEEEEEEEEEEEEEEEEEEE 160
EEEEEEEEEEEEEEEEEEEEEEEEEEEEEEEEEEEEEEEEEEEEEEEEEEEEEEEEEEEE 240
EEEEEEEEEEEEEEEEEEEEEEEEEEEEEEEEEEEEEEEEEEEEEEEEEEEEEEEEEE
```

Fig. 1. An example annotated sequence from the dataset of Pedersen and Nielsen. The 4 occurrences of ATG are underlined. The second ATG is the TIS. The other 3 ATGs are non-TIS. The 100 nucleotides upstream of the TIS are marked by an overline. The 100 nucleotides downstream of the TIS are marked by a double overline. The ".", "i", and "E" are annotations indicating whether the corresponding nucleotide is upstream (.), TIS (i), or downstream (E).

2. Data Set and Evaluation

The vertebrate dataset of Pedersen and Nielsen[658] is used in most of the approaches that we plan to present. So, let us first describe this data set. The sequences contained in this data set are processed by removing possible introns and joining the exons[658] to obtain the corresponding mRNA sequences. From these sequences, only those with an annotated TIS, and with at least 10 upstream nucleotides as well as 150 downstream nucleotides are selected. The sequences are then filtered to remove those belonging to same gene families, homologous genes from different organisms, and sequences which have been added multiple times to the database. Since the dataset is processed DNA, the TIS site is ATG (rather than AUG).

An example entry from this dataset is given in Figure 1. There are 4 ATGs in the example sequence shown. The second ATG is the TIS. The other 3 ATGS are non-TIS. ATGs to the left of the TIS are termed upstream ATGs. So the first ATG in the figure is an upstream ATG. ATGs to the right of the TIS are termed downstream ATGs. So the third and fourth ATGs in the figure are downstream ATGs.

There are a total of 13375 ATG sites in the Pedersen and Nielsen dataset. Of these ATG sites, 3312 (24.76%) are the true TIS, while the other 10063 (75.23%) are non-TIS. Of the non-TIS, 2077 (15.5%) are upstream of a true TIS. At each of these ATG sites, a sequence segment consisting of 100 nucleotides upstream and 100 nucleotides downstream is extracted. These 13375 sequence segments are the inputs upon which TIS recognition is performed. If an ATG does not have enough upstream or downstream context, the missing context is padded with the appropriate number of dont-care symbols.

With the exception of the perceptron method presented in Section 3 and the ribosome-scanning method presented in Section 9, the selection of features and the training and testing of the machine learning methods are all performed with a 3-fold cross validation process on this dataset of Pedersen and Nielsen. That is, the dataset is divided into three equal parts, and each part is in turn reserved for testing the classification trained on the features selected from the other two parts of the data.[328]

The results of cross validation are evaluated using standard performance measures defined as follows. Sensitivity measures the proportion of TIS that are correctly recognized as TIS. Specificity measures the proportion of non-TIS that are correctly recognized as non-TIS. Precision measures the proportion of the claimed TIS that are indeed TIS. Accuracy measures the proportion of predictions, both for TIS and non-TIS, that are correct.

3. Recognition by Perceptrons

One of the earliest attempts at predicting TIS using supervised learning methods is the work of Stormo *et al.*[805] Unlike most of the work surveyed here which deals with TIS recognition for eukaryotes, the problem in Stormo *et al.* was to distinguish TIS in *E. coli*. They trained a perceptron[569] model to recognize TIS for sequences of different lengths. A perceptron is a simple artificial neural net with no hidden units and a single output unit. The output unit is activated—that is, recognizes the TIS—when $W \cdot X \geq T$ (and vice versa, $W \cdot X < T$ is not a TIS), where X is a vector encoding the candidate sequence and T is a chosen threshold value. The perceptron is trained to find a suitable weight vector W by iterating the following steps in Figure 2.

The encoding of a nucleotide is as four bits rather than the less redundant two bit encoding—*i.e.*, "A" as 1000, "C" as "0100", "G" as 0010, and "T" as 0001. This encoding is used because a major limitation of perceptrons is that they can only be used to distinguish linearly separable functions. In particular, it cannot be used to encode the "exclusive or" (XOR) function. Suppose instead that the 2-bit binary representation was used with the following encoding,

"A" as 00, "C" as 01, "G" as 10, and "T" as 11.

It would not be possible then to train the perceptron to recognize XOR which in this encoding is the rule

A or T versus C or G.

TEST:	choose a sequence X from the training set
	if X is a TIS and $W \cdot X \geq T$, goto TEST
	if X is a TIS and $W \cdot X < T$, goto ADD
	if X is not a TIS and $W \cdot X < T$, goto TEST
	if X is not a TIS and $W \cdot X \geq T$, goto SUBTRACT
ADD:	replace W by $W + S$, goto TEST
SUBTRACT:	replace W by $W - S$, goto TEST

Fig. 2. The training of a perceptron is by iterating the steps above. The perceptron convergence theorem guarantees that the solution is found in a finite number of steps if its solution exists. Alternatively, if the changes to the weight vector W plateaus, the training can also be stopped.

Stormo *et al.*[805] try using the 4-bit encoding and sequences of windows of sizes 51, 71, and 101 bases roughly centered around the position of the TIS to find the weights for the perceptron model. Of the three sizes, Stormo *et al.* find that the window of 101 bases is the best. Not surprisingly, the initiation codon and the Shine-Dalgarno region is found to have the greatest significance for the weights W. The authors have also worked on a different rule-based consensus model[804] and the perceptron approach is found to be more accurate and also more precise.

4. Recognition by Artificial Neural Networks

In Pedersen and Nielsen,[658] an artificial neural network (ANN) is trained on a 203 nucleotide window centered on the AUG. It is a feed-forward ANN with three layers of neurons which can be thought of as generalizing the earlier work of Stormo *et al.* to ANNs with hidden units. As with Stormo *et al.*, the nucleotides are encoded using a redundant 4-bit representation. While ANNs with hidden units can overcome the limitations of perceptrons, the redundant representation has the advantage of not introducing encoding bias. The output layer has two neurons. The first neuron predicts if the input is a TIS. The second neuron predicts if the the input is a non-TIS. Whichever of these neurons gives the bigger score wins.

According to Pedersen and Nielsen,[658] the number of neurons in the hidden layer of the ANN does not significantly affect the performance of the ANN. They obtain results of 78% sensitivity on start AUGs and 87% specificity on non-start AUGs on their vertebrate dataset described in Section 2, giving an overall accuracy of 85%. Their system is available on the Internet as the NetStart 1.0 prediction server accessible at http://www.cbs.dtu.dk/services/NetStart.

Pedersen and Nielsen[658] also carry out additional analysis to try to uncover features in their sequences that are important for distinguishing TIS from non-TIS. In one of the analysis, they supply their neural network with input windows which cover the aforementioned 203 nucleotides, except for one position—a "hole"— from which input is disregarded. The hole is shifted along the input window in a series of runs of the neural network and the impact of the hole in each position is noted. This experiment reveals that position −3 is crucial to TIS recognition, as the accuracy of the neural network drops precipitously when a hole is present in this position.

Pedersen and Nielsen[658] also analyse the positions of non-translation initiating ATGs that are misclassified by their neural network as TIS. In this analysis, they discover that ATGs that are in-frame to the TIS are more likely to be misclassified as TIS regardless of whether they are upstream or downstream of the TIS.

5. Recognition by Engineering of Support Vector Machine Kernels

Zien et al.[940, 941] work on the same vertebrate dataset from Pedersen and Nielsen by using support vector machines (SVM) instead. The same 203 nucleotide window is used as the underlying features to be learnt. Each nucleotide is encoded using the same sparse binary encoding as Pedersen and Nielsen.

Homogeneous polynomial kernels[757] of degree d,

$$k(X,Y) = (X \cdot Y)^d = \sum_{i_1} \ldots \sum_{i_d} X[i_1] \times \ldots \times X[i_d] \times Y[i_1] \times \ldots \times Y[i_d]$$

are commonly used in SVM. Due to the encoding used for nucleotides, the position of each bit that is set indicates it is A, C, G, or T. Consequently, the dot product $(X \cdot Y)$ is equivalent to a count of the number of nucleotides that coincide in the two sequences represented by vectors X and Y. Similarly, $(X \cdot Y)^d$ is equivalent to a correlation of the nucleotide frequencies at any d sequence positions. Zien et al.[940] report that SVM achieves TIS recognition performance comparable to Pedersen and Nielsen's ANN using this standard type of kernels on the dataset of Pedersen and Nielsen described in Section 2.

In the polynomial kernel $(X \cdot Y)^d$ above, the correlation of nucleotide frequencies at any d sequence positions is used. However, there are a number of biological reasons that suggest we should only consider sequence positions that are not too far apart. In particular, each amino acid is coded by a triplet of adjacent nucleotides and the region upstream of a TIS is non-coding but the region downstream of a TIS is coding. Thus a kernel that zooms into such localized correlations may be appropriate for TIS recognition.

Inspired by this reasoning, Zien *et al.*[940] show how to obtain improvements by appropriate engineering of the kernel function—using a locality-improved kernel with a small window on each position. The locality-improved kernel emphasizes correlations between sequence positions that are close together, and a span of 3 nucleotides up- and down-stream is empirically determined as optimal. The locality-improved kernel is thus defined as

$$k(X,Y) = \sum_{p=1}^{l} win_p(X,Y)$$

where

$$win_p(X,Y) = \left(\sum_{j=-3}^{3} w_j \times (X =_{p+j} Y) \right)^4$$

$$= \sum_{j_1=-3}^{3} \cdots \sum_{j_4=-3}^{3} w_{j_1} \times (X =_{p+j_1} Y) \times \ldots \times w_{j_4} \times (X =_{p+j_4} Y)$$

Here, w_j are appropriate weights that are increasing from the boundaries to the center of the window, and

$$(X =_{p+j} Y) = \begin{cases} 1, & \text{if the nucleotides at position } p+j \text{ of} \\ & X \text{ and } Y \text{ are the same} \\ 0, & \text{otherwise} \end{cases}$$

With the locality-improved kernel, Zien *et al.*[940] obtain an accuracy of 69.9% and 94.1% on start and non-start AUGs respectively, giving an overall accuracy of 88.1% on the dataset described in Section 2.

Zien *et al.*[941] further improve their previous results by engineering a more sophisticated kernel—a so-called Salzberg kernel. The Salzberg kernel is essentially a conditional probabilitistic model of positional di-nucleotides. The Salzberg kernel gives an overall accuracy of 88.6% on the dataset described in Section 2.

6. Recognition by Feature Generation, Feature Selection, and Feature Integration

Wong *et al.*[515, 928] show that good performance comparable to the best results can be obtained by a methodology based on these three steps:

(1) generate candidate features from the sequences,
(2) select relevant features from the candidates, and

(3) integrate the selected features using a machine learning method to build a
system to recognize specific properties—in this case, TIS—in sequence data.

We present these three steps in the next three subsections

6.1. Feature Generation

The sequence segments prepared in the previous section are not suitable for direct
application of most machine learning techniques, as these techniques rely on explicit signals of high quality. It is necessary to devise various signals and sensors
for these signals, so that given a sequence segment, a score is produced to indicate the possible presence of such a signal in that sequence segment. The obvious
strategy to devising these signals and sensors is to generate a large number of candidate features, and to evaluate them against an annotated dataset to decide which
ones are signals and which ones are noise.

Wong et al.[928] make use of the general technique of k-grams and a few refinements for producing candidate features. A k-gram is simply a pattern of k consecutive letters, which can be amino acid symbols or nucleic acid symbols. K-grams
can also be restricted those in-frame ones. Each k-gram and its frequency in the
said sequence fragment becomes a candidate feature. Another general technique
for producing these candidate features is the idea of position-specific k-gram. The
sensor for such a feature simply reports what k-gram is seen in a particular position in the sequence fragment.

For ease of discussion, given a sequence segment, we refer to each position in
the sequence segment relative to the target ATG of that segment. The "A" in the
target ATG is numbered as +1 and consecutive downstream positions—that is, to
the right—from the target ATG are numbered from +4 onwards. The first upstream
position—that is, to the left—adjacent to the target ATG is −1 and decreases for
consecutive positions towards the 5' end—that is, the left end of the sequence
fragment.

Let us use the sequence segment centered at the second ATG in Figure 1 and
comprising 100 nucleotides upstream and 100 nucleotides downstream of this
ATG for illustration of the various k-gram features described by Wong et al..[928]
These upstream and downstream nucleotides are marked using overline and double overline in the figure.

For the basic k-grams, k is the length of a nucleotide pattern to be generated.
Some typical values for k are 1, 2, 3, 4, and 5. Since there are 4 possible letters
for each position, there are 4^k possible basic k-grams for each value of k. For
example, for k = 3, one of the k-grams is ATG and the frequency of this k-gram is
4 in our example sequence segment. The candidate feature (and value assignment)

corresponding to this is "ATG=4".

The upstream region of a TIS is non-coding and the downstream region of a TIS is coding. It can therefore be expected that they have some different underlying features. So it is wise to introduce additional classes of k-grams to attempt to capture these differences. These are the upstream and downstream k-grams.

For the upstream k-grams, Wong et al.[928] count only occurrences of the corresponding patterns upstream of the target ATG. Again, for each value of k, there are 4^k upstream k-grams. For example, for k = 3, some k-grams are: ATG, which has frequency 1 in this context; GCT, which has frequency 5 in this context; and TTT, which has frequency 0 in this context. The candidate features and value assignments corresponding to these k-grams are "upstream ATG=1", "upstream GCT=5", and "upstream TTT=0".

For the downstream k-grams, Wong et al.[928] count only occurences of the corresponding patterns downstream of the target ATG. Again, for each value of k, there are 4^k downstream k-grams. For example, for k = 3, some k-grams are: ATG, which has frequency 2 in this context; GCT, which has frequency 3 in this context; and TTT, which has frequency 2 in this context. The candidate features and value assignments corresponding to these k-grams are "downstream ATG=2", "downstream GCT=3", and "downstream TTT=2".

The biological process of translating from nucleotides to amino acids is to have 3 nucleotides—the so-called codons—codes for one amino acid, starting from the TIS. Therefore, 3-grams in positions ..., -9, -6, -3, +4, +7, +10, ... are aligned to the TIS. Wong et al.[928] call those 3-grams in positions ..., -9, -6, and -3, the in-frame upstream 3-grams, and those 3-grams in positions +4, +7, +10, ..., the in-frame downstream 3-grams. As these 3-grams are in positions that have biological meaning, they are also good candidate features. There are 2×4^3 such 3-grams. In our example sequence fragment, some in-frame downstream 3-grams are: GCT, which has frequency 1 in this context; TTT, which has frequency 1 in this context; ATG, which has frequency 1 in this context. The corresponding candidate features and value assignments are "in-frame downstream GCT=1", "in-frame downstream TTT=1", and "in-frame downstream ATG=1". Some in-frame upstream 3-grams are: GCT, which has frequency 2 in this context; TTT, which has frequency 0 in this context; ATG, which has frequency 0 in this context. The corresponding candidate features and value assignments are "in-frame upstream GCT=2", "in-frame upstream TTT=0", and "in-frame upstream ATG=0".

Another type of features are what Wong et al.[928] call the position-specific k-grams. For this type of k-grams, they simply record which k-gram appears in a particular position in the sequence segment. It is sufficient to consider only 1-grams, that is, k-grams for k = 1. Since our sequence segment has 100 nucleotides flanking each side of the target ATG, there are 200 position-specific 1-grams. In our example sequence segment, some position-specific 1-grams are: at position +4 is G and at position −3 is G. The corresponding candidate features and value assignments are "position+4=G" and "position−3=G".

Combining all the features discussed above, for $k = 1, ..., 5$, each sequence segment is coded into a record having $(\sum_{k=1}^{5} 4^k + 4^k + 4^k) + 2 \times 4^3 + 200 = 4436$ features. For illustration, our example sequence segment is coded into this record: {..., "ATG=4", ..., "upstream ATG=1", "upstream GCT=5", "upstream TTT=0", ..., "downstream ATG=2", "downstream GCT=3", downstream TTT=2", ..., "in-frame downstream GCT=1", "in-frame downstream TTT=1", "in-frame downstream ATG=1", ..., "in-frame upstream GCT=2", "in-frame upstream TTT=0", "in-frame upstream ATG=0", ..., ..., "position−3=G", ..., "position+4=G", ...}. Such a record is often called a feature vector.

These 4436 features as described above are generated for each of the 13375 sequence segments corresponding to the 13375 ATG sites in the Pedersen and Nielsen dataset.[658] We note that other techniques for generating candidate features are possible. For example, we can compute a specific statistic on the sequence segment, such as its GC ratio. Specific biological knowledge can also be used to devise specialized sensors and features, such as CpG island,[271] Kozak consensus pattern,[452] etc. An exhaustive exposition is outside the scope of this chapter. In the next two subsections, we show how to reliably recognize TIS on the basis of a small subset of our 4436 candidate features.

6.2. Feature Selection

This number of candidate features is clearly too many. Most of them can be expected to be noise that can confuse typical machine learning algorithms. So the next step in the methodology proposed by Wong et al.[928] is to apply a feature selection technique to pick those features that are most likely to help in distinguishing TIS from non-TIS.

Any feature selection techniques can be used, inclusing signal-to-noise measure,[297] t-test statistical measure,[133] χ^2 statistical measure,[514] entropy measure,[242] information gain measure,[692] information gain ratio,[693] Fisher criterion score,[251] Wilcoxon rank sum test,[742] correlation-based feature selection method (CFS),[316] and so on. All of these methods are described in Chapter 3.

Wong et al.[928] apply CFS to the feature vectors derived from the Pedersen and Nielsen dataset as described in Section 2 and Subsection 6.1 in a 3-fold cross validation setting. It turns out that in each fold, exactly the same 9 features are selected by CFS, viz.

(1) "position–3",
(2) "in-frame upstream ATG",
(3) "in-frame downstream TAA",
(4) "in-frame downstream TAG",
(5) "in-frame downstream TGA",
(6) "in-frame downstream CTG",
(7) "in-frame downstream GAC",
(8) "in-frame downstream GAG", and
(9) "in-frame downstream GCC".

These 9 features are thus very robust differentiators of TIS from non-TIS. Furthermore, there are good biological reasons for most of them.

"Position–3" can be explained by the known correspondence to the well-known Kozak consensus sequence, GCC[AG]CCAUGG, for vertebrate translation initiation sites.[403, 450] Hence having a "A" or "G" in this position indicates that the target ATG is more likely to be a TIS. This is the same feature deduced as important by Pedersen and Nielsen[658] in their "hole-shifting" experiment discussed in Section 4.

"In-frame upstream ATG" can be explained by the ribosome scanning model.[9, 451] The ribosome scans the mRNA in a 5'-to-3' (that is, left-to-right) manner until it encounters the first ATG with the right context for translation initiation. Thus a ATG that is closer to the 5' end have a higher probability to be a TIS. Consequently, the presence of an in-frame ATG upstream of the target ATG indicates that the target ATG is less likely to be a TIS. This is also consistent with the observation by Rogozin *et al.*[718] that a negative correlation exists between the strength of the start context and the number of upstream ATGs. This is also a feature deduced by Pedersen and Nielsen[658] in a detailed analysis on the erroneous predictions made by their neural network.

"In-frame downstream TAA", "in-frame downstream TAG", and "in-frame downstream TGA" can be explained as they correspond to in-frame stop codons downstream from the target ATG. These 3 nucleotide triplets—TAA, TAG, TGA—do not code for amino acids. They are called the stop codons. The biological process of translating in-frame codons into amino acids stops upon encountering an in-frame stop codon. Thus the presence of any of these three features means there is an in-frame stop codon within 100 nucleotides downstream of the target ATG. Consequently, the protein product corresponding to the sequence is no more than 33 amino acids. This is smaller than most proteins. Hence the target ATG is not likely to be a TIS. This group of stop codon features are not reported by Pedersen and Nielsen,[658] Zien *et al.*,[940, 941] nor Hatzigeorgiou,[335] presumably the complex and/or low-level nature of their systems prevented them from noticing this important group.

Wong et al.[928] do not have a clear biological explanation for the remaining 4 selected features, other than that of codon biases.

6.3. *Feature Integration for Decision Making*

In order to show that the 9 features identified in the previous step are indeed relevant and reliable differentiators of TIS from non-TIS, Wong et al.[928] demonstrate in a 3-fold cross validation setting that practically any machine learning methods can be trained on these 9 features to produce TIS recognizers of extremely competitive accuracy. In particular, among those classification methods described in Chapter 3, Wong et al.[928] give the results obtained on Naive Bayes (NB),[471] Support Vector Machine (SVM),[855] and C4.5.[693]

According to Wong et al.,[928] NB trained on the 9 features selected above yields an effective TIS recognizer with sensitivity = 84.3%, specificity = 86.1%, precision = 66.3%, and accuracy = 85.7%. SVM trained on these 9 features yields an accurate TIS recognizer with sensitivity = 73.9%, specificity = 93.2%, precision = 77.9%, and accuracy = 88.5%. C4.5 trained on these 9 features also yields an accurate TIS recognizer with sensitivity = 74.0%, specificity = 94.4%, precision = 81.1%, and accuracy = 89.4%.

7. Improved Recognition by Feature Generation, Feature Selection, and Feature Integration

As every 3 nucleotides code for an amino acid, it is legitimate to investigate if an alternative approach to generating features based on amino acids can produce effective TIS recognizers. Also, in the previous sections, Wong et al.[928] use features selected by CFS, hence it is legitimate to investigate if features selected by other methods can produce effective TIS recognizers. Li et al.[494] and Liu and Wong[515] pursue these two alternatives. In this section, we discuss their results.

For generating features, Li et al.[494] and Liu and Wong[515] take the sequence segments of 100 nucleotides upstream and 100 nucleotides downstream of the target ATG as before. Then they consider 3-grams that are in-frame—that is, those 3-grams that are aligned to the ATG at positions ..., -6, -3, +4, +7, ... Those in-frame 3-grams that code for amino acids are converted into the corresponding amino acid letters. Those in-frame 3-grams that are stop codons are converted into a special letter symbolizing a stop codon. From the conversion above, Li et al.[494] generate the following types of k-grams:

(1) up-X, which counts the number of times the amino acid letter X appears in

the upstream part, for X ranging over the standard 20 amino acid letters and the special stop symbol.

(2) down-X, which counts the number of times the amino acid letter X appears in the downstream part, for X ranging over the standard 20 amino acid letters and the special stop symbol.

(3) up-XY, which counts the number of times the two amino acid letters XY appear as a substring in the upstream part, for X and Y ranging over the standard 20 amino acid letters and the special stop symbol.

(4) down-XY, which counts the number of times the two amino acid letters XY appear as a substring in the upstream part, for X and Y ranging over the standard 20 amino acid letters and the special stop symbol.

Li *et al.*[494] also generate the following Boolean features from the original sequence fragments: up-ATG, which indicates that an in-frame ATG occurs in the upstream part; up3-AorG, which indicates that an "A" or a "G" appears in position −3 ; down4-G, which indicates that a "G" appears in position +4. These last two features are inspired by the Kozak consensus sequence, GCC[AG]CCAUGG, for vertebrate translation initiation sites.[403, 450] A total of $2 \times 21 + 2 \times 21^2 + 3 = 927$ features are thus generated as described above.

For selecting features, they use the entropy measure to rank the relevance of each of these 927 candidate features in a 3-fold cross validation setting. In each fold, the top 100 features are selected. The following features are consistently among the top 10 features in each of the 3 folds: up-ATG, down-STOP, down-L, down-D, down-E, down-A, up3-AorG, up-A, down-V. Up-M also appears among the top features in each fold, but we exclude it as it is redundant given that up-ATG is true if and only if up-M > 0. The detailed ranking of these features in each fold is given in Figure 3.

Interestingly, most of these features, except up-A and down-V, correspond to those selected by CFS on the original nucleotide sequence fragments in Section 6.2. Specifically, up-ATG corresponds to "in-frame upstream ATG"; down-stop corresponds to "in-frame downstream TAA", "in-frame downstream TAG", and "in-frame downstream TGA"; up3-AorG corresponds to "position−3"; down-L corresponds to "in-frame downstream CTG"; down-D corresponds to "in-frame downstream GAC"; down-E corresponds to "in-frame downstream GAG"; and down-A corresponds to "in-frame downstream GCC".

For validating whether accurate systems for recognizing TIS can be developed using features based on amino acids, Liu and Wong[515] test the C4.5, SVM, and NB machine learning methods in 3-fold cross validations. The top 100 features selected by the entropy measure are used in each fold.

Fold	up ATG	down STOP	up3 AorG	down A	down V	up A	down L	down D	down E
1	1	2	4	3	6	5	8	9	7
2	1	2	3	4	5	6	7	8	9
3	1	2	3	4	5	6	8	9	7

Fig. 3.   Ranking of the top 9 features selected by the entropy measure method as relevant in each of the 3 folds.

For C4.5, they obtain sensivity = 74.88%, specificity = 93.65%, precision = 79.51%, and accuracy = 89.00%. This is comparable to the performance of C4.5 using the 9 features selected by CFS in Section 6.2.

For SVM, they obtain sensitivity = 80.19%, specificity = 96.48%, precision = 88.24%, and accuracy = 92.45%. This is significantly better than the performance of SVM using the 9 features selected by CFS in Section 6.2. In fact, it is the best reported results on the Pedersen and Nielsen dataset[658] that we know of.

For NB, they obtain sensitivity = 70.53%, specificity = 87.76%, precision = 65.47%, and accuracy = 83.49%. This is considerably worse than the other results. The increase from 9 features in Subsection 6.3 to 100 features here has apparently confused NB.

Li *et al.*[494] use just the top 10 features selected by the entropy measure in their study. They use the PCL (Prediction by Collective Likelihood of emerging patterns) classifier,[497] described in Chapter 3, to integrate these top 10 entropy features. They report that PCL achieves 84.7% sensitivity, 88.7% specificity and 87.7% overall accuracy.

8. Recognition by Linear Discriminant Function

Linear discriminant analysis provides a linear function that separates two classes while minimizing misclassification. Assume a p-feature variable X is given as a vector, then the linear discriminant function

$$y = \sum_{i=1}^{p} \alpha[i] \times X[i]$$

classifies X into the first class if $y \geq c$ and into the second class if $y < c$. The optimal selection of the vector of feature weights $\alpha[\_]$ and the decision threshold

c is typically determined by maximizing the ratio of between-class-variation to within-class-variation.

The ATGpr program of Salamov *et al.*[737] uses a linear discriminant function that combines several statistical features derived from training sequences. Specifically, ATGpr uses the following characteristics extracted from their training sequences:

(1) Positional triplet weight matrix around an ATG. For each triplet nucleotides $i = 1, 2, \ldots, 64$, and position $j = -14, -13, \ldots, +4, +5$, the frequencies for true TIS $f_{TIS}(i, j)$ and all candidate ATGs $f_{totalATG}(i, j)$ are calculated. Then the propensity for a particular triplet i to be in a specific position j relative to the initiation codon is given as:

$$P_{triplet}(i, j) = \frac{f_{TIS}(i, j)}{f_{totalATG}(i, j)}$$

To use these propensities, the total score around each candidate ATG region is added together for the window -14 to $+5$.

(2) In-frame hexanucleotide weight matrix. For each hexanucleotide $k = 1, 2, \ldots, 4^6$, the frequencies of it appearing in an in-frame position upto 300 nucleotide downstream of the candidate ATG for true TIS $f_{coding}(k)$ and downstream of false TIS in a noncoding region $f_{noncoding}(k)$ are calculated. Then the propensity for a particular hexanucleotide is given as:

$$P_{hexamer}(k) = \frac{f_{coding}(k)}{f_{noncoding}(k)}$$

(3) 5' UTR-ORF hexanucleotide difference. A true TIS has a higher value of the average hexamer coding propensities in the potential 5'UTR region $[-1, -50]$ and potential coding region $[+1, +50]$, where the positions are relative to the candidate ATG. So Salamov *et al.* also uses the difference between the average hexanucleotide coding propensities between these two regions as a feature.

(4) Signal peptide characteristic. Within a 30 amino acid window down-stream of each ATG, the most hydrophobic 8-residue peptide was identified. Hydrophobicity is calculated using the hydropathy scale of Kyte and Doolittle.[466]

(5) Presence of another upstream in-frame ATG. This is a simple Boolean-valued feature. If an extra ATG is found upstream of the candidate ATG without encountering an in-frame stop codon, the likelihood of the ATG being an TIS is down-weighted.

(6) Upstream cytosine nucleotide characteristic. The frequency of cytosine in the region $[-7, -36]$ upstream of a candidate ATG is counted, as it has been observed that 5' UTRs of human genes are cytosine rich.[524]

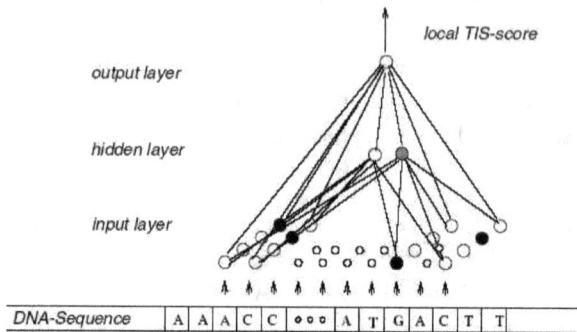

Fig. 4. The consensus ANN of DIANA-TIS. A window of 12 nucleotides is presented to the trained ANN. A high score at the output indicates a possible TIS. (*Image credit: Artemis Hatzigeorgiou.*)

In a more recent work,[618] an improved version of ATGpr called ATGpr_sim is developed, which uses both statistical information and similarities with other known proteins to obtain higher accuracy. ATGpr can be accessed at http://www.hri.co.jp/atgpr/. When searching TIS in a given cDNA or mRNA sequence, the system outputs several ATGs—5 by default—in the order of decreasing confidence. If we always take the ATG with highest confidence as TIS, then for the 3312 sequences in the Pedersen and Nielsen dataset, ATGpr can predict correctly true TIS in 2941 (88.80%) of them. Note that by taking only the ATG with the highest confidence as the TIS, only 1 prediction is made per sequence. Hence this figure is not directly comparable to the 3-fold cross validation figures reported on other methods earlier.

9. Recognition by Ribosome Scanning

Hatzigeorgiou[335] reports a highly accurate TIS prediction program, DIANA-TIS, using artificial neural networks trained on human sequences. Their dataset contains full-length cDNA sequences which has been filtered for errors. An overall accuracy of 94% is obtained using an integrated method which combines a consensus ANN with a coding ANN together with the ribosome scanning model.

The consensus ANN assesses the candidate TIS and its immediate surrounding comprising a window from positions −7 to +5 relative to the candidate TIS. The consensus ANN is a feed-forward ANN with short cut connections and two hidden units, as shown in Figure 4. The coding ANN is used to assess the coding potential of the regions upstream and downstream of a candidate TIS. The coding ANN works on a window of 54 nucleotides. As every three nucleotides form a codon

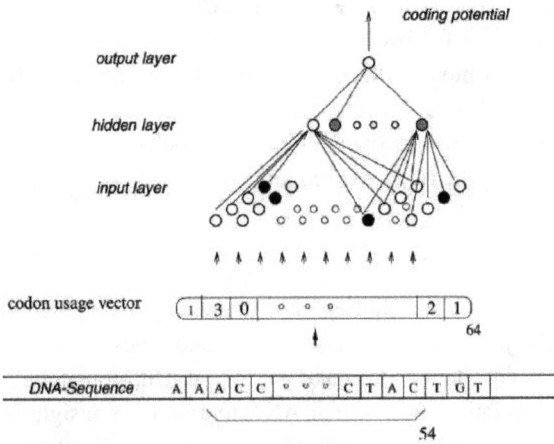

Fig. 5. The coding ANN of DIANA-TIS. A window of 54 nucleotides is presented to the trained ANN. A high score at the output indicates a coding nucleotide. (*Image credit: Artemis Hatzigeorgiou.*)

that translates into an amino acid, there are 64 possible codons. To assess the coding potential of the window of 54 nucleotides, this window is transformed into a vector of 64 units before feeding into the coding ANN. Each unit represents one of the 64 codons and gives the normalized frequency of the corresponding codon appearing in the window. The coding ANN is a feed-forward ANN and has two hidden units, but no short cut connection, as shown in Figure 5. Both the consensus ANN and the coding ANN produce a score between 0 and 1.

The two ANNs are basically integrated as follows. Given a candidate ATG, the consensus ANN is applied to to a window of 12 nucleotides at positions −7 to +5 to calculate a consensus score s_1. Then the coding ANN is applied to the in-frame 60 positions before the ATG by sliding along a window of 54 nucleotides and summing the output of the coding ANN at each position to obtain an upstream coding score s_2. The coding ANN is also applied to the in-frame 60 positions after the ATG to obtain a downstream coding score s_3 in a similar fashion. The final score for the ATG is then obtained as $s_1 \times (s_3 - s_2)$. The score calculations above are applied to all the ATGs in a mRNA sequence one after another, and the first ATG to score at above 0.2 is taken as the TIS of the mRNA sequence. This preference to accept the first ATG in a favour context as the TIS is the so-call ribosome scanning model.

Note that in the ribosome scanning model,[9, 451] an mRNA sequence is scanned from left to right, testing each ATG in turn until one of them is classified as TIS; all the ATGs to the right of this ATG are skipped and classified as non-TIS. In short, exactly one prediction is made per mRNA under the ribosome scanning model. Hence, accuracy figures based on the ribosome scanning model should not be compared with models that tests every ATG. In addition, Hatzigeorgiou uses a dataset that is different from Pedersen and Nielsen. So her results cannot be directly compared to results obtained on the Pedersen and Nielsen dataset or obtained without using the ribosome scanning model.

10. Remarks

The approach of Pedersen and Nielsen[658] is interesting in that their inputs are extremely low level—just a string of nucleotides—and relies entirely on their ANN to learn high-level correlations to make prediction. Unfortunately, it is not easy to extract these correlations out of their ANN to gain more insight into sequence features that distinguish TIS from non-TIS. Nevertheless, by more extensive experiments and analysis, Pedersen and Nielsen are able to suggest that position –3 is crucial to distinguishing TIS from non-TIS.

The approach of Zien *et al.*[940, 941] and Hatzigeorgiou[335] are also very interesting in that they show us how to perform sophisticated engineerings of SVM kernel functions and ANNs. Unfortunately, it is also not easy to extract more insight into sequence features that distinguish TIS from non-TIS from their systems. Nevertheless, the improved results of the locality-improved kernel in Zien *et al.*[940] over that of standard polynomial kernels suggest that local correlations are more important than long-ranged correlations in distinguishing TIS from non-TIS.

The approach of Wong *et al.*[928] and Li *et al.*[494] is interesting in that they focus on deriving high-level understandable features first, and then use these features to distinguish TIS from non-TIS. In fact, by applying a decision tree induction method such as C4.5[693] on the selected features, highly meaningful rules such as "if up-ATG = Y and down-STOP > 0, then prediction is false TIS"; "if up3-AorG = N and down-STOP > 0, then prediction is false TIS"; and "if up-ATG = N and down-STOP ≤ 0 and up3-AorG = Y, then prediction is true TIS" are also extracted.

Finally, we summarize the TIS-recognition performance of the various methods described in this chapter in Figure 6.

I. Accuracy of some TIS recognition methods. Pedersen and Nielsen,[658] Zien *et al.*,[940, 941] and Li *et al.*[494] are directly comparable to results in Parts II and III. Hatzigeorgiou[335] is not directly comparable as she uses a different dataset and also the ribosome scanning model. Salamov *et al.*[737] is also not directly compatible as we have derived its result using the ribosome scanning model.

Classifier	Sensitivity	Specificity	Accuracy
Pedersen and Nielsen[658]	78.0%	87.0%	85.0%
Zien *et al.*[940]	69.9%	94.1%	88.1%
Zien *et al.*[941]	-	-	88.6%
Hatzigeorgiou[335]	-	-	94.0%
Salamov *et al.*[737]	-	-	88.8%
Li *et al.*[494]	84.7%	88.7%	87.7%

II. Accuracy of NB, SVM, and C4.5 reported by Wong *et al.*[928] on the Pedersen and Nielsen dataset based on 9 features selected using CFS.

Classifier	Sensitivity	Specificity	Precision	Accuracy
NB	84.3%	86.1%	66.3%	85.7%
SVM	73.9%	93.2%	77.9%	88.5%
C4.5	74.0%	94.4%	81.1%	89.4%

III. Accuracy of NB, SVM, and C4.5 reported by Liu and Wong[515] on the Pedersen and Nielsen dataset based on the 100 translated features selected using the entropy measure.

Classifier	Sensitivity	Specificity	Precision	Accuracy
NB	70.53%	87.76%	65.47%	83.49%
SVM	80.19%	96.48%	88.24%	92.45%
C4.5	74.88%	93.65%	79.51%	89.00%

Fig. 6. Accuracy results of various feature selection and machine learning methods for TIS recognition.

CHAPTER 5

HOW NEURAL NETWORKS FIND PROMOTERS USING RECOGNITION OF MICRO-STRUCTURAL PROMOTER COMPONENTS

Vladimir B. Bajić

Institute for Infocomm Research
bajicv@i2r.a-star.edu.sg

Ivan V. Bajić

University of Miami
ivan_bajic@ieee.org

Finding regulatory components in genomic DNA by computational methods is an attractive and complex research field. Currently, one of the important targets is finding protein coding genes in uncharacterized DNA. One of the significant aspects of gene recognition is the determination of locations of specific regulatory regions—promoters—that usually occupy the position at the beginning of a gene. Promoters are responsible for the initiation of the DNA transcription process. Current computer methods for promoter recognition are still either insufficiently accurate or insufficiently sensitive. We discuss in this chapter some of the general frameworks and conceptual issues related to the use of artificial neural networks (ANNs) for promoter recognition. The scenario discussed relates to the case when the main promoter finding algorithms rely on the recognition of specific components contained in the promoter regions of eukaryotes. Some specific technical solutions are also presented and their recognition performances on an independent test set are compared with those of the non-ANN based promoter recognition programs.

ORGANIZATION.

Section 1. We first discuss the motivation for finding promoters of genes in uncharacterized DNA sequences. Then we provide some background material on the use of ANN for this purpose.

Section 2. Next we describe some of the characteristic motifs of eukaryotic promoters, as well as problems associated with them. This gives us an understanding of the challenges in arriving at a general model for eukaryotic promoter recognition.

Section 3. A few attempts have been made in eukaryotic promoter recognition that use information based on the more common promoter region motifs—such as their position weight matrices—and their relative distances. We mention some of them next. We also review the study of Fickett and Hatzigeorgiou,[249] which reveals the high level of false positives in some of these programs. The role that enhancers may have played in false positive recognition of promoters is then discussed.

Section 4. Some details of some of the principles used in designing ANNs for recognizing eukaryotic promoters are presented in this and the next sections. In particular, we begin with discussing representations of nucleotides for ANN processing.

Section 5. Then we describe the two main forms of structural decomposition of the promoter recognition problem by ANN, *viz.* parallel vs. cascade composition of feature detectors. The promoter recognition part of GRAIL[549] and promoter 2.0[439] are then used to illustrate these two forms. We also discuss the construction of the underlying feature detectors.

Section 6. After that, we introduce time-delay neural networks, which are used in the NNPP program[702, 703, 705] for promoter recognition. We also discuss the issue of pruning ANN connections in this context.

Section 7. Finally, we close the chapter with a more extensive discussion and survey on the performance of promoter recognition programs.

1. Motivation and Background

Advances in genetics, molecular biology, and computer science have opened up possibilities for a different approach to research in biology—the computational discovery of knowledge from biological sequence data. Numerous methods aimed at different facets of this goal are synergized in a new scientific discipline named "bioinformatics". This new field has the potential to unveil new biological knowledge on a scale and at a price unimaginable 2-3 decades ago. We present here an overview of capabilities of the artificial neural network (ANN) paradigm to computationally predict, in uncharacterised long stretches of DNA, special and important regulatory regions of DNA called promoters. For references on ANNs, see for example Bose and Liang,[96] Caudill,[135] Chen,[143] Fausett,[241] Hercht-Nielsen,[338] Hertz *et al.*,[350] Kung,[465] and Rumelhart and McClelland.[730]

This chapter focuses on some of the basic principles that can be used in constructing promoter prediction tools relying on ANNs that make use of information about specific short characteristic components—the so-called "motifs"—of eukaryotic promoters' micro-structure. These motifs include subregions such as the TATA-box,[75, 116, 173, 624, 661] the CCAAT-box,[75, 223] Inr,[392, 422, 423, 640, 661, 774, 776, 872] the GC-box,[116] and numerous other DNA sites that bind a particular class of proteins known as transcription factors (TFs).[476, 878] In this approach the ANN system attempts to recognize the presence of some of these motifs and bases its final prediction on the evidence of such presences. This is conceptually different from the macro-structural approach, where the recognition of the promoter region is based primarily on the recognition of features of larger sections of the promoter and neighboring regions,[46−49, 51, 190, 321, 383, 674, 753] such as CpG-islands.[81, 180, 181, 271, 474]

Although the techniques to be presented here relate to the recognition of a particular group of regulatory regions of DNA—the promoter regions—the techniques discussed are far more general and are not limited to promoter recognition only. They could well be used in the large-scale search for other important sections of DNA and specific genomic signals, such as enhancers, exons, splice sites, *etc.*

1.1. *Problem Framework*

Probably the most fascinating aspects of bioinformatics is the computational investigation, discovery, and prediction of biological functions of different parts of DNA/RNA and protein sequences. One of the important practical goals of bioinformatics is in reducing the need for laboratory experiments, as these are expensive and time consuming.

The worldwide effort aimed at sequencing the human genome—the so-called Human Genome Project[132, 171, 467, 635, 859, 873, 874]—is virtually finished, although it was initially planned for finalization by the year 2004.[544] To illustrate the quantity of information contained in the human genome, note that it contains approximately 3 billion bp[893] and within it about 35,000–65,000 genes, while the number of regulatory regions and other functional parts in the human genome still remain to a large extent unclear.

An open question is our ability to computationally locate all important functional parts of the human genome, as well as to computationally infer the biological functions of such segments of genetic information. Recognition of constituent functional components of DNA, and consequently annotation of genes within a genome, depend on the availability of suitable models of such components. This relies on our understanding of the functionality of DNA/RNA and related cell products. This problem is very difficult and complex, as our current knowledge and understanding of DNA/RNA functioning is not complete.

Hence, one of the general problems that has to be solved is the accurate complete annotation of different genomes. This involves the identification and location of genes, as well as their associated promoters. Currently, the dominant interest is in finding genes that code for proteins, and the recognition of related promoters is a part of the problem. There are many techniques aimed at recognizing genes or some of their main features in long stretches of DNA—see Burge *et al.*,[121−123] Claverie,[166] Dong and Searls,[210] Eckman *et al.*,[217] Fickett *et al.*,[247, 248]

Gelfand *et al.*,[282, 283] Gish and States,[290] Guigo *et al.*,[305, 306] Hatzigeorgiou *et al.*,[333, 334] Hayes and Borodovsky,[336] Henderson *et al.*,[341] Hutchinson and Hayden,[378] Jiang and Jacob,[397] Kleffe *et al.*,[435] Krogh,[453, 454] Lukashin and Borodovsky,[529] Mathe *et al.*,[546] Milanesi and Rogozin,[566] Murakami and Takagi,[587] Quandt *et al.*,[690] Roytberg *et al.*,[727] Salzberg *et al.*,[738, 739, 741] Snyder and Stormo,[781] Solovyev *et al.*,[784, 785] Sze *et al.*,[811, 812] Tiwari *et al.*,[831] Uberbacher and Mural,[848] Ureta-Vidal *et al.*,[851] Xu *et al.*,[912-914] and Zhang.[930]

Some of these techniques utilize ANNs as parts of the solution at different levels of the problem and for different features and purposes—for example, see Brunak *et al.*,[114] Hatzigeorgiou *et al.*,[329, 331-334] Hayes and Borodovsky,[336] Lapedes *et al.*,[473] Rampone,[697] Snyder and Stormo,[781] and Uberbacher *et al.*[848, 914] However, numerous problems remain unsolved and, so far, the overall results of different types of predictions are not yet satisfactory.[127, 167, 304, 334] Within the problems that still await successful solutions is an accurate recognition of promoters, which remains one of the crucial components of the complete gene recognition problem.

1.2. *Promoter Recognition*

There are several main reasons why we are interested in searching for promoters in genomic DNA.[249, 656, 681] For example:

- Promoters have a regulatory role for a gene. Thus, recognizing and locating promoter regions in genomic DNA is an important part of DNA annotation.
- Finding the promoter determines more precisely where the transcription start site (TSS) is located.
- We may have an interest in looking for specific types of genes and consequently for locating specific promoters characteristic for such genes.

The problem of promoter recognition is not a simple one and it has many facets, such as:

- Determination of the promoter region, without any attempt to find out what such regions contain.
- Determination of the location of different binding sites for numerous TFs that participate in the initiation of the transcription process.
- Determination of the TSS, which is an important reference point in the context of transcription initiation.
- Determination of the functional classes of promoters, *etc.*

Thus, many techniques exist for promoter recognition and location. More details on such methods for prokaryotic organisms can be found in Alexandrov and Mironov,[19] Demeler and Zhou,[196] Grob and Stuber,[302] Hirst and Sternberg,[356] Horton and Kanehisa,[365] Lukashin *et al.*,[528] Mulligan and McClure,[584] Nakata *et al.*,[605] O'Neil,[638] Reese,[702] Rosenblueth *et al.*,[719] Staden,[795] etc.

The techniques for the recognition of eukaryotic promoters are much less efficient. The eukaryotic promoters are far more complex and possess very individual micro-structures that are specialized for different conditions of gene expression. It is thus much more difficult to devise a general promoter recognition method for eukaryotes. As in the case of prokaryotic promoter recognition, different techniques have been used to deal with the recognition of eukaryotic promoters of different classes—see Audic and Claverie,[38] Bajic *et al.*,[43, 44, 46−49, 51] Bucher,[116] Chen *et al.*,[150] Claverie and Sauvaget,[168] Davuluri *et al.*,[190] Down and Hubbard,[211], Fickett and Hatzigeorgiou,[249] Frech and Werner,[262] Hannenhali and Levy,[321] Hatzigeorgiou *et al.*,[330] Hutchinson,[377] Ioshikhes and Zhang,[383] Kondrakhin *et al.*,[446] Mache and Levi,[535] Matis *et al.*,[549] Milanesi *et al.*,[565] Ohler *et al.*,[626−628, 630] Ponger and Mouchiroud,[674] Prestridge,[679, 681] Quandt *et al.*,[690, 691] Reese *et al.*,[703, 705] Scherf *et al.*,[753] Solovyev and Salamov,[784] Staden,[796] and Zhang.[931] Recent evaluation studies[49, 249] of some publicly available computer programs have revealed that computer tools for eukaryotic promoter recognition are not yet mature.

1.3. *ANN-Based Promoter Recognition*

Some of the techniques mentioned are based on the use of artificial neural networks. Results on the recognition of promoters by ANNs in prokaryotes can be found in Demeler and Zhou,[196] Hirst and Sternberg,[356] Horton and Kanehisa,[365] Lukashin *et al.*,[528] Reese;[702] and those for eukaryotic organisms in Bajic,[42] Bajic *et al.*,[43, 44, 46−49, 51] Hatzigeorgiou *et al.*,[330] Knudsen,[439] Mache and Levi,[535] Matis *et al.*,[549] Milanesi *et al.*,[565] Ohler *et al.*,[627] and Reese *et al.*[702, 703, 705]

Results of several programs based on ANNs are available for comparison: NNPP;[702, 703, 705] Promoter2.0,[439] which is an improved version of the program built in GeneID package;[306] SPANN;[42, 43] SPANN2;[44] McPromoter;[627] Dragon Promoter Finder;[48, 49, 51] and Dragon Gene Start Finder.[46, 47] The programs mentioned operate on different principles and use different input information. For example, some of ANN-based programs are designed to recognize specific characteristic subregions of eukaryotic promoters, as in the case of NNPP,[705] Promoter2.0,[439] the promoter finding part of the GRAIL program,[549] the program of Hatzigeorgiou *et al.*,[330] and that of Wang *et al.* in Chapter 6 of this book. On the other hand, SPANN,[42, 43] SPANN2,[44] Dragon Promoter Finder,[48, 49, 51] Dragon Gene Start Finder,[46, 47] and McPromoter[627] use primarily integral information about the promoter region. The scores achieved by some of these programs are shown later and indicate that ANN-based methods for eukaryotic promoter recognition rate favorably with regard to non-ANN based programs.

2. Characteristic Motifs of Eukaryotic Promoters

Promoters are those parts of genomic DNA that are intimately related to the initiation of the so-called transcription process. The starting point of transcription—*i.e.*, the TSS—is generally contained within the promoter region and located close to, or at, its 3' end. A promoter in eukaryotes can be defined somewhat loosely as a portion of the DNA sequence around the transcription initiation site.[677] Eukaryotic promoters may contain different subregions—sometimes also called components or elements—such as TATA-box, CCAAT-box, Inr, GC-box, DPE, together with other different TF binding sites.

The problem with these subregions in eukaryotic promoters is that they vary considerably from promoter to promoter. They may appear in different combinations. Their relative locations with respect to the TSS are different for different promoters. Furthermore, not all of these specific subregions need to exist in a particular promoter. The high complexity of eukaryotic organisms is a consequence of high specialization of their genes, so that promoters in eukaryotes are adjusted to different conditions of gene expression, for example, in different tissues or in different cell types.

Thus, the variability of internal eukaryotic promoter structures can be large. Consequently, the characteristics of the eukaryotic promoter are rather individual for the promoter, than common for a larger promoter group.[216, 402, 476, 557, 570, 621, 681, 774, 799, 872, 878, 887] For this reason it is not easy to precisely define a promoter structure in eukaryotic organisms. This is also one of the reasons why at this moment there is no adequate computer tool to accurately detect different types of promoters in a large-scale search through DNA databases.

The simplistic version of the process of the initiation of transcription implies a possible model for eukaryotic promoters: It should have a number of binding sites. However,

- there is a large number of TFs—see TRANSFAC database details in Matys *et al.*;[553]
- TF binding sites—for one of their databases see Ghosh[287]—for different promoters may be at different relative distances from the TSS;[216, 402, 570, 887]

- for functional promoters the order of TF binding sites may be important;
- for different promoters, not all of the TF binding sites need to be present;[886] and
- the composition of TF binding sites for a particular promoter is essentially specific and not shared by a majority of other promoters.

It is thus very difficult to make a general computer tool for promoter recognition that uses information based on TF binding sites and their composition within the eukaryotic promoters.[43, 656]

There are three types of RNA polymerase molecules in eukaryotes that bind to promoter regions. Our specific interest is in RNA Polymerase II and their corresponding promoters—*viz.* Pol II promoters—whose associated genes provide codes for proteins. Many eukaryotic Pol II promoters have some specific subregions that possess reasonably high consensus. A typical example is the TATA-box. The TATA-box is a short region rich with thymine (T) and adenine (A) and located about −25 bp to −30 bp upstream of the TSS. But there are also other frequently present components like the CCAAT-box, Inr, DPE, *etc.*

Arriving at a general model for an eukaryotic promoter is difficult. Nevertheless, suitable models can be derived for specific classes of promoters. For example, the mammalian muscle-actin-specific promoters are modelled reasonably well for extremely accurate prediction.[261] Such modelling of specific narrow groups of promoters makes a lot of sense in a search for specific genes. The point is, however, that computer tools for the general annotation of DNA are aimed at the large-scale scanning and searching of DNA databases so as to recognize and locate as many different promoters as possible, and not to make large numbers of false recognitions. Obviously, it is difficult to make such tools based on highly specific structures of very narrow types of promoters.

3. Motif-Based Search for Promoters

The problems of initiation and control of transcription processes in eukaryotes have a major importance in the biochemistry of cells[656] and are the subject of intensive research.[125, 138, 249, 406, 448, 621, 685, 717, 775, 799, 860, 878] The accurate prediction of promoter location, including that of the TSS, can significantly help in locating genes. We have indicated previously that there are a number of components within the eukaryotic promoter region that may possibly serve as a basis for promoter recognition. Such methods have to take into account the great variability of the internal eukaryotic promoter structure,[216, 402, 476, 557, 570, 621, 681, 774, 799, 872, 887] which contributes to promoter complexity,[926] and to the complexity of the computational promoter recognition.

A reasonable approach in devising techniques for promoter recognition is to identify those patterns that are common to very large groups of promoters, and then to search for such patterns and their mutual distances. Such algorithms would reflect the biochemical background of the transcription process, and, in principle, should result in the least number of false recognition. Unfortunately, constructing such algorithms depends crucially on the detailed knowledge of the biochemistry of promoter's activity which is not yet fully available. In prokaryotic promoters very common patterns exist—for example, the −10 and −35 regions have reasonably high consensus and a very consistent distance between them. It is thus not surprising that a number of techniques have been developed to deal with prokaryotic promoter recognition.[19, 196, 302, 356, 365, 528, 584, 605, 638, 702, 719, 795] Some of these techniques are based on the use of ANNs.[196, 356, 365, 528, 702]

Due to the high structural complexity and the absence of a greater number of strong common motifs in eukaryotic promoters, the existent techniques aimed at computational recognition of eukaryotic promoters are much less accurate. The previously mentioned very individual micro-structure of eukaryotic promoters that are specialized for different conditions of gene expression complicates enormously the development of adequate techniques for promoter recognition. However, there are certain motifs within eukaryotic promoters that are present in larger promoter groups. Many of the eukaryotic promoter recognition programs base their algorithms on searching for some of these motifs, frequently by using some additional information such as the relative distances between the motifs.

A number of attempts have been made in this promoter recognition task that are aimed at utilizing information from some of these more common promoter region components. About 30% of eukaryotic Pol II promoters contain TATA-like motifs. About 50% of vertebrate promoters contain CCAAT-box motifs. Inr is also a very common subregion in eukaryotic promoters. It is found that combination of some of these specific subregions are crucial in the determination of the correct TSS location—such as the combination of TATA-box and Inr.[621] Also, the position weight matrices (PWMs) for the TATA-box, the CCAAT-box, the GC-box, and the cap site have been determined in Bucher.[116] In spite of the fact that the consensus of the TATA-box is not very strong, the PWM of the TATA-box from Bucher[116] appears to be a very useful tool for recognition of a larger group of eukaryotic Pol II promoters. This weight matrix is normally used in combination with the other methods[679] due to the fact that when it is used alone it produces a large number of false recognition of the order of 1 false recognition per 100–120 bp on non-promoter sequences.[679, 682]

It is possible to use only one target motif in the attempt to recognize promoters that contain such a motif and to achieve relatively good success—see Chapter 6. However, most methods that aim at recognizing eukaryotic promoters do not base their algorithms on locating only one of many possible micro-structural promoter components. They rather look for the existence of a suitable combination of such elements which is then assessed and used in the prediction task. For example, in Prestridge,[679] the prediction of eukaryotic Pol II promoters is based on the prediction of the TF binding sites and then combined with an assessment of the PWM score for the TATA-box. The TF binding sites that are used are those corresponding to the TF database from Ghosh.[287] The method is based on the assumption that the distributions of the TF binding sites in promoter and non-promoter regions are different. The resulting program, Promoter Scan, can predict both TATA-containing and TATA-less promoters and has shown a reduced level of false recognition compared with the other promoter-finding programs.[249] The last version of Promoter Scan[681] has an improved and extended functionality compared with the original version.

A sort of an extension of the method used initially for developing Promoter Scan has been made in TSSG and TSSW programs.[784] These programs are extended by the addition of a linear discriminant function that values (1) the TATA-box score; (2) the sequence composition about the TSS—*viz.* triplet preferences in the TSS region; (3) hexamer preferences in the three upstream regions—*viz.* [−300, −201], [−200, −101], [−100, −1]; and (4) potential TF binding sites. The programs use different TF databases.

Also, as in the case of Promoter Scan, a part of the AutoGene program—the program FunSiteP, which is responsible for finding promoters[446]—contains an algorithm devised on the assumption of different distributions of TF binding sites in the promoter regions and in non-promoter sequences. The database source for FunSiteP is based on a collection of binding sites from Faisst and Meyer.[237]

The other group of methods that explicitly use eukaryotic promoter microstructure components—at least as a part of the algorithm—exploit the modelling and generalization capabilities of ANNs.[44, 46−49, 51, 330, 439, 535, 549, 565, 627, 702−707] Some of them utilize, in one way or other, the fact that combinations of some of the specific subregions—such as the combination of the TATA-box and Inr—helps in the determination of the TSS location. Certain programs also use explicit information on relative distances between such specific subregions. In the next sections we see in more details some of the principles that may be used in designing ANN systems for eukaryotic promoter recognition.

3.1. *Evaluation Study by Fickett and Hatzigeorgiou*

A recent evaluation study[249] of publicly available computer programs has indicated different degrees of success of these programs and revealed that tools for promoter recognitions do require a lot of additional development. On the specific evaluation set used,[249] where only the ability of programs to locate the TSS is considered, the rate of success is in the range of 13%–54% of true positive predictions (TP), while false positive predictions (FP) are in the range of $1/5520$ bp in the best case and up to $1/460$ bp in the worst case.

TP predictions are correct predictions of the TSS location within the prespecified bounds arround the actual TSS location. FP predictions are those reported as predicted TSS locations at positions out of the above mentioned bounds. The interpretation of the FP score of, say, $1/200$ bp means the promoter recognition system produces on an average 1 FP prediction of promoters every 200 bp.

The general observation is that the level of TP predictions is directly correlated to the level of FP predictions. So, the best program in correct positive predictions—NNPP, which is based on neural networks—produces 54% TP predictions, and FP at the level of $1/460$ bp. On the other hand, the Promoter Scan program,[679] which is not neural network based, produces a score of 13% TP predictions, but achieved the lowest proportion of FP at the level of $1/5560$ bp.

It should be indicated that TP and FP as measures of success in prediction programs are not very convenient for comparison of prediction programs that produce different TPs and FPs. Thus, to be able to make a reasonable comparison of different programs on a unified basis, a more convenient measure of success scores from Bajic[50] is used later. It shows a rational ranking of promoter prediction programs and corroborates our statement that ANN-based programs for promoter prediction exhibit comparable or better performance to non-ANN promoter prediction programs.

3.2. *Enhancers May Contribute to False Recognition*

Closely associated with promoter regions in eukaryotes is another class of transcriptional regulatory domains in DNA—the enhancers. Enhancers cooperate with promoters in the initiation of transcription. They are located at various distances from the TSS and sometimes may be several thousands nucleotides away. They can also be located either upstream or downstream of the TSS.

As in promoters, enhancers also contain different TF binding sites. Thus, as pointed out in Fickett and Hatzigeorgiou,[249] one of the reasons for having a high level of FPs produced by programs for finding eukaryotic promoters can be that most of these techniques are mainly based on finding specific TF binding sites within the promoter region, or in the assessment of the density of TF binding sites in the promoter and non-promoter sequences. On this basis it seems that enhancers could frequently be recognized as promoters. Thus, it would be of interest to develop methods to discriminate between the promoter and enhancer regions in uncharacterized DNA, and in this way to contribute to the reduction of the FP scores of some of promoter recognition programs.

4. ANNs and Promoter Components

4.1. *Description of Promoter Recognition Problem*

The best way to understand a possible role of neural networks in promoter prediction in eukaryotes is to examine the essence of the problem that needs to be solved by neural networks. Let us assume that there may exist n specific subregions R_j, $j = 1, 2, ..., n$, some of which we may use for promoter recognition. Let R_j^k denote the region R_j in the k-th promoter. We use the superscript to indicate that the form—the actual composition, length, and relative position with respect to the TSS—of the subregion R_j in the k-th promoter may be, and usually is, different from the form of the same region in another, say, i-th promoter. Let s_j^k and e_j^k denote the starting position and the ending position of the region R_j^k in the k-th promoter. These positions are counted from the 5' end of the DNA string and represent relative distances from an adopted reference position. Thus, $s_j^k < e_j^k$.

Let us also assume that the starting points of these subregions are at distances d_j^k, $j = 1, 2, ..., n$, from the TSS. The values of d_j^k are taken from the set

$$\mathbb{Z}_\perp = \mathbb{Z} \cup \{\perp\}$$

where $\mathbb{Z}$ is the set of integers. Note that, due to the possible absence of a subregion R_j^k from the k-th promoter, it may be that d_j^k cannot be defined. This is the reason for having the special symbol $\perp$ in the definition of $\mathbb{Z}_\perp$. Note also that we use negative values of d_j^k for the locations of R_j^k upstream of the TSS, we use positive values of d_j^k for the downstream locations. Thus the sign of d_j^k determines only the direction from the TSS.

In an uncharacterized genomic sequence we do not know the location of the TSS—it has yet to be determined. Thus, we cannot use the information of distances d_j^k in the search process, even though we can do this during the training process as it may be assumed that in the training set the TSS and subregions of interest are known. In fact, it makes sense to develop specialized ANN systems aimed at searching for specific promoter components, where these systems use information on distances d_j^k directly. This may be the case in tasks when the location of the TSS is known, but when there is not enough information about the promoter structure.[681] However, we mainly consider here the problem of uncharacterized genomic sequences.

The direct use of distances d_j^k can be circumvented if we use relative distances between the subregions. We assume that in one promoter two functional subregions R_i^k and R_j^k do not overlap. Thus

$$D_{ij}^k = \begin{cases} s_j^k - e_i^k - 1, & \text{if } s_j^k > e_i^k \\ \\ s_i^k - e_j^k - 1, & \text{if } s_i^k > e_j^k \end{cases}$$

denotes the mutual distance of subregions R_i^k and R_j^k in the k-th promoter. This distance does not include the ending point e_i^k (respectively, e_j^k) of the first subregion R_i^k (respectively, R_j^k)—counted in the direction from 5' toward the 3' end—nor the starting point s_j^k (respectively, s_i^k) of the subsequent subregion R_j^k (respectively, R_i^k). Alternatively, one can express these distances by means of subregion length c_i^k as

$$D_{ij}^k = \begin{cases} s_j^k - s_i^k - c_i^k, & \text{if } s_j^k \geq \left(s_i^k + c_i^k\right) \\ \\ s_i^k - s_j^k - c_j^k, & \text{if } s_i^k \geq \left(s_j^k + c_j^k\right) \end{cases}$$

Further, the characteristics of the subregion R_j that we are trying to identify—and by which we attempt to recognize its presence in the k-th promoter—may be varied. So, let us assume that we are interested in identifying a feature F_j of the subregion R_j in all promoters under investigation. That is, we try to identify the feature F_j for all R_j^k, $k = 1, 2, ..., n_p$, where n_p is the number of promoters in the group we analyze. The feature F_j may be expressed, say, as a set of probabilities of specific motifs appearing at appropriate positions relative to the reference indicator, or it may be given in the form of a suitably defined discrepancy function $discrep(R_j^k, R_j^{ref})$—for example, the distance of R_j^k from R_j^{ref} in a suitable space—where R_j^{ref} may represent a prespecified motif, consensus sequence, *etc.* Although we can consider several features associated with a subregion, for simplicity we restrict our consideration to only one such feature. We need to highlight the fact that for virtually any of the specific subregions there is no unique subregion description. As an example, many different compositions of nucleotides may represent the same subregion although the subregion is characterized by a strong consensus signature.

The order of subregions R_j in a specific promoter may be of importance for the functionality of the promoter. Thus the ordering of subregions R_j^k is also a candidate as an input parameter for assessment.

nucleotide	code
A	1000
C	0100
G	0010
T	0001

Fig. 1. Binary code that can be used to represent four nucleotides in DNA.

An additional problem that is encountered in dealing with the locations of sub-regions is related to the fact that domains of location for two functional subregions R_i and R_j in a set of sequences may overlap, although in a particular promoter they are separate. The overlapping of possible locations comes from considering a group of promoters containing the same subregions. A typical example is in the case of the Inr and TATA-box in eukaryotic Pol II promoters, where Inr can be located within the −14 bp to +11 bp region relative to the TSS, while the TATA-box can be located in the −40 bp to −11 bp domain. Thus, a number of constraints of this type may be of importance when formulating input information to the neural network system.

So, we can summarize the problems that appear:

- generally, the model describing a subregion and its selected feature(s) does describe a set of more or less similar sequences, but does not determine them uniquely;
- not all subregions need to exist in a particular promoter;
- relative distances of subregions from the TSS are variable—which implies that the relative mutual distances of subregions are variable too;
- order of subregions may be of importance; and
- possible overlapping of subregion domains can occur.

4.2. *Representation of Nucleotides for Network Processing*

The DNA sequence is a sequence of 4 different nucleotides denoted in the customary way as "A", "T", "C", and "G". If no biological or physical properties of these nucleotides are taken into account, then a suitable representation of these nucleotides may be by the code given in Figure 1.

This code has the same Hamming distance between any two coding vectors for the A, C, G, and T nucleotides, which is considered desirable so as not to contribute to biased learning. This code representation has been used in a number of ANN applications to DNA and protein analysis—for example, see Brunak *et al.*,[114] Demeler and Zhou,[196] Farber,[240] Knudsen,[439] O'Neill,[637] Qian and Sejnowski,[687] and Reese and Eeckman.[705]

However, it should be mentioned at this point that this is not the only possible numerical representation of nucleotides. If we want to incorporate some of the physical properties that characterize nucleotides and base our further analysis on such a representation, then, for example, the electron-ion interaction potential (EIIP),[857,858] may also be used with success in promoter recognition algorithms.[42−49,51]

It is difficult to determine at this stage of computational genomics research which coding system is more effective. One can argue that the A, T, C, and G nucleotides generally form two groups with specific chemical characteristics— purines (A, G) and pyrimidines (C, T)[547]—so that their numerical representation for the purpose of computer analysis should reflect such similarities in order to more adequately mimic the real-world situation. Consequently, it seems that it is not the right approach to use the binary coding as mentioned above. Also, the ordering of nucleotides is crucial in determining the function of a particular section of DNA. Since the biochemical functions of DNA segments depend on that order—for example, in a particular context several successive purine nucleotides can have different biochemical properties than if their positions are occupied by pyrimidine nucleotides—it seems more logical to use a coding system that reflects physical or biological properties of the nucleotides. Thus we favor numerical coding of nucleotides via physical characteristics that they may have, over the essentially artificial allocation of binary numerical representation, such as the binary coding presented above.

5. Structural Decomposition

The problem of eukaryotic promoter recognition allows several possible structural decomposition forms. We comment on two of such decomposition structures. The first one basically uses parallel composition of feature detectors (PCFD), while the other uses cascade composition of feature detectors (CCFD). Both structures comprise hierarchical systems.

5.1. *Parallel Composition of Feature Detectors*

It is possible to build a neural network system so that, on the first hierarchical level, ANNs attempt to recognize the individual features F_j either as independent

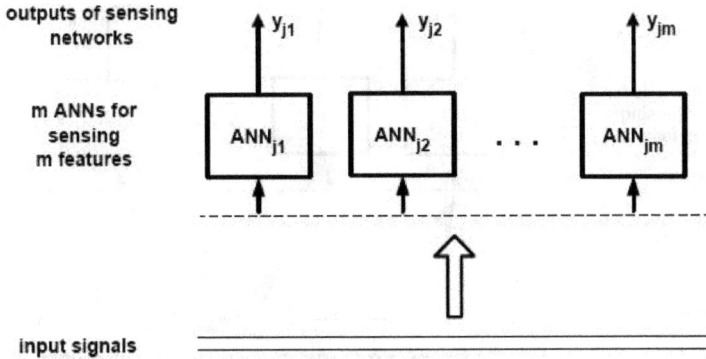

outputs of sensing networks y_{j1} y_{j2} y_{jm}

m ANNs for sensing m features ANN_{j1} ANN_{j2} ... ANN_{jm}

input signals

Fig. 2. Conceptual structure of the first hierarchical level of an ANN system aimed at feature recognition of promoter subregions.

features, or in specific combinations—see Figure 2. Naturally, the independent recognition of individual features is far simpler and more practical.

One possible way to realize this first hierarchical level of ANNs is depicted in Figure 3. Let us assume that there are m subregions that we intend to identify, R_j, $j \in \{j_1, j_2, ..., j_m\}$; that each subregion R_j is characterized by a feature F_j; and that we have m ANNs, ANN_j, $j \in \{j_1, j_2, ..., j_m\}$, for the independent recognition of features F_j, $j \in \{j_1, j_2, ..., j_m\}$. We assume that the neural network ANN_j for the recognition of feature F_j requires information gathered by reading data through a data window w_j that slides along the DNA strand from its 5' end towards its 3' end. Then the process of supplying the ANN system with input information is equivalent to independently sliding m required windows w_j along the DNA sequence and feeding each feature-sensing ANN with information from the appropriate window. The network ANN_j then produces output signal y_j.

Depending on the choice of neural networks, these output signals may be continuous or discrete. If they are continuous, then usually their amplitudes are related to the sensing accuracy—the larger the amplitude, the higher the certainty that the subregion is identified, *i.e.*, the greater the chance that the searched-for feature is detected. If the networks for feature sensing are designed to function like classifiers,[96, 135, 143, 241, 350, 465, 513, 730] then they perform the classification of input patterns into appropriate output categories. For example, they may have outputs at 1 to correspond to the "identified" subregion or feature, *vs.* 2 to correspond to the case when the subregion or feature is not identified at the given position. Depending on the problem, the classifier networks may have to learn

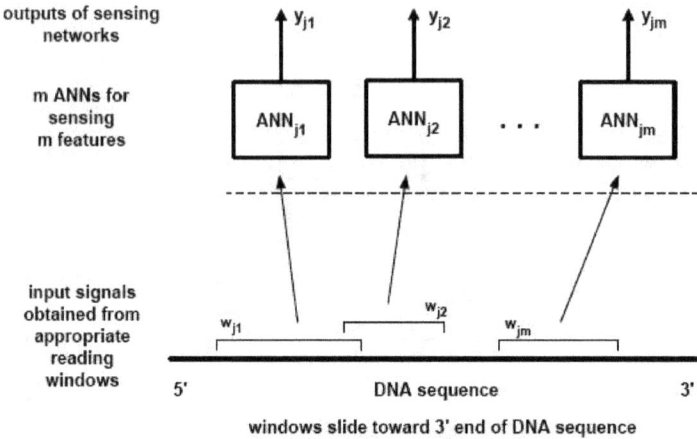

Fig. 3.   Input signals for networks on the first hierarchical level are obtained from the appropriate reading windows $w_{j_1}, w_{j_2}, ..., w_{j_m}$.

very complicated multidimensional decision boundaries so as to be able to conduct classification properly. Generally, ANN classifiers are capable of learning such complicated decision surfaces, but these capabilities and final success depend on the type of network, as well as the training procedure and the training set.[465]

In our case it is convenient to deal with 2 output categories for each feature-sensing networks, but this need not always be the case. Some of the typical types of ANNs that produce discrete output signals convenient for this category are probabilistic neural networks (PNNs).[788, 789, 791] These networks perform input pattern classification with a decision boundary that approaches asymptotically the Bayesian optimal decision surface. That is, they asymptotically converge to the Bayesian classifier. Another useful type of networks, whose decision boundary also approximates the theoretical Bayesian decision surface, contains the learning vector quantization (LVQ) networks,[442, 443] which may be considered as a variant of self-organizing-map (SOM) networks[440, 442, 444] adjusted for supervised learning. Also useful are the radial basis function based classifiers,[591, 726] or the multilayered perceptrons (MLPs).[885] MLPs are capable of approximating arbitrary discrimination surfaces,[513] although their approximations of these surfaces have some topological constraints.[289] This only means that one may need a very large MLP with a sufficient number of hidden layers in order to sufficiently well approximate the smooth decision boundary. The size of such an MLP may lead to problems of training. The situation can be relaxed somewhat if a nonlinear pre-processing of MLP inputs is made.[884]

For practical purposes it is more convenient to have networks with continuous outputs at this level, such as the radial basis function networks (RBFNs),[53, 109, 152, 412, 576, 577, 667, 670, 676] the generalized regression networks (GRNN),[790, 791] or some of the many other forms of ANNs that produce continuous output signals.[96, 135, 143, 241, 350, 465, 730] As an example, MLPs can be associated with the probabilities of detecting a time-evolving feature[97] so as to produce continuous output signals and can be used in the context of promoter subregion sensing.

It is important to note that the networks used on this level can produce a large number of FPs. This is intimately related to the problem discussed in Trifonov,[835] where it has been shown on some "hard-to-believe" examples—to use the terminology from Trifonov[835]—that sequences quite different from consensus sequences functionally perform better than those consensus sequences themselves. This is one of the general problems related to what consensus sequences represent and how "characteristic" they are for the pattern that is searched for. The same difficulty appears with the usage of the PWM that characterizes a specific subregion. This highlights the real problem of what is a suitable definition of similarity between the signature of the pattern R_j^{ref} (the template pattern) we look for and the potential candidate sequence R_j^k that is tested in the search process. In other words, the problem is how to express mathematically the discrepancy function $discrep(R_j^k, R_j^{ref})$ in the most effective way.

On the higher hierarchical level, a neural network system can be built to assess the identified combination of features in association with their mutual distances and their ordering, as depicted in Figure 4. This problem belongs to the class of the so-called multi-sensor fusion/integration problems,[712, 814, 924] but their implementation is complicated by the necessity to cater at this level for spatial/temporal patterns and their translation invariance. If the information on the relative mutual distance between the sensed feature at the lower hierarchical level is not presented to the higher hierarchical level network, then the network on the higher hierarchical level has to have the capability to learn and recognize the spatial/temporal events so as to be able to assess simultaneously the feature signals y_i obtained from the networks at the lower level and their spatial/temporal differences, so that it can estimate the overall combination and produce the system output signal y_0. Different classes of ANNs can be used for this purpose, but dynamic (recurrent) networks seem to be best suited for this task.[229, 866, 867, 870] If, however, information about mutual relative distances of the sensed features is contained in the input

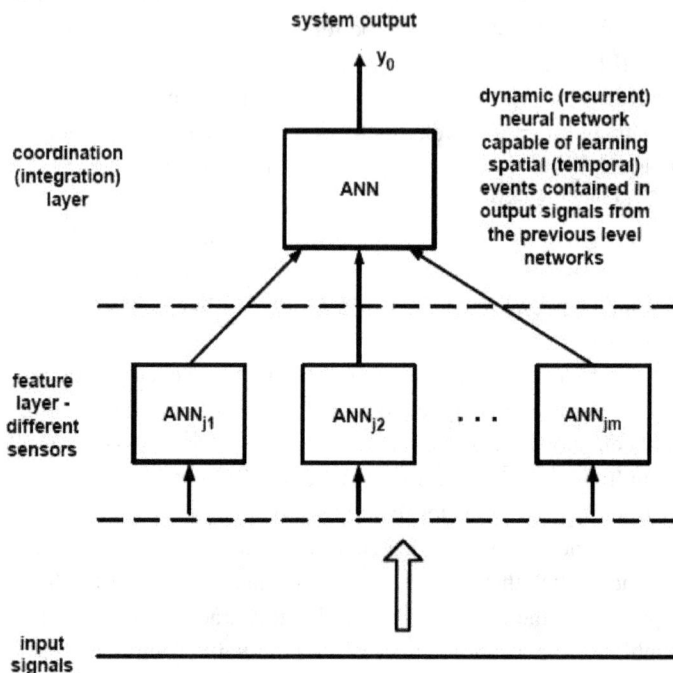

Fig. 4.   One possible general ANN structure for eukaryotic promoter recognition.

signals to the ANN on the coordination level, then such coordination ANN need not be recurrent.

In principle, suitable and sufficiently complex structure of neural network systems can allow a non-hierarchical approach. However, the problems of training such networks may be considerable, either from the viewpoint of the time required for the training, or from the viewpoint of network parameter convergence to reasonably good values, or both. For these reasons, in the case of recognition based on sensing different promoter subregion features, it is pragmatic to apply a hierarchical approach in designing neural network systems.[376, 482, 588, 712] With this basic idea in mind, one can build several different hierarchical structures of ANNs to suit eukaryotic promoter recognition. The structures presented in Figure 4 and Figure 5 show two of many possibilities.

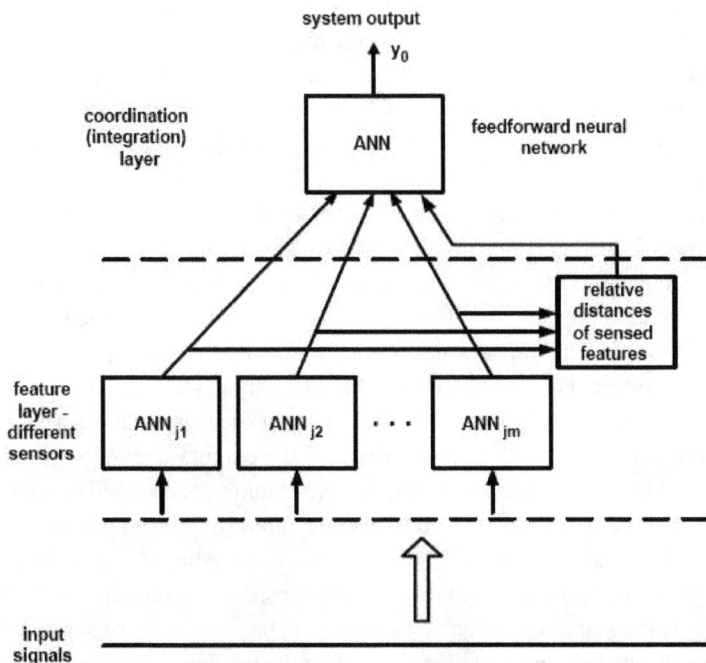

Fig. 5.   Another structure of ANN-based system which uses also distances between the signals for promoter recognition.

5.2. First- and Second-Level ANNs

At the first level we have the "feature sensing" networks, *viz.* those that are trained to sense the presence of specific subregions of promoters on the basis of selected subregion features. These networks can be any static neural networks.[96, 135, 143, 241, 350, 465, 730] Although some solutions[535, 702−707] utilize time-delay neural networks (TDNN) proposed in Waibel *et al.*,[866] it is not necessary to have dynamic networks on this level. Moreover, dynamic ANNs may not be more accurate as feature sensors. Anyway, whatever choice of static neural networks is made for dealing with information processing at this level, the general problem to be further handled is explained by an example of two feature-sensing networks in what follows.

Let us assume that the system is based on the recognition of subregions R_1 and R_2, whose identified features are F_1 and F_2, respectively. Let the first layer neural networks ANN_1 and ANN_2, serve the purpose of identifying features F_1

and F_2, respectively, and produce continuous output signals y_1 and y_2. These output signals are assumed to be confined to the interval $[0, 1]$, where values close to 1 denote a strong signal—a high certainty that at the given position of the reading window the feature is detected. For lower values of the signal, this certainty reduces; with values close to 0 the chances that at the given window position the feature exists are slim.

In many simpler cases the system has to make the choice regarding the levels of output signals at which, and above which, the features are considered detected. These are frequently determined by the cut-off (threshold) value for each output signal. Note that there may be different cut-off values for each of the output signals, although this is not always necessary. In our case depicted in Figure 6 we consider two different cut-off values, one for each output signal.

Note also that the concept of cut-off values is not necessary, although it simplifies the problem to an extent. One can essentially use output values of the first layer networks and leave decisions about whether the features are detected or not to the higher-level coordination ANN. In this case the input signals of the coordination ANN may have a large number of components corresponding to signals sensed at different positions of the data window. Another possibility to avoid cut-off values is to use only the maximum value of a signal from the output of one sensor within the examined data window, and then to consider the maximum value of the output signal of another sensor, and the relative distance between them. This however is constrained by a serious conceptual problem, to be shown later.

Figure 6 shows a possible situation with the measured signals. According to Figure 6, any position from x_1 up to x_2, including these two positions, determines a valid candidate for the predicted existence of F_1. Analogously, the existence of F_2 is predicted on positions starting with x_3 and ending at x_4. One can raise the legitimate question: Why are there so many predicted possible locations for a "detected" feature? The answer is that in the feature domain the overlapping of characteristics of the correct and wrong subregions is huge which leads to a great number of wrong guesses of the ANNs.

The relevant combinations of positions from $[x_1, x_2]$ and from $[x_3, x_4]$ are subject to some constraints on distance $dist(R_1, R_2)$ between the subregions R_1 and R_2. Generally, for many subregions in a group of promoters we know some of the constraints regarding their locations. These constraints on $dist(R_1, R_2)$ can be used to define the logic to handle signals y_1 and y_2 in the context of their mutual distance. Obviously, the minimal and the maximal spatial distances between R_1 and R_2 for the used cut-off values, as determined by the ANNs, are given by

$$\min\left[dist(R_1, R_2)\right] = x_3 - x_2 + 1,$$

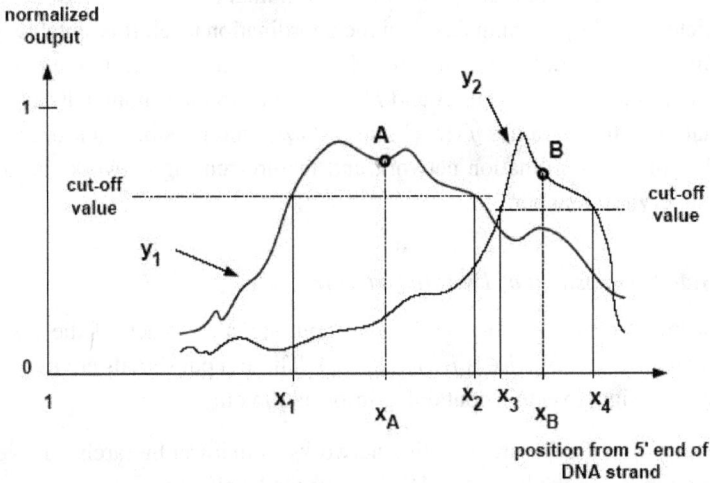

Fig. 6.   Presentation of possible output signals y_1 and y_2 from sensor networks at the lower hierarchical level, their cut-off values, and reference positions x_1, x_2, x_3, and x_4 on the basis of which the spatial range (or time interval) for occurring features is to be determined.

$$\max\left[dist(R_1, R_2)\right] = x_4 - x_1 + 1.$$

If the general constraint on $dist(R_1, R_2)$ is given by $dist(R_1, R_2) \in [\alpha_1, \alpha_2]$, then for an arbitrary pair of points A and B, where A belongs to the candidate region $[x_1, x_2]$ and B to the candidate region $[x_3, x_4]$, only some combination of signal value pairs $(y_A(x_A), y_B(x_B))$ are allowed. Those that are allowed are determined by the condition

$$(x_B - x_A) \in [\alpha_1, \alpha_2]$$

Notice the ordering of features F_1 and F_2 contained in the condition above. The logic block operating at the input of the ANN at the higher hierarchical level should handle this problem, or the solution can be implemented via the ANN design.

In addition, it is not always the case that the strongest signals in the detected intervals $[x_1, x_2]$ and $[x_3, x_4]$ are the correct or the best candidates that characterizes the correct detection of features F_1 and F_2. This is a consequence of the same reasons highlighted in the discussion in Trifonov[835] and mentioned in the preceding sections. This is one of the most serious criticisms of solutions that use the maximum output values of the feature sensing ANNs to "recognize" the presence of the sensed feature. The decision of what are the best or successful combinations of the obtained values of output signals and their mutual distance in space or time has to be determined by training ANN on the coordination level. It is suitable that the coordination network be dynamic in order to be capable of learning the most successful combinations of points A and B—i.e., to learn their mutual distance— in combination with the values $y_A(x_A)$ and $y_B(x_B)$, and possibly their ordering. The training of the coordination network and feature-sensing networks is to be done in a supervised manner.

5.3. Cascade Composition of Feature Detectors

To explain this scenario, denote by X_j the input signal for each of the feature detection networks $ANN_j, j \in \{j_1, j_2, ..., j_m\}$. These input signals are assumed to be composite signal vectors containing information on:

- the output Y_j of the feature detection networks from lower hierarchical levels;
- information Z_j on the basic raw DNA sequence level; and
- possibly other information—represented by vector G_j—acquired either during the training process, or from the biological constraints related to the data used.

Output Y_j of ANN_j is obtained by postprocessing the raw output y_j of ANN_j. This structure is depicted in Figure 7. Note that at each of the hierarchical levels a particular information can be used or not. Thus,

$$X_j = \langle \lambda_{j-1} \times Y_{j-1}, \lambda_{j-2} \times Y_{j-2}, \ldots, \lambda_1 \times Y_1, \lambda_{j,z} \times Z_j, \lambda_{j,g} \times G_j \rangle$$

Switches λ_i, $i = 1, ..., j - 1$, and $\lambda_{j,z}, \lambda_{j,g}$, have value of 1, if the respective information they are associated with is used at the jth hierarchical level, or they have value of 0 otherwise. Training of feature detector networks is done successively in a partial hierarchy, where the network at the level j is trained by having all lower level networks included in the structure. The training process can take many forms. Such cascade hierarchical structures are expected to have good filtering characteristics, but it cannot be concluded that they are advantageous over the PCFD structures.

5.4. Structures Based on Multilayer Perceptrons

MLPs can be used to detect individual promoter components, as well as to combine accurately such evidence into higher hierarchical ANNs so as to provide the final prediction of promoter existence. We present two such solutions that are parts of larger packages aimed at gene recognition. The first one a part of the GRAIL package.[549] The other one named Promoter2.0[439] is an improved version of the

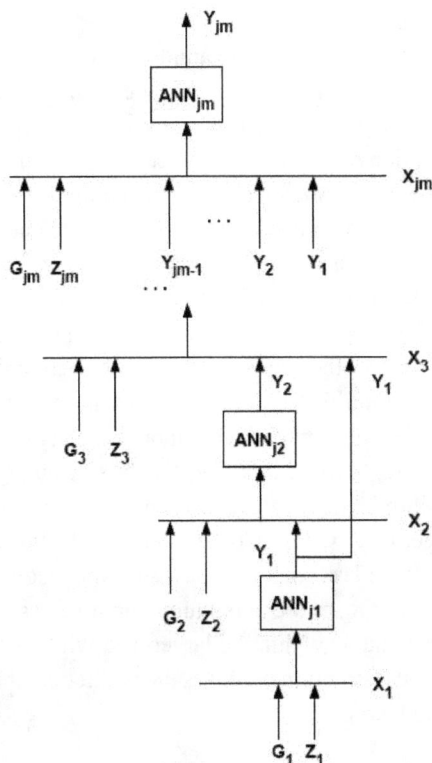

Fig. 7. The cascade structure of feature detectors.

promoter prediction part of the GeneID package.[306] Both of these solutions rely on the sensors for different promoter components and relative distances between them.

The solution presented in Matis *et al.*[549] for the GRAIL package is an example of a parallel composition of feature dectectors (PCFD) structure—*i.e.*, a conventional feedforward structure—which is in principle a two-level hierarchical structure. The MLP at the second hierarchical level receives information from different sensors related to promoter components, such as the Inr, TATA-box, GC-box, CCAAT-box, as well as from the translation start site, and the constraints on the relative distances between them. Such information is nonlinearly processed and the final prediction is produced. The sensors for different components can be based on ANNs, but this need not necessarily be the case. It is emphasized that this solution uses information about the presence of the translation start site, which is not part of promoter region.

The Promoter2.0 program[439] is an example of a cascade composition of feature detectors (CCFD) structure—*i.e.*, a cascade feedforward structure—and presents a different hierarchical solution that uses essentially four hierarchical levels for the final prediction of promoter presence. Based on the explanation provided in Knudsen,[439] the functioning of the system is roughly as follows. ANNs at each of the four levels consist of only one hidden neuron and one output neuron. In each of the ANNs, both neurons receive, as part of the input signals, information about the DNA composition in a binary form from a window of 6 nucleotides. In addition, they also receive in a floating point form the input signals representing the maximum outputs of the other networks on lower hierarchical levels multiplied by a separation function that relates to the normalized distances between the sensed components. The four promoter components that the ANNs are trained for are TATA-box, cap-site, CCAAT-box, and GC-box. Training of this network structure is done by a simplified genetic algorithm. The scanning of the DNA sequences is done in larger data windows of 200–300 nucleotides, within which a smaller window of 6 nucleotides slides, and the results are recorded for each of the four ANNs. After the recording of the maximum output values and the respective positions of the smaller window within the larger data window is finished for all four ANNs, the hierarchical structure produces the prediction of the presence of a promoter in the examined larger data window.

6. Time-Delay Neural Networks

A popular choice of dynamic networks that can inherently comprise the structure presented in Figure 4 is the time-delay neural networks (TDNN). Due to their special architecture, TDNNs are capable of learning to classify features that are invariant regarding their spatial (or temporal) translation. The basic unit (neuron) of a TDNN is a time-delay (TD) neuron, illustrated in Figure 8. These networks are initially used in the problem of phoneme recognition[866, 867] and in word recognition.[470] There are some similarities in the problems of phoneme recognition and in the recognition of promoters, where the latter is based on detection of the micro-promoter components. The features that need to be detected, if they are found in a promoter, may be on different mutual spatial locations. TDNN provides a convenient mechanism to make the neural network system insensitive to mutual distances of relevant events.

A TD neuron, depicted in Figure 8, has s inputs u_i, $i = 1, 2, ..., s$, and each of these inputs is also delayed by a maximum of N time units. Each of the inputs and its delayed version has its own weight $w_{i,j}$, $i = 1, 2, ..., s$, $j = 0, 1, ..., N$.

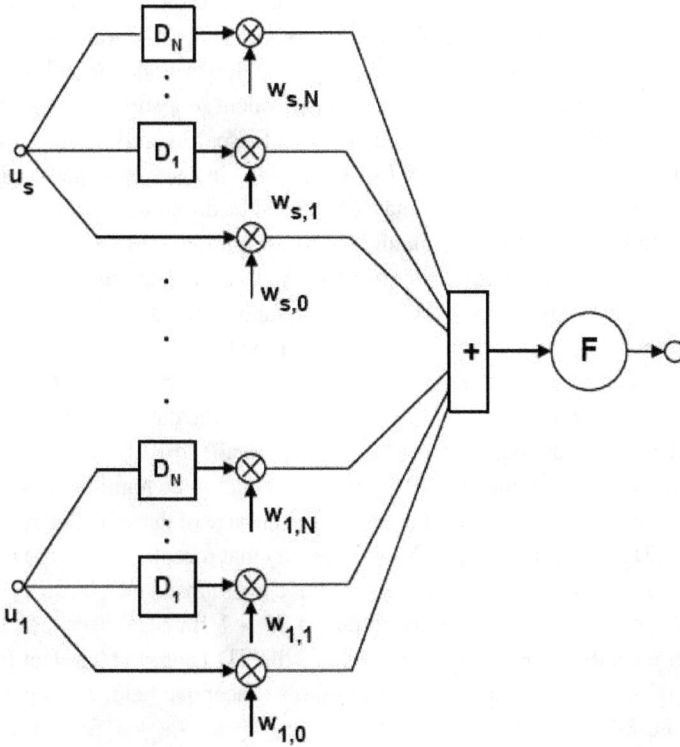

Fig. 8. The TD neuron in this figure has s inputs and each of these inputs is delayed a number of time (position) units (in our case N units). The weighted sum of all delayed inputs and the non-delayed inputs is passed through the nonlinear transfer element F.

This means that there are $s \times (N + 1)$ weighted signals that are summed so as to make the input to the nonlinear transfer element characterized by function F. This transfer element is sigmoid or threshold. Thus, using the notations $u_i(k)$ to mean the value of input u_i at instance k and $y(k)$ to mean the value of output y at instance k, the value of the TD neuron output y at instant k is given by

$$y(k) = F\left(\sum_{i=1}^{s} \sum_{j=0}^{N} w_{i,j} \times u_i(k - j) \right)$$

Without aiming to be rigorous, we explain in what follows some of the main aspects of information processing in TD neurons and TDNNs. Particular applications may have many variations of these basic ideas. Let us assume that the input data are of the form of 1-D signals, either scalar or vector. The inputs that the TDNN receives are collections of frames of this data, and these frame collections are supplied to the TDNN in a sequential manner. However, due to a TD neuron input-delaying functionality, this process is equivalent to a sort of inherent rearranging of the 1-D input signals to the 2-D events; see Figure 9. Let us assume that a feature F_0 that we are looking for is contained in a subsequence contained in a data frame of length s which may be located at different relative positions with regard to some reference point, and that the variation of this position is limited to the data window of length w. In other words, the data window of length w contains the subsequence characterized by F_0 and comprises all of its possible positions relative to the reference point. We mention here that the reference point may be a position of another feature F_1 detected by a feature-sensing ANN.

The data window slides along the DNA string and the data it contains at each position of the window can be considered rearranged, so that all window data make—after the rearrangement—a 2-D form. The maximum number of N delays of individual inputs to the TD neuron determines the size of the so-called receptive field of the TD neuron. This size is $N+1$. The way that receptive fields are formed is explained in Figure 9. Data from s consecutive positions in a DNA string serve as inputs to the TD neurons. The consecutive $N + 1$ frames correspond to the TD neuron receptive field—i.e., at moment k, the TD neuron reads data from a collection of $N + 1$ data frames that belong to its receptive field. As depicted in Figure 9, the 2-D form of the window data has $q = w - s + 1$ frames of data, where data in the first frame corresponds to $U(k) = \langle u_1, u_2, ..., u_s \rangle$, in the second frame to $U(k-1) = \langle u_2, u_3, ..., u_{s+1} \rangle$, and so on. Here, the counting of positions starts at the rightmost position of a frame on the DNA string. In order to be able to capture a feature irrespective of its relative position within the window, a layer of TD neurons should have at least $q - N + 1$ TD neurons. This however can be altered and the task of grasping spatial or temporal invariance can be transferred to a higher-level layer of TD neurons.

One of the most crucial ingredients of the translation-invariance of TDNNs is that weights related to different frame positions are copies of the set of weights that correspond to the first set of the receptive field frames. The learning procedure for the TDNN can be based on back-propagation.[728,729] Different learning procedures can be found in Waibel et al..[867] The TDNN is exposed to a sequence of learning patterns so as to be able to learn invariance in the space or time translation of the relevant patterns. In the training process, all the weights are treated as independent—i.e., they can change individually for one learning iteration step— but after that the related weights are averaged and these values are allocated to the relevant connections.[866,867] The process is repeated through a great number

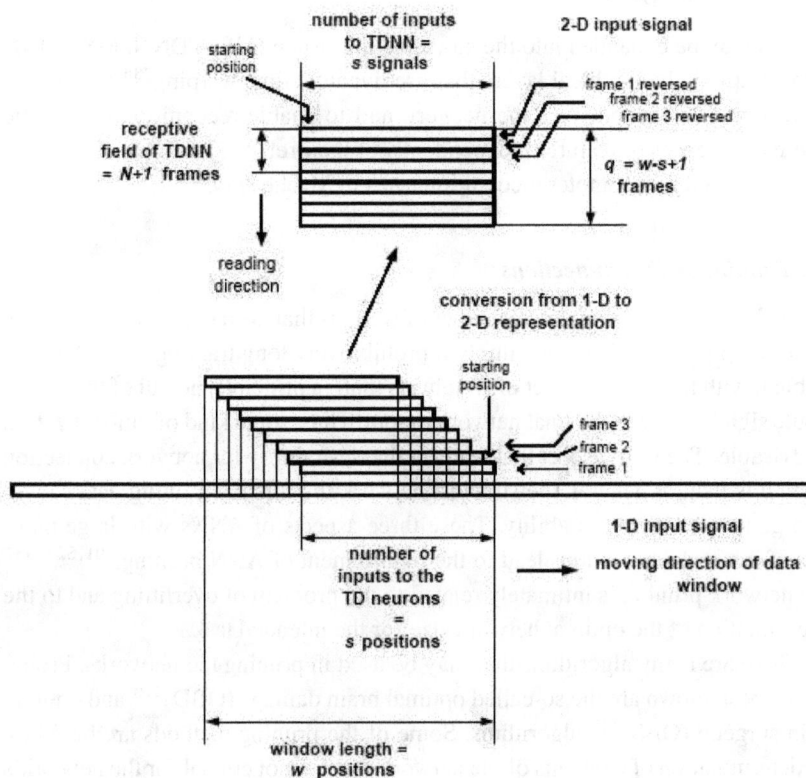

Fig. 9. An equivalent conversion of a 1-D form of input signal to a 2-D form for processing by TDNN. A window of length w positions slides along the input signal. The extracted window part of the signal is rearranged so that it becomes a collection of $q = w - s + 1$ data frames of length s, where s is the number of input signals to the TD neuron.

of learning iteration steps until the network performance function achieves the desired value.

The training phase of the TDNNs can be long and tedious even for shorter time patterns.[866, 867] However, it may achieve a very good classification performance. Owing to the great number of training parameters of TDNNs it is not possible to use many of the efficient training procedures based on second-order methods, such as Levenberg-Marquardt, *etc*. In addition, the backpropagation learning suffers the problem of determination of the proper learning rate. This problem can be circumvented by using adaptive optimized learning rates.[766-768]

6.1. *Multistate TDNN*

TDNNs may be expanded into the so-called multistate (MS) TDNN. MS-TDNN have additional hierarchical layers that use dynamic time warping[736] to enhance the recognition capability of the network and to enable recognition of specific ordered sequences of features independently of their relative locations. These networks are used for promoter recognition tasks in Mache *et al.*[535]

6.2. *Pruning ANN Connections*

A TDNN may have a very large number of weights that need to be adjusted during the training process. This may imply a prohibitively long training time. Another problem with the large number of weights is that, in principle, not all of them contribute significantly to the final network output. Thus, some kind of rationalization is advisable. The third aspect that leads to the desirability for network connection pruning is that, in general, the simplest network that fits the training data should have good generalization ability. These three aspects of ANNs with large numbers of internal connections lead to the requirement of ANN pruning.[701, 824, 937] The network pruning is intimately related to the problem of overfitting and to the determination of the optimal network size for the intended task.

There are many algorithms that may be used in pruning the networks. Probably the best known are the so-called optimal brain damage (OBD)[479] and optimal brain surgeon (OBS)[325] algorithms. Some of the pruning methods are based on the determination of elements of the network that are not crucial for the network's successful operation by calculating the sensitivity of the error function with regards to the change of these elements.[415, 479, 581, 762] There are other approaches, such as the penalty-term methods that result in weight decays, where the error function is modified so that the training algorithm drives to zero the nonessential weights;[140, 385, 395, 622, 669, 876, 877] interactive methods;[771, 772] local and distributed bottleneck methods;[459, 460] genetic algorithm based pruning;[883] *etc.*

The problem to be solved by network pruning is how to simplify the network structure by eliminating a number of internal connections so that the pruned network remains good and achieves, after retraining, improved performance. Essentially, the large initial network size is gradually reduced during the training by eliminating parts that do not contribute significantly to the overall network performance. After elimination of the parts with insignificant contributions, the network is retrained; and the process is repeated until a sufficiently good network performance is achieved. The pruning of the TDNN in the NNPP program for promoter recognition during the training process appears to be crucial for the performance of NNPP.[703−706]

7. Comments on Performance of ANN-Based Programs for Eukaryotic Promoter Prediction

In this section we comment on the reported results of some ANN-based programs for eukaryotic promoter recognition that use recognition of individual promoter components as a part of the solution, *viz.* the NNPP program,[705] the Promoter2.0 program,[439] and the SPANN2 program.[44]

The NNPP program is based on the TDNN architecture that belongs to PCFD structures, similar to that presented in Figure 9. It contains two feature-sensing TDNNs that are used to react in the presence of two promoter subregions: TATA-box and Inr. There is also a coordination TDNN that processes outputs of the feature-sensing networks and produces the final output. The feature-sensing TDNNs are trained independently, and all three networks are pruned during the training process.[705] Promoter2.0 is based on a CCFD hierarchical structure as commented on before. The SPANN2 program is based on preprocessing transformation and clustering of input data. The data in each cluster are processed by a structure similar to PCFD one. In total, 11 ANNs are used in the SPANN2 system. Note that SPANN2 system combines the assessment of promoter region and the signal of the presence of the TATA motif.

In order to obtain some reasonable assessment of the capabilities of ANN-based systems for eukaryotic promoter recognition, we use the results obtained in an evaluation study of nine programs for the prediction of eukaryotic promoters, as presented in Fickett and Hatzigeorgiou.[249] These programs are listed in Figure 10 and denoted by program numbers 1 to 9. In addition, we use the results of three other programs that make strand-specific searches and whose results are reported after the study[249] on the same data set. These three programs are indicated in Figure 10 as IMC[626] (program 10), SPANN[42, 43] (program 11), and SPANN2[44] (program 12). The original result achieved by program 3 is replaced by a new result reported in Knudsen,[439] as it scores better. For details on programs 1 to 9 see Fickett and Hatzigeorgiou[249] and references therein. Since different measures of prediction accuracy are available, which produce different rankings of the achieved scores of prediction, we use the average score measure (ASM) as proposed in Bajic[50] to obtain a balanced overall ranking of different promoter prediction programs.

Without entering into the details of the evaluation test, we mention only that the data set from Fickett and Hatzigeorgiou[249] contains 18 sequences of a total length of 33120 bp and 24 TSS locations. A prediction is counted as correct if it is within −200 nucleotides upstream of the real TSS location and +100 nucleotides downstream of the TSS location. For obtaining relative ranking of the above men-

Program	program No.	TP	FP
Audic[38]	1	5	33
Autogene[446]	2	7	51
Promoter2.0	3	10	43
NNPP	4	13	72
PromoterFind[377]	5	7	29
PromoterScan	6	3	6
TATA[116]	7	6	47
TSSG[784]	8	7	25
TSSW[784]	9	10	42
IMC	10	12	39
SPANN	11	12	44
SPANN2	12	8	16

Fig. 10. List of programs whose performance is compared.

tioned 12 programs, we use the ASM as a balanced method for reconciling different ranking results produced by different measures of prediction success. Eleven different measures of prediction success have been used in Bajic[50] to produce the final ASM based ranking and the results are given in Figure 11. The ranking of performances is given in ascending order, so that the best overall performance got the rank position 1.

The ASM, which is a balanced measure, ranks the NNPP program[705] which has best absolute TP score, only at position 9 in the total ranking due to a very large number of FPs. On the other hand, the Promoter Scan program of Prestridge[679]—this program is not based on ANNs—although achieving the least absolute TP score, is ranked much better at 4th overall position. Another evaluated ANN-based program, Promoter2.0,[439] ranks overall at 5th position. For illustration purpose only we observe that the other two ANN-based programs, SPANN[42, 43] and SPANN2,[44] which use very different mechanism of promoter recognition, rank well in the overall system at positions 3 and 1, respectively.

This comparison indicates that ANN-based systems exhibit prediction performances comparable to other non-ANN based algorithms. However, since the content of the data set from Fickett and Hatzigeorgiou[249] is not very representative of eukaryotic promoter sequences, no conclusions about the absolute values of the

Rank Position	Program No.
1	12
2	10
3	11
4	6
5	8
6	9
7	3
8	5
9	4
10	1
11	2
12	7

Fig. 11. Ranking prediction performances of different programs based on ASM which uses 11 different performance measures.

compared programs should be made based on the comparison results in Figure 11.

The comparison analysis of Fickett and Hatzigeorgiou[249] is by now a bit outdated. A new generation of ANN-based programs[46-49, 51, 627] has been developed and evaluated on the complete human chromosomes 21 and 22. The results are presented in Figure 12 together with the results of two other non-ANN based systems, Eponine[211] and FirstEF.[190] Figure 12 gives ranking based on the ASM, as well as on the correlation coefficient (CC). Again, we find that some of the ANN-based promoter prediction programs have superior preformance compared to other non-ANN based systems. On the other hand, we also find that some of the ANN-based programs are considerably inferior to the non-ANN based solutions. This suggests that the real problem is not in the selection of the technology on which promoter predictions are based, but rather on the information used to account for the presence of promoters.

In summary, we have presented in this chapter some of the basic ideas that may be used, or which are already being used, in building ANN systems for the recognition of eukaryotic promoters in the uncharacterized DNA strings. These ANNs are trained to recognize individual micro-structural components—*i.e.*, specific motifs—of the promoter region. The results achieved by ANN-based pro-

V. B. Bajić & I. V. Bajić

Program	TP	FP	Rank by ASM	Rank by CC
Dragon Gene Start Finder[46]	198	52	1	1
Dragon Promoter Finder[48]	250	868	5	5
Eponine[211]	128	19	2	3
FirstEF[190]	234	300	4	4
McPromoter[627]	162	65	3	2
NNPP2[703]	249	3539	6	6
Promoter2.0[439]	135	1869	7	7

Fig. 12. Comparison of several promoter prediction programs on human chromosomes 21 and 22. The total length of sequences is 68,666,480 bp and there are in total 272 experimentally determined TSS.

grams are comparable with those of programs that do not use ANNs. However, the performance of promoter recognition programs are not satisfactory yet. The problem of eukaryotic promoter recognition represents, and remains, a great challenge in the general field of pattern recognition.

CHAPTER 6

NEURAL-STATISTICAL MODEL OF
TATA-BOX MOTIFS IN EUKARYOTES

Haiyan Wang

Max-Planck Institute for Molecular Genetics
whyinsa@yahoo.com

Xinzhong Li

Imperial College
xinzhong@doc.ic.ac.uk

Vladimir B. Bajić

Institute for Infocomm Research
bajicv@i2r.a-star.edu.sg

The TATA-box is one of the most important binding sites in eukaryotic Polymerase II promoters. It is also one of the most common motifs in these promoters. The TATA-box is responsible mainly for the proper localization of the transcription start site (TSS) by the biochemical mechanism of DNA transcription. It also has very regular distances from the TSS. Accurate computational recognition of the TATA-box can improve the accuracy of the determination of the TSS location by computer algorithms. The conventional recognition model of the TATA-box in DNA sequence analysis is based on the use of a position weight matrix (PWM). The PWM model of the TATA-box is widely used in promoter recognition programs. This chapter presents a different, nonlinear, recognition model of this motif, based on a combination of statistical and neural network modelling. The resulting TATA-box model uses "statistical filtering" and two LVQ neural networks. The model is derived for a sensitivity level that corresponds to approximately 67.8% correct recognition of TATA motifs. The system is tested on an independent data set used in the evaluation study by Fickett and Hatzigeorgiou, and it performs better in promoter recognition than three other methods, including the one based on the matching score of the TATA-box PWM of Bucher.

ORGANIZATION.

Section 1. We first give some background about promoter recognition via the recognition of the TATA-box.

Section 2. Then we provide a statistical analysis of TATA motifs and their surroundings.

Section 3. Next we describe the use of the LVQ ANNs in modelling TATA motifs.

Section 4. Finally, we present a new multistage LVQ ANN system that models the TATA motifs. The results of application of this system to promoter recognition are also presented.

1. Promoter Recognition via Recognition of TATA-Box

Extracting new biological knowledge in a computational manner from recorded biological sequence databases is one of the key issues of bioinformatics.[26] One of the currently most important general problems in bioinformatics is the annotation of uncharacterized biological sequences. It presumes finding and locating functionally active segments of biological sequences. Once these are found, their positions, functions, and other relevant information are recorded and stored in databases.

At present, the primary target of annotation of sequences originated from eukaryotes is the location of protein coding genes.[123, 127, 166, 247, 304, 587] However, correctly recognizing the starting and ending points of different genes is not a simple task, and methods for this purpose are not sufficiently accurate yet.[123, 127, 166, 247, 304, 334, 587] The starting end of genes can be more precisely determined through the location of promoters, since promoters are usually located before the respective gene, so that recognizing a promoter allows for a more precise determination of the gene's 5' end.[38, 249, 306]

Computational promoter finding has received more attention in the last decade.[38, 43, 44, 46–48, 116, 150, 168, 190, 211, 249, 262, 321, 330, 377, 383, 439, 446, 535, 549, 565, 626, 627, 629, 630, 674, 679, 681, 690, 691, 703, 753, 784, 796, 901, 931] Current knowledge of promoter functionality relies on extensive experimental work.[249, 656] In simplified terms, the promoter region of protein coding genes of eukaryotes— shortly called eukaryotic Pol II promoters—represents a section of DNA to which RNA Polymerase II enzyme and different transcription factors (TFs) bind, forming the so-called transcription preinitiation complex that makes the initiation of the DNA transcription possible. Different transcription factors bind to different subsections of the promoter region. These docking sites, called transcription factor binding sites, are recognized by transcription factors via the bio-chemical machinery of the cell. One of the most important transcription factor binding sites in eukaryotes is the so-called TATA-box.[75, 116, 173, 314, 624, 661, 773, 889]

The eukaryotic promoters are far more complex that the prokaryotic ones, and possess very individual structures which are not common to large groups of promoters. It is thus difficult to design a general algorithm for recognition of eukaryotic promoters. There exists a number of computer programs to aid in promoter identification; see recent overviews by Fickett and Hatzigeorgiou[249] and Prestridge.[681]

Although much progress has been achieved in recent years in developing such general type promoter-recognition software, the general conclusion of the study by Fickett and Hatzigeorgiou[249] is that the performances of the publicly available promoter recognition programs are not very satisfactory. That study has demonstrated that on the specific test set these programs recognize the existing promoters in the range between 13% to 54%, producing false positives in the range of about 1/460bp down to 1/5520bp. The general tendency is an increased number of false predictions as the number of correct predictions increases. A recently reported result[753] that aims at a low level of false recognition recognizes only 29% of the true promoters on the same test set, although it makes only 1 false prediction per 2547bp.

Although there are a number of transcription factor binding sites in eukaryotic Pol II promoters—such as Inr, CCAAT-box, and GC-box—that are shared among larger promoter groups, the most common transcription factor binding site among them seems to be the TATA-box. Some estimates are that it is present in 70%–80% of eukaryotic Pol II promoters.[681] This motivates using the recognition of TATA-box as a part of eukaryotic promoter prediction algorithms.[44, 116, 439, 549, 679, 784]

Recognition of this element is heavily dependent on finding good matches to the TATA-like motif.[116, 661] This is frequently done by using the position weight matrix (PWM) description[322, 795, 801, 803, 805] of the TATA-box.[116] The very first attempts of using the PWM models in biological sequence analysis are in modeling the transcription start site and translation initiation site in *E. coli*.[322, 805] However, the recognition performance based on PWM score only is not very good and too many false recognitions are produced if a high sensitivity level of recognition is required. It thus makes sense to search for a better characterization of the TATA motif, as it may improve recognition accuracy of both the TATA-box and the promoter.

One such possible approach is presented in this chapter. It should be mentioned that it is a general opinion[38] that the recognition of eukaryotic promoters cannot be made very accurate if it relies on the detection of the presence of only one of the transcription factor binding sites—even if that is the TATA-box—and that other promoter characteristics or binding sites should be used simultaneously in promoter recognition algorithms. Although we are aware of this, we still test the new neural-statistical model of the TATA motif in the recognition of eukaryotic Pol II promoters. Our result outperforms three other previously tested programs, including the one based on the original TATA-box PWM from Bucher.[116]

We develop in this chapter a statistical nonlinear characterization of TATA-box motifs of eukaryotic Pol II promoters. Development of this model is based on the representation of nucleotides by the electron-ion-interaction potential (EIIP),[857, 858] the PWM of the TATA motif,[116] domain characteristics of neighboring segments of the TATA-like motifs, positional information of the motif, and Artificial Neural Network (ANN) based modeling. The new model is obtained

- by finding regularities in the DNA segments around the TATA-box and in the distribution of the TATA-box motifs based on the PWM matching score,
- by developing a new system for TATA motif recognition that combines the LVQ ANNs[441, 443, 445] augmented with statistical analysis, and
- by using a genetic algorithm (GA)[254] to optimize the initial weights of the LVQ ANNs in an attempt to improve the prediction accuracy of the model.

2. Position Weight Matrix and Statistical Analysis of the TATA-Box and Its Neighborhood

In this section we develop some results based on the PWM of TATA motifs, and combine these with the properties of local neighborhoods of TATA-boxes. These are used together with information on motif position in subsequent sections to support TATA motif modelling by ANNs. The model development in this section is backed by the statistical and biological regularities of the TATA motif and local surrounding regions.

2.1. *TATA Motifs as One of the Targets in the Search for Eukaryotic Promoters*

Promoters perform a crucial role in the initiation of the DNA transcription process. They indicate and contain the starting point of transcription near its 3' end. Eukaryotic Pol II promoters contain numerous different transcription factor binding sites that are not always present in each of the promoters, whose location with respect to the transcription start site (TSS) may change in different promoters, as may also their ordering. These facts result in a huge number of microorganizational combinations of these functional regions, and consequently in a large number of possible eukaryotic promoter structures, which is also reflected in the high complexity of eukaryotic organisms. Hence, it is difficult to create a unique eukaryotic promoter model precisely.

The recognition of TATA-box is frequently used as a part of computer algorithms[44, 116, 439, 549, 211, 679, 784] that aim at recognizing eukaryotic Pol II promoters. Other binding sites are also used in many solutions.[150, 151, 259, 260, 439, 446, 549, 678–680, 683, 689, 705, 784] This provides the motivation for improving the quality of models of these different short DNA motifs, so that their recognition becomes more accurate. Our interest in this study is focussed on the TATA motif as it is a frequent binding site in eukaryotic Pol II promoters. The TATA-box is a hexamer sequence with a consensus given by TATAAA, though it is more accurate to describe it with a PWM.[116]

The TATA motif is found in a majority of protein coding genes, with the position of its first nucleotide generally falling in the range of -25 to -35 upstream from the TSS and centered about position -30; see Figure 1. In the transcription process, the TATA-box serves as the binding site to TFIID[656] and, when it is present in the promoter, it is responsible for the correct determination of the TSS by the transcription machinery.[656]

2.2. Data Sources and Data Sets

The data used for building a model of the TATA motif is collected from some public nucleotide databases. All extracted sequences were from vertebrate organisms. For the core promoter regions the data source was the Eukaryotic Promoter Database (EPD),[115, 662] which contains an annotated collection of experimentally mapped TSS and surrounding regions. We have extracted 878 vertebrate core promoter sequences. All promoter sequences are of length 43bp from position -45 to -3 relative to the TSS. The location of the TSS is assumed to be between the nucleotide at position -1 and the nucleotide at position $+1$. In this notation there is no nucleotide in position 0 and the first transcribed nucleotide is the one at position $+1$. The other two group of sequences—belonging to the exon and intron regions—are extracted from the GenBank database, and they are also taken from vertebrate organisms. All exon and intron sequences are divided into non-overlapping segments of length 43bp.

From the extracted promoter sequences, we randomly select 640 which make up the training set P_{tr} for promoters, while the promoter test set P_{tst} contains the remaining 238 promoter sequences. From the set of exon sequences we randomly select 8000 sequences and this set is denoted as S_{cds}. Analogously, from the set of intron sequences, another 8000 sequences are randomly selected to form the set S_{int}. We used the whole S_{cds} set as the negative training set, and the S_{int} set as a negative test set.

2.3. *Recognition Quality*

Since we are developing a new model of TATA-box motifs, we also have to test
the quality of recognition of the TATA-like sequences based on this new model.
Thus we have to use proper measures to express the success of such recognition.

There are several measures that can be used to assess the quality of motif recognition. The four customary numbers TP, TN, FP, and FN, denoting
"true positive", "true negative", "false positive", and "false negative" recognition
respectively, are generally used in different combinations to describe the basic
recognition quality. These four numbers are expressed as percentages in this chapter. However, it is often more convenient to use measures expressed by a single
number that relate to the recognition quality. For a recent review on these, see
Bajic.[50]

Any way, during model development, we use the correlation coefficient (CC)
to measure the success of prediction, although this measure does not sufficiently
penalize the large FP prediction. However, for the final assessment of the quality
of the developed model we use the average score measure (ASM)[50] so as to be
able to make the relevant comparison of the promoter prediction results obtained
by the new TATA motif model and the results obtained by other algorithms.

2.4. *Statistical Analysis of TATA Motifs*

The DNA sequence is composed of 4 different nitrogenous bases denoted in the
customary way as "A", "T", "C", and "G", relating to adenine, thymine, cytosine, and guanine respectively. Before any analysis of a DNA sequence is made,
we convert it to a numeric sequence by replacing each of the nucleotides by its
electron-ion-interaction potential (EIIP)[857, 858]. After that, the values of EIIP for
A and T are increased by tenfold. The collection of the original values of EIIP for
C and G, and the modified EIIP values for A and T, are denoted as the modified
EIIP values (MEIIPVs).

Given an uncharacterized genomic DNA sequence, the goal is to recognize
the TATA-box motif by moving along the sequence. If accurate recognition of the
TATA-box can be made, then one can also make a recognition of the promoter. In
this study, promoter recognition is made only on the basis of the detected presence of TATA motifs. The TATA-box is modeled based on the PWM determined
in Bucher[116] from a set of 502 unrelated eukaryotic Pol II promoters. In order
to better represent the core TATA motif, we also use some features of the short
regions 6bp before and 6bp after the core motif.

Thus, due to the overlapping of the PWM length and the three hexamer regions, the data window analyzed in the search for the TATA motif is only 20bp

wide. This information is finally combined with the position information of the TATA motif relative to the TSS. All these data serve as input information to the ANN system developed in the subsequent sections.

2.4.1. *PWM*

Computer models of specific biological signals normally form part of the tools for nucleotide sequence analyses. Since a large amount of data is available in databases, proper statistical analysis is possible in many cases and can help in developing predictive models of specific biological signals. A very convenient method for such modeling is based on the position weight matrix (PWM) of specific motifs.[322, 795, 801, 803, 805] This has found broad and very successful applications in many motif search computer programs.

PWM is a statistical motif descriptor. It is derived from the base-frequency matrices that represent the probabilities of a given nucleotide occurring at a given position in a motif. Since it basically relates to probabilities, the PWM attempts to describe in a statistical fashion some characteristics of the motif found in a collection of sequences. For the proper determination of the PWM it is important that enough data is available.

Bucher[116] has enhanced the basic algorithm for the determination of the PWM by introducing an optimization criterion based on a measure of local over-representation. It is possible to estimate in this way the cut-off matching score to the weight matrix, as well as its width and location of the preferred region of occurrence. Due to the way that it is determined, the PWM for DNA motifs are represented by a rectangular matrix that has 4 rows, where each row corresponds to one of the 4 bases, and its number of columns is equal to the length of the motif. The PWM of the TATA-box motif from Bucher[116] is of size 4×15, giving the motif length of 15 nucleotides, with the start of the core motif at column 2 of the PWM.

This PWM can be used to scan a sequence for the presence of the TATA motif. A window of length 15 nucleotides slides along the sequence. The matching score for the motif is calculated based on the nucleotides found and the PWM. The matching score for the window is given by

$$x = \sum_{i=i}^{15} w_{b_i,i} \tag{1}$$

where $w_{b_i,i}$ is the weight of base b_i at the ith position of the motif, and b_i is the ith base of the sequence in a window; b_i relates to A, C, G, or T. To determine which of the matching scores suggests the presence of the TATA-like motifs, we compare the matching score with the threshold τ. If $x > \tau$, where the cut-off value τ is usually determined either experimentally or statistically, then the data window suggests the presence of the TATA-box.

2.4.2. *Numerical Characterization of Segments Around TATA-Box*

DNA is an organized high capacity biochemical structure—a macromolecule— that stores information. It thus seems logical that the nucleotides in the DNA sequence are ordered following certain regularities. The mechanism of recognition of binding sites in promoters by transcription factors is not completely known. However, one may assume that a binding site and its immediate surrounding regions—the region before the binding site and the region after the binding site— have some properties that enable easy recognition of the site by the respective transcription factors.

Based on this assumption, we expect certain regularities in the case of the TATA-box, in the segments before and after the core TATA motif, which enable easier recognition of the motif. We also expect that these regularities for the functional TATA motifs are different from those relating to non-functional ones.

Consequently, we focus our attention on the distribution of the bases A, C, G, and T around TATA motifs. To describe the potential features of the segments around the TATA-box, we consider three hexamer segments of DNA associated with this motif in the direction from the 5' toward the 3' end of the sequence: The segment S_1 immediately before the TATA-box, the segment S_2 representing the TATA-box hexamer, and the segment S_3 representing the hexamer immediately after the core TATA-box. All three segments are 6bp in length each. However, in addition to these three hexamer we also use the PWM of the TATA motif to calculate the matching score. Since the core TATA motif is represented in the PWM starting from position 2, we need to consider in the modeling process a subregion of the total length 20bp relating to the TATA motif . We use certain characteristics of these three segments in the development of our TATA-box model.

Consider a sequence s of length 43 nucleotide positions. Assume it contains a TATA-box. Let AV_1, AV_2, and AV_3 denote the average of the MEIIPVs of the nucleotides contained in the segments S_1, S_2, and S_3 respectively. Let e_j, $j = -45, -44, ..., -3$, represent the MEIIPV of each of the nucleotides in the sequence s. Further assume that the 5' end—the starting nucleotide—of the TATA-box is at the position i relative to the TSS. Then AV_1, AV_2, and AV_3 are given by

$$AV_1 = \frac{1}{6} \times \sum_{k=1}^{6} e_{i-k}, \quad AV_2 = \frac{1}{6} \times \sum_{k=0}^{5} e_{i+k}, \quad AV_3 = \frac{1}{6} \times \sum_{k=6}^{11} e_{i+k} \quad (2)$$

2.4.3. *Position Analysis of TATA Motifs*

In order to obtain initial information on the location of the TATA motif, we scan each sequence from P_{tr} using a window of 15 nucleotides. The first window starts at position –40 relative to the TSS, and the last one starts at position –17. We calculate according to Equation 1 the matching scores for each of the windows with the PWM for the TATA-box as determined by Bucher.[116] Then we assume that the position of the window with the maximum matching score for a sequence corresponds to the window which contains a hypothetical TATA motif. This computational determination of putative TATA motifs is not the best one since the highest score does not imply biological relevance. But in the absence of experimentally determined TATA motifs for all sequences in P_{tr}, we adopt this unified way of determining the putative motifs. It is also assumed that at most one of the TATA motifs can exist in any of the sequences from P_{tr}. With a selected threshold of $\tau = 0.45$, a total of 475 sequences from P_{tr} are extracted. This amounts to about 75% of the sequences in P_{tr} which can be regarded as those that contain the TATA-box. The extracted sequences form the set P_{TATA}.

The sequences from P_{TATA} are clustered into different groups G_i, according to the distance of the 5' end of the TATA-box hexamer from the TSS. In the PWM of Bucher[116] the first nucleotide of the TATA-box hexamer is in position 2, so that the detected TATA motifs by using the PWM begin from position –39 to –16, making in total 24 possible positions for the distribution of the motif. Let $G_i, i = -39$, $-38, ..., -16$, denote the group of promoter sequences containing the TATA-box hexamers which all start at position i. Thus, $P_{TATA} = \bigcup_{i=-39}^{-16} G_i$. The number of sequences $N(G_i)$ in different G_i's is not the same. The total number of the computationally found TATA-box motifs in P_{tr} with the threshold $\tau = 0.45$—*i.e.*, the number of sequences in the set P_{TATA}—is $N_{TATA} = \sum_{i=-39}^{-16} N(G_i) = 475$.

The distribution of the putative TATA motifs according to the position of their starting nucleotide with respect to the TSS is shown in Figure 1. This distribution can be approximately modelled by

$$Y = 0.192 \times e^{-0.125 \times (i+30)^2}$$

where Y is the probability that the TATA hexamer starts at position $i = -3, ...,$ -16. This model distribution is also depicted in Figure 1 as the dotted curve. Note that the model distribution is based on the P_{tr} promoter set and, in principle, may change if the set is changed. However, it resembles well the previously observed positional distribution of TATA-like motifs in eukaryotic Pol II promoters.

Figure 1 indicates that the starting point of the TATA-box hexamer in the promoter set P_{tr} concentrates around the position –30. The middle point of the TATA-box hexamer is by convention considered to be represented by the second T in the

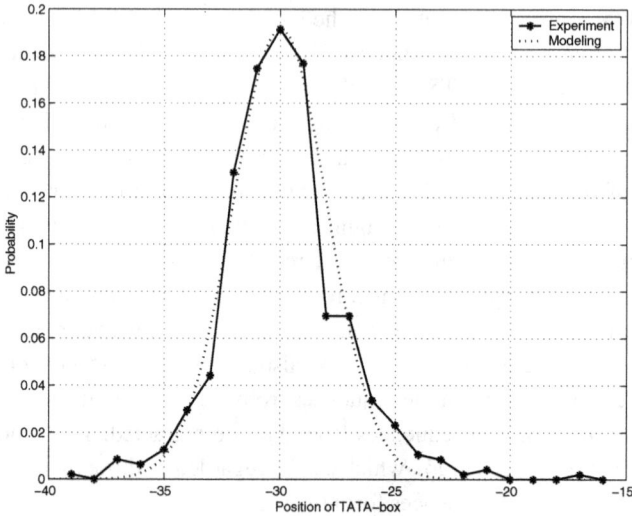

Fig. 1.   Experimental and modelled distributions of TATA-box motifs found in P_{tr} based on PWM matching score greater than the threshold $\tau = 0.45$.

TATAAA consensus. This implies that the middle point of the motif is concentrated around position –28. This is close to experimental results, which suggests the location of the TATA-box to be about –25bp to –30bp upstream of the TSS. The distribution of the TATA-box can be regarded as one of the features of the motif, as it expresses a very regular statistical property.

2.4.4. *Characteristics of Segments S_1, S_2, and S_3*

We want to find some regularities that stem from the positional distributions of the TATA motif. For this purpose we use the P_{TATA} set of promoter sequences, and a part of the sequences from S_{cds} that are selected as follows. For each of the sequences from S_{cds} the maximum matching score is obtained with the PWM for TATA-box motifs. Then if such score is greater than $\tau = 0.45$, the sequence passes the filter, otherwise it is rejected. We obtain 1040 filtered sequences from S_{cds} in this way to make the set S_{cds}^f. Further, we find for each sequence from $P_{TATA} \cup S_{cds}^f$, based on the TATA PWM, the maximum matching score among the scores that correspond to the positioning of the 5' end of the TATA hexamer at –39, ..., –16, relative to the TSS. Each such position is considered to determine a putative TATA motif. Then we calculate the respective AV_1, AV_2, and AV_3 values of the segments S_1, S_2, and S_3 using MEIIPVs.

The following observations follow from the analysis of AV_i values:

- there are ranges of AV_i values produced only by sequences from P_{TATA};
- analogously, there are ranges of AV_i values produced only by sequences from S_{cds}^f; and
- some ranges of AV_i values contain a majority of data produced by either P_{TATA} or by sequences from S_{cds}^f.

These observations are used in analyzing the distributions of values of two additional parameters $D_1 = AV_2 - AV_1$, and $D_2 = AV_2 - AV_3$, that describe the contrasts between the neighboring segments S_1 vs. S_2, and S_2 vs. S_3 respectively. The observations about the ranges of numerical parameters AV_1, AV_2, AV_3, as well as the derived parameters D_1 and D_2, suggest that restricting the values of these parameters can serve as a sort of filtering of TATA-containing sequences, and thus could be used in possible recognition of the TATA motifs and consequently the promoters.

In order to illustrate the possible effects of such filtering on TATA motif (and promoter) recognition, we slide a window of 20 nucleotides in length along each sequence in P_{tr}, P_{tst}, S_{cds}, and S_{int}. For each position of the window, the AV_1, AV_2, and AV_3, as well as D_1 and D_2, are calculated based on the content from positions 1 to 18 within the window. The PWM matching score for TATA motif is calculated based on content from position 6 to 20 of the window.

If the AV_2, D_1, and D_2 values fall into pre-selected data ranges, and the PWM score is higher than the selected threshold τ, the sequence is considered as a TATA-box containing sequence (and thus a promoter). The position of S_2 for which this is observed relates to the position of the found TATA motif. Such filtering possibilities, by restricting the values of D_1, D_2, and AV_2, are illustrated in Figure 2, where the results of promoter recognition via the TATA motif recognition are given by using some ranges of D_1, D_2, and AV_2 values, together with the threshold $\tau = 0.45$ of the PWM matching score, as a filtering criteria.

For example, Figure 2 indicates that $FP = 4.8\%$ is obtained on the set of S_{int} and $FP = 2.6\%$ on the set of S_{cds}, when $TP = 58.1\%$ on P_{tr} and $TP = 45.4\%$ on P_{tst}. The results in Figure 2 are not all that good by themselves. Nevertheless, they illustrate the positive effect of restricting the values of numerical parameters D_1, D_2, and AV_2. This is utilized in combination with the other information in the construction of our TATA motif model.

Case	Value Intervals		$P_{tr}(\%)$	$P_{tst}(\%)$	$S_{cds}(\%)$	$S_{int}(\%)$
	D_1	D_2				
1	$[0.3, 1.3]$	$[0.39, 1.2]$	58.1	45.4	2.6	4.8
2	$[0.5, 1.3]$	$[0.43, 1.2]$	43.1	33.2	1	1.5
3	$[0.5, 1.3]$	$[0.58, 1.2]$	41.4	31.5	0.51	0.55
4	$[0.7, 1.3]$	$[0.6, 1.2]$	22.4	15.1	0.3	0.19
5	$[0.95, 1.3]$	$[-0.1, 1.2]$	13.1	6.7	0.025	0
6	$[0.97, 1.3]$	$[-0.1, 1.2]$	11.1	5.8	0	0

$$AV_2 \in [1.07, 1.097] \cup [1.275, 1.4] \text{ for all cases}$$

Fig. 2.   Recognition of promoters based on attempted recognition of the TATA motif by constraining numerical parameters that characterize S_1, S_2, and S_3 and using threshold $\tau = 0.45$

2.5. *Concluding Remarks*

Up to now we have described some statistical regularities of TATA-like motifs in eukaryotic Pol II promoters. These properties are based on the PWM and are associated with certain data ranges obtained by the analysis of numerical representation of the TATA-box hexamer and its immediate neighboring hexamers. Finally, strong regularities are also detected in the positional distribution of the motif. These regularities reflect different aspects of the statistical, and partly biological, features of the TATA-containing eukaryotic Pol II promoter sequences.

Based on these regularities one can derive a set of information data for each promoter sequence as follows. For a promoter sequence from P_{tr} of length 43bp spanning positions –45 to –3, we slide a window of length 15 nucleotides from the 5' end toward the 3' end of the sequence. We calculate 24 matching scores p_i against the TATA-box PWM, associated with the starting positions of the window $i = -40, -39, ..., -17$, relative to the TSS. For the sequence, the window that achieves the maximum score is considered to be most likely the one which contains the TATA-box. If the position of the first nucleotide in the window with the maximum matching score p_i is i, then the position of the first nucleotide in the TATA-box hexamer is $j = i + 1$.

The following 8 data are then generated and associated with the detected

TATA-box hexamer

$$\left.\begin{array}{ll} x_1 = \max_i p_i & x_2 = \dfrac{j + 40}{24} \\ x_3 = 0.192 \times e^{-0.125 \times (j+30)^2} & x_4 = AV_1 \\ x_5 = AV_2 & x_6 = AV_3 \\ x_7 = AV_2 - AV_1 & x_8 = AV_2 - AV_3 \end{array}\right\} \quad (3)$$

These data reflect the statistical and distributional information of the TATA-box sequence and will be used in the training of ANNs for TATA motif and promoter recognition.

3. LVQ ANN for TATA-Box Recognition

There are obviously many approaches that can be used to recognize the presence of a TATA-box in genomic uncharacterized sequences. These may be based on the consensus sequence,[801] PWM, other statistical methods including Hidden Markov Models, Factor Analysis, *etc.*, or even simple feature extraction. However, a well-known approach to pattern recognition based on multidimensional feature space is the use of artificial neural networks (ANNs). ANNs often perform pattern recognition with a high recognition rate. Moreover, the use of ANNs represents a complementary approach to conventional methods based on explicitly defined feature extraction, so that systems for pattern recognition based on ANNs can be combined with other methods in a "mixture of experts" approach.

Many ANN structures and learning algorithms can be successfully employed in TATA motif recognition. However, we concentrate here on using the so-called Learning Vector Quantization (LVQ) ANNs,[441, 443, 445] for this task. The LVQ ANNs have shown to be very successful in pattern recognition.[204, 572] Thus we expect that them to perform well on this problem.

In this section we explain:

- how data preprocessing is done to enable easier training of LVQ ANNs,
- how LVQ ANNs can be used for the TATA motif recognition, and
- how initial parameters of the LVQ ANN can be optimized by a genetic algorithm for more efficient subsequent network training.

These explanations provide the background for the design of the final model of the TATA-box motif to be presented in the next section.

3.1. *Data Preprocessing: Phase 1*

For successful training of ANNs, a high quality training data set is extremely important. The method of training data selection, data coding, as well as data information content, always play a significant role. Although training an ANN can be relatively fast for moderate sized training data sets given current capabilities of computers and learning algorithms, the training with a large number of training data is still usually slow. This speed also depends on the size of the feature vectors that represent input patterns. This is a typical situation when dealing with the DNA sequences because, depending on the problem in question, these sequences may have feature vectors of high dimension and the number of sequences in the training set can be huge.

It is generally ineffective to feed a large amount of raw sequence data to an ANN and to expect it to be able to capture accurately the essential transformation from the input feature space to the target output space. In order to make the learning of the ANNs more efficient, as well as to enhance their generalization capabilities, we normally perform a number of preprocessing transformations of raw data to make it more suitable for ANN processing, as well as to emphasize more important information contained in the raw data. However, the coding of the raw DNA data may prevent proper data preprocessing, as is the case of binary coding of DNA sequences.

The data we select for the training and test sets, as well as the numerical representation of DNA sequences, have been described earlier. We have also explained, based on the preliminary statistical analysis in the last section, how we generate a set of 8 numerical data to represent a TATA-box motif. These are given in Equation 3. In principle, a vector $X = \langle x_1, x_2, ..., x_8 \rangle$ may be used as an input to the LVQ ANN. However, we proceed differently below.

Due to the statistical nature of the matching score for TATA-like motifs obtained by the PWM, and the distribution range of D_1 and D_2 values, we first use these to perform the primary filtering of the training feature vectors. This procedure considerably reduces the quantity of negative data that represent non-TATA motifs, although it sacrifices to an extent the positive data. Since our ultimate goal is to obtain a system with high recognition accuracy of TATA-box motifs, we change the level of filter threshold and examine its effect. The idea is to find the most suitable threshold—one that makes the greatest difference between the correct and false recognition—and to leave most of the further recognition task to the ANN.

At first, we set a threshold τ for the matching score p_i. Only those feature vectors whose first element x_1 is greater than τ are selected as potentially representing the TATA-box motif, and are used for further analysis. We also use the distributions of x_7 and x_8 as constraining conditions for the initial filtering of input data. These conditions are based on the statistical analysis of the last section.

The set of sequences that satisfies these first filtering conditions is defined by

$$S_{F1} = \{X \mid x_1 > \tau,\ x_7 \in [0.3, 1.3],\ x_8 \in [0.3, 1.2]\} \qquad (4)$$

The initial training set T_{tr} is formed by taking the union of P_{tr} and S_{cds} and labeling the sequences with their class labels. The sequences from this initial training set is then prefiltered based on the conditions in Equation 4.

The efficiency of such initial filtering is depicted in Figure 3, showing the effects of different threshold values τ combined with the value range filtering for x_7 and x_8. The recognition quality represented by TP and FP varies with the change of the threshold, while the ranges for x_7 and x_8 are fixed as mentioned above. The curve "TP" is given for the set P_{tr}, while the curve "FP" is given for S_{cds}. The discrimination factor, $TP - FP$, is above 50% when $\tau \in [0.34, 0.46]$. This filter is denoted as SF_1.

Another prefiltering that we use in the stage of training LVQ ANNs is, when values ranges for x_7 and x_8 are changed, so that the sequences satisfying these new filtering conditions are given by

$$S_{F2} = \{X \mid x_1 > \tau,\ x_7 \in [0.1, 1.3],\ x_8 \in [0.1, 1.2]\}$$

This filter is denoted as SF_2 and its efficiency agains threshold τ is presented in Figure 4.

As can be seen, both filters have a maximum discrimination for τ of around 0.42. However, the filter SF_2 allows for more feature vectors to pass the filtering condition. Also, the maximum discrimination of SF_2 is about 63%, while that of SF_1 is 59.5%.

Since these filters operate only at the initial stage of the process of the TATA-box recognition, we select $\tau = 0.4$, which is a bit smaller than the τ with the best discrimination ability for any of the filters, to allow more promoter sequences to pass the filter and to make further "filtering" by the ANN system. The value of $\tau = 0.4$ allows about 67% of TP and 8% FP for SF_1, as well as 77% of TP and 17% of FP for SF_2. Note that the filter SF_1 is used in the final TATA motif model, while the filter SF_2 is used in the process of training LVQ ANNs. So, these two "statistical filters" are never used simultaneously.

Feature vectors X that pass such initial filtering are subjected to further processing, as explained in the next subsection on Phase 2 of data preprocessing. This converts vectors X into the transformed feature vectors Y that are input to the LVQ ANN. The overall structure of the preprocessing phases and the LVQ ANN that makes use of these data is presented in Figure 5 for both the training phase and the testing phase.

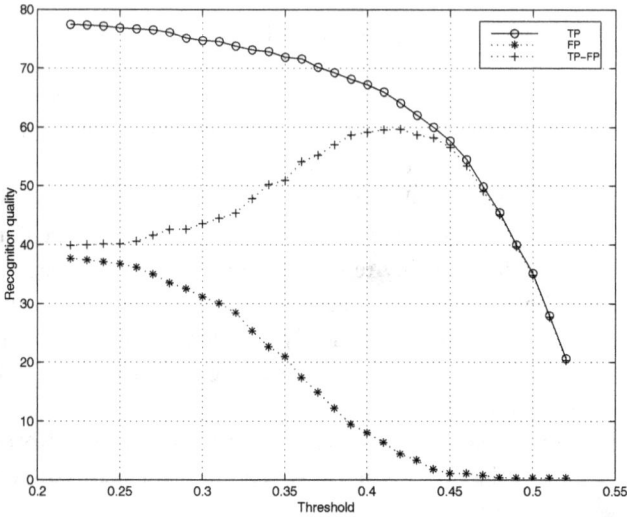

Fig. 3.   The efficiency of the combined threshold and data range restriction filtering—*i.e.*, "statiscal filter"—with $x_1 > \tau$, $x_7 \in [0.3, 1.3]$, and $x_8 \in [0.3, 1.2]$. This filter is named SF_1.

3.2. *Data Preprocessing: Phase 2 — Principal Component Analysis*

After sequences are initially filtered by the statistical filter SF_2, we are faced with the question of whether the filtered data perform efficiently as input data to the LVQ ANN. We need only the relevant information to be fed to the LVQ ANN and that non-crucial information be neglected. One of the well-known statistical techniques that ranks the information contained in a data set according to their information content is the so-called Principal Component Analysis (PCA).[154, 326]

This technique is very useful for the preprocessing of the raw data. It helps eliminate components with insignificant information content, and in this way it performs dimension reduction of the data. This can significantly enhance the ability of ANNs to learn only the most significant features from the data set. PCA has been widely used in data analysis, signal processing, *etc*.[154, 326] PCA is a method based on linear analysis aimed at finding the direction in the input space where most of the energy of the input lies. In other words, PCA performs specific feature extraction. The training data set Y_{tr} that we use for the LVQ ANN is obtained by first filtering the sequences from $P_{tr} \cup S_{cds}$ by the statistical filter SF_2, after which normalization and PCA are applied as follows. Let us assume that after statistical filtering the promoter sequences that pass the filter is the set P_f. Then the

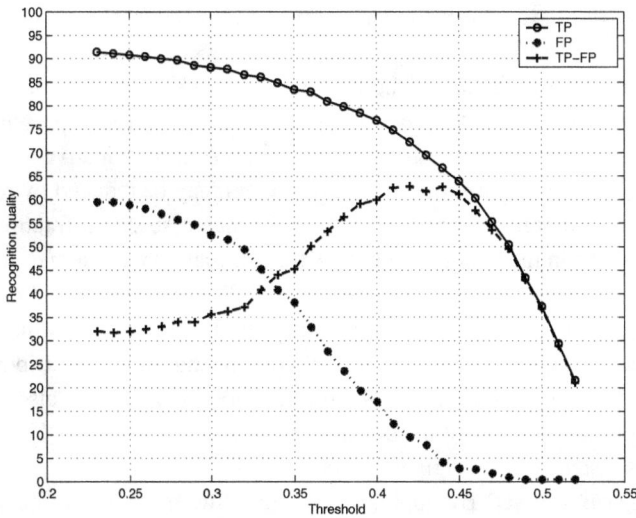

Fig. 4. The efficiency of the "statiscal filter" with $x_1 > \tau$, $x_7 \in [0.1, 1.3]$, and $x_8 \in [0.1, 1.2]$. This filter is named SF_2.

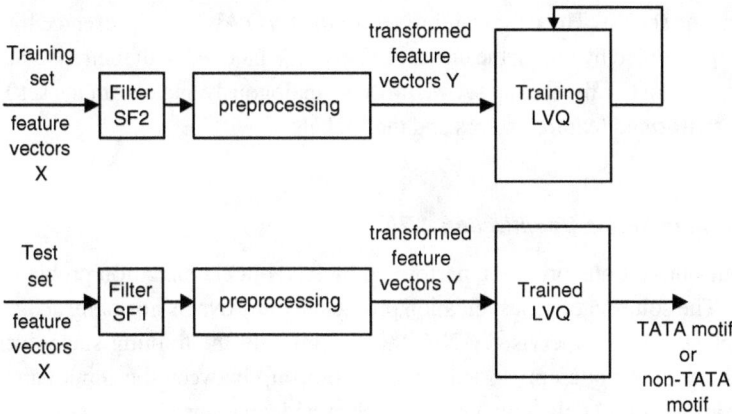

Fig. 5. Statistical filtering, preprocessing, and LVQ ANN.

following steps are made:

- **Step 1: Normalization.** We generate for each sequence in P_f a feature vector

X as explained before. All these vectors constitute the set X_{pf}. This set is normalized to have zero mean and a standard deviation of one, resulting in the set X_{pfN}.

- **Step 2: PCA.** Application of the PCA method to X_{pfN} transforms the normalized feature data so that the transformed feature vectors represent points in the space spanned by the eigenvectors of the covariance matrix of the input data. In addition, the size of the feature vectors may be reduced by retaining only those components that contribute more than a specified fraction of the total variation in the data set. The PCA transformation matrix M_{PCA} is generated to transform X_{pfN} to the new set Y_p. We set the required minimum contribution of components in the transformed data to be 0.01, which means that after the PCA transformation only components which contribute more than 1% to the total variation of the data set are retained. This also result in the reduction of the size of the transformed input feature vectors Y in Y_p to 6 elements instead of the original 8.

- **Step 3: Training set.** By applying statistical filtering SF_2 on the set S_{cds} we obtain filtered non-promoter sequences which make the set N_f. Then, for each sequence in N_f we calculate the feature vector X. All feature vectors for sequences in N_f make the set X_{Nf}. Using the mean of X_{pf}, the standard deviation of X_{pf}, and M_{PCA} obtained in Steps 1 and 2 for the promoter sequences, we transform feature vectors in X_{Nf} to new feature vectors that make the set Y_N. Thus, the training set for the LVQ ANN is represented by the set Y_{tr}, formed by taking the union of Y_p and Y_N labelled with their associated class labels. For the testing set we proceed analogously and obtain the set Y_{tst} of transformed feature vectors and their labels.

3.3. *Learning Vector Quantization ANN*

The input-output transformation performed by ANN in classification problems is discrete. The solution of a classification problem by an ANN is normally achieved in two phases. In a supervised ANN, there is initially the training stage during which the network attempts to learn the relationship between the input patterns and the desired output class by means of adjusting its parameters, and possibly its structure. Such learning procedures are in most cases iterative and attempt to tune the network parameters and structure until the discrepancy measure between the responses produced by the network and the desired responses is sufficiently small. After that phase, one considers the network training completed, and the values of network parameters remain fixed after that. In the testing phase, a test set of new examples, which are not contained in the training set, are presented to the network and compared to the desired network responses. In this way the test of the quality of the trained ANN can be evaluated and its generalization ability assessed.

An ANN system for TATA-box recognition can be considered as a classifier system. For classification problems, many types of ANN can be selected, such as feedforward ANNs, radial basis ANNs, decision based ANNs, *etc.* We use a Learning Vector Quantization (LVQ) ANN for this purpose. LVQ-competitive networks are used for supervised classification problems.[441, 443, 445] Each codebook vector—*i.e.*, input vector—is assigned to one of several target classes. Each class may have many codebook vectors. A pattern is classified by finding the codebook vector nearest to it and assigning the pattern to the class corresponding to that codebook vector. Thus, the LVQ ANN performs a type of nearest-neighbor classification. The standard vector quantization can be used for supervised classification and such methods can provide universally consistent classifiers[204] even in the cases when the codebook vectors are obtained by unsupervised methods. The LVQ ANNs attempt to improve this approach using the adaptation of the codebook vectors in a supervised way.

3.4. The Structure of an LVQ Classifier

An LVQ network has two layers. The first one is a competitive layer, while the second one is a linear layer. The role of the competitive layer is to learn to classify input feature vectors X into C_{hidden} clusters in the C_{hidden} space. After the patterns are clustered into C_{hidden} clusters by the competitive layer, the linear layer transforms the competitive layers' clusters into final target classes C_{tar} in the C_{tar} space. One neuron per class/cluster is used in both layers. For this reason the competitive layer can learn to classify up to C_{hidden} clusters. In conjunction with the linear layer, this results in C_{tar} target classes. The role of the linear layer is to convert the competitive layer intermediate classification into target classes. Thus, since the outputs of the competitive layer are vectors with only one element equal to 1, while all others are equal to 0, the weights of the linear layer can be fixed to appropriate 1s and 0s after the ANN initialization and need not be adaptively changed during the training process.

Let us assume that the input patterns are represented by N-dimensional feature vectors that have to be classified into K target classes, by using M intermediate classes at the output of the hidden layer. Then the LVQ ANN has N nodes in the input layer, M nodes in the hidden layer, and K nodes in the output layer. Let the weight matrix $W_{M \times N}$ describe the connection from the input layer to the hidden layer, while the matrix $V_{K \times M}$ describes the connection from the hidden layer to the output layer. For any input pattern, all the nodes in the hidden layer output 0s, except one node that outputs 1. Thus the output vectors of the hidden layer are composed of all 0s and only one element which is equal to 1. Also, only one node in the output layer produces 1 at its output, while all other nodes in the output layer produce 0s. The output vector $\langle 1, 0, ...0 \rangle$ represents the first class, $\langle 0, 1, 0...0 \rangle$ represents the second class, and so on. A structure of a trained LVQ ANN with two output classes is shown in Figure 6.

Because $V_{K \times M}$ is constant in LVQ ANN, we need only to tune $W_{M \times N}$ based on the training data. The purpose of the learning process is to place the code-book vectors in the input space in a way to describe the boundaries of the classes by taking into account data from the training set. Thus the LVQ algorithm attempts an optimal placement of the codebook vectors in the input space in order to optimally describe class boundaries. This is achieved via the adaptive iteration process. Class boundaries are segments of hyperplanes placed at the mid-distance of two neighboring codebook vectors that belong to different classes.

We know that for a classification problem the number of input and output nodes are determined by the nature of data and the problem setting. However, there are no clear and precise methods for determining the number of hidden nodes and learning rates, especially in the absence of prior knowledge about the class probability densities. Too few hidden nodes may not be enough for good class separation. Too many lead to a very long training phase and does not always produce good generalization. Thus, a compromise has to be achieved through experimentation so as to achieve the best result.

3.5. *LVQ ANN Training*

The training data are obtained as before, using statistical filtering, normalization, and PCA preprocessing. Each of the input vectors is obtained so as to correspond to the hypothetical TATA-like motif found in the original promoter sequence.

At the beginning of LVQ ANN training the weights should be initialized to small values either randomly or by some heuristic procedure. A proportion of neurons in the hidden layer, corresponding to the classes of feature vectors representing the TATA motif and those not representing this motif, is given as 4/6, so that 40% of the neurons in the competitive layer relate to subclusters of Class 1 (TATA motif class) and 60% of the neurons in the competitive layer relate to subclusters for Class 2 (non-TATA motif class). Thus, the weight matrix V from the hidden subclasses layer to the output target classes is also defined. The number of nodes in the hidden layer is set to 100, and the number of learning epochs is set to 1000. The LVQ ANN is trained by the $lvq2$ learning method.[445]

The structure of the trained LVQ ANN is depicted in Figure 6. The input vec-

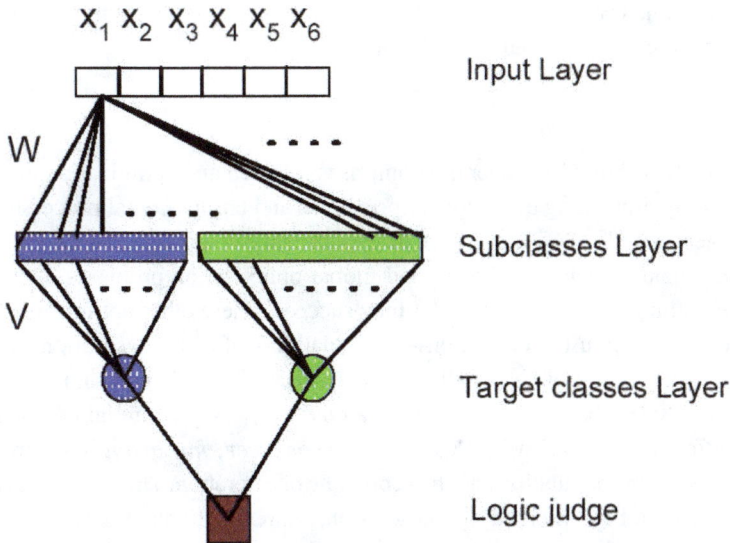

Fig. 6.   A structure of a trained LVQ ANN.

tors are classified into two categories in the output target class layer, and the outputs are finally produced by the "logic judge." When the network is initialized, the neurons which relate to the different classes in the competitive layer are randomly distributed. However, through the process of learning they become properly organized to reflect the distribution of the classes in the input data, possibly forming clusters that relate to data subclasses within each of the output classes.

3.6. *Initial Values for Network Parameters*

An LVQ ANN usually performs efficiently if the initial weight vectors are close to their optimized final values. If the initialization of weights of the LVQ ANN is random, then the initial weights are far from the optimized values. When different initial values of the weights are attempted, and the ANN trained produce very different results in the recognition quality of the TATA motif. Thus it is decided to make a systematic search for the good initial weights that can ultimately produce an accurate recognition system. In the ANN field some heuristic procedures are frequently used for weight initialization, such as k-means for Kohonen's SOM ANN, *etc*. A good approach to the problem of optimizing initial weights in an LVQ ANN is to attempt to optimize these globally. This leads to a solution that employs a genetic algorithm for the determination of the optimized initial weights.

After the weights are reasonably well tuned, we apply the $lvq2$ training algorithm to fine tune the decision boundaries in the LVQ ANN.

3.6.1. *Genetic Algorithm*

A genetic algorithm (GA) generally comprises a search and optimization method developed by mimicking evolutionary principles and chromosomal processing in natural genetics.[254] GAs are a part of evolutionary computing, and become a promising field associated with AI and global optimization problems. GAs are iterative optimization procedures that may succeed where other optimization approaches fail. They use a collection—a population—of possible solutions to the problem in each iteration. Normally, in the absence of knowledge from the problem domain, a GA begins its search with a randomly assigned initial population. Then, different "genetic" operators, such as *crossover, mutation, etc.*, are applied to update the population and thus complete one iteration. The iteration cycles are repeated until a sufficiently good solution, expressed by the fitness criterion, is achieved.

For determination of the initial weights in an LVQ ANN we define a chromosome—*i.e.*, an individual—as the whole set of the weights in the LVQ ANN. The mutation operator is defined as in the conventional GA. From generation to generation the population evolves, improving slowly the fitness criterion. The best solution in each iteration is copied without changes to the next generation in order to preserve the best chromosome. Consequently, the best solution can survive to the end of the process.

3.6.2. *Searching for Good Initial Weights*

To find the good initial weights of the LVQ ANN we proceed as follows. From each sequence in the set $P_{tr} \cup S_{cds}$, we determine a vector of 8 parameters related to a hypothetical TATA-box motif. These vectors are filtered by SF_2—*i.e.*, the threshold $\tau = 0.4$, and bounds for data regions for D_1 and D_2 selected as $D_1 = [0.1, 1.3]$, $D_2 = [0.1, 1.2]$. Then these vectors are normalized and PCA transformed, resulting in the set Y_{tr} that contains feature vectors of size 6 and their labels. In total 492 data vectors that originate from P_{tr} pass the filter SF_2, while 1359 data vectors that originate from S_{cds} pass the filter SF_2.

Then an LVQ ANN is constructed with 6 nodes in the input layer and 100 neurons in the competitive layer. Of these 100 neurons, 40 are related to Class 1 (TATA motif class) and 60 are related to Class 2 (non-TATA motif class). The ratio 4/6 of the hidden layer neurons does not reflect properly the input data distribution 492/1359. However, if the 492/1359 proportion is used, then the network

τ	P_{tr} (TP%)	S_{cds} (FP%)	$CC_{training}$	P_{tst} (TP%)	S_{int} (FP%)	CC_{test}
0.4	67.5	2.4	0.6586	58.8235	2.875	0.4525

Fig. 7.   Recognition using LVQ ANN1 after the initial weights are determined by GA

performance appears to be very poor for the reason that there is an insufficient number of neurons in the competitive layer associated with the TATA related feature vectors. On the other hand, if we want to increase sufficiently the number of neurons for Class 1 to achieve a good TP level, and at the same time keep the proportion of neurons relating to Class 1 and Class 2 correct, the total number of neurons in the competitive layer becomes very big and the training time for GA appears to be extremely long. Thus we decide to retain the ratio of dedicated neurons in the competitive layer as indicated. After the LVQ ANN is specified, we use GA to search for the good initial weights for the LVQ ANN. The GA is run for 21 epochs. After 21 epochs, the fitness has already reached about 87%. Thus it can be regarded as corresponding to relatively good initial weights.

Note that we filter feature vectors by SF_2 before the GA is applied to search for the optimized initial weights of the LVQ ANN. However, in order to test how good the determined initial weights are, we are not going to apply the complete SF_2 filtering, but only filtering by the threshold value of $\tau = 0.4$, and no restriction to the values of numerical parameters x_7 and x_8 is applied. This allows more feature vectors—in both the training and test sets—to be processed by the LVQ ANN than if the SF_2 filtering is applied. We apply the normalization and PCA transformation of feature vectors for all sequences from $P_{tr} \cup S_{cds}$ and $P_{tst} \cup S_{int}$, that pass the threshold filtering, using the preprocessing parameters as determined previously. Then such transformed feature vectors are presented to the LVQ ANN and their ability to recognize the TATA motif (and promoters) is evaluated. The results of recognition are presented in Figure 7. Then, the LVQ ANN is further trained with the $lvq2$ algorithm to provide fine tuning of the decision boundaries for the network. The learning rate $LR = 0.01$ and 5500 epochs are used for training. The recognition quality has improved as can be seen from results presented in Figure 8. However, the results are not sufficiently good for constructing a system for the recognition of TATA-like sequences in long stretches of DNA. Such a system is generated in what follows.

τ	P_{tr} (TP%)	S_{cds} (FP%)	$CC_{training}$	P_{tst} (TP%)	S_{int} (FP%)	CC_{test}
0.4	66.8750	1.2125	0.7198	58.8235	2.55	0.4712

Fig. 8.   Recognition results with LVQ ANN1, whose initial weights are determined by GA and then fine-tuned by the $lvq2$ algorithm.

4. Final Model of TATA-Box Motif

4.1. *Structure of Final System for TATA-Box Recognition*

The system to be used for the recognition of the TATA-box is depicted in Figure 9. It consists of a statistical filter SF_1, a block for data normalization and PCA transformation, and a block with two LVQ ANNs (LVQB). The whole system is denoted as a multi-stage LVQ ANN (MLVQ) system. The feature vectors X that contain 8 numerical components are fed to the input of the statistical filter SF_1.

The feature vectors that pass the filter enter the normalization and PCA transformation block, at the output of which we get transformed feature vectors Y that contain only 6 numerical components. These transformed feature vectors enter the LVQB. At the output of the MLVQ system the input feature vectors X are classified into those that represent a TATA motif or those which do not.

The LVQB depicted in Figure 10 consists of two LVQ ANNs that process data sequentially. The input of the LVQ ANN1 are transformed feature vectors Y. If the feature vector Y is classified as the one representing the TATA motif, it is passed as input to the LVQ ANN2 that makes the final assessment whether Y represents the TATA motif or not.

4.2. *Training of the ANN Part of the Model*

The training of the LVQ ANNs within the MLVQ system is as follows.

- **The first stage LVQ ANN1.** This ANN is selected with 100 neurons in the hidden layer, and with the 50% of the neurons in the hidden subclasses layer corresponding to Class 1 (TATA motif class) and 50% of the neurons to relate to Class 2 (non-TATA motif class). The training set for this ANN includes feature vectors X of all sequences from P_{tr}, those of 640 sequences from S_{cds}, and the respective class labels. First we apply GA to search for the optimized initial weights of the network as explained in the last section. The

MLVQ SYSTEM FOR THE
TATA MOTIF RECOGNITION

feature
vectors X
of size 8 → [Statistical filter SF1] → [normalization + PCA transformation] → transformed feature vectors Y of size 6 → [LVQB] →

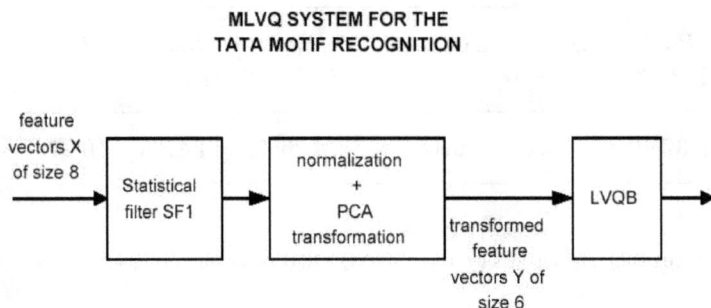

Fig. 9.   The system for the TATA motif recognition.

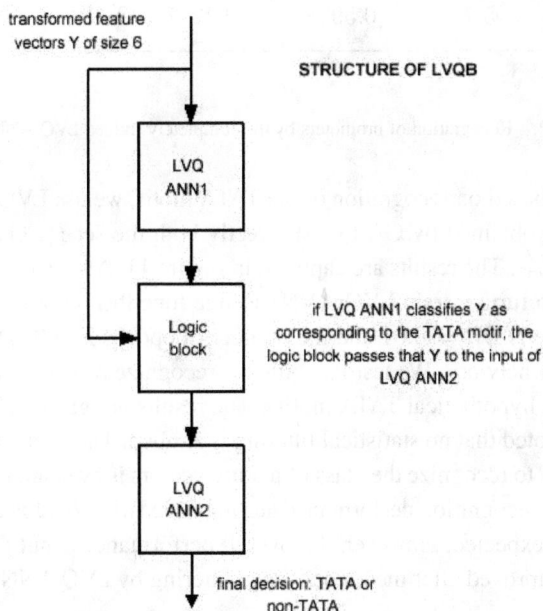

transformed feature
vectors Y of size 6

STRUCTURE OF LVQB

[LVQ ANN1]

[Logic block]   if LVQ ANN1 classifies Y as corresponding to the TATA motif, the logic block passes that Y to the input of LVQ ANN2

[LVQ ANN2]

final decision: TATA or non-TATA

Fig. 10.   The structure of LVQB. The transformed input feature vectors Y are classified at the output of the LVQ ANN as representing the TATA motif or not.

only reason for the restriction of the of the number of sequences from S_{cds} to 640 is the speed of GA search. After 51 generations, the fitness reaches 85.03%. In order to see its performance in attempted recognition of promoter

P_{tr} (TP%)	S_{cds} (FP%)	$CC_{training}$	P_{tst} (TP%)	S_{int} (FP%)	CC_{test}
79.3750	11.9500	0.4707	74.3697	14.4125	0.27

Fig. 11. Recognition of promoters by means of LVQ ANN1 using initial weights obtained by GA.

P_{tr} (TP%)	S_{cds} (FP%)	$CC_{training}$	P_{tst} (TP%)	S_{int} (FP%)	CC_{test}
76.4063	5.75	0.5918	69.3277	8.1125	0.3436

Fig. 12. Recognition of promoters by the completely trained LVQ ANN1

sequences, based on recognition of the TATA motif, we use LVQ ANN1 with the weights obtained by GA to test directly both the set $P_{tr} \cup S_{cds}$ and the set $P_{tst} \cup S_{int}$. The results are captured in Figure 11. Also, we apply the $lvq2$ algorithm to further train LVQ ANN1 to fine tune the class boundaries. The learning rate is $LR = 0.01$ and the number of epochs is 500. This trains the LVQ ANN1 network. We test its ability to recognize promoter sequences by recognizing hypothetical TATA motifs. The results are given in Figure 12. It should be noted that no statistical filtering is applied. Thus, only the ability of LVQ ANN1 to recognize the class of feature vectors is evaluated. One can notice that the recognition performance has improved after $lvq2$ is applied. This is what we expected. However, the overall performance is not good enough. It will be improved after the second stage filtering by LVQ ANN2.

- **The second stage LVQ ANN2.** After filtering of feature vectors that correspond to $P_{tr} \cup S_{cds}$ by the LVQ ANN1, 489 feature vectors corresponding to sequences from P_{tr} ($TP = 76.4063\%$) and 460 feature vectors corresponding to sequences from S_{cds} ($FP = 5.75\%$) remain. These, together with the associated class labels serve as the training set for LVQ ANN2. The number of neurons in the hidden layer for LVQ ANN2 is also 100, and the distribution of neurons in the hidden layer is 50% for Class 1 and 50% for Class 2, as

P_{tr} (TP%)	S_{cds} (FP%)	$CC_{training}$	P_{tst} (TP%)	S_{int} (FP%)	CC_{test}
56.25	0.8000	0.6722	48.3193	1.4375	0.4767

Fig. 13.   Promoter recognition results with LVQB, where LVQ ANN2 was tuned only by GA.

P_{tr} (TP%)	S_{cds} (FP%)	$CC_{training}$	P_{tst} (TP%)	S_{int} (FP%)	CC_{test}
58.1250	0.7625	0.6885	51.6807	0.6885	0.5377

Fig. 14.   Promoter recognition results with the completely tuned LVQB.

the positive and negative data for training are approximately of the same size. GA is used to search for the initial weights for LVQ ANN2. After 101 generations, the set of optimized initial weights is obtained, producing the fitness of 79.66%. The ability of LVQB (LVQ ANN1 and LVQ ANN2 together) to recognize promoters by recognizing the presence of the TATA motif is now tested. The results are given in Figure 13. After the optimized initial weights for LVQ ANN2 are obtained, the LVQ ANN2 is further fine tuned using the $lvq2$ algorithm with the same parameters as for LVQ ANN1. Then, the LVQB is considered trained, and the results of its recognition accuracy testing are given in Figure 14. Compared with recognition results of LVQ ANN1, FP for S_{cds} has reduced 7.5 times, FP for S_{int} reduced 11.7 times, while TP for the training set falls about 31% and for the test set about 35%. Thus, the obtained improvement in accuracy is rather significant, which is also reflected through the increase in the CC values for training and for the test set results.

4.3. *Performance of Complete System*

The complete system for TATA motif recognition includes the statistical filter SF_1, preprocessing block (normalization and PCA transformation of feature vectors), and the tuned LVQB. The only global parameter that can influence the recog-

H. Wang, X. Li, & V. B. Bajić

τ	P_{tr} (TP%)	S_{cds} (FP%)	$CC_{training}$	P_{tst} (TP%)	S_{int} (FP%)	CC_{test}
0.3600	57.9688	0.7625	0.6874	50.8403	1.0875	0.5312
0.3700	57.9688	0.7625	0.6874	50.8403	1.0875	0.5312
0.3800	57.9688	0.7625	0.6874	50.8403	1.0875	0.5312
0.3900	57.9688	0.7625	0.6874	50.8403	1.0875	0.5312
0.4000	57.9688	0.7125	0.6911	50.8403	1.0000	0.5411
0.4100	**57.6563**	**0.6625**	**0.6925**	50.8403	1.0000	0.5411
0.4200	57.5000	0.6625	0.6914	50.8403	0.9250	0.5500
0.4300	57.0313	0.6125	0.6918	**50.8403**	**0.9125**	**0.5515**
0.4400	56.2500	0.6000	0.6871	49.1597	0.9125	0.5384
0.4500	54.6875	0.5875	0.6767	47.4790	0.7625	0.5442
0.4600	52.6563	0.5875	0.6617	43.6975	0.6750	0.5254
0.4700	49.6875	0.5875	0.6393	39.4958	0.6625	0.4913
0.4800	46.2500	0.5000	0.6197	35.7143	0.4750	0.4867
0.4900	41.2500	0.4625	0.5825	30.6723	0.3000	0.4716
0.5000	36.2500	0.4625	0.5396	23.9496	0.1375	0.4408

Fig. 15.   Recognition results of the complete tuned system.

nition accuracy and that is also simple for evaluation is the threshold level τ. From previous experimental results we expect the best recognition accuracy of the system for τ at around 0.42. The results of promoter recognition by means of hypothetical TATA motif recognition for the complete system with the threshold of the statistical filter variable, are presented in Figure 15.

It can be observed that a rather good recognition quality appears at the threshold $\tau = 0.43$ for the test set, providing $CC = 0.5515$. Also, the best recognition quality for the training set is achieved for $\tau = 0.41$, producing $CC = 0.6925$. Both of these threshold values are close to the expected best values of around $\tau = 0.42$. Note that the recognition level of $TP = 50.84\%$ on the test set corresponds to the correctly recognized promoter sequences based on the recognition of the hypothetical TATA-box. Since it was estimated in Section 2 that about 75% of promoter sequences used in this study contain a hypothetical TATA-box, the level of correct recognition of TATA motifs is $TP = 67.79\%$.

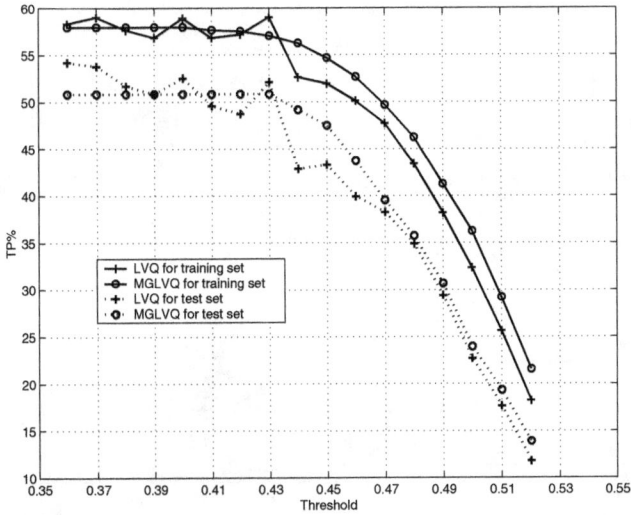

Fig. 16. Comparison of $TP\%$, as a function of τ, between the single LVQ system and the MLVQ system.

4.3.1. *Comparison of MLVQ and System with Single LVQ ANN*

It is interesting to compare the performance of the complete MLVQ system and the one consisting of the statistical filter SF_1, preprocessing unit, and a single LVQ ANN. The comparison results are depicted in Figure 16, Figure 17, and Figure 18. Figure 16 shows that the TP level of both systems are roughly similar as τ changes. However, a noticeable difference appears in the level of FP as indicated in Figure 17. For most of the τ values the FP for the test set of the MLVQ system is considerably lower than for the single LVQ system, implying a much better generalization ability of the MLVQ system. On the other hand, the single LVQ system performs better on the training set, indicating that it has been overtrained. These observations are also visible from the curves of CC changes as given in Figure 18. Although compared to the MLVQ system, the single LVQ system performs better on the training set, its performance is much weaker on the test set.

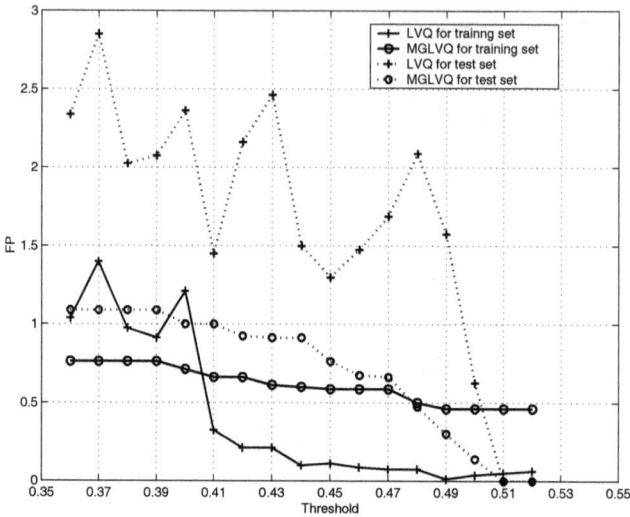

Fig. 17.   Comparison of $FP\%$, as a function of τ, between the single LVQ system and the MLVQ system.

4.4. The Final Test

The final test of the neural-statistical model of the TATA-box implemented in the MLVQ system is made in an attempt to evaluate how well the system performs on longer DNA sequences. As the test set, we use a collection of 18 mammalian promoter sequences selected and used for the evaluation of promoter recognition programs in Fickett and Hatzigeorgiou.[249] The total length of these sequences is 33120bp, and the set contained 24 TSS. By applying the MLVQ system to this set we can assess the ability of the system to recognize TSS locations by means of recognizing potential TATA-box motifs.

The threshold is used with the values of $\tau = 0.41$ and $\tau = 0.43$. The former one corresponds to the highest CC value obtained previously on the training set, while the latter corresponds to the highest CC value obtained on the test set. We slide along each sequence from its 5' end toward the 3' end a window of 43 nucleotides in length. For 24 successive positions within the window we find the hypothetical TATA-like motif and calculate the 8 parameters associated with them as described previously. This data is fed to the MLVQ system and predictions of the location of the TATA-box are made. The predictied location of the TSS is counted as being 30bp downstream of the predicted TATA-box location. The same

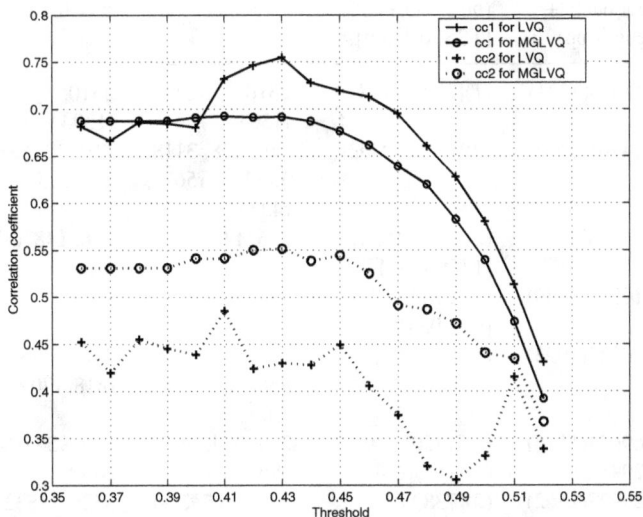

Fig. 18. Comparison of the correlation coefficient CC, as a function of τ, between the single LVQ system and the MLVQ system.

is done with the reverse complement of each of the sequences.

The same criterion, as used in Fickett and Hatzigeorgiou,[249] is applied to determine if a predicted TSS is correct: It is counted as correct if it falls within 200bp upstream or up to 100bp downstream of the actual TSS. Figure 19 shows the test result, where the "+" after a position indicates the correct prediction. The position indicated in bold is the one that counts as correct according to the adopted criteria. Positions with r in front are made on the reverse complement strand.

As can be observed with the threshold $\tau = 0.41$, the MLVQ system identifies 8 (33%) of the known promoters and makes 47 false positives (1/705bp). When the threshold is $\tau = 0.43$, the system finds 7 TSS (29%), whilst making 41 false positive predictions (1/808bp). Figure 20 shows the recognition quality of some other programs tested in Fickett and Hatzigeorgiou.[249] On this test set the MGLVQ system performs better than Audic, Autogene, and the PWM of the TATA-box. The results for comparison are given in Figure 20, where the last column shows the ranking results of different programs based on the average score measure from Bajic.[50] The average score measure is obtained based on 9 different measures of prediction quality and thus can be considered a balanced overall measure of prediction success for different programs. The ranking is in ascend-

No.	Seq. name + Length(bp)	(number) + [TSS location]	$\tau = 0.41$	$\tau = 0.43$
1	L47615(3321)	(1)[2078,2108]	2510, r2456, r1304	2510, r2456, r1304
2	U54701(5663)	(2)[935], [2002]	119, 228, 3118, 5369, r5619 r4272	119, 228, 3118, 5369, r5619
3	Chu(2003)	(2)[1483,1554], [1756,1783]	246, **1487(+)**	246, **1487(+)**
4	U10577(1649)	(2)[1169,1171], [r1040,r1045]	354	354
5	U30245(1093)	(1)[850,961]	194, 446, **690(+)**, r818, r709	446, **690(+)**, r818, r709
6	U69634(1515)	(1)[1450]	299, 934, r645	299, 934, r645
7	U29912(565)	(1)[143,166]	None	None
8	U29927(2562)	(2)[738,803], [1553,1717]	332, **1532(+)**, 2114, r1658	332, **1532(+)**, 2114
9	Y10100(1066)	(1)[1018,1033]	159, 781, r666, r349, r181	159, 781, r349, r181
10	Nomoto(2191)	(1)[1793,1812]	486, 867, 1191, 1375, **1643(+)**	486, 867, 1191, 1375, **1643(+)**
11	U75286(1984)	(1)[1416,1480]	r863	None
12	U52432(1604)	(1)[1512,1523]	217, **1380(+)**, r1451, r138	217, r1451, r138
13	U80601(632)	(1)[317,400]	94	94
14	X94563(2693)	(1)[1163,1200]	489, 2527, r2176, r227	489, 2527, r227
15	Z49978(1352)	(3)[855], [1020], [1150]	**1011(+)**, **1024(+)**, r1160, r1122, r405	**1011(+)**, r1160, r1122, r405
16	U49855(682)	(1)[28,51]	r162	r162
17	X75410(918)	(1)[815,835,836]	473, **685(+)**, **732(+)**, **786(+)**, r521	473, **685(+)**, **732(+)**, **786(+)**, r521
18	U24240(1728)	(1)[1480]	r1529, r1071	r1529, r1071

Fig. 19. Results obtained by MLVQ system on the Fickett and Hatzigeorgiou data set.

ing order with the best performed program on position 1. One can observe that the MLVQ system performs much better than if the recognition is based on the PWM of the TATA-box from Bucher.[116] Since the goal is to develop a more accu-

#	Prog. name	$TP\#$	$TP\%$	$FP\#$	FP per bp	Rank
1	Audic[38]	5	24%	33	1/1004bp	9
2	Autogene[446]	7	29%	51	1/649bp	10
3	Promoter2.0	10	42%	43	1/770bp	4
4	NNPP	13	54%	72	1/460bp	6
5	PromoterFind[377]	7	29%	29	1/1142bp	5
6	PromoterScan	3	13%	6	1/5520bp	2
7	PWM of TATA-box[116]	6	25%	47	1/705bp	11
8	TSSG[784]	7	29%	25	1/1325bp	1
9	TSSW[784]	10	42%	42	1/789bp	3
10	MGLVQ ($\tau = 0.41$)	8	33%	47	1/705bp	8
11	MGLVQ ($\tau = 0.43$)	7	29%	41	1/808bp	7

Fig. 20.   Results of other promoter finding programs tested on the Ficket and Hatizigoergiou data set.

rate model of the TATA motif, this improved accuracy is indirectly demonstrated through the better performance in promoter recognition.

One should note that the Audic program uses Markov chains to characterize the promoter region, while the part of the Autogene system that recognizes promoters (the FunSiteP program) bases its logic on the different densities of the transcription factor binding sites in the promoter and non-promoter regions. It it thus surprising in a way, that the MLVQ system that uses only one transcription factor binding site, the TATA-box motif, performs better than these programs. However, it would be interesting to evaluate the performance of MLVQ on a larger promoter set, since the one from Fickett and Hatzigeorgiou[249] is not very big and is not statistically structured to properly represent the eukaryotic Pol II promoters.

5. Summary

The chapter presents an application of ANN to the recognition of TATA-like motifs in eukaryotic Pol II promoters and, based on that, the recognition of promoter sequences. A short region of DNA of 20bp containing the hypothetical TATA-like motif is represented by 8 numerical parameters obtained as a combination of the TATA-box PWM matching score and a statistical analysis. Five of these numerical parameters are derived from the modified EIIP values of nucleotides in the TATA-box hexamer and the neighboring hexamers. These parameters show some regularities that were used in the design of the recognition system. The MLVQ system for TATA motif recognition contains a statistical filter, a normalization and PCA transformation block, and two LVQ ANNs. The initial weights of the ANNs in MGLVQ have been obtained by GA. These helped to obtain finally trained system with improved recognition quality. It is demonstrated on an independent data set of Fickett and Hatzigeorgiou[249] that the new neural-statistical model of the TATA-box (contained in the MLVQ system) achieves improved recognition accuracy as compared with the recognition based only on the use of the matching score of the TATA PWM. Moreover, the MLVQ system performed better than the other two promoter-finding programs that use other promoter characteristics in the search for promoters. Since our system is aimed at improved modeling of the TATA-like motifs and not for promoter recognition, we can conclude that a much better model of the TATA-box has been developed by the combined statistical filtering and ANN system.

CHAPTER 7

TUNING THE DRAGON PROMOTER FINDER SYSTEM FOR HUMAN PROMOTER RECOGNITION

Vladimir B. Bajić

Institute for Infocomm Research
bajicv@i2r.a-star.edu.sg

Allen Chong

Institute for Infocomm Research
achong@i2r.a-star.edu.sg

Discovery of new genes through the identification of their promoters in anonymous DNA and the study of transcriptional control make promoter and transcription start site recognition an important issue for Bioinformatics. The biological process of transcription activation is very complex and is not completely understood. Hence computer systems for promoter recognition may not perform well if we do not pay attention to their tuning.

This chapter explains the tuning of a computer system for the recognition of functional transcription start sites in promoter regions of human sequences. The system is called Dragon Promoter Finder, and it can be accessed at `http://sdmc.i2r.a-star.edu.sg/promoter`. The tuning of this complex system is set up as a multi-criteria optimization problem with constraints. The process is semi-automatic, as it requires an expert assessment of the results.

ORGANIZATION.

Section 1. We briefly discuss the importance and challenges of promoter and transcription start site recognition. Then we briefly describe the performance of several existing systems for this recognition problem.

Section 2. Good performance in promoter recognition requires careful tuning of the recognition system. We use Dragon Promoter Finder, which is one of the best performing system on this problem, to illustrate this tuning process. So, we zoom into a more extensive exposition of the architecture of Dragon Promoter Finder in this section.

Section 3. After that, we identify the parameters of Dragon Promoter Finder which should be tuned. These parameters include the parameters of a nonlinear signal processing block and a threshold on the output node of an artificial neural network.

Section 4. In order to determine the optimal parameter values, carefully chosen tuning data should be used. The selection of our tuning data set, comprising samples of both promoters and non-promoters, is described in this section.

Section 5. We are then ready to dive into the details of the tuning process of Dragon Promoter Finder. This process is a multi-criteria optimization that is cast as a Gembicki goal attainment optimization.

Section 6. Finally, we present the fruit of this tuning process. Specifically, we demonstrate the significant superiority of Dragon Promoter Finder's performance compared to several systems for finding promoters and transcription start sites. On a test set of 1.15Mbp of diverse sequences containing 159 transcription start sites, Dragon Promoter Finder attains several folds less false positives than other systems at the same level of sensitivity.

1. Promoter Recognition

Promoters are functional regions of DNA that control gene expression. The biochemical activity in this region—involving the interaction of DNA, chromatin and transcription factors—determines the initiation and the rate of gene transcription. In eukaryotes, the promoter region is usually located upstream of, or overlaps, the transcription start site.[249, 656, 878] A gene has at least one promoter.[656, 878] One can thus find a gene by first locating its promoter.[879] Discovering new genes through the identification of their promoters in anonymous DNA and the study of transcriptional control make promoter recognition an extremely important issue for Bioinformatics.

Even though promising solutions have been proposed—by Bajic *et al.*,[48, 49, 51] Davuluri *et al.*,[190] Down and Hubbard,[211] Hannenhalli and Levy,[321] Ioshikhes and Zhang,[383] and Scherf *et al.*[753]—computational recognition of promoters still has not yet achieved a satisfactory level of confidence.[249, 802] The reason is that the biological process of transcription activation is very complex and hierarchical, and is not completely understood.[878] There are numerous and functionally diverse transcription factors that individually bind to specific DNA consensus sequences—called transcription factor binding sites—in the promoter to activate the transcriptional machinery in concert with RNA polymerases.

Many simplistic approaches have been taken in computational promoter recognition.[681] Unfortunately, at significant levels of true positive recognition, these have produced a significant number of false positive recognition.[249, 681, 708] A false positive prediction is a prediction that indicates the presence of a promoter at a location where the promoter does not exist. A true positive prediction is one that correctly identifies the location of a promoter. Promoter recognition systems for large-scale DNA screening require an acceptable ratio of true positive and false positive predictions. That is, these systems should maximize true positive recognition while minimizing false positive recognition.

While the boundaries of promoter region are loosely defined, each promoter has at least one strong reference site: the transcription start site (TSS). Promoter search can thus focus either on locating the promoter region,[190,321, 383,753] or on pinpointing the TSS.[48,49,51,211,439,628,707] The system in this chapter is a TSS finder. Existing TSS-finders—like NNPP2.1,[707] Promoter2.0,[439] and McPromoter[628]—produce a lot of false positive predictions,[628,656,681,708] making them unsuitable for locating promoters in large genomic sequences.

A recently reported system, Eponine,[211] has demonstrated very good performance. However, its predictions are very much related to CpG-island associated promoters[81,180,271,474] and to G+C rich promoters. The G+C-content of a DNA segment is the proportion of the total number of G and C nucleotides relative to the length of that segment. CpG islands are unmethylated segments of DNA longer than 200bp, with the G+C content of at least 50dinucleotides—that is, a C followed by a G—being at least 60the G+C content of the segment.[81,180,271,474] CpG islands are found around gene starts in approximately half of mammalian promoters[180,474] and are estimated to be associated with about 60% of human promoters.[181] For this reason it is suggested by Pedersen *et al.*[656] that CpG islands could represent a good global signal to locate promoters across genomes. At least in mammalian genomes, CpG islands are good indicator of gene presence. The G+C content is not uniformly distributed over the chromosomes and the CpG island density varies according to the isochores' G+C content.[673]

Several promising systems have been developed in the last two years. PromoterInspector[753] has been reported to produce a considerably reduced level of false positive recognition compared to other publicly available promoter recognition programs. After its introduction, three other systems with similar performance have also been reported.[190,321,383] These four systems predict regions that either overlap promoter regions, or are in close proximity to the promoter. The localization of TSS is not considered in Hannenhalli and Levy[321] and Ioshikhes and Zhang.[383] However, if we wish to develop a promoter model that can efficiently search for genes of a specific genetic class through the recognition of relevant promoter features for the targeted class of genes, it is necessary to pinpoint the TSS and thus, localize the promoter region.

2. Dragon Promoter Finder

It is well known that TSS-finding systems produce a high number of false positives.[249,628,681,708] The TSS-finding system that we present in this chapter considerably reduces the number of false positives. In contrast to solutions which are aimed at the recognition of specialized classes of promoters—such as CpG-island related promoters,[321, 383] or G+C-rich promoters with a TATA-box[211]— our system is aimed at analyzing and identifying general human polymerase II promoters. While the design details of our system, Dragon Promoter Finder, have been published elsewhere,[48, 49, 51] the tuning process of this complex system has not described previously. We thus present here the details of the tuning process for Dragon Promoter Finder system. The tuning process is performed using Gembicki's goal attainment optimization process.[99, 284, 754]

Our system is at `http://sdmc.i2r.a-star.edu.sg/promoter`. It is based on a hierarchical multi-model structure with models specialized for

(a) different promoter groups, and
(b) different sensitivity levels.

To the best of our knowledge, this is the first reported composite-model structure used in promoter recognition systems based on (a) and (b) above. First, the short DNA segments around TSS representing the promoter data are separated into the G+C-rich and G+C-poor groups. This separation of data and subsequent development of models for both of these promoter groups, as well as the sophisticated tuning of the models, has resulted in considerably enhanced system performance. The resulting system combines:

- multiple hierarchically organized models optimally tuned for different sensitivity requirements,
- specialization of models to G+C-rich or G+C-poor promoter groups,
- sensor-integration,
- nonlinear signal processing, and
- artificial neural networks (ANN).

This makes it conceptually different from the approaches used in other promoter-finding systems,[249, 681] including those that use several region sensors.[488, 628, 753]. The system is shown to be capable of successfully recognizing promoters that are CpG-island related and those that are not, as well as promoters in G+C-rich and in G+C-poor regions. This makes it quite universal as opposed to solutions which are specialized in recognizing CpG-islands related promoters,[321, 383] or to the one in Down and Hubbard.[211]

The practical significance of Dragon Promoter Finder is in its use for identification and annotation of promoters in anonymous DNA, as well as in the enhancement of gene hunting by more accurate determination of the 5' end of the gene and parts of the gene's regulatory regions.

3. Model

The description of the system is presented by Bajic and colleagues.[48, 49, 51] The system possesses three sensors for promoters, coding exons, and introns. Let the produced signals of the promoter, coding exon, and intron sensors be denoted respectively by σ_p, σ_e, and σ_i. These signals enter a nonlinear signal processing block where they are transformed according to

$$s_E = f(\sigma_p - \sigma_e, a_e, b_e, c_e, d_e)$$
$$s_I = f(\sigma_p - \sigma_i, a_i, b_i, c_i, d_i)$$
$$s_{EI} = f(\sigma_p - \sigma_{ei}, a_{ei}, b_{ei}, c_{ei}, d_{ei})$$

In Ver. 1.2 of Dragon Promoter Finder, the function f is defined by

$$f = blin = \begin{cases} c \times x, & \text{if } x > a \\ x, & \text{if } b \le x \le a \\ d \times x, & \text{if } b > x \end{cases}$$

In Ver. 1.3 of Dragon Promoter Finder, the function f is defined by

$$f = sat = \begin{cases} a, & \text{if } x > a \\ x, & \text{if } b \le x \le a \\ b, & \text{if } b > x \end{cases}$$

The parameters a_k, b_k, c_k, d_k, for $k = e, i, ei$, are part of the tunable system parameters. Also, the signals s_E, s_I, and s_{EI} are subject to whitening in Ver.1.2, and to principal component transform in Ver.1.3. The transformed signals—z_E, z_I, and z_{EI}—are inputs to the feed-forward ANN. The ANN is trained by the Bayesian regularization method[82] for the best separation between the classes of input signals, with initial parameters $a_k = +\infty$ and $b_k = -\infty$ for $k = e, i, ei$. The trained ANN is then used as a part of the system in the final tuning.

4. Tuning Data

In order to determine the optimal parameter values, a set of tuning data composed of promoter and non-promoter sequences must be created. This section briefly describes our tuning data set.

4.1. *Promoter Data*

For the promoter data of the tuning data set, we used 793 different vertebrate promoter sequences from the Eukaryotic Promoter Database[662]. These sequences are extracted from the window $[-250, +50]$ relative to the TSS position. Note that, by convention, there is no nucleotide position "0". The nucleotide at the TSS is assigned the position "1" and the nucleotide immediately preceding the TSS is assigned the position "−1". Additionally, we used $[-250, +50]$ sequence segments of 20 full-length gene sequences with known TSS, whose promoters are not included by the Eukaryotic Promoter Database.

4.2. Non-Promoter Data

For the non-promoter data of the tuning data set, we randomly collect from Genbank[76] Rel. 121 a set of non-overlapping human coding exons and intron sequences, each 250bp in length. We also selected non-overlapping human sequences from the 3'UTR regions taken from the UTRdb.[665] All sequences in these three groups are checked for similarity using the BLAST2Sequences program[820] to ensure that any two sequences within the group have less than 50% identity relative to each other. In total, 1300 coding exon sequences, 4500 intron sequences, and 1600 3'UTR sequences are selected. Additionally, from the 20 gene sequences mentioned earlier, we include as non-promoter data all 250bp segments that do not overlap the $[-250, +50]$ regions.

5. Tuning Process

The Dragon Promoter Finder requires careful tuning to achieve the best performance of the system in recognition of TSS in a blind promoter search. The general goal of tuning is to maximize the level of true positives versus false positives over the entire range of sensitivity settings. Different models are trained and each is tuned for the best performance at a predefined sensitivity level. This means that we aim at making the highest positive predictive value (ppv)—sometimes denoted as specificity in bioinformatics[50]—for the predefined sensitivity level. The sensitivity and ppv are given by

$$S_e = \frac{TP}{TP + FN}$$
$$ppv = \frac{TP}{TP + FP}$$

where FN stands for false negatives and equals the number of true promoters not predicted by the promoter prediction programs; TP stands for true positives and equals the number of true promoters predicted by the promoter prediction programs; and FP stands for false positives and equals to the number of non-promoters incorrectly claimed by the promoter prediction programs as promoters.

The tuning process can thus be considered as an optimization process with two goals expressed by

$$\max S_e$$

$$\max ppv$$

However, TP, FP, and FN can be only positive integers. Therefore, the formulation of the tuning of the above-mentioned optimization problem cannot take full advantage of the sophisticated optimization algorithms with continuous criteria functions. So, we need to reformulate the optimization problem for the tuning purpose.

For Ver. 1.3, the set of tunable parameters p of the system consists of the ANN threshold τ and a_k, b_k, c_k, d_k, for $k = e, i, ei$. For Ver. 1.2, additional sensor signal thresholds are also used. These parameters have to be adjusted so that the tuned system achieves the desired performance. The tuning process is conducted 10 times for each selected level of sensitivity, and different models are produced in the process. Then, from all of the models, the selection of the representative model for each sensitivity level is made.

In the tuning process, a sequence from the tuning set is presented to the system and this produces an output signal s. This signal is compared to the desired target value t, which is 1 for promoters and -1 for non-promoters, and the error $e = s - t$ is calculated. The error serves as the signal that determines the change in the system's tunable parameters by means of a feedback process. This tuning process can be considered as a multi-criteria optimization with constraints, more specifically as a Gembicki's goal attainment optimization.[99, 284, 754] The details of which is described in the remainder of this section.

We define several objectives for which we want to achieve the predefined goals in the tuning process. These objectives are captured in the vector F of objectives of the system produced on the tuning set as

$$F = \langle E_p, E_e, E_i, E_{utr}, E_g, S_e, ppv \rangle$$

where

$$E_k = \frac{1}{N_k} \times \sum_{j=1}^{N_k} |e_j^k|, \text{ for } k = p, e, i, utr, g$$

$$e_j^k = s_j^k - t_j^k,$$

$$t_j^k = \begin{cases} 1, & \text{if } k = p \\ -1, & \text{otherwise} \end{cases}$$

Here, N_k is the number of presented sequences of a specific class k; s_j^k and t_j^k are the system output signal and target values respectively, for group k when the

j-th sequence is presented to the system; p, e, i, utr, g stand for promoter, coding exon, intron, 3'UTR, and the sequences corresponding to non-promoter positions in the selected 20 genes, respectively.

The goal attainment process is defined as

$$\min_{\gamma \in R,\ p \in \Omega} \gamma$$

subject to the following constraints:

$$E_j - w_j \times \gamma \leq 1, \text{ for } j = p, e, i, utr, g$$

$$\frac{1}{ppv} - w_f \times \gamma \leq 1$$

$$\frac{1}{S_e} - w_t \times \gamma \leq \frac{1}{L_s}$$

Here, L_s is the predefined sensitivity level for which the model is tuned, Ω is the overall parameter space, and w_j, w_t, and w_f are the slack weights in the optimization. The tuning process is repeated 10 times with the tuning parameters randomly initialized.

After the collection of models are produced for the all selected sensitivity levels, then the selection of the appropriate models for each level is made. This second phase of the process is not automated and requires manual selection. The goals in the model selection process are

- the change of the parameter values for the selected models for different successive sensitivity levels has to be gradual, and
- the best possible models, expressed in terms of E_{TP} and E_{FP}, should be selected.

Sometimes it is not possible to choose models that satisfy the gradual change in the parameters. Then the tuning for the critical sensitivity levels is repeated sufficient number of times until this can be done.

6. Discussions and Conclusions

This tuning has resulted in a superior recognition ability of Dragon Promoter Finder. For the purpose of illustration, we present in Figure 1 some of the performance comparison results.

The Dragon Promoter Finder is compared with the Promoter2.0 program,[439] the NNPP2.1 program,[707] and the PromoterInspector program.[753] The first two programs are TSS-finding programs, while the third one is a promoter region

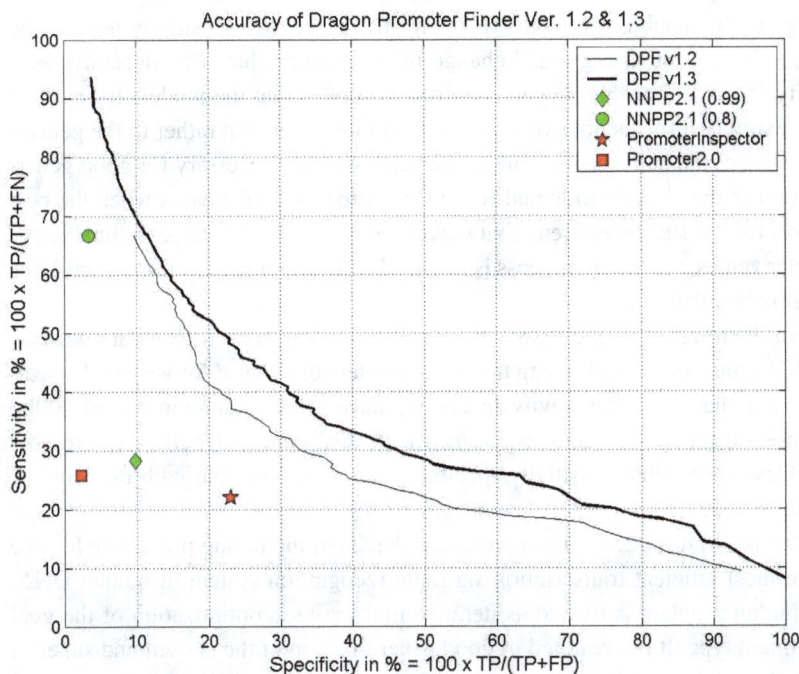

Fig. 1. Comparison of performances of three TSS prediction programs—Dragon Promoter Finder Ver.1.2 and Ver.1.3, Promoter2.0 and NNPP2.1—on the test set from Bajic *et al.*[51] which has a total length of 1.15Mbp and comprises 159 TSS.

finder. The details of the comparisons, as well as of the test set, are given in Bajic *et al.*[48, 49, 51], and at http://sdmc.i2r.a-star.edu.sg/promoter. In these tests, Dragon Promoter Finder has produced several folds smaller number of false positive predictions than the other promoter prediction systems.

Due to the very complex structure of the system and its many tunable parameters, it is not possible to make sequential tuning of parameters in the model. A reasonable solution to this problem is to use a general optimization approach. So, as described in the previous section, we have opted for the Gembicki's goal attainment optimization. Choosing random initialization values for parameters 10 times we obtained different models at each of the predefined sensitivity levels. This is necessary as, in general, the model parameters converge to different values. In total, we have generated models for 85 levels of sensitivity, spanning from $S_e = 0.1$ up to $S_e = 0.94$.

One of the obstacles in the final selection of the models is the non-gradual change in the tunable parameter values for the successive sensitivity levels. The reason for requesting a gradual change in parameter values for successive sensitivity levels is that this property indirectly implies that the models for each of the sensitivity levels is not overfitted to the training data, but rather to the general properties of the data classes. These characteristics are necessary for good generalization in the classification and recognition process. In the cases when the best models for the successive sensitivity levels showed abrupt change in tunable parameter values, the tuning process is repeated sufficient number of times until this criterion is satisfied.

For the lower sensitivity range from 0.1 to 0.27, we have used a data-window of 250bp since the models using this sequence length allowed for very high specificity. For the higher sensitivity levels, we have used a data-window of 200bp because—from our previous experience with this window length—we are able to achieve reasonable sensitivity/specificity ratios not possible with the window length of 250.

Let us summarize. We have presented details of the tuning procedure for one of the most efficient transcription start site recognition system in human DNA. The tuning problem is treated as iterative multi-criteria optimization of the goal attainment type. It has resulted in good generalization of the system and superior prediction ability. This tuning approach has opened a way for a more general method of tuning complex prediction systems for computational biology.

CHAPTER 8

RNA SECONDARY STRUCTURE PREDICTION

Wing-Kin Sung

National University of Singapore
ksung@comp.nus.edu.sg

Understanding secondary structures of RNAs helps to determine their chemical and biological properties. Only a small number of RNA structures have been determined currently since such structure determination experiments are time-consuming and expensive. As a result, scientists start to rely on RNA secondary structure prediction. Unlike protein structure prediction, predicting RNA secondary structure already has some success. This chapter reviews a number of RNA secondary structure prediction methods.

ORGANIZATION.

Section 1. We begin by briefly introducing the relevance of RNA secondary structures for applications such as function classification, evolution study, and pseudogene detection. Then we present the different basic types of RNA secondary structures.

Section 2. Next we provide a brief description of how to obtain RNA secondary structure experimentally. We discuss physical methods, chemical methods, and mutational analysis.

Section 3. Two types of RNA secondary structure predictions are then described. The first type is based on multiple alignments of several RNA sequences. The second type is based on a single RNA sequence. We focus on the predicting secondary structure based on a single RNA sequence.

Section 4. We start with RNA structure prediction algorithms with the assumption that there is no pseudoknot. A summary of previous key results on this topic is given. This is then followed by a detailed presentation of the algorithm of Lyngso, Zuker, and Pedersen.[531]

Sections 5– 7. Then we proceed to some of the latest works on RNA structure prediction that allow for pseudoknots. In particular, we present the $O(n^4)$-time algoritm of Akutsu[15] for a restricted kind of pseudoknots and the 1/3-approximation polynomial time algorithm of Ieong *et al.*[381] for general pseudoknots.

1. Introduction to RNA Secondary Structures

Due to the advance in sequencing technologies, many RNA sequences have been discovered. However, only a few of their structures have been deduced. The chemical and biological properties of many RNAs—like tRNAs—are determined primarily by their secondary structures. Therefore, determining the secondary structures of RNAs is becoming one of the most important topics in bioinformatics. We list a number of applications of RNA secondary structures below:

- Function classification. Many RNAs that do not have similar sequences do have similar functions.[437] An explanation is that they have similar secondary structure. For example, RNA viruses have a high mutation rate. Distant groups of RNA viruses show little or no detectable sequence homology. In contrast, their secondary structures are highly conserved. Hence, researchers classify RNA viruses based on their secondary structure instead of their sequences.
- Evolutionary studies. Ribosomal RNA is a very ancient molecule. It evolves slowly and exists in all living species. Therefore, it is used to determine the evolutionary spectrum of species.[891] One problem in the evolution study is to align the ribosomal RNA sequences from different species. Since the secondary structures of ribosomal RNAs are highly conserved, researchers use the structure as the basis to get a highly accurate alignment.
- Pseudogene detection. Given a DNA sequence that is highly homologous to some known tRNA gene, such a sequence may be a gene or a pseudogene. A way to detect if it is a pseudogene is by computing its secondary structure and checking if that looks similar to some tRNA secondary structure.[525]

Before studying the structures of RNAs, we need to understand the interactions between a pair of RNA nucleotides. RNA consists of a set of nucleotides that can be either adenine (A), cytosine (C), guanine (G), or uracil (U). Each of these nucleotides is known as the base and can be bonded with another one via hydrogen bonds. When this bonding happens, we say that the two bases form a base-pair. There are two types of base-pairs: canonical base-pair and wobble base-pair. The canonical base-pair are formed by a double hydrogen bond between A and U, or a triple hydrogen bond between G and C. The wobble base-pair is formed by a single hydrogen bond between G and U. Apart from these two types of base-pairs, other base pairs like U-C and G-A are also feasible, though they are relatively rare. To simplify the study, we assume only canonical and wobble base-pairs exist.

Unlike DNA, which is double stranded, RNA is single stranded. Due to the extra hydrogen bond in each RNA base, RNA bases in a RNA molecule hybridize with itself and form complex a 3D structure. Biologists describe RNA structures in three levels: primary structure, secondary structure, and tertiary structure. The primary structure of an RNA is just its sequence of nucleotides. The secondary structure of an RNA specifies a list of canonical and wobble base-pairs that occur in the RNA structure. The tertiary structure is the actual 3D structure of the RNA.

Although the tertiary structure is more useful, such a tertiary structure is difficult to predict. Hence, many researchers try to get the secondary structure instead, as such a secondary structure can already explain most of the functionalities of the RNA. This chapter focuses on the secondary structure of RNAs. Consider a RNA polymer $s_1 s_2 \ldots s_n$ of length n. Generally, the secondary structure of the RNA can be considered as a set S of base pairs (s_i, s_j) where $1 \leq i < j \leq n$ that satisfies the following two criteria:

(1) Each base is paired at most once.
(2) Nested criteria: if $(s_i, s_j), (s_k, s_l) \in S$, we have $i < k < j \iff i < l < j$.

Actually, a RNA secondary structure may contain base pairs that do not satisfy the two criteria above. However, such cases are rare. If criteria (1) is not satisfied, a base triple may happen. If criteria (2) is not satisfied, a pseudoknot becomes feasible. Figure 1 shows two examples of pseudoknots. Formally speaking, a pseudoknot is composed of two interleaving base pairs (s_i, s_j) and (s_k, s_l) such that $i < k < j < l$.

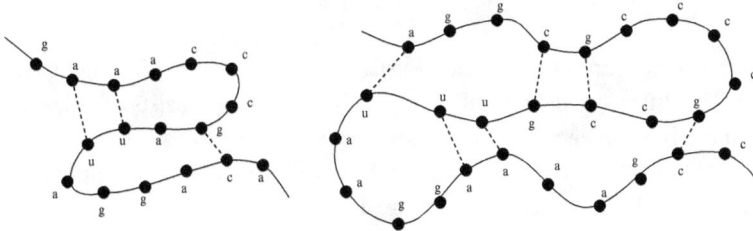

Fig. 1.   Pseudoknots.

When no pseudoknot appears, the RNA structure can be described as a planar graph. See Figure 6 for an example. Then, the regions enclosed by the RNA backbone and the base pairs are defined as loops. Based on the positions of the base pairs, loops can be classified into the following five types:

- Hairpin loop—a hairpin loop is a loop that contains exactly one base-pair. This can happen when the RNA strand folds into itself with a base-pair holding them together as shown in Figure 2.

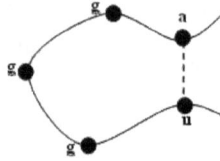

Fig. 2.   Hairpin loop.

- Stacked pair—a stacked pair is a loop formed by two base-pairs that are adjacent to each other. In other words, if a base-pair is (i, j), the other base-pair that forms the stacked pair could be $(i + 1, j - 1)$. Figure 3 shows an example.

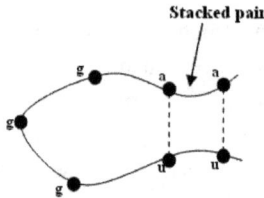

Fig. 3.   Stacked pair.

- Internal loop—an internal loop consists of two base-pairs like the stacked pair. The difference between them is that internal loop consists of at least one unpaired base on each side of the loop between the two base-pairs. In short, the length of the two sides of the RNA between the two base-pairs must be greater than 1. This is shown in Figure 4.

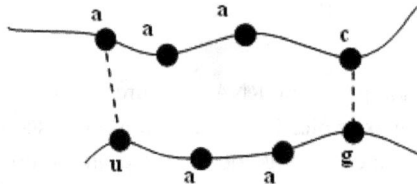

Fig. 4.   Internal loop.

- Bulge—a bulge has two base-pairs like the internal loop. However, only one side of the bulge has unpaired bases. The other side must have two base-pairs adjacent to each other as shown in Figure 5.

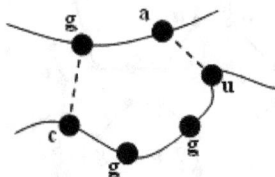

Fig. 5. Bulge.

- Multi-loop—any loop with 3 or more base-pairs is a multi-loop. Usually, one can view a multi-loop as a combination of multiple double-stranded regions of the RNA. Figure 6 shows examples of various loop types.

2. RNA Secondary Structure Determination Experiments

In the literature, there are several experimental methods for obtaining the secondary structure of an RNA, including physical methods, chemical/enzymatic methods, and mutational analysis.

- Physical methods—the basic idea behind physical methods is to infer the structure based on the distance measurements among atoms. Crystal X-ray diffraction is a physical method that gives the highest resolution. It reveals distance information based on X-ray diffraction. For example, the structure of tRNA is obtained using this approach.[431] However, the use of this method is limited since it is difficult to obtain crystals of RNA molecules that are suitable for X-ray diffraction. Another physical method is Nuclear Magnetic Resonance (NMR), which can provide detail local conformation based on the magnetic properties of hydrogen nuclei. Currently, NMR can only resolve structures of size no longer than 30–40 residues.
- Chemical/enzymatic methods—enzymatic and chemical probes[224] which modify RNA under some specific constraints can be used to analyse RNA structure. By comparing the properties of the RNA before and after applying the probes, we can obtain some RNA structure information. Note that the RNA structure information extracted are usually limited as some segments of an RNA polymer is inaccessible to the probes. Another issue is the experiment temperatures for various chemical or enzymatic digestions. A RNA

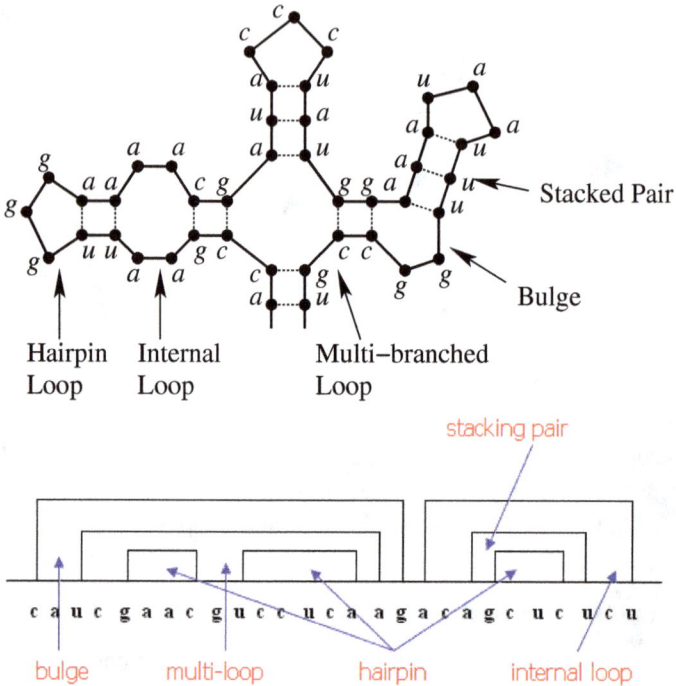

Fig. 6.   Different loop types.

polymer may unfold due to high experiment temperature. Caution is required in interpreting such experimental results.

- Mutational analysis—this method makes specific mutation to the RNA sequence. Then, the binding ability between the mutated sequence and some proteins is tested.[817] If the binding ability of the mutated sequence is different from the original sequence, we claim that the mutated RNA sequence has structural changes. Such information helps us deduce the secondary structure.

3. RNA Structure Prediction Based on Sequence

Based on laboratory experiment, a denatured RNA renatures to the same structure spontaneously *in vitro*. Hence, it is in general believed that the structure of RNAs are determined by their sequences. This belief motivates us to predict the secondary structure of a given RNA based on its sequence only. There is a growing body of research in this area, which can be divided into two types:

(1) structure prediction based on multiple RNA sequences which are structurally similar; and

(2) structure prediction based on a single RNA sequence.

For the first type, the basic idea is to align the RNA sequences and predict the structure. Sankoff[743] considers the case of aligning two RNA sequences and inferring the structure. The time complexity of his algorithm is $O(n^6)$. (In Computer Science, the big-O notation is used to expressed the worst-case running time of a program. It tries to express the worst-case running time by ignoring the constant factor. For example, if your program runs in at most $10 \times n^2$ steps for an input of size n, then we say the time complexity of your program is $O(n^2)$. Throughout this chapter, we use this notation to express running time.) Corpet and Michot[174] present a method that incrementally adds new sequences to refine an alignment by taking into account base sequence and secondary structure. Eddy and Durbin[220] build a multiple alignment of the sequences and derive the common secondary structure. They propose covariance models that can successfully compute the consensus secondary structure of tRNA. Unfortunately, their method is suitable for short sequences only. The methods above have the common problem that they assume the secondary structure does not have any pseudoknot. Gary and Stormo[274] proposes to solve this problem using graph theoretical approach.

This chapter focuses on the second type. That is, structure prediction based on a single RNA sequence. Section 4 studies RNA structure prediction algorithms with the assumption that there is no pseudoknot. Then, some latest works on RNA structure prediction with pseudoknots are introduced in Sections 5 to 7.

4. Structure Prediction in the Absence of Pseudoknot

This section considers the problem of predicting RNA secondary structure with the assumption that there is no pseudoknot. The reason for ignoring pseudoknots is to reduce computational time complexity. Although ignoring psuedoknots reduces accuracy, such an approximation still looks reasonably good as psuedoknots do not appear so frequently.

Predicting RNA secondary structure is quite difficult. The most naive approach relies on an exhaustive search to find the lowest free-energy conformation. Such an approach fails because the number of conformations with the lowest free-energy is numerous. Identifying the correct conformation likes looking for a needle in the haystack.

Over the past 30 years, researchers try to find the correct RNA conformation based on simulating the thermal motion of the RNA—*e.g.*, CHARMM[108] and AMBER.[653] The simulation considers both the energy of the molecule and the net force experienced by every pair of atoms. In principle, the correct RNA conformation can be computed in this way. However, such an approach fails because of the following two reasons:

(1) Since we still do not fully understand the chemical and physical properties of atoms, the energies and forces computed are merely approximated values. It is not clear whether the correct conformation can be predicted at such a level of approximation.
(2) The computation time of every simulation iteration takes seconds to minutes even for short RNA sequences. Unless CPU technology improves significantly, it is impossible to compute the structure within reasonable time.

In 1970s, scientists discover that the stability of RNA helices can be predicted using thermodynamic data obtained from melting studies. Those data implies that loops' energies are approximately independent. Tinoco et al.[829, 830] rely on this finding and propose the nearest neighbour model to approximate the free energy of any RNA structure. Their model makes the following assumptions:

(1) The energy of every loop—including hairpin, stacking pair, bulge, internal loop, and multi-loop–is independent of the other loops.
(2) The energy of a secondary structure is the sum of all its loops.

Based on this model, Nussinov and Jacobson[623] propose the first algorithm for computing the optimal RNA structure. Their idea is to maximize the number of stacking pairs. However, they do not consider the destabilising energy of various loops. Zuker and Stiegler[944] then give an algorithm that accounts for the various destabilising energies. Their algorithm takes $O(n^4)$ time, where n is the length of the RNA sequence. Lyngso, Zuker, and Pedersen[531] improves the time complexity to $O(n^3)$. Using the best known parameters proposed by Mathews et al.,[548] the predicted structure on average contains more than 70% of the base pairs of the true secondary structure. Apart from finding just the optimal RNA structure, Zuker[943] also prposes an approach that can compute all the suboptimal structures whose free energies are within some fixed range from the optimal.

In this section, we present the best known RNA secondary structure prediction algorithm—which is the one proposed by Lyngso, Zuker, and Pedersen.[531]

4.1. *Loop Energy*

RNA secondary structure is built upon 5 basic types of loops. Mathews, Sabina, Zuker, and Turner[548] have derived the 4 energy functions that govern the formation of these loops. These energy functions are:

- $eS(i, j)$—this function gives the free energy of a stacking pair consisting of base pairs (i, j) and $(i + 1, j - 1)$. Since no free base is included, this is the only loop which can stabilize the RNA secondary structure. Thus, its energy is negative.

- $eH(i, j)$—this function gives the free energy of the hairpin closed by the base pair (i, j). Biologically, the bigger the hairpin loop, the more unstable is the structure. Therefore, $eH(i, j)$ is more positive if $|j - i + 1|$ is large.

- $eL(i, j, i', j')$—this function gives the free energy of an internal loop or a budge enclosed by base pairs (i, j) and (i', j'). Its free energy depends on the loop size $|i' - i + 1| + |j' - j + 1|$ and the asymmetry of the two sides of the loop. Normally, if the loop size is big and the two sides of the loop are asymmetric, the internal loop is more unstable and thus, $eL(i, j, i', j')$ is more positive.

- $eM(i, j, i_1, j_1, \ldots, i_k, j_k)$—this function gives the free energy of a multi-loop enclosed by base pair (i, j) and k base pairs $(i_1, j_1), \ldots, (i_k, j_k)$. The mutli-loop is getting more unstable when its loop size and the value k are big.

4.2. First RNA Secondary Structure Prediction Algorithm

Based on the nearest neighbor model, the energy of a secondary structure is the sum of all its loops. By energy minimization, the secondary structure of the RNA can then be predicted. To speedup the process, we take advantage of dynamic programming. The dynamic programming can be described using the following 4 recursive equations.

- $W(j)$—the energy of the optimal secondary structure for $S[1..j]$.
- $V(i, j)$—the energy of the optimal secondary structure for $S[i..j]$ given that (i, j) is a base pair.
- $VBI(i, j)$—the energy of the optimal secondary structure for $S[i..j]$ given that (i, j) closes a bulge or an internal loop.
- $VM(i, j)$—the energy of the optimal secondary structure for $S[i..j]$ given that (i, j) closes a multi-loop.

We describe the detail of the 4 recursive equations below.

4.2.1. $W(j)$

The recursive equations for $W(j)$ are given below. When $j = 0$, $W(j) = 0$ as the sequence is null. For $j > 0$, we have two cases: either $S[j]$ is a free base; or there exists i such that $(S[i], S[j])$ forms a base pair. For the first case, $W(j) =$

$W(j-1)$. For the second case, $W(j) = \min_{1 \le i < j}\{V(i,j) + W(i-1)\}$. Thus, we have the following recursive equations.

$$W(j) = \begin{cases} 0 & \text{if } j = 0 \\ \min\{W(j-1), \min_{1 \le i < j}\{V(i,j) + W(i-1)\}\} & \text{if } j > 0 \end{cases}$$

4.2.2. $V(i,j)$

$V(i,j)$ is the free energy of the optimal secondary structure for $S[i..j]$ where (i,j) forms a base pair. When $i \ge j$, $S[i..j]$ is a null sequence and we cannot form any base pair. Thus, we set $V(i,j) = +\infty$. When $i < j$, The base pair (i,j) should belong to one of the four loop types: hairpin, stacked pair, internal loop, and multi-loop. Thus the free energy $V(i,j)$ should be the minimum of $eH(i,j)$, $eS(i,j)+V(i+1,j-1)$, $VBI(i,j)$, and $VM(i,j)$. Hence we have the following equations.

$$V(i,j) = \begin{cases} +\infty, & \text{if } i \ge j \\ \min \left\{ \begin{array}{ll} eH(i,j) & \text{Hairpin;} \\ eS(i,j) + V(i+1,j-1) & \text{Stacked pair;} \\ VBI(i,j) & \text{Internal loop;} \\ VM(i,j) & \text{Multi-loop.} \end{array} \right\}, & \text{if } i < j \end{cases}$$

4.2.3. $VBI(i,j)$

$VBI(i,j)$ is the free energy of the optimal secondary structure for $S[i..j]$ where the base pair (i,j) closes a bulge or an internal loop. The bulge or the internal loop is formed by (i,j) together with some other base pair (i',j') where $i < i' < j' < j$. The energy of this loop is $eL(i,j,i',j')$. The energy of the best secondary structure for $S[i..j]$ with (i,j) and (i',j') forms an internal loop is $eL(i,j,i',j') + V(i',j')$. By trying all possible (i',j') pairs, the optimal energy can be found as:

$$VBI(i,j) = \min_{i < i' < j' < j}\{eL(i,j,i',j') + V(i',j')\}$$

4.2.4. $VM(i,j)$

$VM(i,j)$ is the free energy of the optimal secondary structure for $S[i..j]$ where the base pair (i,j) closes a multi-loop. The multi-loop is formed by (i,j) together with k base pairs $(i_1, j_1), \ldots, (i_k, j_k)$ where $k > 1$ and $i < i_1 < j_1 < i_2 < j_2 < \ldots < i_k < j_k < j$—see Figure 7 for an example. Similar to the calculation of

$VBI(i,j)$, we get the following:

$$VM(i,j) = \min_{i<i_1<j_1<\ldots<i_k<j_k<j} \left\{ eM(i,j,i_1,j_1,\ldots,i_k,j_k) + \sum_{h=1}^{k} V(i_h,j_h) \right\}$$

4.2.5. Time Analysis

Based on the discussion above, computing the free energy of the optimal secondary structure for $S[1..n]$ is equivalent to finding $W(n)$. Such a computation requires us to fill in 4 dynamic programming tables for the 4 recursive equations $W(\cdot)$, $V(\cdot,\cdot)$, $VBI(\cdot,\cdot)$, and $VM(\cdot,\cdot)$. The optimal secondary structure can then be obtained by backtracking. We give below the time analysis for filling in the 4 tables.

- $W(i)$—it is an array with n entries. Each entry requires finding the minimum of n terms, $V(i,j) + W(i-1)$ for i varying from 1 to $j-1$. So, each entry needs $O(n)$ time. As a result, it costs $O(n^2)$ time in total.
- $V(i,j)$—it is an array with n^2 entries. Each entry requires finding the minimum of 4 terms, $eH(i,j)$, $eS(i,j)+V(i+1,j-1)$, $VBI(i,j)$, and $VM(i,j)$. Since each entry can be filled in $O(1)$ time, this matrix can be computed in $O(n^2)$ time.
- $VBI(i,j)$—it is an array with n^2 entries. Each entry requires finding the minimum of n^2 terms: $eL(i,j,i',j') + V(i',j')$ for $i < i' < j' < j$, where both i' and j' vary from 1 to n at most. So, each term needs $O(n^2)$ time. As a result, it costs $O(n^4)$ time in total.
- $VM(i,j)$—it is an array with n^2 entries. Each entry requires finding the minimum of exponential terms: $eM(i,j,i_1,j_1,\ldots,i_k,j_k) + \sum_{h=1}^{k} V(i',j')$ for $i < i_1 < j_1 < \ldots < i_k < j_k < j$. So in total, it costs exponential time.

In summary, the execution time of the algorithm is exponential. The major problem is on those computations pertaining to multi-loops and internal loops, which require time that is exponential and quartic in n respectively. For multi-loops, we assume that the energy of multi-loops can be approximated using an affine linear function, through which we can reduce the time cost of $VM(\cdot,\cdot)$ from exponential time to $O(n^3)$ time. For internal loops, we reduce the overhead of $VBI(\cdot,\cdot)$ to $O(n^3)$ time by using the approximation equation suggested by Ninio.[645] Therefore, we can reduce the overall complexity to $O(n^3)$ time from the original exponential time. The two speed-up methods are discussed in detail in Subsections 4.3 and 4.4.

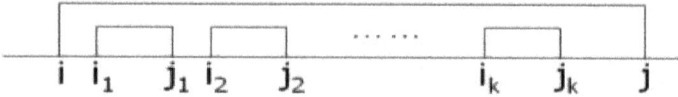

Fig. 7.   Structure of a multi-loop.

4.3. Speeding up Multi-Loops

4.3.1. Assumption on Free Energy of Multi-Loop

To make the problem tractable, the following simplified assumption is made. Consider a multi-loop formed by base pairs (i, j), (i_1, j_1), ..., (i_k, j_k) as shown in Figure 7. The energy of the multi-loop can be decomposed into linear contributions from the number of unpaired bases in the loop, the number of base pairs in the loop, and a constant, that is

$$eM(i, j, i_1, j_1, \ldots, i_k, j_k) = a + b \times k + c \times \left(\begin{array}{c} (i_1 - i - 1) + \\ (j - j_k - 1) + \\ \sum_{h=1}^{k-1}(i_h + 1 - j_h - 1) \end{array} \right)$$

where a, b, c are constants; k is the number of base pairs in the loop; and $((i_1 - i - 1) + (j - j_k - 1) + \sum_{h=1}^{k-1}(i_h + 1 - j_h - 1))$ is the number of unpaired bases in the loop.

4.3.2. Modified Algorithm for Speeding Up Multi-Loop Computation

Given the assumption above, the RNA structure prediction algorithm can be speeded up by introducing a new recursive equation $WM(i, j)$. $WM(i, j)$ equals the energy of the optimal secondary structure of $S[i..j]$ that constitutes the substructure of a multi-loop structure. Here, inside the multi-loop substructure, a free base is penalized with a score c while each base pair belonging to the multi-loop substructure is penalized with a score b. Thus, we have the following equation.

$$WM(i, j) = \min \begin{cases} WM(i, j - 1) + c, & j \text{ is free base;} \\ WM(i + 1, j) + c, & i \text{ is free base;} \\ V(i, j) + b, & (i, j) \text{ is pair;} \\ \min_{i < r \leq j} \left\{ \begin{array}{l} WM(i, r - 1) + \\ WM(r, j) \end{array} \right\}, & \begin{array}{l} i \text{ and } j \text{ not free, and} \\ (i, j) \text{ is not pair} \end{array} \end{cases}$$

Given $WM(i, j)$, $VM(i, j)$ can be modified as

$$VM(i, j) = \min_{i+1 < r \leq j-1} \{WM(i + 1, r - 1) + WM(r, j - 1) + a\}$$

We can find an r between $i + 1$ and $j - 1$ that divides $S[i..j]$ into 2 parts. The sum of the two parts' energy should be minimal. Then the energy penalty a of the multi-loop is added to the sum to give $VM(i,j)$.

4.3.3. *Time Complexity*

After making these changes, we need to fill in 5 dynamic programming tables, *viz.* $W(i)$, $V(i,j)$, $VBI(i,j)$, $VM(i,j)$, and $WM(i,j)$.

The time complexity for filling tables $W(i)$, $V(i,j)$, and $VBI(i,j)$ are the same as the analysis in Section 4.2.5. They cost $O(n^2)$, $O(n^2)$, and $O(n^4)$ time respectively.

For the table $WM(i,j)$, it has n^2 entries. Each entry can be computed by finding the minimum of 4 terms: $WM(i, j-1)+c$, $WM(i+1, j)+c$, $V(i, j)+b$, and the minimum of $WM(i, k-1) + WM(k, j)$ for $i < k \le j$. The first 3 terms can be found in $O(1)$ time while the last term takes $O(n)$ time. In total, filling in the table $WM(i, j)$ takes $O(n^3)$ time.

For the table $VM(i, j)$, it also has n^2 entries. But now each entry can be evaluated by finding the minimum of the n terms: $WM(i+1, k-1)+WM(k, j-1) + a$ for $i + 1 < k \le j - 1$. Thus, filling table $VM(i, j)$ also takes $O(n^3)$ time.

In conclusion, the modified algorithm runs in $O(n^4)$ time.

4.4. *Speeding Up Internal Loops*

4.4.1. *Assumption on Free Energy for Internal Loop*

Consider an internal loop or a bulge formed by two base pairs (i, j) and (i', j') with $i < i' < j' < j$. We assume its free energy $eL(i, j, i', j')$ can be computed as the sum:

$$eL(i, j, i', j') = size(n_1 + n_2) + stacking(i, j)+ \\ stacking(i', j') + asymmetry(n_1, n_2)$$

where

- $n_1 = i' - i - 1$ and $n_2 = j - j' - 1$ are the number of unpaired bases on both sides of the internal loop, respectively;
- $size(n_1 + n_2)$ is an energy function depending on the loop size;
- $stacking(i, j)$ and $stacking(i', j')$ are the energy for the mismatched base pairs adjacent to the two base pairs (i, j) and (i', j'), respectively;
- $asymmetry(n_1, n_2)$ is energy penalty for the asymmetry of the two sides of the internal loop.

To simplify the computation, we further assume that when $n_1 > c$ and $n_2 > c$, it is the case that $asymmetry(n_1, n_2) = asymmetry(n_1 - 1, n_2 - 1)$. In practice, $asymmetry(n_1, n_2)$ is approximated using Ninio's equation,[645] *viz.*

$$asymmetry(n_1, n_2) = \min\{K, |n_1 - n_2| \times f(m)\}$$

where $m = \min\{n_1, n_2, c\}$, K and c are constants, and $f(m)$ is an arbitrary penalty function that depends on m. Note that $asymmetry(n_1, n_2)$ satisfies the above assumption and c is proposed to be 1 and 5 in two literatures.[645, 664]

The above two assumptions imply the following lemma which is useful for devising an efficient algorithm for computing internal loop energy.

Lemma 1: *Consider* $i < i' < j' < j$. *Let* $n_1 = i' - i - 1$, $n_2 = j - j' - 1$, *and* $l = n_1 + n_2$. *For* $n_1 > c$ *and* $n_2 > c$, *we have*

$$eL(i, j, i', j') - eL(i + 1, j - 1, i', j') = size(l) - size(l - 2) + \\ stacking(i, j) - stacking(i + 1, j - 1)$$

Proof: This follows because $eL(i, j, i', j') - eL(i + 1, j - 1, i', j') = (size(l) + stacking(i, j) + stacking(i', j') + asymmetry(n_1, n_2)) - (size(l - 2) + stacking(i + 1, j - 1) + stacking(i', j') + asymmetry(n_1 - 1, n_2 - 1))$. By the assumption that $asymmetry(n_1, n_2) = asymmetry(n_1 - 1, n_2 - 1)$, we have $eL(i, j, i', j') - eL(i + 1, j - 1, i', j') = size(l) - size(l - 2) + stacking(i, j) - stacking(i + 1, j - 1)$ as desired. ☐

4.4.2. Detailed Description

Based on the assumptions, $VBI(i, j)$ for all $i < j$ can be found in $O(n^3)$ time as follows.

We define new recursive equations VBI' and VBI''. $VBI'(i, j, l)$ equals the minimal energy of an internal loop of size l closed by a pair (i, j). $VBI''(i, j, l)$ also equals the minimal energy of an internal loop of size l closed by a pair (i, j). Moreover, VBI'' requires the number of the bases between i and i' and the number of the bases between j and j', excluding i, i', j, and j', to be more than a constant c. Formally, VBI' and VBI'' are defined as follows.

$$VBI'(i, j, l) = \min_{\substack{i < i' < j' < j, \\ i' - i - 1 + j - j' - 1 = l}} \{eL(i, j, i', j') + V(i', j')\}$$

$$VBI''(i, j, l) = \min_{\substack{i < i' < j' < j, \\ i' - i - 1 + j - j' - 1 = l, \\ i' - i - 1, j - j' - 1 > c}} \{eL(i, j, i', j') + V(i', j')\}$$

Together with Lemma 1, we have

$$VBI''(i,j,l) - VBI''(i+1,j-1,l) = size(l) - size(l-2)+$$
$$stacking(i,j) + stacking(i+1,j-1)$$

$$VBI'(i,j,l) = \min \begin{cases} VBI''(i+1,j-1,l)+ \\ size(l) - size(j-2)+ \\ stacking(i,j) - stacking(i+1,j-1), \\ \min_{1 \le d \le c} \left\{ \begin{array}{c} V(i+d,j-l-d)+ \\ eL(i,j,i+d,j-l+d-2) \end{array} \right\}, \\ \min_{1 \le d \le c} \left\{ \begin{array}{c} V(i+l+d,j-d)+ \\ eL(i,j,i+l-d+2,j-d) \end{array} \right\} \end{cases}$$

The last two entries of the above equation handle the cases where this minimum is obtained by an internal loop, in which d is less than a constant c, especially a bulge loop when c is equal to 1, that is at $j' = i + 1$ or $j' = j - 1$. By definition, we have $VBI(i,j) = \min_l \{VBI'(i,j,l)\}$.

4.4.3. *Time Analysis*

The dynamic programming tables for $VBI'(\cdot,\cdot,\cdot)$ and $VBI''(\cdot,\cdot,\cdot)$ have $O(n^3)$ entries. Each entry in $VBI'(\cdot,\cdot,\cdot)$ and $VBI''(\cdot,\cdot,\cdot)$ can be computed in $O(c)$ and $O(1)$ time respectively. Thus, both tables can be filled in using $O(c \times n^3)$ time. Given $VBI'(\cdot,\cdot,\cdot)$, the table $VBI(\cdot,\cdot)$ can be filled in using $O(n^2)$ time.

Together with filling the tables $W(\cdot)$, $V(\cdot,\cdot)$, $VM(\cdot,\cdot)$, $WM(\cdot,\cdot)$, the time required to predict secondary structure without pseudoknot is $O(n^3)$.

5. Structure Prediction in the Presence of Pseudoknots

Although pseudoknots are not frequent, they are very important in many RNA molecules.[183] For examples, pseudoknots form a core reaction center of many enzymatic RNAs, such as RNAseP RNA[522] and ribosomal RNA.[447] They also appear at the 5'-end of mRNAs, and act as a control of translation. Therefore, discovering pseudoknots in RNA molecules is very important.

Up to now, there is no good way to predict RNA secondary structure with pseudoknots. In fact, this problem is NP-hard.[15, 381, 530] Different approaches have been attempted to tackle this problem. Heuristic search procedures are adopted in most RNA folding methods that are capable of folding pseudoknots. Some examples include quasi-Monte Carlo searches by Abrahams *et al.*,[4] genetic algorithms by Gultyaev *et al.*,[307] Hopfield networks by Akiyama and Kanehisa,[14] and stochastic context-free grammar by Brown and Wilson.[110]

These approaches cannot guarantee that the best structure is found and are unable to say how far a given prediction is from the optimal. Other approaches are based on maximum weighted matching,[274, 813]. They report some successes in predicting pseudoknots and base triples.

Based on dynamic programming, Rivas and Eddy,[715] Lyngso and Pedersen,[530] and Akutsu[15] propose three polynomial time algorithms that can find optimal secondary structure for certain kinds of pseudoknots. Their time complexities are $O(n^6)$, $O(n^5)$, and $O(n^4)$, respectively. On the other hand, Ieong et al.[381] propose two polynomial time approximation algorithms that can handle a wider range of pseudoknots. One algorithm handle bi-secondary structures—i.e., secondary structures that can be embedded as a planar graph—while the other algorithm can handle general secondary structure. The worst-case approximation ratios are $1/2$ and $1/3$, respectively.

To illustrate the current solutions for predicting RNA secondary structure with pseudoknots, the next two sections present Akutsu's $O(n^4)$-time algorithm and Ieong et al.'s $1/3$-approximation polynomial time algorithm.

6. Akutsu's Algorithm

6.1. Definition of Simple Pseudoknot

This section gives the definition of a simple pseudoknot.[15] Consider a substring $S[i_0..k_0]$ of a RNA sequence S where i_0 and k_0 are arbitrarily chosen positions. A set of base pairs M_{i_0,k_0} is a simple pseudoknot if there exist j_0, j_0' such that

(1) each endpoint i appears in M_{i_0,k_0} once;
(2) each base pair (i,j) in M_{i_0,k_0} satisfies either $i_0 \leq i < j_0' < j \leq j_0$ or $j_0' \leq i < j_0 < j \leq k_0$; and
(3) if pairs (i,j) and (i',j') in M_{i_0,k_0} satisfy $i < i' < j_0'$ or $j_0' \leq i < i'$, then $j > j'$.

The first two parts of the definition divides the sequence $S[i_0..k_0]$ into three segments: $S[i_0..j_0]$, $S[j_0..j_0']$, and $S[j_0'..k_0]$. For each base pair in M_{i_0,k_0}, one of its end must be in $S[j_0..j_0']$ while the other end is either in $S[i_0..j_0]$ or $S[j_0'..k_0]$. The third part of the definition confines the base pairs so that they cannot intersect each other. Part I of Figure 8 is an example of a simple pseudoknot. Parts II, III, and IV of Figure 8 are some examples that are not simple pseudoknot.

With this definition of simple pseudoknots, a RNA secondary structure with simple pseudoknots is defined as below. A set of base pairs M is called a RNA secondary structure with simple pseudoknots if $M = M' \cup M_1 \cup \ldots \cup M_t$ for some non-negative integer t such that

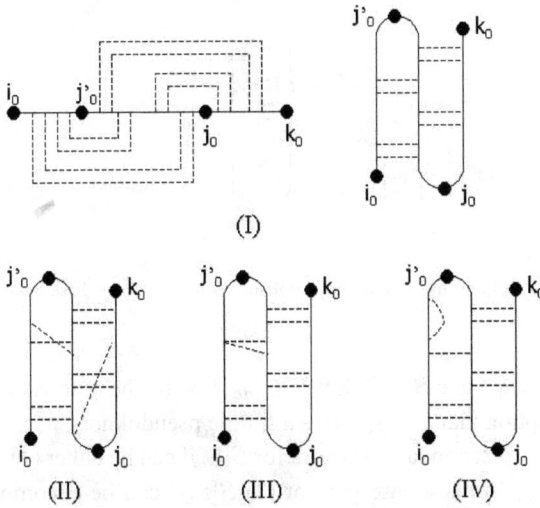

Fig. 8.   An illustration of simple pseudoknots.

(1) For $h = 1, 2, \ldots, t$, M_h is a simple pseudoknot for $S[i_h..k_h]$ where $1 \leq i_1 < k_1 < i_2 < k_2 < \ldots < i_t < k_t \leq n$.

(2) M' is a secondary structure without pseudoknot for string S' where S' is obtained by removing segments $S[i_h..k_h]$ for all $h = 1, 2, \ldots, t$.

6.2. RNA Secondary Structure Prediction with Simple Pseudoknots

This section presents an algorithm which solves the following problem.

Input: A RNA sequence $S[1..n]$

Output: A RNA secondary structure with simple pseudoknots that maximizes the score.

Score: In this section for simplicity, the score function used is different from that of the RNA secondary structure prediction without pseudoknot. The score function here is the number of the base pairs in $S[1..n]$. In short, we maximize the number of base pairs. Note that the score function can be generalized to some simple energy function.

A dynamic programming algorithm is designed to solve the problem above. Let $V(i, j)$ be the optimal score of an RNA secondary structure with simple pseu-

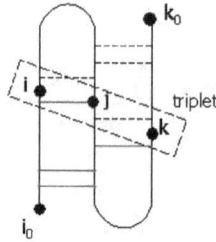

Fig. 9. (i, j, k) is a triplet in the simple pseudoknot. Note that all the base pairs in solid lines are below the triplet.

doknots for the sequence $S[i..j]$. Let $V_{pseudo}(i, j)$ be the optimal score for $S[i..j]$ with the assumption that $S[i..j]$ forms a simple pseudoknot.

For $V(i, j)$, the secondary structure for $S[i..j]$ can be either (1) a simple pseudoknot, (2) (i, j) forms a base pair, or (3) $S[i..j]$ can be decomposed into two compounds. Therefore, we get the following recursive equation.

$$V(i, j) = \max \left\{ \begin{array}{l} V_{pseudo}(i, j), \\ V(i + 1, j - 1) + \delta(S[i], S[j]), \\ \max_{i < k \le j}\{V(i, k - 1) + V(k, j)\} \end{array} \right\}$$

where $V(i, i) = 0$ for all i. Also, $\delta(S[i], S[j]) = 1$ if $\{S[i], S[j]\} = \{a, u\}$ or $\{c, g\}$; otherwise, $\delta(S[i], S[j]) = -\infty$.

For $V_{pseudo}(i_0, k_0)$, its value can also be computed using a dynamic programming algorithm. To explain the algorithm, we first give some notations. Recall that, in a simple pseudoknot $S[i_0..k_0]$, the sequence is partitioned into three segments $S[i_0..j_0']$, $S[j_0'..j_0]$, and $S[j_0..k_0]$ for some unknown positions j_0 and j_0'. We denote the three segments as left, middle, and right segments, respectively. See Part I of Figure 8 for an example. For a triplet (i, j, k) where $i_0 \le i \le j_0'$, $j_0' < j \le j_0$, and $j_0 < k \le k_0$, we say a base pair (x, y) is below the triplet (i, j, k) if either $x \le i$ and $y \ge j$, or $x \ge j$ and $y \le k$. Figure 9 is an example illustrating this concept of "below". All the base pairs in red color are "below" the triplet (i, j, k).

For a triplet (i, j, k), $S[i]$, $S[j]$, and $S[k]$ should satisfy one of the following relations: (1) (i, j) is a base pair, (2) (j, k) is a base pair, or (3) both (i, j) and (j, k) are not base pair. Below, we define three variables, based on the above three relationships, which are useful for computing $V_{pseudo}(i_0, k_0)$.

- $V_L(i, j, k)$ is the maximum number of base pairs below the triplet (i, j, k) in a pseudoknot for $S[i_0..k_0]$ given that (i, j) is a base pair.

- $V_R(i, j, k)$ is the maximum number of base pairs below the triplet (i, j, k) in a pseudoknot for $S[i_0..k_0]$ given that (j, k) is a base pair.
- $V_M(i, j, k)$ is the maximum number of base pairs below the triplet (i, j, k) in a pseudoknot for $S[i_0..k_0]$ given that both (i, j) and (j, k) are not a base pair.

Note that $\max\{V_L(i, j, k), V_R(i, j, k), V_M(i, j, k)\}$ is the maximum number of base pairs below the triplet (i, j, k) in a pseudoknot for $S[i_0..k_0]$. Then $V_{pseudo}(i_0, j_0)$ can be calculated as:

$$V_{pseudo}(i_0, k_0) = \max_{i_0 \leq i < j < k \leq k_0} \{V_L(i, j, k), V_M(i, j, k), V_R(i, j, k)\}$$

6.2.1. $V_L(i, j, k)$, $V_R(i, j, k)$, $V_M(i, j, k)$

We define below the recursive formulae for the 3 variables $V_L(i, j, k)$, $V_R(i, j, k)$, and $V_M(i, j, k)$.

$$V_L(i, j, k) = \delta(S[i], S[j]) + \max \left\{ \begin{array}{l} V_L(i-1, j+1, k), \\ V_M(i-1, j+1, k), \\ V_R(i-1, j+1, k) \end{array} \right\}$$

$$V_R(i, j, k) = \delta(S[i], S[j]) + \max \left\{ \begin{array}{l} V_L(i, j+1, k-1), \\ V_M(i, j+1, k-1), \\ V_R(i, j+1, k-1) \end{array} \right\}$$

$$V_M(i, j, k) = \max \left\{ \begin{array}{l} V_L(i-1, j, k), V_M(i-1, j, k), \\ V_L(i, j+1, k), V_M(i, j+1, k), V_R(i, j+1, k), \\ V_M(i, j, k-1), V_R(i, j, k-1) \end{array} \right\}$$

Here we provide an intuitive explanation for the formulae above. For both $V_L(i, j, k)$ and $V_R(i, j, k)$, the first term represents the number of base pairs on the triplet (i, j, k) while the second term represents the number of base pairs below (i, j, k). For $V_M(i, j, k)$, since there is no base pair on the triplet (i, j, k), the formula only consists of the number of base pairs below the triplet. Note that the two variables, $V_R(i-1, j, k)$ and $V_L(i, j, k-1)$, do not appear in the formula for $V_M(i, j, k)$. This is because $V_M(i, j, k)$ indicates that both (i, j) and (j, k) are not base pair.

6.2.2. *To Compute Basis*

To compute $V_L(i,j,k)$, $V_R(i,j,k)$, $V_M(i,j,k)$, some base values are required.

$$V_R(i_0 - 1, j, j+1) = \delta(S[j], S[j+1]), \quad \text{for all } j \tag{1}$$
$$V_R(i_0 - 1, j, j) = 0, \qquad\qquad \text{for all } j \tag{2}$$
$$V_L(i, j, j) = \delta(S[i], S[j]), \qquad \text{for all } i_0 \le i < j \tag{3}$$
$$V_L(i_0 - 1, j, k) = 0, \qquad\qquad \text{for all } k = j+1 \text{ or } k = j \tag{4}$$
$$V_M(i_0 - 1, j, k) = 0, \qquad\qquad \text{for all } k = j+1 \text{ or } k = j \tag{5}$$

The base case (3) can be explained by Part I of Figure 10. The base cases (1) and (2) can be explained by Parts II and III of Figure 10 respectively. The base cases (4) and (5) are trivial since they are out of range.

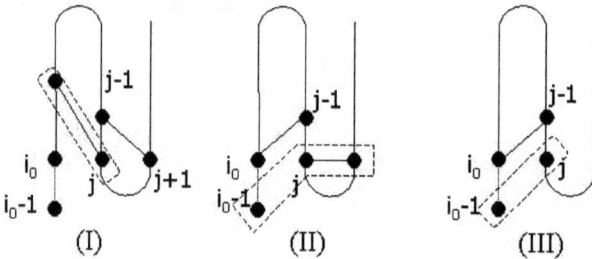

Fig. 10.   Basis for the recursive equations V_L, V_M, and V_R.

6.2.3. *Time Analysis*

This section gives the time analysis. First, we analyse the time required for computing V_{pseudo}.

Lemma 2: $V_{pseudo}(i_0, k_0)$ *for all* $1 \le i_0 < k_0 \le n$ *can be computed in* $O(n^4)$ *time.*

Proof: Observe that the base cases for $V_L(\cdot, \cdot, \cdot)$, $V_M(\cdot, \cdot, \cdot)$, $V_R(\cdot, \cdot, \cdot)$ only depend on i_0. Thus for a fixed i_0, the values for the base cases of $V_L(\cdot, \cdot, \cdot)$, $V_M(\cdot, \cdot, \cdot)$, and $V_R(\cdot, \cdot, \cdot)$ can be computed in $O(n^2)$ time. Then the values of tables $V_L(\cdot, \cdot, \cdot)$, $V_R(\cdot, \cdot, \cdot)$, and $V_M(\cdot, \cdot, \cdot)$ are independent of k_0, and can be computed in $O(n^3)$ time since each table has n^3 entries and each entry can be computed in $O(1)$ time.

Based on the definition of V_{pseudo}, we have the following recursive equation:

$$V_{pseudo}(i_0, k+1) = \max \left\{ \begin{array}{l} V_{pseudo}(i_0, k), \\ \max_{i_0 < i < j < k+1} \left\{ \begin{array}{l} V_L(i,j,k+1), \\ V_R(i,j,k+1), \\ V_M(i,j,k+1) \end{array} \right\} \end{array} \right\}$$

Thus for a fixed i_0, $V_{pseudo}(i_0, k_0)$ for all k_0 can be computed in $O(n^3)$ time. Since there are n choices for i_0, $V_{pseudo}(i_0, k_0)$ for all $1 \leq i_0 < k_0 \leq n$ can be computed in $O(n^4)$ time.

By the lemma above, the RNA secondary structure with simple pseudoknots for a sequence S can be predicted in $O(n^4)$ time.

Proposition 3: *Consider a sequence $S[1..n]$. We can predict the RNA secondary structure of S with simple pseudoknots in $O(n^4)$ time.*

Proof: Based on the lemma above, $V_{pseudo}(i, j)$ for all i, j can be computed in $O(n^4)$ time. Then, we need to fill in n^2 entries for the table $V(\cdot, \cdot)$ where each entry can be computed in $O(n)$ time. Hence, the table $V(\cdot, \cdot)$ can be filled in using $O(n^3)$ time. In total, the problem can be solved in $O(n^4)$ time.

7. Approximation Algorithm for Predicting Secondary Structure with General Pseudoknots

The previous algorithm can only handle some special types of pseduoknots. This section addresses general pseudoknots. As the problem of predicting secondary structure with general pseudoknots is NP-hard, we approach the problem by giving an approximation algorithm.[381]

Given a RNA sequence $S = s_1 s_2 \ldots s_n$, this section constructs a secondary structure of S that approximates the maximum number of stacking pairs with a ratio of $1/3$. First, we need some definition. We denote a stacking pair $\{(s_i, s_j), (s_{i+1}, S_{j-1})\}$ as $(s_i, s_{i+1}; s_{j-1}, s_j)$. For a consecutive of q ($q \geq 1$) stacking pairs $(s_i, s_{i+1}; s_{j-1}, s_j), (s_{i+1}, s_{i+2}; s_{j-2}, s_{j-1}), \ldots, (s_{i+q-1}, s_{i+q}; s_{j-q}, s_{j-q+1})$, it is denoted as $(s_i, s_{i+1}, \ldots, s_{i+q}; s_{j-q}, \ldots, s_{j-1}, s_j)$. The approximation algorithm uses a greedy approach. Figure 11 shows the algorithm $GreedySP(\cdot, \cdot)$.

In the following, we analyze the approximation ratio of the algorithm. The algorithm $GreedySP(S, i)$ will generate a sequence of consecutive stacking pairs SP's. Let $SP_1, SP_2, \ldots, SP_h$ be the generated sequence. We have the following fact.

// Let $S = s_1 s_2 \ldots s_n$ be the input RNA sequence.

// Initially, all s_j are unmarked.

// Let E be the set of base pairs output by the algorithm.

// Initially, $E = \emptyset$.

$GreedySP(S, i)$ // $i \geq 3$

(1) Repeatedly find the leftmost i consecutive stacking pairs SP—i.e., find $(s_p, \ldots, s_{p+i}; s_{q-i}, \ldots, s_q)$ such that p is as small as possible—formed by unmarked bases. Add SP to E and mark all these bases.

(2) For $k = i - 1$ downto 2,
    Repeatedly find any k consecutive stacking pairs SP formed by unmarked bases. Add SP to E and mark all these bases.

(3) Repeatedly find the leftmost stacking pair SP formed by unmarked bases. Add SP to E and mark all these bases.

Fig. 11.   The 1/3-approximation algorithm.

Fact 4: For any SP_j and SP_k, $j \neq k$, the corresponding stacking pairs in SP_j and SP_k do not overlap.

For each $SP_j = (s_p, \ldots, s_{p+t}; s_{q-t}, \ldots, s_q)$, we define two intervals of indices, $\mathcal{I}_j$ and $\mathcal{J}_j$, as $[p..p + t]$ and $[q - t..q]$ respectively. We want to compare the number of stacking pairs formed with that in the optimal case, so we have the following definition.

Definition 5: Let $\mathcal{P}$ be an optimal secondary structure of S with a maximum number of stacking pairs. Let $\mathcal{F}$ be the set of all stacking pairs of $\mathcal{P}$. For each SP_j computed by $GreedySP(S, i)$, we define the set $\mathcal{X}_\beta$, where $\beta = \mathcal{I}_j$ or $\mathcal{J}_j$, as follows.

$$\mathcal{X}_\beta = \left\{ \begin{pmatrix} s_k, s_{k+1}; \\ s_{w-1}, s_w \end{pmatrix} \in \mathcal{F} \;\middle|\; \text{at least one of indices } k, k+1, w-1, w \in \beta \right\}$$

Next, we observe that

Lemma 6: Let $SP_1, SP_2, \ldots, SP_h$ be the sequence of SP's computed by $GreedySP(S, i)$. Then $\bigcup_{1 \leq j \leq h} \{\mathcal{X}_{\mathcal{I}_j} \cup \mathcal{X}_{\mathcal{J}_j}\} = \mathcal{F}$.

Proof: We prove it by contradiction. Suppose that there exists a stacking pair $(s_k, s_{k+1}; s_{w-1}, s_w) \in \mathcal{F}$ but not in any of $\mathcal{X}_{\mathcal{I}_j}$ and $\mathcal{X}_{\mathcal{J}_j}$. By Definition 5, none of the indices $k, k+1, w-1, w$, is in any of $\mathcal{I}_j$ and $\mathcal{J}_j$. This contradicts with Step 3 of the algorithm $GreedySP(S, i)$. $\qquad\square$

Note that $\mathcal{X}_\beta$'s may not be disjoint.

Definition 7: For each $\mathcal{X}_{\mathcal{I}_j}$, we define $\mathcal{X}'_{\mathcal{I}_j}$ to be $\mathcal{X}_{\mathcal{I}_j} - \bigcup_{k<j}\{\mathcal{X}_{\mathcal{I}_k} \cup \mathcal{X}_{\mathcal{J}_k}\}$; and define $\mathcal{X}'_{\mathcal{J}_j}$ as $\mathcal{X}_{\mathcal{J}_j} - \bigcup_{k<j}\{\mathcal{X}_{\mathcal{I}_k} \cup \mathcal{X}_{\mathcal{J}_k}\} - \mathcal{X}_{\mathcal{I}_j}$.

Let $|SP_j|$ be the number of stacking pairs represented by SP_j. Let $|\mathcal{I}_j|$ and $|\mathcal{J}_j|$ be the number of indices in the intervals $\mathcal{I}_j$ and $\mathcal{J}_j$ respectively.

Lemma 8: *Let N be the number of stacking pairs computed by the algorithm $GreedySP(S, i)$ and N^* be the maximum number of stacking pairs that can be formed by S. If for all j, we have $|SP_j| \geq \frac{1}{r} \times |(\mathcal{X}'_{\mathcal{I}_j} \cup \mathcal{X}'_{\mathcal{J}_j})|$, then $N \geq \frac{1}{r} \times N^*$.*

Proof: By Definition 7, $\bigcup_k\{\mathcal{X}_{\mathcal{I}_k} \cup \mathcal{X}_{\mathcal{J}_k}\} = \bigcup_k\{\mathcal{X}'_{\mathcal{I}_k} \cup \mathcal{X}'_{\mathcal{J}_k}\}$. By Fact 4, $N = \sum_j |SP_j|$, so $N \geq \frac{1}{r} \times |\bigcup_k\{\mathcal{X}_{\mathcal{I}_k} \cup \mathcal{X}_{\mathcal{J}_k}\}|$. Now by Lemma 6, we conclude $N \geq \frac{1}{r} \times N^*$ as desired. $\qquad\square$

This brings us to the main approximation result:

Proposition 9: *For each SP_j computed by $GreedySP(S, i)$, we have*

$$|SP_j| \geq \frac{1}{3} \times |\mathcal{X}'_{\mathcal{I}_j} \cup \mathcal{X}'_{\mathcal{J}_j}|$$

Proof: There are 3 steps of the $GreedySP(S, i)$ algorithm to be considered. For each SP_j computed by $GreedySP(S, i)$ in Step 1, we know that $SP_j = (s_p, \ldots, s_{p+i}; s_{q-i}, \ldots, s_q)$ is the leftmost i consecutive stacking pairs—*i.e.*, p is as small as possible. By definition, $|\mathcal{X}'_{\mathcal{I}_j}|, |\mathcal{X}'_{\mathcal{J}_j}| \leq i + 2$. We further claim that $|\mathcal{X}'_{\mathcal{I}_j}| \leq i+1$. Then, as $i \geq 3$, $|SP_j|/|\mathcal{X}'_{\mathcal{I}_j} \cup \mathcal{X}'_{\mathcal{J}_j}| \geq i/((i+1)+(i+2)) \geq 1/3$.

We prove the claim by contradiction. Assume that $|\mathcal{X}'_{\mathcal{I}_j}| = i + 2$. That is, for some integer t, $\mathcal{F}$ has $i + 2$ consecutive stacking pairs $(s_{p-1}, \ldots, s_{p+i+1}; s_{t-i-1}, \ldots, s_{t+1})$. Furthermore, none of the bases $s_{p-1}, \ldots, s_{p+i+1}, s_{t-i-1}, \ldots, s_{t+1}$ are marked before SP_j is being chosen. Otherwise, suppose one of such bases, says s_a, is marked when the algorithm chooses SP_l for $l < j$, then the stacking pairs adjacent to s_a do not belong to $\mathcal{X}'_{\mathcal{I}_j}$ and they belong to $\mathcal{X}'_{\mathcal{I}_l}$ or $\mathcal{X}'_{\mathcal{J}_l}$ instead. Therefore, $(s_{p-1}, \ldots, s_{p+i-1}; s_{t-i+1}, \ldots, s_{t+1})$ is the leftmost i consecutive stacking pairs formed by unmarked bases before SP_j is chosen. As SP_j is not the leftmost i consecutive stacking pairs, this contradicts with the algorithm. The claim is thus proved.

For each SP_j computed by $GreedySP(S, i)$ in Step 2, let $|SP_j| = k \geq 2$ and let $SP_j = (s_p, \ldots, s_{p+k}; s_{q-k}, \ldots, s_q)$. By definition, $|\mathcal{X}'_{\mathcal{I}_j}|, |\mathcal{X}'_{\mathcal{J}_j}| \leq k + 2$. We claim that $|\mathcal{X}'_{\mathcal{I}_j}|, |\mathcal{X}'_{\mathcal{J}_j}| \leq k+1$. Then $|SP_j| / |\mathcal{X}'_{\mathcal{I}_j} \cup \mathcal{X}'_{\mathcal{J}_j}| \geq k/((k+1)+(k+1))$, which is at least $1/3$ as $k \geq 2$.

We can prove the claim $|\mathcal{X}'_{\mathcal{I}_j}| \leq k + 1$ by contradiction. Assume $|\mathcal{X}'_{\mathcal{I}_j}| = k + 2$. Thus, for some integer t, there exist $k + 2$ consecutive stacking pairs $(s_{p-1}, \ldots, s_{p+k+1}; s_{t-k-1}, \ldots, s_{t+1})$. Similar to Step 1, we can show that none of the bases $s_{p-1}, \ldots, s_{p+k+1}, s_{t-k-1}, \ldots, s_{t+1}$ are marked before SP_j is chosen. Thus, $GreedySP(S, i)$ should select some $k+1$ or $k+2$ consecutive stacking pairs instead of the chosen k consecutive stacking pairs. Thus, we arrive at a contradiction. We can prove the claim $|\mathcal{X}'_{\mathcal{J}_j}| \leq k + 1$ in a similar way.

For each SP_j computed by $GreedySP(S, i)$ in Step 3, SP_j is a leftmost stacking pair—that is, $SP_j = (s_p, s_{p+1}; s_{q-1}, s_q)$. Using the same approach as in Step 2, we can show that $|\mathcal{X}'_{\mathcal{I}_j}|, |\mathcal{X}'_{\mathcal{J}_j}| \leq 2$. We further claim that $|\mathcal{X}'_{\mathcal{I}_j}| \leq 1$. Then $|SP_j| / |\mathcal{X}'_{\mathcal{I}_j} \cup \mathcal{X}'_{\mathcal{J}_j}| \geq 1/(1 + 2) = 1/3$.

To verify the claim $|\mathcal{X}'_{\mathcal{I}_j}| \leq 1$, we consider all possible cases with $|\mathcal{X}'_{\mathcal{I}_j}| = 2$ while there are no 2 consecutive stacking pairs. The only possible case is that for some integers r, t, both $(s_{p-1}, s_p; s_{r-1}, s_r)$ and $(s_p, s_{p+1}; s_{t-1}, s_t)$ belong to $\mathcal{X}'_{\mathcal{I}_j}$. However, this means that SP_j is not the leftmost stacking pair formed by unmarked bases. This contradicts the algorithm and completes the proof. $\square$

By Lemma 8 and Proposition 9, we derive the following corollary.

Corollary 10: *Given a RNA sequence S. Let N^* be the maximum number of stacking pairs that can be formed by any secondary structure of S. Let N be the number of stacking pairs output by $GreedySP(S, i)$. Then $N \geq \frac{1}{3} \times N^*$.*

We remark that by setting $i = 3$ in $GreedySP(S, i)$, we can already achieve the approximation ratio of $1/3$. The following lemma gives the time and space complexity of the algorithm.

Lemma 11: *Given a RNA sequence S of length n. The algorithm $GreedySP(S, i)$, where i is a constant, can be implemented in $O(n)$ time and $O(n)$ space.*

Proof: Recall that the bases of a RNA sequence are chosen from the alphabet $\{a, u, g, c\}$. If i is a constant, there are only constant number of different patterns of consecutive stacking pairs that we have to consider. For any $1 \leq k \leq i$, there are only 4^k different strings that can be formed by the four characters $\{a, u, g, c\}$. So for all possible values of k, the locations of the occurrences of these possible strings in the RNA sequence can be recorded in an array of linked lists indexed by the pattern of the string using $O(n)$ time preprocessing. There are at most 4^k linked lists and there are only n entries in all linked lists.

Now, fix a constant k. In order to locate all k consecutive stacking pairs, we scan the RNA sequence from left to right. For each substring of k consecutive characters, we look up the array to see if we can form k consecutive stacking pairs. By a simple bookkeeping procedure, we can keep track which bases have been used already. Each entry in the linked lists is thus scanned at most once. So the whole procedure takes only $O(n)$ time. Since i is a constant, we can repeat the whole procedure for i different values of k and the total time complexity is still $O(n)$ time. □

CHAPTER 9

PROTEIN SUBCELLULAR LOCALIZATION PREDICTION

Paul Horton

National Institute of Industrial Science and Technology
horton-p@aist.go.jp

Yuri Mukai

National Institute of Industrial Science and Technology
yuri-mukai@aist.go.jp

Kenta Nakai

University of Tokyo
knakai@ims.u-tokyo.ac.jp

This chapter discusses various aspects of protein subcellular localization in the context of bioinformatics and reviews the twenty years of progress in predicting protein subcellular localization.

ORGANIZATION.

Section 1. We first provide the motivation for prediction of protein subcellular localization sites, as well as discuss changes being brought about by progress in proteomics.

Section 2. After that, we describe the biology of protein subcellular location. In particular, we explain the principle of protein sorting signals.

Section 3. Then we present several experimental techniques for determining protein subcellular localization sites. The techniques surveyed include traditional methods such as immunofluorescence microscopy, as well as large-scale methods such as green fluorescent protein.

Section 4. Next we mention some of the general issues involved in predicting protein subcellular localization, such as what are the sites? how many sites per protein? how good are the predictions? and so on. We also discuss the distinction between features that reflect causal influences on localization versus features that merely reflect correlation with localization.

Section 5. Lastly, we offer a survey of computational methods and approaches for predicting protein subcellular localization. These include discriminant analysis of amino acid
composition, localization process modeling, machine learning, feature discovery, and
literature analysis.

1. Motivation

The prediction of the subcellular localization sites of proteins from their amino
acid sequences is a fairly long-standing problem in bioinformatics. Nishikawa
and Ooi observed that amino acid composition correlates with localization sites
in 1982.[617] Around that time early work on the characterization and prediction
of secretory signal peptides began.[556, 861, 862] Over a decade later, the publication
of a signal peptide prediction program SignalP[611, 614] was awarded a "Hot Paper
in Bioinformatics" Award,[613] indicative of the high level of interest in such predictions. The main driving force behind this redoubled interest in predicting protein localization from sequence has been the need to annotate massive amounts
of sequence data coming from various genome projects.[227] The importance of
the biological phenomenon underlying protein localization was underscored by
the 1999 Nobel Prize in physiology or medicine, awarded to Günter Blobel for the
discovery that "proteins have intrinsic signals that govern their transport and localization in the cell". More recently, the emergence of proteomic technologies—see
Chapter 13—has given birth to terms such as "secretome".[218] Indeed recent experimental studies have determined the localization sites of a large fraction of all
the proteins in yeast.[374, 463, 721] Several excellent reviews on protein localization
and prediction are available.[232, 244, 599]

   Each compartment in a cell has a unique set of functions, thus it is reasonable to assume that the compartment or membrane in which a protein resides is
one determinant of its function. This assertion is supported by the fact that localization correlates with both protein-protein interaction data[357, 734, 735, 760] and
with gene expression levels.[212, 374] Thus protein localization is a valuable clue to
protein function, especially when homologs of known function cannot be found
by sequence similarity—a situation that is still common today. A recent study
of 134 bacterial genomes and several eukaryotic genomes shows that standard
sequence similarity reveals useful functional information in only about half of
all proteins, although there is significant variance in that proportional between
species.[809] Some applications are interested in a particular localization site. Many
industrial applications are concerned with the efficiency of the secretion of non-
native proteins in micro-organisms.[854] Proteins on the cell membrane are attractive as potential drug targets because they are accessible from outside the cell.

   Recent progress[374, 463] in large-scale experiments to determine protein localization indicates that in the foreseeable future the localization sites of a large percentage of proteins for some model organisms may be experimentally determined. For other organisms, the ability to infer localization by sequence similarity, an approach quantatively analyzed by Nair and Rost[595], will increase significantly. This clearly reduces the practical value of prediction schemes. So you may want to skip to the next chapter...

Still here? Good, because we think there are many excellent reasons for studying localization prediction. For one reason, the rose-colored scenario stated above is not quite upon us yet. There is still less than full coverage and significant error in the proteomic localization data, some of which may be systematic—see Section 3. For example, a recent study[374] covers 75% of the proteome and shows roughly 20% disagreement with data from the *Saccharomyces* Genome Database,[206] which contains data from two other large-scale experiments.[463, 721] Also, so far these experiments have been done on yeast, which has many fewer protein encoding genes than, for example, humans—approximately 6,000 for yeast[206, 293] versus perhaps 30,000 for human.[467] Many human proteins do not have close homologs in yeast. As mentioned in Chapter 13, the accumulated knowledge from decades of small scale experiments may contain fewer errors than recent large-scale experiments but the coverage is low. We found that, as of October 2003, only around 0.25% of SWISS-PROT entries include an explicit firm localization site assignment. (We searched only the "CC" fields and excluded assignments marked as "potential", "probable", or "by similarity".) However, do see Section 5.6 for automated methods to gain more localization information from SWISS-PROT annotations. Moreover the error rate is still significant. For example, a study of chloroplast signal peptides found roughly 10% of cleavage sites to be incorrect or based on insufficient evidence.[231, 614] Thus there is still some utility in predicting localization from amino acid sequence.

This notwithstanding, the coverage of experimental methods is certainly increasing rapidly and the accuracy of information derived from large scale experiments can be increased by comparing the results of multiple experiments—see Chapter 13. Prediction methods can play a role here in identifying outliers that may indicate experimental error. As the number of proteins whose localization has been determined experimentally increases, the role of prediction schemes is certain to change. A black box prediction program for native proteins is going to be of little use in the near future—even if it attains a high accuracy. We believe two qualities are to be demanded of prediction schemes in the future, *viz.*

(1) a high explanatory power, and
(2) the ability to accurately predict the localization of non-native, mutant, and

artificial proteins.

The first is because as scientists we are not satisfied with simply knowing that protein A is localized to site B. We also want to know how it gets there. Prediction schemes encoding our hypothesis about the process can help us gauge how much we really know, and machine learning techniques can help us generate new hypotheses. The second is because when designing new proteins we would like to be able to do experiments *in silico* on proteins that do not yet exist—and therefore clearly do not have experimentally determined localization sites.

Regrettably, most readers of this book will not become specialists in protein localization. Therefore perhaps a more compelling reason to read this chapter is the fact that localization prediction has been intensively studied with a variety of computational techniques and is an excellent vehicle for discussing several issues that apply to many areas of bioinformatics. More specifically there has been significant cross-fertilization of ideas between localization prediction and the related topics of predicting of membrane protein structure and post-translational modifications, *e.g.* lipidification, of proteins. Many groups working on protein localization prediction have also consistently published in these areas.[226, 394, 559, 600, 724]

2. Biology of Localization

Living organisms are classified into two categories: prokaryotes and eukaryotes. Unlike prokaryotes, eukaryotic cells are equipped with many kinds of membrane-bound compartments called organelles—*e.g.*, the nucleus, mitochondrion, endoplasmic reticulum (ER), and vacuole. We also consider some other subcellular sites such as the cytoplasm, plasma membrane, and cell wall; see Figure 1. Each organelle plays some specific cellular roles thanks to the presence of specifically localized proteins.

It is well known that proteins are synthesized based on the genetic information encoded in DNA. Although some information is encoded in the DNA within mitochondria—and chloroplasts in plants—most information is encoded in the nuclear DNA. Even most mitochondrial and chloroplast proteins are encoded in the nuclear DNA. The proteins encoded in the nuclear DNA are first synthesized within the cytoplasm and then specifically transported to each final localization site. Note that the transportation across the membrane of the ER generally starts in a pipelined fashion before synthesis of the amino acid chain is finished. The study of molecular mechanisms on how the final localization site of a protein is recognized and transported—often called "protein sorting"—is one of the central themes in modern cell biology. General textbooks on molecular cell biology typically devote many pages to this topic. As a recent example, we recommend the textbook of Alberts *et al.*[18]

The most important principle of protein sorting is that each protein has the information of its final localization site as a part of its amino acid sequence. In many cases, proteins are first synthesized as precursors having an extra stretch of polypeptide that function as a "sorting signal". They are specifically recognized and transported with some molecular machinery. After they are localized at their final destination, these sorting signals are often cleaved off. Therefore, it should be possible to predict the subcellular localization site of a protein if we can specifically recognize its sorting signal, as the cellular machinery does. This attempt is still challenging because our knowledge is incomplete.

Like all principles in biology, the sorting signal hypothesis allows some exceptions. That is, some proteins do not have sorting signals within their amino acid sequences, but instead are localized by binding with another protein that has the information. Fortunately for the developers of prediction methods, this "hitch-hiking" strategy does not seem to be common—probably because it is difficult for protein complexes to go through the organellar membranes. The nucleus is somewhat special in this respect because its membrane has large nuclear pores that allow small—up to about 60 kDa—proteins to diffuse into and out of the nucleus and also makes hitch-hiking relatively easier.

In many cases, the information of sorting signals is encoded within a limited length of the polypeptide. However, there are some examples where sorting information is encoded by sequence patches that are only recognizable in the 3D structure. The sequence features of sorting signals are variable. Some are represented as relatively well-conserved sequence motifs. Others appear more ambiguous—such as hydrophobic stretches—to our eyes, at least. Usually, at least to some extent, sorting signals can be discriminated with appropriately employed pattern recognition algorithms without the knowledge of their 3D structures.

We should keep in mind that the sorting signal for each localization site is not necessarily unique. For example, many mitochondrial proteins have the mitochondrial transit peptide on their N-terminus, but many others do not have this kind of signal and are instead localized by some different pathways. Indeed, recent developments of cell biology have enriched our knowledge of protein sorting greatly for certain proteins. Nevertheless, such knowledge is often applicable to a very limited set of proteins and is not general enough to raise overall prediction accuracy significantly.

Generally speaking localization signals appear to be well conserved across species. For example, Schneider *et al.* have studied mitochondrial targeting peptides in mammals, yeast, *Neurospora crassa*, and plant proteins. Although they observe some differences between plant and non-plant species, the clustering of

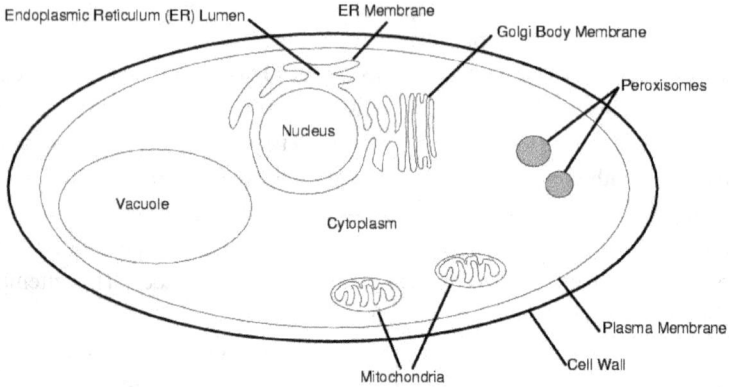

Fig. 1.   A cartoon of some of the comparments of a yeast cell is shown.

targeting peptides produced by an unsupervised learning algorithm—Kohonen self-organizing map—do not produce clusters based on species. We expect that prediction methods trained on one species and applied to another species can give reasonable results, even for distant pairs such as yeast and human. Indeed most predictive studies have used training data that include proteins from multiple species. Some differences do exist and have been analyzed at some length.[222, 598]

3. Experimental Techniques for Determining Localization Sites

3.1. *Traditional Methods*

Two traditional methods for determining the cellular localization sites of proteins are immunofluorescent staining and gradient centrifugation. We briefly outline each technique here.

3.1.1. *Immunofluorescence Microscopy*

Immunofluorescence microscopy can be used to determine the localization of a target protein. Cells—treated with detergent to help solubolize their plasma membrane—on a cover glass slide are treated with two antibodies. The first antibody is chosen to selectively bind to the target protein. The second antibody, which is tagged with a fluorescent marker, binds to the immunoglobulin epitope of the first antibody. The advantage of this scheme is that the second antibody does not depend on the target protein.

To more accurately determine the localization site of the target protein, co-localization with a reference marker—which is known to localize and fluoresce at a specific localization site—can be measured. The reference marker is chosen so that it emits a different wavelength of light when fluorescing than the second antibody used to detect the target protein. If the spatial pattern of the two wavelengths of light coincide then one may conclude that the target protein has the same localization site as the reference marker. The reference marker does not need to be a protein. Some well known reference markers are DAPI, MitoTracker®, and LysoTracker®; which are specific for the nucleus, mitochondrion, and lysosome respectively. Unfortunately, it is difficult to accurately measure the quantitative distribution of the target protein with immunofluorescence alone.

3.1.2. Gradient Centrifugation

Cell homogenation is the process of disrupting the plasma membrane of cells by mechanical means; for example with a rotating rod placed in a test tube. If carefully done, cells can be homogenated with the nucleus and most organelles intact. Fortunately these different compartments have different densities, allowing them to be separated by density gradient centrifugation.

The contents of each fraction, corresponding to a particular localization site, obtained by centrifugation can be analyzed with a "Western blot"—in which SDS-PAGE is used to separate the fraction on a electrophoretic gel and immunofluorescence is used to detect which band corresponds to the target protein. Once the band is identified it is possible to accurately measure the amount of protein it contains based on its size and darkness. Thus although this approach still requires the preparation of an antibody specific to the target protein, it has the advantage of allowing measurement of the quantity of the target protein at each localization site.

3.2. Large-Scale Experiments

3.2.1. Immunofluorescent Microscopy

Traditional immunofluorescent microscopy methods for determining localization required an antibody specific to each target protein. Unfortunately the development of such antibodies is a difficult and costly process. Kumar[463] and colleagues employed a scheme which uses immunofluorescent microscopy but does not require target specific antibodies. For each target protein they constructed vectors that expressed a fusion protein consisting of the target protein fused at its C-terminal to the V5 epitope. The localization of the fused protein could thus be determined with fluorescently labeled V5 antibodies. They also performed similar experiments using a haemagglutinin (HA) tag and HA antibodies instead of V5. They have determined the localization of 2022 and 1083 proteins using V5 and HA respectively. The union of the two sets contains 2744 proteins.

3.2.2. Green Fluorescent Protein

Green Fluorescent Protein (GFP) is a valuable tool[286] for studying localization. GFP is a 238 amino acid protein, with known 3D structure,[916] that naturally occurs in the bioluminescent jellyfish *Aequorea victoria*. GFP emits green light upon excitation with blue light. Genetically-engineered variants of GFP with different emission wavelengths such as Yellow Fluorescent Protein, Red Fluorescent Protein (RFP), and Cyan Fluorescent Protein are also available. An important property of GFP and its variants is that, unlike typical bioluminescent molecules, it does not require any co-factors. GFP is used in studying protein localization by creating fusion proteins consisting of the target protein and GFP fused together, usually with GFP connected to the target protein C-terminus. The location of the fused protein can be traced with fluorescence microscopy. The fluorescence of GFP is relatively stable over time and thus lends itself to quantitative measurements. One way to introduce these fusion protein into cells is by transfection with an expression vector containing DNA coding for the fusion protein co-expressed with a gene that can be selected for, such as resistance to a particular drug.

Recently GFP fusion proteins have been combined with homologous recombination in the largest experiment on localization performed so far. Huh *et al.*[374] transfected cells with vectors specific to each yeast protein coding gene (more precisely ORF). The vector for each target gene contains specific sequences that allows the vector to be inserted into the chromosome in the native position with the native promoter of the target gene. This technique solves the problem of over or under expression mentioned in Section 3.2.3. Of a total of 6,234 ORFs, fluorescence from 4,156 fusion proteins was detected and divided by localization site into: cell periphery, bud, bud neck, cytoskeleton & microtubule, cytoplasm, nucleus, mitochondrion, endoplasmic reticulum, vacuole, vacuolar membrane, punctate, and ambiguous. Many proteins showed multiple localization. The localization of some proteins was further investigated using proteins with known localization tagged with RFP. The comparison of the resulting pattern of green and red fluorescence gives detailed information about the localization of the target protein.

3.2.3. Comments on Large-Scale Experiments

The large-scale experiments described here are extremely impressive and have radically increased what we know about localization. However GFP is a large tag and may interfere with localization in some cases. Most sorting signals are located near the N-terminal of proteins, but in some cases the C-terminal region can also affect localization. For example, the peroxisomal translocation signal PTS1[230, 574] and ER retention signal "KDEL"[659] are found in C-terminal region. Other signals which are not specific to the C-terminal region may also sometimes occur there.[709, 828] It is also at least conceivable that a GFP tag can interfere with the diffusion of small proteins into and out of the nucleus. The smaller tags used in immunofluoresence microscopy studies have not been expressed with native promoters and may exhibit abnormal expression levels. Over-expression potentially causes false positives in the cytoplasm by overloading localization mechanisms. Conversely, under-expression may lead to false negatives in various compartments due to missing small amounts of the protein. Thus there are still some potential sources of systematic error that may explain part of the 20% disagreement between localization data gained with the large-scale GFP experiment[374] and compilations of previous experimental data.

4. Issues and Complications

Before we delve into specific prediction schemes we mention some of the general issues involved in localization prediction.

4.1. *How Many Sites are There and What are They?*

Predictive studies have divided the cell into anywhere from two to around 12 sites. In a sense the classification of proteins as integral membrane proteins versus soluble or peripheral membrane proteins is a kind of localization prediction. Many tools have been developed for the prediction of membrane spanning regions in proteins.[182, 353, 457, 575] A binary classification problem that is at the heart of understanding protein localization is the prediction of signal peptides. The development of the SignalP program[611] is an example of an influential paper which focused on the binary classification problems of predicting the presence or absence of signal peptides, and distinguishing between cleaved signal peptides versus their uncleaved counterparts (N-terminal signal anchor sequences). PSORT[602] classifies eukaryotic proteins into 14 (17 for plants) localization classes. PSORTII[368] classifies eukaryotic proteins into 10 classes.

There is in fact no single accepted "correct" scheme for defining localization classes. Recent experimental techniques have allowed the localization site of proteins in yeast cells to be determined at the resolution of roughly 22 distinct sites.[374] This allows some sites, such as the nuclear periphery and the endoplasmic reticulum, that have been lumped together by most predictive studies to be distinguished. On the other hand, the fluorescence microscopy used in their study does not allow them to distinguish between lumen versus membrane for the mitochondria and endoplasmic reticulum sites. So in this area, the annotation accumulated from small-scale experiments in SWISS-PROT often offers higher resolution.

Indeed the annotations regarding subcellular location in SWISS-PROT[86] strongly reflect the historic lack of a canonical scheme for defining localization sites. Considering only entries containing a subcellular location annotation in the "CC" field and canonicalizing for capitalization and the use of white space, one finds 3214 distinct descriptions for subcellular location. The top 45 such descriptions are shown in Figure 4. Note that the descriptions vary both in terms of localization site and the firmness of the evidence upon which the annotation is based. In fact many annotations—*e.g.*, those containing "by similarity"—do not represent experimental verifications.

The large diversity in annotations has been a practical difficulty to overcome when devising classification schemes for prediction programs. However, less common descriptions often add useful specific information such as conditions—*e.g.*, temperature—under which a localization is observed, or the topology of integral membrane proteins, *etc.* See also Section 5.4. The recent increased use of controlled vocabularies, such as the cellular component vocabulary of the Gene Ontology[35] should make localization site definitions and dataset preparation easier in the future.

4.2. Is One Site Per Protein an Adequate Model?

All of the studies that we are aware of use something close to a one-site-per-protein model. Some work has gone a little beyond this; for example, the PSORT-B[272] database includes eight localization classes for gram-negative bacteria, three of which correspond to proteins that localize to two different sites—such as outer membrane and extracellular.

It is known that the function of some transcription binding factors is partially regulated by selective localization to either the nucleus or the cytoplasm—where their action is obviously blocked.[71,568,636] Kumar *et al.*[463] find that approximately 25% of proteins that localize to an organelle also show significant cytoplasmic staining upon immunofluoresence analysis. The distribution of localization sites from another large-scale study[374] gives the dual localization of nucleus and cytoplasm as the number one localization site, as shown in Figure 5. One of the difficulties in accurately predicting the localization of mitochondrial and chloroplast proteins in plants is that their localization signals are very similar, and in fact—in rare cases—the same protein can be localized to both organelles.[137,777]

Although most work to date on eukaryotes has used the one-site-per-protein model, this is a drastic simplification of reality. Some proteins localize to multiple sites simultaneously, some proteins change their localization in a regulated way, and some proteins constantly move—such as proteins found in vesicle membranes which shuttle between the Golgi body and the endoplasmic reticulum membrane. Some simplification of this complicated reality seems necessary to make the problem tractable.

However, given the data in Figure 5, we believe that adding multiple localization to the nucleus and cytoplasm explicitly to prediction schemes is worth considering. On the other hand, for the vast majority of proteins, multiple localization generally appears to be limited to a few pairs of sites. So it is probably unnecessary to require a model to allow proteins to multiply localize to arbitrary combinations of sites.

4.3. How Good are the Predictions?

The prediction of integral membrane proteins appears to be the easiest one, with percent accuracies in the high 90s being reported by several studies.[182, 353, 457, 575] A much higher resolution prediction for gram negative bacteria also seems to be relatively easy. Horton and Nakai[368] claim an accuracy of 94% with PSORTII in classifying 336 *E. coli* proteins into 7 localization classes (unfortunately this accuracy estimate is buried in the discussion of their paper). Recently, slightly lower accuracy has been reported by PSORT-B[272] with a much larger data set of 1443 sequences.[a]

Localization in eukaryotic cells has proven harder to predict. For example, PSORTII[368] only achieved a somewhat disappointing accuracy of 60% for ten yeast localization sites in 1997. Interestingly, most of the mistakes are between the cytoplasm and the nucleus. Some of those "mistakes" are likely to have been proteins that, depending on certain conditions, can localize to either site. Some advances have been made since then, and several methods have been published with much higher estimated accuracy for eukaryotic cells.

Some of these methods have been compared on a common dataset by Emanuelsson,[232] who found TargetP to classify plant and non-plant eukaryotic cells proteins into four sites with an accuracy of 85% and 90% respectively. Nair and Rost[595] compared TargetP,[232] SubLoc,[369] and NNPSL[710] on a common dataset. They obtained high coverage (99% and 93%) for extracellular and mitochondrial proteins with TargetP but with low precision (51% and 46%). For

---

[a]PSORT, PSORTII, and PSORT-B are three distinct prediction programs, although the feature vector and class definitions used by PSORTII are mainly taken from PSORT.

cytoplasmic and nuclear proteins, the study found SubLoc to yield coverages of 67% and 82% with precisions of 60% and 76% respectively.

Many methods not covered in these studies have also claimed high predictions accuracies with various datasets and localization site definitions. We have no intention of doubting any particular estimates, but refer the reader to Section 4.4 for caveats. An independent comparison[562] of signal peptide predictions has compared the accuracy of weight matrix, neural network, and hidden Markov models. Due to different balances of recall versus precision, the results of Table 1 in their study appear insufficient to declare which method is best. However, it seems that an overall accuracy of around 90% is possible.

4.3.1. *Which Method Should I Use?*

Of course we encourage researchers to look at several methods and admit that our coverage in this chapter is only partial. Figures 2 and 3 show some public prediction servers and datasets. For most users we recommend trying the TargetP[232] server because it has a good reported prediction accuracy and the programs it is based on, such as SignalP,[611, 614] appear to have been designed by researchers with a thorough understanding of the current state of knowledge regarding sorting signals and processes. One drawback of SignalP is that it is a commercial piece of software and the program is not freely available to certain non-profit research organizations. However, use of the public server is free. Another minor drawback is that it may not be well suited for proteins that are localized through mechanisms other than signal peptides.[610] These are relatively rare, and we expect such proteins to be hard for any prediction program. Although some of the various PSORT programs are in serious need of updating, such updates are planned and we believe the PSORT web site will continue to be a valuable resource.

In large genome projects, gene finding programs often mispredict the N-terminal region of proteins;[265] see also Chapter 5. Thus methods—many of which are shown in Figure 8—which do not rely on N-terminal signals are especially useful because they can be expected to be relatively robust against start site errors.[369] In any case, since gene finding programs typically can make use of similar sequences to increase accuracy, the accuracy of gene finding is likely to increase along with the rapidly increasing availability of sequences from various organisms.

4.4. *A Caveat Regarding Estimated Prediction Accuracies*

We note here that the prediction accuracy reported in various studies, including our own, may be a bit optimistic. One obvious important difficulty in comparing

Program	URL	Ref.
PSORT	`psort.ims.u-tokyo.ac.jp`	61,597
PSORTII	`psort.ims.u-tokyo.ac.jp`	368,597
NNPSL	`www.doe-mbi.ucla.edu/~astrid/` `astrid.html`	710
TargetP	`www.cbs.dtu.dk/services/` `TargetP`	232
LOC3d	`cubic.bioc.columbia.edu/db/` `LOC3d`	596
PLOC	`www.genome.ad.jp/SIT/ploc.html`	648
SubLoc	`www.bioinfo.tsinghua.edu.cn/` `SubLoc`	369
ProtComp	`www.softberry.com/berry.phtml?` `topic=proteinloc`	
Predotar	`www.inra.fr/predotar`	

Fig. 2. Some public localization prediction servers are shown with references to the literature where available.

prediction accuracies is the variety of localization site definitions and datasets that have been used in different studies. The majority of the works have used annotations in SWISS-PROT[86] to train and test their methods however, which leads to another more subtle problem in estimating prediction. Strictly speaking, estimating the accuracy based on performance on a test set is only valid if the test set data is used just once.

Consider two arbitrary classifiers—on a particular data set one may classify more accurately than the other simply by chance. Many works have been published using various subsets of the same data. Moreover, since each work generally reflects the results of testing multiple classifiers or one classifier with many parameter settings, the effect is amplified and the same data has been used many times for testing. Thus the results of even rigorous cross-validation studies should be taken with this in mind.

In the machine learning community, the UCI Repository[84] contains datasets—including two for protein localization—that can be used to test classification algorithms. Despite the fact that the repository contains more than 100 datasets, there

Dataset	URL	Ref.
SWISS-PROT	www.ebi.ac.uk/swissprot	86
NLSDB	cubic.bioc.columbia.edu/db/ NLSdb	593
MITOP	mips.gsf.de/proj/medgen/mitop	749
YEAST GFP LOCALIZATION DB	yeastgfp.ucsf.edu	374
YPL.db	genome.tugraz.at/ypl.html	312
TRIPLES	ygac.med.yale.edu/triples/ triples.htm	463
MIPS CYGD	mips.gsf.de/genre/proj/yeast/ index.jsp	564

Fig. 3. Some public localization datasets are shown with references to the literature.

is a serious concern that its repeated use has lead to inaccurate conclusions on the general accuracy of classifiers tested on it.[740]

4.5. *Correlation and Causality*

An important issue in evaluating prediction schemes for localization is the distinction between sequence features that reflect causal influences on localization versus those which merely reflect correlation with localization. Figure 6 uses the localization of a transcription factor to the nucleus to illustrate this point. Nuclear localization signals (NLS)[299] in proteins cause them to be selectively imported into the nucleus by importins. Indeed a classic study by Goldfarb *et al.*[295] shows that not only is nuclear localization impaired by mutations in the NLS, but also that non-nuclear proteins are imported into the nucleus when modified to include artificial NLS's. Thus the presence of an NLS naturally correlates with nuclear localization.

The vast majority of the DNA in a eukaryotic cell is found in the nucleus. Thus proteins whose function is to interact with DNA are generally imported to the nucleus, and therefore DNA binding motifs such as the zinc finger binding motif[438] also correlate with nuclear localization. There is however a fundamental difference between these two correlations. The zinc finger binding motif is not believed

Freq.	Description	Freq.	Description
7307	cytoplasmic (by similarity)	309	integral membrane protein. inner membrane (by similarity)
6380	cytoplasmic		
5166	secreted	296	integral membrane protein. inner membrane (probable)
4181	integral membrane protein		
3655	nuclear	270	membrane-bound
3251	integral membrane protein (potential)	226	attached to the membrane by a gpi-anchor
2177	cytoplasmic (probable)	216	periplasmic (by similarity)
1862	cytoplasmic (potential)	212	mitochondrial inner membrane
1542	chloroplast		
1241	type i membrane protein	211	membrane-bound. endoplasmic reticulum
1114	nuclear (potential)		
1029	nuclear (probable)	204	cytoplasmic and nuclear (by similarity)
869	nuclear (by similarity)		
775	mitochondrial	202	attached to the membrane by a lipid anchor (potential)
721	integral membrane protein (probable)		
		199	membrane-associated (by similarity)
495	integral membrane protein. inner membrane (potential)		
		193	inner membrane-associated (by similarity)
484	mitochondrial matrix		
435	integral membrane protein. mitochondrial inner membrane	180	periplasmic (potential)
		175	type i membrane protein (potential)
428	secreted (by similarity)		
393	extracellular	165	attached to the membrane by a lipid anchor (probable)
381	periplasmic		
373	integral membrane protein. inner membrane	162	nuclear; nucleolar
		161	mitochondrial (by similarity)
355	chloroplast thylakoid membrane	156	type ii membrane protein
		156	secreted (probable)
352	integral membrane protein (by similarity)	153	integral membrane protein. chloroplast thylakoid membrane
312	secreted (potential)	148	lysosomal

Fig. 4.   The 45 most frequent descriptions of subcellular localization in SWISS-PROT.

to exert a causal influence on nuclear localization. For example, Mingot *et al.*[568] has created a mutant form of a nuclear protein in which DNA binding is abolished but nuclear localization is retained.

The correlation between zinc finger binding motifs and nuclear localization is

Freq.	Description	Freq.	Description
827	cytoplasm, nucleus	13	bud neck, cell periphery
823	cytoplasm	12	cytoplasm, nucleolus, nucleus
496	nucleus	11	punctate composite, early
485	mitochondrion		Golgi
266	ER	11	early Golgi
157	ambiguous	11	ambiguous, bud neck,cell pe-
121	vacuole		riphery, bud
73	punctate composite	10	microtubule
73	nucleolus, nucleus	10	cell periphery, bud
70	nucleolus	10	bud neck, cytoplasm,cell pe-
57	cell periphery		riphery
54	vacuolar membrane	10	bud neck, cytoplasm
53	nuclear periphery	10	ambiguous, bud neck, cyto-
39	spindle pole		plasm, cell periphery, bud
34	endosome	9	ER, cytoplasm
33	late Golgi	8	nucleus, spindle pole
27	actin	8	mitochondrion, punctate com-
21	peroxisome		posite
21	cell periphery,vacuole	8	cytoplasm, nucleolus
19	lipid particle	8	ambiguous, bud neck, cyto-
18	cytoplasm, punctate compos-		plasm, bud
	ite	8	ER, vacuole
18	cytoplasm, mitochondrion	6	ER to Golgi
18	Golgi, early Golgi	5	mitochondrion, nucleus
15	bud neck	5	early Golgi, late Golgi
15	Golgi	5	cytoplasm, vacuole

Fig. 5. The 45 most frequent descriptions of subcellular localization in the yeast GFP[374] database.

real and useful for the prediction of native proteins. Especially since NLS's are relatively difficult to detect from primary sequence information—Nair and Rost reported that NLS's were detected in only 1% of eukaryotic proteins, using sequence analysis with a strict precision requirement.[596] However it should be kept in mind that non-causal correlations such as the one between DNA binding motifs and nuclear localization may not be robust when applied to mutant or non-native proteins. The understanding of which correlations reflect causal influences is critical to the ability to design novel protein sequences with some desired behavior. A classic example of this issue from another area of bioinformatics is the use of codon usage in gene finding. Indeed the question of how to treat causal and

NLS   Zn Finger

Nuclear Localization   DNA Binding

Transcription Regulation

High Evolutionary Fitness

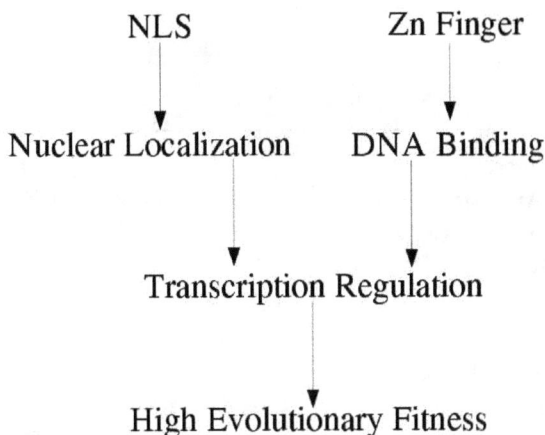

Fig. 6. A schematic diagram illustrating some causal influences is shown. NLS is the nuclear localization signal. Many omitted variables also exert causal influence on function and fitness.

non-causal correlations is not specific to protein localization, but is important in any application of machine learning or statistical analysis. We refer the reader to Pearl[652] for an in depth analysis of causality in the context of statistical inference.

5. Localization and Machine Learning

An impressive number of learning and knowledge representation techniques have been applied to the problem of predicting protein subcellular localization. Indeed the list is fairly representative of the techniques from AI, pattern recognition, and machine learning that have been applied across the entire field of bioinformatics. We have attempted to organize the work in this field by classifier (Figure 7) and by the kind of input used (Figure 8) to classify. We must admit that these two tables are incomplete and imperfect. In particular, many works that are specific to localization to a particular site are omitted—although many are included—and the categorization of classifier and input type is rather arbitrary in some cases. For papers that cover a variety of classifiers or input techniques we are likely to have made some errors of omission as well. Nonetheless we believe that these figures are a useful way to organize the existing body of work on localization prediction.

Many of the methods in Figure 7 are briefly described in Chapter 3. We do not describe them in detail here. Instead we outline some of the common approaches and we give our thoughts on the strengths and weaknesses of these methods when applied to localization prediction.

Prediction Schemes: Classifier Technique Used	Ref.
Rule-Based Expert System	228, 601, 602
Discriminant Analysis	136, 159, 165, 245, 604, 617, 923
Principle Component Analysis	25, 923
Bayesian Network	272
Naïve Bayes	213, 368
Decision Trees	368
Nearest-Neighbor Methods	368, 594, 595
Support Vector Machines	131, 161, 272, 369, 648
Feed-Forward Neural Network	89, 134, 165, 231, 232, 389, 589, 611, 710, 755
Kohonen Self-Organizing Map	755
(Hidden) Markov Models	134, 267, 272, 612, 922
Human-Designed Structured Model	134, 367, 601, 602
(Generalized) Weight Matrix	160, 862, 863
Feature Discovery/Data Mining	61, 272, 366

Fig. 7. A partial list of classifying methods and programs that employ them to predict localization is shown. The Bayesian network framework is general enough to encode any probability distribution, thus including many other categories, but here we reserve the Bayesian network category for authors who presented their methods as Bayesian networks. Weight Matrix includes methods that couple a few fixed columns to allow for interdependence between sites. We include Markov chain models in (hidden) Markov models. Discriminant analysis is used as a catch all for discriminant methods that don't fall into more specific categories.

5.1. *Standard Classifiers Using (Generalized) Amino Acid Content*

The correlation between amino acid composition and protein localization was observed as early as 1982.[617] As can be seen in Figure 8, many methods have been developed based on amino acid composition or slightly generalized features such as the composition of amino acid pairs separated by 0-3 amino acids, including recent studies[369, 648] using support vector machines.[178, 856]

These methods achieve competitive accuracy. Moreover they have the advantage that they may be accurate even for proteins that have sorting signals which are too subtle to be found with current prediction techniques, or that localize without the direct use of sorting signals—see Section 2.

As mentioned above, for genome projects in which putative amino acid chains

Prediction Schemes: Information Used as Input	Ref.
(Generalized) Amino Acid Composition	131, 136, 165, 245, 369, 604, 710, 798, 922, 272, 648
N-terminal Sequence Region	164, 165, 367, 368, 601, 602, 611, 862, 863, 61, 134, 160, 231, 272, 755
Entire Amino Acid Sequence	134, 367, 368, 601, 602
Sequence Periodicity	159, 648
Sequence Similarity	272, 595
Sequence Motifs	61, 213, 231, 366
Protein Signatures	130, 161, 272
Physiochemical Properties	61, 164, 213, 245, 755
mRNA Expression Data	213
Knockout lethality	213
Integral α-helix Transmembrane Region Prediction	134, 272, 367, 368, 601, 602
Surface Residue Composition	25
Text Descriptions	228, 594, 798
Fluorescence Microscope Images	89, 589
Meta Localization Features	272, 367, 368, 601, 602

Fig. 8. A partial list of types of input information and references for programs that use that input to predict localization is shown. Generalized amino acid composition may include slightly higher-order inputs such as the composition of adjacent pairs of amino acids. Protein Signatures—some of which can also be classified as sequence motifs—are taken from PROSITE[238], SBASE-A[590], and InterPro.[582] Meta localization features are the results of localization prediction programs; for example, PSORT(II) uses a modified version of McGeoch's signal peptide prediction program[863] as an input feature.

may have incorrectly predicted N-terminal regions, the lack of a strong dependence on N-terminal sorting signals is also a practical advantage. However, we speculate that some of the amino acid composition bias utilized by these methods reflects an adaptation to functioning effectively in the different chemical environments found in different compartments, as discussed in Andrade *et al.*,[25] rather than a causal factor in their localization.

We do not dismiss all of the bias as being mere correlation. For example, hydrophobicity can certainly affect the integration into or transition through membranes. Still we feel that this approach is relatively prone to relying on non-causal correlations. The potential drawbacks of such a reliance is discussed in Section 4.4.

5.2. *Localization Process Modeling Approach*

PSORT[601, 602] uses a tree-based reasoning scheme designed to roughly reflect the localization process. Rules are supplied for each decision node in the tree in which a feature is primarily chosen from biological knowledge. For example, a weight matrix designed to detect signal peptides is used at the node representing transport through the endoplasmic reticulum membrane.

Unlike the decision trees mentioned in Section 5.5, the tree architecture is designed from prior knowledge rather than induced from statistical properties of a training set. The advantage of this system is that it is not only a prediction scheme but is also a kind of knowledge base. To the extent that biologists understand the mechanisms of localization it should be possible to design prediction schemes that model the process. Given the dependence on prior knowledge we expect that most of the correlations PSORT uses are causal in nature.

One disadvantage of this approach is the labor intensive process of updating the rules to reflect the ever growing body of knowledge regarding localization processes. Another disadvantage is the lack of a good way—such as cross-validation, but see Section 4.4—to estimate the accuracy on unkown sequences. It is of course not feasible to remove the influence of a set of randomly chosen test sequences on the body of knowledge regarding localization. Finally this method has less ability to fully leverage all correlations between sequence features and localization to maximize prediction accuracy.

5.3. *Sequence Based Machine Learning Approaches with Architectures Designed to Reflect Localization Signals*

A series of works by Nielsen, Emanuelsson, and colleagues have taken an approach in which sophisticated sequence-based classifiers are used—but knowledge of localization is employed to select the classifier architecture and input sequence region.[231, 232, 611, 612, 614] In particular, many of those works use feed-forward neural networks[57, 82, 687, 730] to predict sorting signals. Although provably optimal learning procedures are not known, in principle feed-forward neural networks can learn non-linear interactions between distant amino acids that affect localization.

Applied without careful analysis of the learned weights, neural networks generally have a tendency to produce "black box" classifiers. However the works describe in this section do provide some such analysis and their restriction of the sequence region input should greatly reduce reliance on spurious correlations. In other words, they use prior knowledge to limit the complexity of the input but rely on machine learning techniques to determine the actual function of the input that is used for classification. We feel this general approach should be effective for many bioinformatics applications.

5.4. *Nearest Neighbor Algorithms*

PSORTII[368] uses the PSORT features. But instead of using the PSORT reasoning tree, it uses the k nearest neighbors classifier (k-NN).[214] To classify a sequence, PSORTII simply considers the k sequences in the training data whose feature vector most closely matches, by euclidean distance, the feature vector of the sequence to be classified. This classifier was found to be more effective than decision tree induction, Naïve Bayes, and a structured probabilistic model roughly based on the PSORT reasoning tree[368]. Predicting localization by sequence similarity search for close homologs of known localization is another commonly employed method,[272, 595] which amounts to a kind of nearest neighbor classification.

Nearest neighbor classifiers do not summarize the training data, and thus in a machine learning sense have very little explanatory power. They do however naturally provide the particular examples in the training data which are most similar to the data to be classified. This is useful in problems such as localization where much valuable *ad hoc* annotation information can be given in addition to the predicted class. For example, even if the localization site definition used does not give nucleolar proteins their own distinct class, if the nearest neighbors to a sequence are annotated as "nuclear; nucleolar" in SWISS-PROT, that gives a valuable clue that the sequence may localize to the nucleolus.

We are currently designing a successor to PSORTII but plan to retain at least some form of nearest neighbor classification. It has been reported[603] that the accuracy of PSORTII can be improved by using more sophisticated variants of nearest neighbor classifiers such as discriminant adaptive nearest neighbor classifiers.[327]

5.5. *Feature Discovery*

In this section we introduce two studies, Horton[366] and Bannai *et al.*[61], that focused on trying to automatically discover simple sequence features that are relevant to protein localization. Although neither study was able to surpass previous methods in estimated prediction accuracy, we feel their feature discovery approach merits mention. Since the methods used in these studies are less established than many of the classification algorithms mentioned in Figure 7, we briefly describe them here.

Horton[366] first identified substrings in the protein sequences that correlate significantly to localization site and then built decision trees with the standard decision tree algorithm using those substrings as potential features. Decision trees were chosen because decision trees are relatively easy to interpret and the standard decision tree induction algorithm[692] includes feature selection. The leaves of the decision trees are localization sites and the internal nodes are binary nodes that represent tests of the number of occurrences of a particular substring in the input sequence. Such trees were induced on several random subsets of the localization data with the idea that consistently selected substring features would be important. Figure 9 shows one of the induced trees. The prediction accuracy of this tree is far from competitive with the best methods and many discovered features appear to simply reflect amino acid composition bias—but we were pleased that a test for the presence of a carboxy-terminal phenylalanine was consistently selected from an *E.coli* dataset. Although we were not aware of the fact until after the feature discovery experiment was conducted, the presence of a carboxyl-terminal phenylalanine is an experimentally verified factor in localization to the outer membrane in bacteria.[807]

Bannai *et al.*[61] considered a broader class of features, including substrings pattern that only require a partial match and patterns that group amino acids based on published amino acid indexes, many of which reflect various chemical properties of amino acids. Instead of the decision tree induction algorithm they used an extensive search of possible rules to build a decision list[472] which is a special case of a decision tree with a linear structure. Their method was able to "discover" the known fact that mitochondrial targeting signals have an amphiphilic α-helix structure and predict signal peptides with competitive accuracy using only a simple average of hydrophobicity over the appropriate sequence region.

5.6. *Extraction of Localization Information from the Literature and Experimental Data*

This chapter has focused on predicting protein localization from information directly derivable from the amino acid sequence of the protein, which is more or less the same problem nature is faced with. There is of course another source of information available to use—experimental data and the vast literature based upon it. Ultimately all of the methods mentioned in this chapter are based on information gained by human interpretation of experimental data. But recently there has been progress in developing computer programs to automatically extract such information. Eisenhaber and Bork[228] have developed a rule-based method for classify-

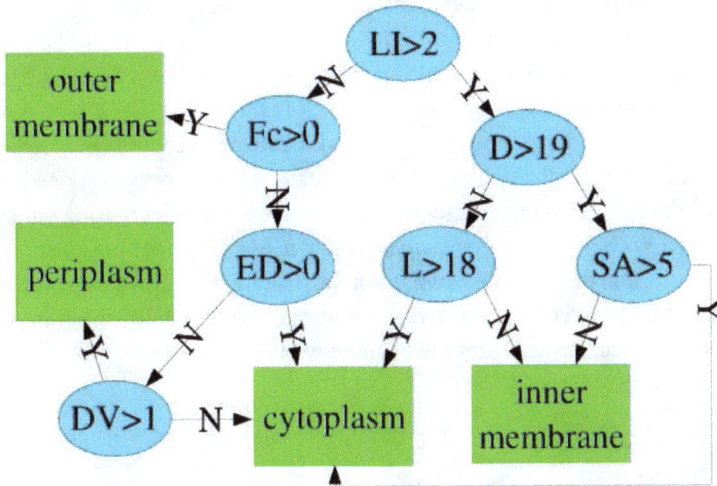

Fig. 9. A decision tree induced for localization in *E.coli* is shown. The leaf nodes (green rectangles) represent localization sites. To simplify presentation, leaves representing the same localization site have been merged. Internal nodes (blue ovals) represent binary tests. The root node test for whether the substring "LI" appears more than twice or not. A lower case "c" represents the C-terminal of the amino acid sequence. Thus the condition "Fc> 0" tests for the presence of a carboxy-terminal phenylalanine. (*Image credit: Adapted from Figure 15.3 of Horton.*[366])

ing SWISS-PROT entries that lack an explicit "subcellular localization" tag but contain sufficient information in their description to determine their localization. For example, they observe that the functional description of "cartilage protein" is sufficient to infer extracellular localization. Support vector machines have also been applied to extracting localization information from the literature.[798] Nair and Rost[594] also used machine learning to predict localization from SWISS-PROT keywords. Murphy and colleagues have developed methods for classifying localization from fluorescence microscopy images,[89, 589] and methods to extract and analyze such images automatically from figures and captions in the literature.[449]

6. Conclusion

In this chapter we have briefly discussed many aspects of protein localization in the context of bioinformatics. We hope that the overview of the biology and some experimental techniques presented here will be valuable background for computer scientists seeking to develop new prediction algorithms. For biologists who are primarily users of such algorithms, we hope that our brief summary of common approaches to prediction will give helpful insights as to what is going on "under the hood" in many commonly used prediction tools, as well as providing a starting point for algorithm developers.

We have stated our views regarding how priorities will shift in the age of proteomics—namely towards methods that reflect the biology in a robust way. As we have mentioned earlier, some aspects of localization cannot be understood statically, since the localization of some proteins is dynamic or conditioned upon the state of the cell. With proteomic-scale experiments vastly increasing our body of knowledge, we believe that more sophisticated models incorporating some form of time or cell state may soon become feasible. The roughly 20 years of research reviewed in this chapter is impressive—but at the current rate of innovation we believe the next decades will prove even more exciting.

CHAPTER 10

HOMOLOGY SEARCH METHODS

Daniel G. Brown

University of Waterloo
browndg@monod.uwaterloo.ca

Ming Li

University of Waterloo
mli@pythagoras.math.uwaterloo.ca

Bin Ma

University of Western Ontario
bma@uwo.ca

Homology search methods have advanced substantially in recent years. Beginning with the elegant Needleman-Wunsch and Smith-Waterman dynamic programming techiques of the 1970s, algorithms have been developed that were appropriate for the data sets and computer systems of their times. As data sets grew, faster but less sensitive heuristic algorithms, such as FASTA and BLAST, became a dominant force in the late 1980s and 1990s. As datasets have grown still larger in the post-genome era, new technologies have appeared to address these new problems. For example, the optimal spaced seeds of PatternHunter increase speed and sensitivity. Using these ideas, we can achieve BLAST-level speed and sensitivity approaching that of slow algorithms like the Smith-Waterman, bringing us back to a full circle. We wish to take you with us on this round trip, with some detours along the way so as to study both global and local alignment. We present methods for general purpose homology that are widely adopted, not individual programs.

ORGANIZATION.

Section 1. We begin with a brief history of the development of homology search methods.

Section 2. Then we introduce the notions of edit distance, alignment, score matrix, and gap penalty.

Section 3. Next, the classic dynamic programming approach to sequence alignment is presented. We discuss both global and local alignments, as well as the issues of gaps and memory usage.

Section 4. Then we proceed to the probabilitistic issues underlying sequence alignment. In particular, we discuss in some details the PAM and BLOSUM scoring matrices and their derivations. We also briefly visit the issue of assessing the significance of an alignment.

Section 5. After that, we introduce the second generation of sequence homology search methods. These methods sacrifice sensitivity for speed by using simple seeds to index into possible matching regions before more expensive dynamic programming alignments are performed.

Section 6. Then we come to the third generation of sequence homology search methods. These methods rely on the more advanced idea of spaced seeds to achieve simultaneously the high sensitivity of the first-generation classical methods and the high speed of second-generation methods. We discuss the optimality of spaced seeds with respect to different sequence models. We also discuss the effectiveness of using multiple spaced seeds.

Section 7. Finally, we show some experiments comparing the sensivity and speed of these different generations of sequence homology search methods.

1. Overview

Two sequences are homologous if they share a common evolutionary ancestry. Unfortunately, this is a hypothesis that usually cannot be verified simply from sequence data. Therefore, our title is really a convenient misuse of terminology. Homology search methods provide evidence about homologies, rather than demonstrating their existence.

Homology search is important as its product—high scoring alignments—is used in a range of areas, from estimating evolutionary histories, to predicting functions of genes and proteins, to identifying possible drug targets. All contemporary molecular biologists use it routinely, and it is used in many of the largest supercomputing facilities worldwide. The NCBI BLAST server for homology search is queried over 100,000 times a day and this rate is growing by 10–15% per month.

The basic homology search problem is so easy that it is usually the first topic in a bioinformatics course. However, the problem is also very hard, as queries and databases grow in size, and the emphasis is on very efficient algorithms with high quality. More programs have been developed for homology search than for any other problem in bioinformatics, yet after 30 years of intensive research, key problems in this area are still wide open.

As is true of many topics in early bioinformatics, the first two important sequence alignment algorithms—the Needleman-Wunsch algorithm for global alignment and the Smith-Waterman algorithm for local alignment—were both identified during the 1970s and early 1980s by a variety of different groups of authors working in different disciplines. However, these algorithms ran in time that was too slow as databases of DNA and protein sequences grew during the 1980s. Since the mid 1980s and 1990s, heuristic algorithms—like FASTA[512] and BLAST[23]—that sacrificed sensitivity for speed became popular. These algorithms offer far faster performance, while missing some fraction of good sequence homologies. The development of good homology search software makes another advance recently, as researchers focus on the cores of alignments that are identified by heuristic search programs. This is seen in local alignment programs—such as PatternHunter[533] and BLAT[427]—that allow substantial improvement in sensitivity at minimal cost. The spaced seeds of PatternHunter, in particular, can be optimized to be highly sensitive for alignments matching a particular model of alignments,[102, 103, 118, 425, 500] which allows substantial improvement in sensitivity. In fact, one can use these spaced models to approach Smith-Waterman sensitivity at BLAST speed.[500] This chapter takes you through this round trip, stopping to note key ideas along the way, rather than specific programs.

2. Edit Distance and Alignments

2.1. *Edit Distance*

The central question in this field is whether two given DNA or protein sequences are homologous. This question cannot be answered precisely without intimate knowledge of the origin of the biological sequences, even if they are very similar in their sequence. However, the number of evolutionary mutations required to change one sequence to the other can be used to estimate the probability that the two sequences are homologous. We use a distance metric—"edit distance"—to measure the evolutionary similarity of two sequences.

When a DNA sequence is copied from a parent to a child, three common types of mutations can be introduced: substitution, insertion, and deletion. Substitution is the change of one nucleotide to another. Insertion inserts a new nucleotide to the sequence, and deletion deletes an existing nucleotide. See Figure 1.

These three operations are similar to the operations used when one edits a text file. Any sequence can be converted to any other sequence using these three edit operations, since we can change any sequence to the empty sequence by deletions, and the empty sequence to any other sequence by insertions. However, such a sequence of operations is rarely the shortest. The minimum number of operations that are required to change a sequence s to a sequence t is called the edit distance of s and t, denoted here by $d(s, t)$.

parent:	A$\underline{C}$GTCT	AGTCT	A$\underline{C}$GTCT
	↓	↓	↓
child:	AGTCT	A$\underline{C}$GTCT	A$\underline{G}$GTCT
	deletion	insertion	substitution

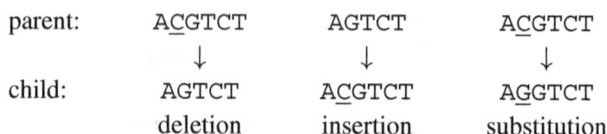

Fig. 1.   The three types of modifications.

Before we examine the algorithm that computes the actual edit distance of two given sequences, we prove that edit distance is a metric.

Theorem 1: Edit distance is a metric. That is, for any three sequences x, y, and z, $d(\cdot,\cdot)$ satisfies the following three properties:

(1) Isolation: $d(x,y) = 0$ if and only if $x = y$.
(2) Symmetry: $d(x,y) = d(y,x)$.
(3) Triangular inequality: $d(x,y) \leq d(x,z) + d(z,y)$.

Proof: For Property 1, it is trivial.

For Property 2, first note that a deletion in one sequence is equivalent to an insertion in the other sequence. Therefore, if d_1 insertions, d_2 deletions and d_3 substitutions convert x to y, then d_3 substitutions, d_2 insertions and d_1 deletions convert y to x.

For Property 3, we can transform x to z in $d(x,z)$ steps, and z to y in $d(z,y)$ steps, for any z. This gives an upper bound of $d(x,z) + d(z,y)$ steps on the edit distance from x to y. □

Another measure that is often used in the comparison of strings is the "Hamming distance". For two strings $s = s_1 s_2 \ldots s_n$ and $t = t_1 t_2 \ldots t_n$ of the same length, their Hamming distance $d_H(s,t)$ is the number of positions i, $1 \leq i \leq n$, such that $s_i \neq t_i$. Unfortunately, despite its simplicity, Hamming distance is not suitable for measuring the similarity of long genetic sequences, because of the existence of insertions and deletions.

2.2. *Optimal Alignments*

Edit distance is defined in the form of the minimum number of edit operations between two sequences, where insertions, substitutions and deletions all count as one mutation. However, edit distance is usually formulated as sequence alignment. Let $s = s_1 s_2 \ldots s_m$ and $t = t_1 t_2 \ldots t_n$ be two sequences over a finite alphabet Σ. We augment the alphabet Σ with a "space" symbol denoted by "−" that is not in

Σ, yielding the alignment alphabet Σ'. Any equal-length sequences S and T over the alphabet $\Sigma \cup \{-\}$ that result from inserting space characters between letters of s and t are called an alignment of the original sequences s and t.

Usually, the objective of the sequence alignment of s and t is to maximize the similarity of S and T, so that characters in the same position in both sequences are closely related or identical. The simplest goal is to minimize the Hamming distance $d_H(S, T)$. Figure 2 shows an optimal alignment of two sequences ACGTCT and AGTACG under this objective.

```
ACGT-CT

A-GTACG
```

Fig. 2.   The optimal alignment of two sequences

Computing the optimal alignment to minimize Hamming distance is equivalent to computing edit distance, as shown in the following theorem:

Theorem 2: Let s and t be two sequences, and $d(s, t)$ be their edit distance. Let S and T be the optimal alignment of s and t that minimizes $d_H(S, T)$. Then $d_H(S, T) = d(s, t)$.

Proof: It is easy to see that in the edit operations that convert s to t, each insertion corresponds to a space symbol in S, each deletion corresponds to a space symbol in T, and each substitution corresponds to a mismatch between S and T.   □

2.3. *More Complicated Objectives*

It is possible to perform biologically meaningful sequence alignments using only the edit distance. However, typical homology search programs optimize somewhat more complicated functions of the alignments. As we will see in Section 4, there is actually a probabilistic basis to these more complicated functions. In the interim, we provide a descriptive overview here.

2.3.1. *Score Matrices*

The edit distance measure causes the same cost to be incurred for every substitution. Yet—especially for protein sequences—some mutations are far less conservative than others, and are less expected in truly homologous sequences. In these cases, a "score matrix" is used to discriminate different types of matches and mismatches. The score matrix value $M(a, b)$ is a real number that is the contribution to the alignment score of aligning a and b, for any $a, b \in \Sigma'$. Given two sequences s and t, and their alignment (S, T), the score of the alignment is defined by $\sum_{i=1}^{k} M(S[i], T[i])$. When a score matrix is used, we usually let $M(a, b) > 0$ if a and b are closely related. Therefore, the objective of the alignment is to find an alignment that maximizes the alignment score $\sum_{i=1}^{k} M(S[i], T[i])$.

In this framework, minimizing Hamming distance can be regarded as a special case, using a score matrix $M(a, a) = 0$ and $M(a, b) = -1$ for $a \neq b$.

2.3.2. Gap Penalties

Another important way of scoring alignments is to look at the lengths of regions consisting entirely of space characters in an alignment. In homologous sequences, such "gaps" correspond to either deletion or insertion mutations. If we use a typical scoring matrix where we score a negative constant $M(a, -)$ for aligning any letter a to a gap, the cost of the gap is proportional to its length. In both theory and practice, this is undesirable. As such, the typical response is to penalize gaps through a length-dependent gap penalty, rather than only through the scoring matrix. The most common type of gap penalty is the affine penalty, where the score of a gap of length i is $o + i \times e$. In this scheme, the gap "opening" penalty o is typically much more negative than the "extension" penalty e paid per gap position. Both costs, however, are typically negative.

3. Sequence Alignment: Dynamic Programming

3.1. *Dynamic Programming Algorithm for Sequence Alignment*

We return to the simplest case of sequence alignment, where we incorporate the costs of insertions and deletions in the score matrix M. Here, the optimal alignment of two sequences is easily computed by dynamic programming.[300, 607]

Let $s = s[1]s[2]\ldots s[m]$ and $t = t[1]t[2]\ldots t[n]$. Let $DP[i, j]$ be the optimal alignment score for $s[1..i]$ and $t[1..j]$. The dynamic programming algorithm computes $DP[i, j]$ recursively, and $DP[m, n]$ ends up as the score of the optimal alignment between s and t. The actual alignment of the two sequences can be computed by a standard backtracking procedure after $DP[i, j]$ is computed for all $0 \leq i \leq m$ and $0 \leq j \leq n$.

Suppose the optimal alignment of $s[1..i]$ and $t[1..j]$ has k columns. There are three possibilities for the last column of the alignment:

(1) $s[i]$ is aligned to "$-$". In this case, the first $k - 1$ columns of the alignment should be an optimal alignment of $s[1..i - 1]$ and $t[1..j]$. Otherwise, we could

replace the first $k - 1$ columns with the optimal alignment of $s[1..i - 1]$ and $t[1..j]$, and get a better alignment for $s[1..i]$ and $t[1..j]$. Therefore, $DP[i, j] = DP[i - 1, j] + M(s[i], -)$.

(2) $t[j]$ is aligned to "$-$". For the same reason as in Case 1, $DP[i, j] = DP[i, j - 1] + M(-, t[j])$.

(3) $s[i]$ is aligned to $t[j]$. For the same reason as in Cases 1 and 2, $DP[i, j] = DP[i - 1, j - 1] + M(s[i], t[j])$.

By now we have almost derived a way to compute $DP[i, j]$ recursively, except that for each (i, j), we must choose the correct case, which is the case that yields the highest score. This gives the following recursive algorithm to compute $DP[i, j]$.

$$DP[i, j] = \max \begin{cases} DP[i - 1, j] + M(s[i], -) \\ DP[i, j - 1] + M(-, t[j]) \\ DP[i - 1, j - 1] + M(s[i], t[j]) \end{cases} \quad (1)$$

By definition, $DP[0, 0]$ is the alignment score of two empty sequences. Therefore, $DP[0, 0] = 0$. Also, $DP[i, 0]$ is the alignment score between $s[1..i]$ and an empty sequence. Therefore, $DP[i, 0] = \sum_{k=1}^{i} M(s[k], -)$. Similarly we can compute $DP[0, j]$.

To fill in the DP matrix, we may calculate Formula 1 for i from 1 to m and for j from 1 to n. Therefore, when we compute $DP[i, j]$, all the needed right-hand side values in Formula 1 are already known, and $DP[i, j]$ can be computed in constant time. Hence, the time complexity of this algorithm is $O(m \times n)$.

When the algorithm is implemented in a programming language, the values of $DP[i, j]$ are usually stored in a two dimensional array, as shown in Figure 3. The first row and column of the array must be initialized. For each of the other entries, its value is determined by the three adjacent entries—*viz.* the upper, left, upper left entries. The arrows in Figure 3 illustrate that the three entries determine the value of the last entry.

After the DP table is filled, the value of $DP[m, n]$ is the score of the optimal alignment of the two sequences. In constant time, we can determine which of the three cases maximizes the value in Formula 1, and thus determine the last column of the alignment. For example, in Figure 3, it is the third case that maximizes the value of the last entry. Therefore, the last column of the alignment is $s[m]$ matching $t[n]$, preceded by the optimal alignment of $s[1..m - 1]$ and $t[1..n - 1]$. These preceding columns can be identified by examining $DP[m - 1, n - 1]$, and so on. Eventually, we trace back to $DP[0, 0]$ and get the optimal alignment of s and t. This procedure is called backtracking. Figure 4 shows the backtracking

		A	C	G	T	C	T
	0	−1	−2	−3	−4	−5	−6
A	−1	1	0	−1	−2	−3	−4
G	−2	0	0	1	0	−1	−2
T	−3	−1	−1	0	2	1	0
A	−4	−2	−2	−1	1	1	0
C	−5	−3	−1	−2	0	2	1
G	−6	−4	−2	0	−1	1	1

Fig. 3.   The DP table for the alignment of two sequences ACGTCT and AGTACG.

through the table in Figure 3, giving the alignment of Figure 2.

During backtracking, it is possible that for an entry, more than one of the three cases give the entry its value. If we want all possible optimal alignments, we must examine all adjacent entries that give this optimal value. Therefore, there are potentially an exponential number of backtracking paths, each corresponding to an optimal alignment with the same value. If we only need one optimal alignment, we can choose an arbitrary one that maximizes the value. Because each step of the backtracking reduces either i or j or both by one, the total number of steps does not exceed $m + n$. Thus, most computation time in computing a single optimal alignment is spent filling the DP table.

3.1.1. *Reducing Memory Needs*

The algorithm we have given requires $O(m \times n)$ space to store the entire DP matrix. However, if all that is desired is the score of the optimal alignment, this can be computed in $O(n)$ space by only keeping track of two columns or rows of the DP matrix—since when computing a row, we need only to know the value of the previous row.

A slightly more complicated trick[354] allows the computation of the optimal alignment in $O(n)$ space as well. Here, we keep track of the most recent two rows of the matrix, and after we have computed row $m/2$ of the matrix, we also remember which cell in the $m/2$ row the alignment path that is optimal in the current cell passed through last. When we compute this for the $[m, n]$ entry of the

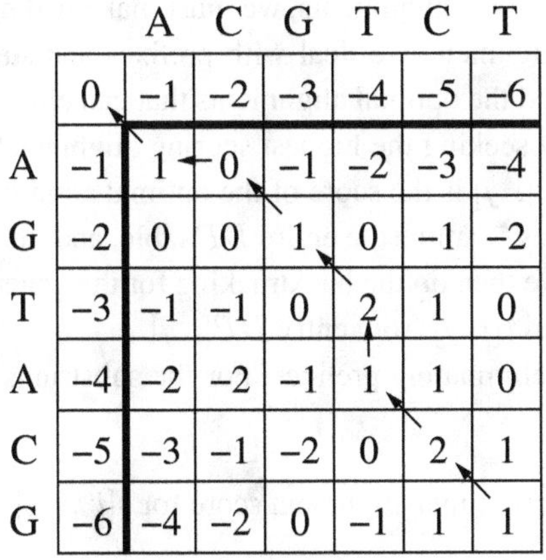

		A	C	G	T	C	T
	0	−1	−2	−3	−4	−5	−6
A	−1	1 ← 0	−1	−2	−3	−4
G	−2	0	0	1	0	−1	−2
T	−3	−1	−1	0	2	1	0
A	−4	−2	−2	−1	1	1	0
C	−5	−3	−1	−2	0	2	1
G	−6	−4	−2	0	−1	1	1

Fig. 4. Backtracking.

DP matrix, this allows us to divide the problem in half and recurse on smaller problems, which preserves both the $O(m \times n)$ runtime and the $O(n)$ space.

Since these heuristics work fine for columns as well, they can actually be used to produce optimal alignments in $O(\min(m, n))$ space.

3.2. *Local Alignment*

The dynamic programming algorithm given in Section 3.1 gives optimal alignments of complete sequences. However, we often want to find closely homologous parts of the two sequences. For example, a protein usually consists of a few domains. Two proteins may have one domain in common, but be otherwise unrelated. In this case, we may want to find the similar domains. Identifying these regions of similarity, such that the alignment score of the two local regions are maximized, is called "local alignment". In contrast, the alignment we described in Section 3.1 is sometimes called "global alignment".

One way to consider local alignment is to think that eliminating prefixes and suffixes of the two sequences is free. Therefore, if the global alignment has negative scores at its either end, we can eliminate the end to get a better local alignment. Consequently, if the same scoring matrix is used, the local alignment score of two sequences is always higher than or equal to the global alignment score. However, the optimal local alignment may not be a part of the global alignment. For example, the optimal local alignment of the two sequences in Figure 2 is the substring ACG of the first sequence aligned with the substring ACG of the second sequence, which is not a part of the alignment shown in Figure 2.

To compute optimal local alignments, we must make two modifications to the algorithm for global alignments, to deal with prefixes and suffixes respectively. First, we note that to find the optimal alignments that can eliminate suffixes of the strings, we are actually seeking the highest scoring alignment of two prefixes of s and t. Recall that $DP[i,j]$ is the score of the optimal alignment between $s[1..i]$ and $t[1..j]$. We must thus examine the entire DP table, and find the (i,j) pair that maximizes $DP[i,j]$. We then do the backtracking for the generation of the actual alignments of $s[1..i]$ and $t[1..j]$ from entry $DP[i,j]$.

Next, we consider eliminating prefixes. For the substrings $s[1..i]$ and $t[1..j]$, let

$$DP'[i,j] = \max_{\substack{i' \leq i+1, \\ j' \leq j+1}} \{\text{optimal alignment score for } s[i' \ldots i] \text{ and } t[j' \ldots j]\}$$

That is, $DP'[i,j]$ is the maximum possible alignment score of two suffixes of $s[1..i]$ and $t[1..j]$. It is possible that both suffixes are empty; the value then is just zero.

Similar to the computation of $DP[i,j]$ in a global alignment, we examine the last column of the optimal alignment of the two optimal suffixes of $s[1..i]$ and $t[1..j]$. The three cases in Section 3.1 still exist. However, there is one more case where both suffixes are empty strings, and $DP'[i,j] = 0$. This gives a new recursive formula for the score:

$$DP'[i,j] = \max \begin{cases} DP'[i-1,j] + M(s[i],-) \\ DP'[i,j-1] + M(-,t[j]) \\ DP'[i-1,j-1] + M(s[i],t[j]) \\ 0 \end{cases} \tag{2}$$

Combining the two modifications, we may fill the DP' matrix using Formula 2. We require that $DP'[0,i] = DP'[j,0] = 0$ for all values of i and j. Lastly, we search the matrix to find the highest value, and backtrace from there until we reach a position where $DP[i,j] = 0$, when we stop.

This algorithm is typically called the Smith-Waterman algorithm,[780] in honor of the authors of one of the original papers introducing it. Its runtime and space complexity are both $O(m \times n)$, and the space complexity can be reduced to $O(\min(m,n))$ using methods analogous to those given in Section 3.1.1.

Exercise: Given two sequences s and t, give an algorithm that identifies a substring t' of t that maximizes the optimal global alignment score of s and t'. This type of alignment is called "fit alignment".

3.3. *Sequence Alignment with Gap Open Penalty*

We now move to consider optimal global alignments when there are specific score penalties imposed for opening gaps, as in Section 2.3.2. Let M be a score matrix, and $o \leq 0$ be the gap open penalty. We also assume that the score $M(a, -)$ of aligning any letter $a \in \Sigma$ to a "$-$" symbol is a constant negative value e.

The dynamic programming algorithm in Section 3.1 cannot be straightforwardly used here to compute the optimal alignment with gap open penalty. The problem is in the first two cases in Formula 1. If the "$-$" symbol is the first one in a gap, then we are opening a gap and should add the gap open penalty o to the score. Otherwise, we should not. The DP table does not provide enough information to differentiate these cases.

To solve this problem, we use three different DP tables—$DP_1[i, j]$, $DP_2[i, j]$ and $DP_3[i, j]$—to record the optimal alignment score of the alignments whose last column is respectively

(1) $s[i]$ matches "$-$",
(2) "$-$" matches $t[j]$, and
(3) $s[i]$ matches $t[j]$.

Consider these three DP tables, using the computation of $DP_1[i, j]$ as an example. The last column must be $s[i]$ matching "$-$". The preceding column can be any of the three cases:

(1) $s[i - 1]$ matches "$-$", and therefore the gap in the last column is not a gap opening. Thus, $DP_1[i, j] = DP_1[i - 1, j] + e$.
(2) "$-$" matches $t[j]$, and therefore the gap in the last column is a gap opening. Thus, $DP_1[i, j] = DP_2[i - 1, j] + e + o$.
(3) $s[i - 1]$ matches $t[j]$, and therefore the gap in the last column is a gap opening. Thus, $DP_1[i, j] = DP_3[i - 1, j] + e + o$.

Because $DP_1[i, j]$ is the best of these three possibilities, we have:

$$DP_1[i, j] = e + \max \begin{cases} DP_1[i - 1, j] \\ DP_2[i - 1, j] + o \\ DP_3[i - 1, j] + o \end{cases}$$

We make a special note that Case 2 indicates that in the alignment, a column of aligning $s[i]$ to "$-$" is followed immediately by a column of aligning "$-$" to $t[j]$. The two columns can obviously be replaced by a column of aligning $s[i]$ to $t[j]$. For many scoring matrices M, the latter alignment is better. Therefore, usually we can discard Case 2 in the computation.

Similarly, we can fill the other two DP tables with the following two formulas:

$$DP_2[i,j] = e + \max \begin{cases} DP_1[i,j-1] + o \\ DP_2[i,j-1] \\ DP_3[i,j-1] + o \end{cases}$$

$$DP_3[i,j] = M(s[i],t[j]) + \max \begin{cases} DP_1[i-1,j-1] \\ DP_2[i-1,j-1] \\ DP_3[i-1,j-1] \end{cases}$$

After appropriate initialization of the three DP tables, we can compute all three tables row-by-row. Then the maximum value of $DP_1[m,n]$, $DP_2[m,n]$, and $DP_3[m,n]$ is the optimal alignment score with gap open penalty. Which of the three is the maximum determines which of the three cases is the last column of optimal alignment. A backtracking from that table entry to $DP_3[0,0]$ gives the actual optimal alignment.

Similarly to Section 3.2, one can define the local alignment with gap open penalty. The extension from global alignment with gap open penalty to local alignment with gap open penalty is similar to what we have done in Section 3.2.

Exercise: Similar to Section 3.2, extend the algorithm for global alignment with gap open penalty to local alignment with gap open penalty.

4. Probabilistic Approaches to Sequence Alignment

Underlying the simple dynamic programming algorithms of the previous sections is a rich mathematical theory.[413, 414, 579] In particular, the score of an alignment can be seen as a measure of the surprisingness of that alignment, given probabilistic models for related and unrelated sequences. The richest part of this theory— which we do not visit in this tour—estimates the probability that an alignment with a particular score or higher would occur in random sequences, to allow researchers to estimate the statistical significance of an alignment.

We begin by considering how to pick the members of a scoring matrix.

4.1. *Scoring Matrices*

What values should be used in a scoring matrix? Intuitively, the value $M[i,j]$ should be more positive for more closely related symbols. But how do we quantify this?

For DNA sequences, the scoring matrix is often very simple. For example, the default score matrix used in the DNA homology search program BLASTN

	A	C	G	T
A	91	−114	−31	−123
C	−114	100	−125	−31
G	−31	−125	100	−114
T	−123	−31	−114	91

Fig. 5. Blastz score matrix.

is simple. Matches between the same base score 1, mismatches between bases score −3, and matches between a "−" symbol and a base score −1. BLAST also includes a gap open penalty, which by default is −5, for the alignments it generates. Other programs—such as Blastz—use a more carefully investigated scoring scheme, proposed in Chiaromonte *et al.*[156] and shown in Figure 5 with gap open penalty −400 and gap extension −30. This matrix comes from a similar origin to the probabilistic explanation given below for protein alignments.

For protein sequences, a more complicated approach is used, based on alignments of sequences known to be homologous. Two complementary approaches are found in PAM matrices[192] (Point Accepted Mutation, or Percent Accepted Mutation) and BLOSUM matrices[346] (BLOcks SUbstitution Matrices). But both use fundamentally similar ideas.

The central idea is to have a scoring matrix where higher-scoring entries are more likely to be aligned in homologous sequences than in random sequences. To encapsulate this, entries in both PAM and BLOSUM matrices are related to the logarithm of the odds ratio—also known as "log-odds"—that a given pair of symbols are aligned due to homology versus by chance.

For two amino acids i and j, let p_i and p_j be the probabilities of amino acids i and j occurring at a random position of a random protein sequence, respectively. Then $p_i \times p_j$ is the probability that i is aligned to j in a random position in the alignment of two unrelated sequences. Let q_{ij} be the probability that a column of a true alignment of two related sequences aligns i to j. The odds ratio that an i matched to a j comes from a true alignment versus a random alignment is thus $q_{ij}/(p_i \times p_j)$, and its logarithm is $\log(q_{ij}/(p_i \times p_j))$.

4.1.1. *PAM Matrices*

PAM matrices—also called Dayhoff matrices in honor of their inventor—are based on a very simple model of protein evolution, in which amino acid sequences evolve by a series of independent single substitution events. PAM matrices describe the frequency with which amino acids "safely" mutate to other amino acids. Because different biomolecules may mutate at different rates, the PAM unit—instead of the time—is used to measure the amount of variation.

One PAM of mutation describes an amount of evolution which changes, on the average, 1% of the amino acids. The PAM-1 mutation matrix has been calculated by Dayhoff*et al.*,[192] by examining the mutations that separate closely related protein sequences. Here we denote the PAM-1 mutation matrix by Q, where $Q[i,j]$ is the probability that a given amino acid i is mutated to j after one PAM of mutations have occurred. Given the assumption that evolution acts as independent mutation events, it is easy to prove that if n PAMs of mutations have occured, the mutation matrix PAM-n is equal to Q^n, the n-th power of the PAM-1 matrix.

After computing the PAM-n matrix Q_n, we usually use it in practice as a log-odds scoring matrix. To do this, we need the probability p_i with which amino acid i appears in a random position of random protein sequences. In a random alignment, i and j are aligned with probability $p_i \times p_j$. But in a random position of an alignment of two sequences separated by n PAMs of mutation, the probability that i and j are aligned is $p_i \times Q_n[i,j]$. Given these two possible hypotheses for why i is aligned to j, the odds ratio is just

$$M_n[i,j] = \frac{p_i \times Q_n[i,j]}{p_i \times p_j} = \frac{Q_n[i,j]}{p_j}$$

This is then converted to a score matrix by taking the logarithm in base 10 and multiplying by 10:

$$M_n'[i,j] = 10 \times \log_{10} \frac{Q_n[i,j]}{p_j}$$

If i is aligned to j more frequently at a random column of the alignment of two homologous proteins than that of two random proteins, $q_{ij} > p_i \times p_j$ and $M_n'[i,j] > 0$. Otherwise, $M_n'[i,j] \leq 0$.

4.1.2. *BLOSUM Matrices*

An alternative scoring approach for protein sequences is the BLOSUM matrices. Recall that PAM matrices are derived from alignments of closely related sequences. By contrast, the BLOSUM matrices invented by Henikoff and Henikoff[346, 347] are derived from contiguous segments of multiple alignments of sequences that are not known to be closely related or even homologous in an evolutionary sense.

The database used is the BLOCKS database, which identifies protein motifs from multiple alignments of amino acid sequences. The database includes thousands of blocks, where each block contains many sequence segments that are similar with each other. For the computation of BLOSUM n matrix, the segments that are identical at $n\%$ or more of the positions within a block are clustered and weighted as a single sequence. This reduces the multiple contributions to amino acid pairs from the most closely related members of a protein family. From these blocks, one can then identify how often two amino acids i and j are aligned to each other in these blocks, versus how often they appear unrelatedly in the BLOCKS database.

If i is aligned to j a total of $a(i, j)$ times in the blocks, out of a total of P aligned pairs in blocks, then the probability of i being aligned to j in a random aligned pair is clearly $a(i, j)/P$. By contrast if the probability of that i occurs at a random position in the database is p_i and the probability of that j occurs is p_j, then the probability of aligning i to j in unrelated sequences is just $p_i \times p_j$. The odds ratio is therefore

$$B[i, j] = \frac{a(i, j)}{p_i \times p_j \times P}$$

Typically, BLOSUM matrices result from doubling the logarithm in base 2 of this B matrix. As such, they are again log-odds matrices.

Reducing the value of n allows the BLOSUM n matrix to be biased toward finding homologies between sequences that are not very similar, ideally allowing their use in searching for homologies that are not very strong. In practice, BLOSUM matrices like BLOSUM 62 or BLOSUM 50 have largely supplanted the use of PAM matrices.

4.1.3. *Weaknesses of this Approach*

A serious issue with these scoring methods is that they are based on encapsulating statistics from existing alignments. In particular, they can be inappropriate in estimating the quality of an alignment between membrane proteins, given that the alignments that are used to develop both PAM and BLOSUM matrices are of largely globular proteins.

4.2. *Probabilistic Alignment Significance*

In addition to the log-odds interpretation of protein scoring matrices, a similar approach can be used to interpret alignment scores. We make a completely unjustified assumption: that positions of an alignment are all independent of one another. Now, consider an alignment A between two unrelated sequences s and t, in a scoring system with no added penalties for opening gaps. If the probability of a symbol i occurring is p_i in a random aligned sequence, then the probability of alignment A appearing randomly is just $\prod_i p_{s_i} \times p_{t_i}$. If, by contrast, the probability that i is aligned to j in sequence that matches a model of homologous sequence is q_{ij}, then the probability of that alignment occuring at random is $\prod_i q_{s_i t_i}$. The odds ratio of these probabilities is just

$$\prod_i \frac{q_{s_i t_i}}{p_{s_i} \times p_{t_i}}$$

If the alignment is scored with a log-odds scoring matrix M, where $M[i,j]$ is the logarithm of the ratio between the probability that i is aligned to j in the known homologous sequences versus at random, then the score of the alignment is just

$$\sum_i \log \frac{q_{s_i t_i}}{p_{s_i} \times p_{t_i}}$$

which is exactly the logarithm of the odds ratio for the alignment.

How do we use this observation? There are two common approaches, from Bayesian and frequentist statistics. In the Bayesian approach, we assess the probability that the alignment is of homologous sequences, given a prior estimate of that probability, using the odds ratio. In the more complicated frequentist approach,[413,414] the probability that a local alignment with score greater than x would appear in unrelated sequences of length $|s|$ and $|t|$ is estimated, and this is used as a measure of the statistical significance of the alignment. This can be extended to consider the sequence content of s and t as well in the estimation.

5. Second Generation Homology Search: Heuristics

Filling in an entire dynamic programming matrix when aligning two sequences is quite time consuming. As a result, in the late 1980s and early 1990s, obvious heuristic methods were proposed. These methods share a common theme: sacrifice sensitivity for speed. That is, they run much faster than full dynamic programming, but they may miss some alignments. The two most popular heuristics are found in FASTA[512] and BLAST.[23] To focus on central ideas, we concentrate on DNA sequence homology search. Protein sequence homology search involves similar strategies.

5.1. FASTA and BLAST

One of the earliest of these heuristics is FASTA. FASTA uses a hashing approach to find all matching k-tuples (between 4 and 6 for DNA), between the query and database. Then nearby k-tuples, separated by a constant distance in both sequences, are joined into a short local alignment. With these short local alignments as seeds, Smith-Waterman dynamic programming is applied to larger gaps between two high scoring pairs still separated by short distances, with the restriction that only the part of the dynamic programming matrix nearest the diagonal is filled in. FASTA outputs only one alignment per query sequence, after the dynamic programming phase, and estimates the probability of the alignment occurring by chance.

More popular has been the BLAST family of heuristics. BLAST works similarly at its beginning, identifying seed matches of length k (= 9–11 for DNA). Each seed match is extended to both sides until a drop-off score is reached. Along the way, seed matches that are being extended in ways that are not typical of truly homologous sequences are also thrown out. BLAST can be set so that two nonoverlapping seed matches may be required before alignments are extended. Newer versions of BLAST[24] allow gapped alignments to be built. BLAST outputs all alignments found, and estimates for each alignment the expected number of alignments of unrelated sequences whose score would be as large. This quantity is estimated in ways described in Section 4.2.

5.2. Large-Scale Global Alignment

Another topic of great interest in the last years has been the identification of global alignments between long stretches of sequenced genomes. Beginning with the work of Delcher *et al.*,[194] a number of other authors have begun to use somewhat different methods than classical dynamic programming ones for this problem.

This problem is of interest partially because portions of these global alignments in some genomes have undergone enough mutations that their homology is no longer significant, yet they may have regulatory or other significant functions that are still conserved. At the same time, researchers identifying homologous blocks that have not been broken by major genome rearrangement events may want to use the non-exonic regions as a tag to identify which of two duplicated blocks is truly homologous to another sequence. As such, the whole-genome alignment problem has been the subject of considerable research in the last few years.

Two methodologies in particular augment and complement the techniques used in heuristic local alignment. The first is the use of seed alignments; but instead of using them to build local alignments, whole genome alignment packages typically use them as anchors that are required to be used in the chosen alignment. The second is the use of variable-length regions as these anchors—most often accomplished by the use of suffix trees or similar data structures, which are used to identify long completely conserved regions as anchors.

6. Next-Generation Homology Search Software

In the post-genome era, supercomputers and specialized hardware implementing sequence alignment methods in digital logic are employed to meet the ever expanding needs of researchers. Pharmaceutical corporations and large scientific funding agencies proudly spend much money to support such supercomputing centers. Unfortunately, the reliability of these solutions must be considered in light of the consistent doubling of sequence databases, as GenBank doubles in size every 18 months.[606]

In the late 1990s, however, several methods have been developed that improve the sensitivity of homology search software to a level comparable to that of full-scale dynamic programming, while avoiding very large runtime complexities. These have largely focused on characterizing the central seeds from which heuristic alignment programs build their local alignments.

6.1. *Improved Alignment Seeds*

BLAST-like heuristics first find short seed matches which are then extended. This technique faces one key problem: As seeds grow longer, we can expect fewer homologies to have the large conserved regions. However, shorter seeds yield many random hits that significantly slow down the computation.

To resolve this problem, a novel seeding scheme has been introduced in PatternHunter.[533] BLAST looks for matches of k—default $k = 11$ in BLASTN and $k = 28$ in MegaBlast—consecutive letters as seeds. PatternHunter instead uses *non*-consecutive k letters as seeds. The relative positions of the k letters is called a "spaced seed model", and k its "weight". For convenience, we denote a model by a 0-1 string, where ones represent required matches and zeros represent "don't care" positions. For example, if we use the weight 6 model 1110111, then the alignment **ACTGCCT** versus **ACTTCCT** matches the seed, as does **ACTGCCT** versus **ACTGCCT**. In this framework, BLAST can be thought of as using models of the form $111\ldots1$.

Let L be the length of a homologous region with no indels, and M be the length of a seed model. Then there are $L - M + 1$ positions that the region may contain a hit; see Figure 6. In a BLAST type of approach, as long as there is one hit in such a region, the region can be detected. Therefore, although the hit probability at a specific position is usually low, the probability that a long region contains a hit can be reasonably high.

```
        TACTGCCTG
        ||||  ||||
        TACTACCTG
  1:  1110101
  2:    1110101
  3:      1110101
```

Fig. 6. There are many positions that a homology may contain a hit. In this figure, the seed model 1110101 hits the region at the second position.

Ma, Tromp, and Li[533] notice that different seed models with identical weight can result in very different probabilities to hit a random homology. For a seed with weight W, the fewer zeros it has, the shorter the seed is, and the more positions it can hit the region at. Therefore, intuitively, BLAST's seed model with W consecutive ones seems to have the highest hit probability among all the weight-W seed models. Quite surprisingly, this is not true. The reason is that the hits at different positions of a region are not independent. For example, using BLAST's seed, if a hit at position i is known, the chance to have a second hit at position $i + 1$ is then very high because it requires only one extra base match. The high dependency between the hits at different positions make the detection of homologies "less efficient".

The same authors observe that the dependency can be reduced by adding some zeros into the seed model. For example, if seed model 1110101 is used and there is a hit at position i, then the hit at position $i + 1$ requires three extra base matches, compared to one extra base match of the BLAST's seed. Thus, hits at different positions are less dependent when spaced seed models are used. On the other hand, spaced seed models are longer than the consecutive seed model with the same weight, and therefore have fewer positions to hit a region at. As a result, the optimal seed must balance these two factors. In the same paper, the authors have developed a method to find the optimal seed model that maximizes the hit probability in a simple model, and the optimal seeds are then used to develop the PatternHunter program.

Some other seeding or related strategies have also been developed before or after PatternHunter's spaced seed model. In the program WABA,[428] Kent proposes the use of a simple pattern in identifying homologous coding sequences. Since these sequences often vary in the third, "wobble", position of codons, WABA ignores these positions when identifying positions that match a seed. In the framework of PatternHunter, this is equivalent to using spaced seeds of form 110110 . . ..

Kent's approach takes advantage of the special properties of the coding region homologies. Kent has also introduced a different approach for detecting non-coding region homologies in his program BLAT.[427] BLAT uses consecutive seeds, but allows one or two mismatches to occur in any positions of the seed. For example, a BLAT hit might require at least ten matches in twelve consecutive positions. This scheme naturally allows more false negatives, but the resultant collection of hits is more enriched for true positives at a given level of false positives than for consecutive seeds where all positions are required to match.

In a random hashing strategy, Buhler[117] uses his experience with identifying sequence motifs using random projection[119] to speed up detection of homology search. This idea is previously used by Indyk and Motwani.[382] Basically, this approach is to find all hits by random hashing over long sequence intervals. A simple probability calculation allows the computation of how many projections are required to ensure a given probability that a homologous alignment has a hit to at least one of these random projections. For good choices of projection weights, this approaches 100% sensitivity. Other than the fact that high-weight random projections are not suitable and not designed for BLAST-type searches, this approach also ignores the possibility of optimizing the choice of those projections.

Of these first three approaches—specific spaced seeds, consecutive seeds allowing a fixed number of mismatches, and random spaced seeds—the first, typified by PatternHunter, allows for optimization. That is, one can use a seed specifically tuned for the types of alignments one expects to see, with the highest sensitivity at a particular false positive rate.

The optimal seed models in the PatternHunter paper[533] are optimized for non-coding regions. Later, Brejová, Brown, and Vinař[102] develop an algorithm for optimizing the seeds in more complicated models, specifically for coding regions. In a later paper,[103] they also propose a unified framework to represent all of the above mentioned seeding methods. In their framework, the seeds are represented by a pair (v, T): a vector v that represents the seed sequence—*e.g.*, the zeros and ones from PatternHunter seeds—and a threshold T that in its simplest case identifies how many "one" positions from the seed vector v must be matching in the two sequences to yield a hit.

6.2. *Optimized Spaced Seeds and Why They Are Better*

Optimized spaced seeds can have substantially greater sensitivity than the consecutive seed models of BLAST. Here we give one example. The simplest model of an alignment is of a region of a fixed length where each position matches with some probability p, independent of all other positions. Figure 7 compares the optimal spaced seed model of weight 11 and length at most 18—*viz.*

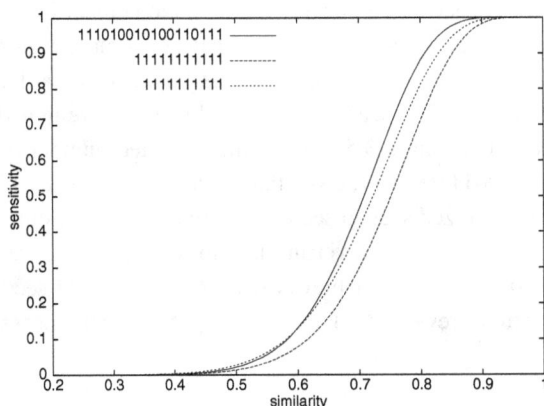

Fig. 7.   1-hit performance of weight 11 spaced model versus weight 11 and 10 consecutive models, coordinates in logarithmic scale.

111010010100110111—with BLAST's consecutive models of weight 11 and 10, for alignments of this type, of fixed length 64. For each similarity rate p shown on the x-axis, the fraction of regions with at least 1 hit is plotted on the y-axis as the sensitivity for that similarity rate. From the figure, one observes that the seemingly trivial change in the seed model significantly increases sensitivity. At 70% homology level, the spaced seed has over 50% higher probability—at 0.47—to have a hit in the region than BLAST weight 11 seed—at probability 0.3.

However, the added sensitivity does not come at the cost of more false positive hits or more hits inside true alignments:

Lemma 3: *The expected number of hits of a weight-W length-M model within a length L region of similarity $0 \leq p \leq 1$ is $(L - M + 1) \times p^W$.*

Proof: The expected number of hits is the sum, over the $(L - M + 1)$ possible positions of fitting the model within the region, of the probability of W specific matches, the latter being p^W. □

Lemma 3 reveals that spaced seeds have fewer expected hits, but have higher probability to hit a homologous region, as shown in Figure 7. This is a bit counter intuitive. The reason is that a consecutive seed often generates multiple hits in a region, because a hit at position i increases the hit probability at position $i + 1$ to p as only one extra base match is required. However, optimized spaced seeds are less likely to have multiple hits in a region because the second hit requires more base matches. Therefore, given many homologous regions, although the total number

of hits generated by a spaced seed is comparable to the number for a consecutive
seed with the same weight, the spaced seed hits can cover more regions.

To quantify this, let $p = 0.7$ and $L = 64$, as for the original PatternHunter
model. Given that the BLASTN seed 11111111111 matches a region, the expected
number of hits in that region is 3.56, while the expected number of hits to the
spaced seed 101101100111001011, given that there is at least one, is just 2.05.

Thus, using an optimized spaced seed, a homology search program increases
sensitivity but not running time. Inverting the above reasoning, we can use an
optimal weight-12 spaced seed to achieve the BLAST weight 11 seed sensitivity,
but generating four times fewer hits. This speeds up the search process by roughly
a factor of four.

6.3. *Computing Optimal Spaced Seeds*

The probability of a seed generating a hit in a fixed length region of a given level
similarity can be computed by dynamic programming[102, 103, 118, 158, 425, 500, 533]
under various assumptions. To choose an optimal seed, we compute the hit probability for all seeds, and pick the one with the highest probability.

Suppose we are given a seed s, of length M and weight W, and a homology
region R, of length L and homology level p, with all positions independent of
each other. In this model, we can compute the probability of s having a hit in R.
We represent R by a random string of zeros and ones, where each position has
probability p of being a one.

We say that seed s has a seed match to R at location i if the L-length substring
of R starting at position i has a one in each position with a one in the seed s. Let A_i
be the event that seed s has a seed match at location i in R, for all $0 \leq i \leq L - M$.
Our goal is to find the probability that s hits $R : \Pr[\cup_{i=0}^{L-M} A_i]$.

For any $M \leq i \leq L$ and any binary string b such that $|b| = M$, we use $f(i, b)$
to denote the probability that s hits the length i prefix of R that ends with b:

$$f(i, b) = \Pr[\cup_{j=0}^{i-M} A_j \mid R[i - l, \ldots, i - 1] = b].$$

In this framework, if s matches b,

$$f(i, b) = 1.$$

Otherwise, we have the recursive relationship:

$$f(i, b) = (1 - p) \times f(i - 1, 0b') + p \times f(i - 1, 1b'),$$

where b' is b deleting the last bit.

Once we have used this dynamic programming algorithm to compute $f(L - M, b)$ for all strings b, we can compute the probability of s hitting the region. It is:

$$\sum_{|b|=M} \Pr[R[i-1, \ldots, i-1] = b] \times f(L - M, b)$$

This is the simplest algorithm, used in the original paper[533] but not published anywhere, to compute the hit probability of a seed. Other algorithms generalize this algorithm to more sophisticated probabilistic assumptions of the region R and improving time complexity.[102, 118, 158, 425, 500]

6.4. *Computing More Realistic Spaced Seeds*

The simple model of homologous regions described above is not very representative of real alignments. Homologous regions are not all of length 64, and vary internally in how conserved they are. Of course, they also include gaps, but we are not going to consider this in producing seeds.

For example, more than 50% of significant alignments in the human and mouse genomes are between exonic regions, and these regions have much more conservation in the first two positions of a codon than in the third, which has traditionally been called the "wobble" position.[428] A seed that takes advantage of this periodicity by ignoring the third position in codons is much more likely to hit than a seed that does not. There is also substantial dependence within a codon: If the second position is not matched, it is quite likely that neither are the first or third.

Similarly, real alignments vary substantially internally in their alignment as well. This is particularly true for coding alignments. Such alignments tend to have core regions with high fidelity, surrounded by less well-conserved regions.

Models that do not account for these variabilities can substantially underestimate the hit probability of a seed.

6.4.1. *Optimal Seeds for Coding Regions*

Two recent papers try to address this need to optimize better models in different ways. Brejová *et al.*[102] use a Hidden Markov Model (HMM) to represent the conservation pattern in a sequence. Their model accounts for both internal dependencies within codons, and also for multiple levels of conservation throughout a protein. They update the dynamic programming model above to this new framework, and show that one can still relatively efficiently compute the probability that a given seed matches a homologous region. Meanwhile, Buhler *et al.*[118] represent the sequences by Markov chains, and present a different algorithm, based on finite automata to compute the probability of a seed hit in that model.

The difference in using seeds tuned to find hits in homologous coding regions versus seeds for general models is quite large. In particular, the optimal seed—111001001001010111—of weight 10 and length at most 18 for noncoding regions is ranked 10,350 among the 24,310 possible seeds in its theoretical sensitivity in the HMM trained by Brejová *et al.*, and ranked 11,258 among these seeds in actual sensitivity on a test set of coding region alignments, matching just 58.5% of them. By contrast, the three optimal coding region seeds—which are also optimal for the theoretical hidden Markov model—match between 84.3% and 85.5% of alignments. These seeds—11011011000011011, 11011000011011011, and 11000011011011011—all ignore the third positions of codons, and also skip entire codons. As such, they model the conservation pattern of real coding sequences much better than the non-periodic seeds optimized for noncoding regions.

Much of the advantage does of course come from the modeling. A previous program, WABA[428] uses three-periodic seeds of the form 110110110... with no formal justification. In practice, this seed has sensitivity close to the optimum, at 81.4%. Still, the optimal seeds give a good improvement over this seed, and also allow the optimization of multiple seeds for still greater sensitivity.

For the homology search in coding regions, an alternative approach is the use of a "translated homology search" program—*e.g.*, tblastx. Such a program first translates DNA sequences to protein sequences, from which the homologies are then found. The translated homology search is supposed to be more sensitive than a DNA-based program for coding regions, however, is substantially slower.

Kisman, Ma, and Li[434] have recently extended the spaced seed idea to translated homology search and developed tPatternHunter, which is both faster and more sensitive than tblastx.

6.4.2. *Optimal Seeds for Variable-Length Regions*

As to the length of regions, it is quite simple for all of these algorithms to incorporate distributions on the length of homologous regions into the model. For a given generative model, we simply compute the probability $\sigma(\ell)$ of a hit in a region of length ℓ, and the distribution $\pi(\ell)$ of lengths of the region; the probability of a hit in a random region is then just $\sum_{\ell} \sigma(\ell) \times \pi(\ell)$.

6.5. *Approaching Smith-Waterman Sensitivity Using Multiple Seed Models*

Another idea that comes directly from the idea of optimal spaced seeds is the one of using multiple seed models, which together optimize the sensitivity. In such an approach, a set of several seed models are selected first. Then all the hits generated by all the seed models are examined to produce local alignments. This obviously increases the sensitivity because more hits than using one seed model are examined. But now, the several seed models need to be optimized together to maximize the sensitivity.

This idea appears in the PatternHunter paper [533] and is further explored in several recent papers.[103, 118, 500] Brejová *et al.*[103] use a heuristic method to design a good pair of seeds. Buhler *et al.*[118] use hill-climbing to locally improve good sets of seeds and a pair of seeds have been designed by their method. Li *et al.*[500] extend the dynamic programming algorithm in Section 6.3 to compute a suboptimal set of seeds greedily.

Li *et al.*[500] show that, in practice, doubling the number of seeds can achieve better sensitivity than reducing the weight of the seeds by one. However, for DNA homology search, the former only approximately doubles the number of hits, whereas the latter increases the number of hits by a factor of four (the size of DNA alphabet). Thus, multiple seeds are a better choice.

It is noteworthy that the multiple-seed approach is only possible when spaced seeds are used—there is only one BLAST-type of consecutive seed with a given weight. The newest version of PatternHunter implements the multiple-seed scheme,[500] having greedily chosen a set of sixteen seeds of weight 11 and length at most 21 in 12 CPU days on a Pentium IV 3GHz PC. When the random region has length 64 and similarity 70%, the first four seeds are: 111010010100110111, 111100110010100001011, 110100001100010101111, 1110111010001111. Multiple seeds for coding regions are also computed and implemented. The experimental results shown later in Section 7 demonstrate that using carefully selected multiple seeds can approach Smith-Waterman sensitivity at BLAST's speed.

6.6. Complexity of Computing Spaced Seeds

Many authors[102, 103, 118, 158, 425, 533] have proposed heuristic or exponential time algorithms for the general seed selection problem: Find one or several optimal spaced seeds so that a maximum number of target regions are each hit by at least one seed. A seemingly simpler problem is to compute the hit probability of k given seeds. Unfortunately, these are all NP-hard problems.[500] Thus the greedy algorithm and the exponential time dynamic programming are the best we can do. Although the proofs are beyond the scope of this tutorial, we to list some of the recent results for these problems. Let $f(n)$ be the maximum number of 0's in each seed, where n is the seed length, the following are true.[500]

(1) If $f(n) = O(\log n)$, then there is a dynamic programming algorithm that computes the hit probability of k seeds in polynomial time; otherwise the

problem is NP-hard.

(2) If $f(n) = O(1)$, one or a constant number of optimal seeds can be computed in polynomial time by enumerating all seed combinations and computing their probabilities; otherwise, even selecting one optimal seed is NP-hard.

(3) If $f(n) = O(1)$, then the greedy algorithm of picking k seeds by enumeration, then adding the seed that most improves the first seed, and so on, approximates the optimal solution within ratio $1 - \frac{1}{e}$ in polynomial time, due to the bound for the greedy algorithm for the maximum coverage problem;[359] otherwise the problem cannot be approximated within ratio $1 - \frac{1}{e} + \epsilon$ for any $\epsilon > 0$, unless $NP = P$.

7. Experiments

Here, we present some experimental comparisons between heuristic search techniques and full Smith-Waterman dynamic programming. As expected, Smith-Waterman dynamic programming is too slow for practical use when the database is large. What is striking is that a good choice of multiple optimal spaced seeds can allow near-total success in detecting alignments, with vastly better runtime.

The results that we show in this section are originally reported by Li *et al.*[500] In the paper, several software packages—SSearch,[654] BLAST, and PatternHunter—are used to find homologies between 29715 mouse EST sequences and 4407 human EST sequences. Those sequences are the new or newly revised mouse and human EST sequences in NCBI's GenBank database within a month before 14 April 2003. After downloading the EST sequences, a simple "repeat masking" is conducted to replace all the sequences of ten or more identical letters to "N"s. This is because they are low complexity regions and their existence generates so many trivial sequence matches that overwhelm the real homologies.

SSearch is a subprogram in the FASTA package and implements the Smith-Waterman algorithm that gurantees to find the optimal alignment of every pair of sequences. Therefore, SSearch's sensitivity is regarded to be 100% in the comparison, and both BLAST and PatternHunter can only find a subset of the homologies found by SSearch. The performance of BLAST version 2.2.6, and PatternHunter version 2.0 are compared against SSearch. Each program uses a score scheme equivalent to:

$$
\begin{array}{rcl}
\text{match} & : & 1 \\
\text{mismatch} & : & -1 \\
\text{gap open penalty} & : & -5 \\
\text{gap extension penalty} & : & -1
\end{array}
$$

All pairs of ESTs with a local alignment of score at least 16 found by SSearch are

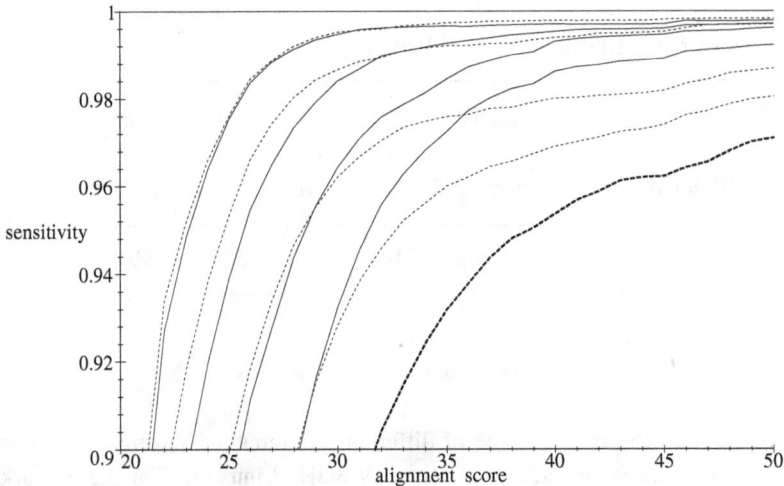

Fig. 8.   The thick dashed curve is the sensitivity of Blastn, seed weight 11. From low to high, the solid curves are the sensitivity of PatternHunter using 1, 2, 4, and 8 weight 11 coding region seeds, respectively. From low to high, the dashed curves are the sensitivity of PatternHunter using 1, 2, 4, and 8 weight 11 general purpose seeds, respectively.

recorded. Also, if a pair of ESTs has more than two local alignments, only the one with the highest score is considered. All of these alignments are kept as being the correct set of homologies, noting of course that some of these alignments may be between sequences that are not evolutionarily related.

As expected, SSearch takes approximately 20 CPU days, while BLASTN takes 575 CPU seconds, both on a 3GHz Pentium IV. SSearch finds 3,346,700 pairs of EST sequences that have local alignment score at least 16, with maximum local alignment score 694.

It is difficult to compare SSearch's sensitivity with BLASTN and Pattern-Hunter. This is because BLASTN and PatternHunter are heuristic algorithms, and need not compute optimal alignments. Thus Li *et al.* have decided—a bit arbitrarily—that if SSearch finds a local alignment with score x for a pair of ESTs, and BLAST (or PatternHunter) finds an alignment with score $\geq x/2$ for the same pair of ESTs, then BLAST (or PatternHunter) "detects" the homology. The successful detection rate is then regarded the sensitivity of BLAST (or PatternHunter) at score x. PatternHunter is run several times with different number of spaced seeds. Two sets of seeds—coding-region seeds and general-purpose seeds—are used, respectively. The results are in Figure 8.

SSearch	Blastn	PatternHunter				
		seeds	1	2	4	8
20 days	575 s	general	242 s	381 s	647 s	1027 s
		coding	214s	357s	575s	996s

Fig. 9.   Running time of different programs.

Figure 9 lists the running time of different programs, with weight 11 seeds for Blastn and PatternHunter, on a Pentium IV 3GHz Linux PC. This benchmark demonstrates that PatternHunter achieves much higher sensitivity than Blastn at faster speed. Furthermore, PatternHunter with 4 coding region seeds runs at the same speed as Blastn and 2880 times faster than the Smith-Waterman SSearch, but with a sensitivity approaching the latter. It also demonstrates that the coding region seeds not only run faster—because there are less irrelevant hits—but are also more sensitive than the general purpose seeds. This is not a surprise because the EST sequences are coding regions.

For now, our brief tour through the world of homology search methods is coming to an end. Early landmarks include the classic global and local alignment algorithms that use full dynamic programming. Later highlights have been heuristic search algorithms, including increasingly sophisticated ones based on optimizing seeds for a particular type of alignment. At the end of it all, we have an algorithm that is almost as sensitive as Smith-Waterman, but requiring 3 orders of magnitude less time. We cannot believe, however, that the tour is entirely over, and encourage our readers to enjoy some time sightseeing on their own.

Acknowledgments

This work was supported in part by the Natural Science and Engineering Research Council of Canada (NSERC), the Ontario PREA award, Canada Council Research Chair program, the Killam Fellowship, and the Human Frontier Science Program. We would like to thank those who provided preprints or unpublished material that was used in this article.

CHAPTER 11

ANALYSIS OF PHYLOGENY: A CASE STUDY ON SAURURACEAE

Shao-Wu Meng

Institute for Infocomm Research
swmeng@i2r.a-star.edu.sg

Phylogenetics is the study of the origin, development, and death of a taxon. It is a useful tool, for example, in the conservation of species. This chapter is an introduction to phylogenetics from the perspective of plant molecular biologists, using a case study on Saururaceae. Based on analysis using integrated data from DNA sequences and morphology, we also draw a surprising conclusion that *Saururus* is not the most primitive genus in Saururaceae, contradicting a long-held tradition of the field. We point out some deficiencies in the older studies.

ORGANIZATION

Section 1. We begin with a brief explanation of what is phylogeny, why do people study phylogeny, and how do people study phylogeny.

Section 2. To illustrate the description of phylogenetics in Section 1, we do a case study on the phylogeny of the relic paleoherb Saururaceae. A background of Saururaceae and the case study is given.

Section 3. A summary of the materials and methods used in the case study is presented. The methods include the steps of collecting plant materials, extracting and sequencing DNA from these materials, aligning these DNA sequences, and performing parsimony analysis of these sequences and morphological data.

Section 4. Then we present the results of each step in detail, focusing especially on the phylogentic trees constructed by parsimony analysis on data from 18S nuclear genes, from *trn*L-F chloroplast DNA sequences, from *mat*R mitochondrial genes, from a combined matrix of these genes, and from morphological data.

Section 5. After that, we dive into an extended discussion on the phylogeny of Saururaceae. On the basis of our analysis, we conclude that Saururacease, *Saururus*, and *Gymnotheca* are monophyly, that *Anemopsis* and *Houttuynia* are sister group, that *Saururus* and *Gymnotheca* are also sister group, and that the *Anemopsis-Houttuynia* form the first clade of Saururaceae. This is in disagreement with several earlier studies that *Anemopsis* and *Houttuynia* are derived from *Saururus*,[631] that *Saururus* is the first to diverged from the ancestral Saururaceous stock,[846] *etc.* We point out some of the deficiencies of these earlier studies.

Section 6. Finally, we provide some advice on reconstructing phylogeny, spanning the aspects of sampling of species, selection of out-group, alignment of sequences, choosing phylogeny reconstruction methods and parameters, dealing with morphological data, and comparing phylogenies from molecular data and morphological data.

1. The What, Why, and How of Phylogeny

1.1. *What is Phylogeny?*

Phylogeny is the evolutionary process of an organism from its initial occurrence to a concrete geological time. It includes the evolutionary process, and also the organism itself and its descendants. Phylogenetics is a research field to study the phylogeny of organisms. It is impossible for a researcher to study the phylogeny of all extinct and existing organisms. Phylogenetists generally study only the origin, development, and death of one taxon.

1.2. *Why Study Phylogeny?*

The aim of studying phylogeny is to reconstruct the evolutionary process and the phylogenetic relationship among the descendants of an organism. It is useful as described in the following two points.

First, the study on phylogeny can satisfy the curiosity of the scientist and the public. *E.g.*, which species is the closest relative of Man, and when did Man separate from it? After careful phylogenetic analysis of extensive data, including DNA-DNA hybridization[129] and mitochondrial DNA sequence,[364] the closest extant relatives of Man were identified as two Chimpanzee species, followed by gorillas, orang utans, and the nine gibbon species.[501] According to microcomplement fixation data, Sarich and Wilson[744] estimated that the divergence time between Human and Chimpanzee was 5 million years ago, rather than 15 million years ago, which was commonly accepted by paleontologists at that time. Molecular phylogenetics has also been greatly pushed forward by answering these questions.

Second, the study on phylogeny can guide us today. *E.g.*, the dusky seaside sparrow, scientifically named *Ammodramus maritimus nigrecens*, had habited the salt marshes in Brevard County, Florida. By 1980, there were only six individuals, and all of them were male. In order to conserve the subspecies, an artificial breeding program was launched. First of all, the key of the program was to find out the phylogenetically closest subspecies to *A. M. nigrecens*. The subsequent steps was to mate females from their closest subspecies with males from *A. M. nigrecens*, then to mate the female hybrids of the first generation with the males of *A.*

M. nigrecens, and then to mate the female hybrids of the second generation with the males of *A. M. nigrecens*, and so on, as long as the males of *A. M. nigrecens* lived.[501] All effort would be useless if the closest subspecies was chosen wrongly. Hence, phylogenetic analysis played an important role in such a program.

In short, phylogeny study is a useful tool, not only in theoretical study, but also in practical use. It is very helpful for a study that needs to know evolutionary history or the phylogenetic relationship among organisms.

1.3. *How to Study Phylogeny*

According to the definition of phylogeny, most contents of phylogeny are invisible because they have been historical. Fortunately, a few extinct organisms have been fossilized and kept in different stratums due to physical or chemical reaction of geology. Moreover, a part of these fossils have been dug out from different stratums by paleontologists. The visible contents of phylogeny—the basis upon which to reconstruct the evolutionary process—are the existing organisms now and those fossils that have been dug out.

Paleontologists reconstruct phylogeny according to fossils, DNA sequences in fossils, and geological accidents.[486] The reconstruction process and its conclusions can be reliable if paleontologists have enough fossils. It is unfortunate that fossils are often rare and not full-scale when compared with the extinct organisms in geological time. Anyway, fossils are very important to reconstruct a time frame of evolution.

Other phylogenetists reconstruct phylogeny mainly according to the characters of extant organisms. Since characters of extant organisms are the results of evolution, they should reflect evolutionary history. Morphological phylogenetists reconstruct phylogeny mainly according to morphological characters from gross morphology, anatomy, embryology, cytology, physiology, chemistry, *etc*.[545, 846] Molecular phylogenetists reconstruct phylogeny mainly according to isozyme, DNA sequences, protein sequences, *etc*.[239, 688, 786]

Recently, phylogenetic reconstruction by comparing sequences of whole genomes gradually becomes fashionable. The advantage of comparing whole genomes is that it can avoid unilateral results from only one or a few types of DNA sequences.[942] Even though this method is important in molecular phylogenetics, comparing whole genomes should not be overwhelming. For example, the sequence difference in genomes between Human and Chimpanzee is less 0.2%. However, the phenotype difference between them is great. What is the reason? It is possible that the DNA sequence difference of less 0.2% is fatal, or there are other mechanisms, such as differences in secondary structure of RNAs and proteins, to decide the phenotype difference between Human and Chimpanzee. Moreover, phenotype is affected not only by genotype, but also by environment. Since a conclusion drawn only from DNA sequence comparison ignores the environmental differences, it could be biased.

Each of the ways above has its own advantages and disadvantages. Hence, they can complement each other. So a better way is to study phylogeny based on integrated data, that come from different subjects and from different levels. In subsequent sections, we present a case study of the phylogeny of Saururaceae on the basis of integrated data from three types of DNA sequences from three genomes and 58 morphological characters.

2. Case Study on Phylogeny of Saururaceae

Saururaceae is a core member of the paleoherbs.[845] It is an ancient and relic family with six species in four genera, viz. *Saururus*, *Gymnotheca*, *Anemopsis*, and *Houttuynia*.[505] They are perennial herbs with simple flowers that bear bracts without perianths. Saururaceae is an East Asian-North American disjunctive family, with *Anemopsis* and *Saururus cernuus* in North America, and *Houttuynia*, *Gymnotheca*, and *Saururus chinensis* in East Asia. Due to its important systematic position and interesting geographical pattern of distribution, Saururaceae has been a hot spot for phylogenetists even though it is a small family having just a few species.

The viewpoints on the phylogeny of Saururaceae are very different based on morphology, including gross morphology, cytology, floral morphogensis, *etc.* Wu and Wang[908] included *Saururus*, *Circaeocarpus*, *Anemopsis*, *Houttuynia*, and *Gymnotheca* in Saururaceae. They thought that *Circaeocarpus* was derived from *Saururus* firstly, *Anemopsis* secondly, *Gymnotheca* thirdly, and *Houttuynia* fourthly. Later, they[909] detected that the newly published genus, *Circaeocarpus*, was in fact a member of Piperaceae, and *Circaeocarpus sauruoides* C. Y. Wu and *Zippelia begoniaefolia* Blume were conspecific. From the point of view of plant biogeography, Wu[906] later thought *Anemopsis* and *Houttuynia* were vicariant genera, and *S. chinensis* and *S. cernuus* were vicariant species. Based on the basic chromosome numbers of some genera in Saururaceae, Okada[631] put forward that *Anemopsis* and *Houttuynia* were respectively derived from *Saururus*, and they were at the same advanced level. Lei *et al.*[484] supported Okada's opinion, and thought that *Gymnotheca* was the most advanced genus. On the basis of a cladistic analysis of morphological and ontogenetic characters, Tucker *et al.*[846] made an estimate that *Saururus* was the first to diverge from the ancestral Saururaceous stock. They also suggested this was followed by *Gymnotheca*, with *Hout-*

tuynia and *Anemopsis* being sister taxa. Combining the data from gross morphology, anatomy, embryology, palynology, cytology, and flower development, Liang[505] proposed that the ancestor of Saururaceae was divided into two branches at early times. One was the *Gymnotheca-Anemopsis*. The other was the *Saururus-Houttuynia*. Some genera of Saururaceae have been represented in recent studies on molecular phylogeny of higher-level within angiosperm.[139, 688, 786, 787] However, they have not studied the phylogeny of Saururaceae.

In short, although several scientists have done a lot of research on Saururaceae, there is still no uniform opinion on the phylogeny of Saururaceae. Hence, we try to construct a more reliable phylogeny of Saururaceae. Our study is based on three types of DNA sequences from all three genomes and 58 stable morphological characters. The three types of DNA sequences are: 18S functional gene from nuclear genome, *trn*L-F DNA sequence from chloroplast genome, and *mat*R functional gene from mitochondrial genome.

18S and *mat*R are generally used for reconstructing higher-level phylogeny, such as relationships of orders, families, or distant genera.[688, 787] *Trn*L-F are commonly used for genera, species, and lower levels. For studying the phylogenetic relationships within Saururaceae, an ancient and relic family, we select the three types of DNA sequences. Meanwhile, we select 58 morphological characters to rebuild the phylogeny of Saururaceae, and to compare with the phylogenies of previous studies on Saururaceae. These morphological characters are stable and come from gross morphology, anatomy, embryology, palynology, cytology, and flower development.

3. Materials and Methods

3.1. *Plant Materials*

We collect from natural populations or cultivated plants all six species of the in-group, *Anemopsis californica*, *Gymnotheca chinensis*, *Gymnotheca involucrata*, *Houttuynia cordata*, *S. cernuus*, and *S. chinensis*; and three designated out-groups, *Peperomia tetraphylla*, *Piper mullesua*, and *Z. begoniaefolia* (all Piperaceae). Then, we deposit vouchers in the herbarium of the Kunming Institute of Botany, Chinese Academy of Sciences, Kunming, Yunnan Province, People's Republic of China.

3.2. *DNA Extraction, PCR, and Sequencing*

A detailed description of the DNA extraction, PCR, and sequencing steps are given in Meng *et al.*[560]

3.3. Alignment of Sequences

We check each DNA sequence with its electrophoretic map using the SeqEd program (Applied Biosystems), and decide the ends of 18S gene by comparing with AF206929 (from GenBank, the same thereafter), the ends of *trn*L-F by comparing with AF200937, and the ends of *mat*R gene by comparing with AF197747, AF197748 and AF197749. Using the Clustal-X program[825] and MEGA2b3,[464] we align all our sequences.

3.4. Parsimony Analysis of Separate DNA Sequences

Parsimony is one of several criteria that may be optimised in building phylogenetic tree. The key idea of parsimony analysis is that some trees fit the character-state data better than other trees. Fit is measured by the number of evolutionary character-state changes implied by the tree. The fewer changes the better.

We analyze the aligned sequences using maximum parsimony in PAUP.[810] More concretely, we use branch-and-bound search with random addition sequence and ACCTRAN character state optimization, and treat gaps as missing data. The number of replicates in bootstrap analysis is 1000.

3.5. Parsimony Analysis of Combined DNA Sequences

The mutation in 18S gene is so slow that the 18S sequences between *Peperomia tetraphylla* and *Peperomia serpens* are almost identical. Hence, we replace *Pe. serpens* by *Pe. tetraphylla* in the alignment of 18S gene. Ditto for *mat*R genes. Therefore, we combine all sequence data into one alignment. Using maximum parsimony analysis and branch-and-bound search, we do a partition-homogeneity test for different parts of the combined data. The partition-homogeneity test checks the homogeneity among different parts of a matrix; it is useful for analyzing a data matrix that is combined from different data. When executing the test, the number of replicates is set to 1000. Then, we analyze the combined data by using the same settings as when analyzing separate DNA alignments. Alignments of each gene or combined DNA sequences are available upon request from us.

3.6. Parsimony Analysis of Morphological Data

We selected 58 morphological characters to reconstruct the phylogeny of Saururaceae. These characters are given in Figures 7–9. The characters are from studies of herbarium specimens and literature.[484, 502−507, 561, 839−844, 846] These characters pertain to gross morphology, anatomy, embryology, palynology, cytology, and flower development. Concretely, 2 characters are from cytology, 11 characters are from vegetative organs, and 45 characters are from reproductive organs. Moreover, we treat 34 of these characters as binary characters and 24 of these characters as multi-state characters, as detailed in Figure 10. We designate *Z. begoniaefolia*, *Piper*, and *Peperomia* as out-group. As before, we used PAUP[810] to analyze the morphological matrix of Saururaceae. All characters are un-weighted and unordered. Other settings are the same as when analyzing the DNA sequences.

3.7. *Analysis of Each Morphological Characters*

Using WINCLADA,[619] we analyzed each morphological character in order to know which one is homologous and which one is homoplasious. We used maximum parsimony analysis and the following setting: heuristics search, 1000 replications, 1 starting tree per replication, multiple TBR and TBR search strategy, 0 random seed, and slow optimization.

Here, TBR is an acronym for tree-bisection-reconnection. It is a heuristic algorithm for searching through treespace. It proceeds by breaking a phylogenetic tree into two parts and then reconnecting the two subtrees at all possible branches. If a better tree is found, it is retained and another round of TBR is initiated. This is quite a rigorous method of searching treespace.

4. Results

4.1. *Phylogeny of Saururaceae from 18S Nuclear Genes*

Alignment of 18S gene sequences produces a matrix of 1567 positions. 64 of these positions are variable-uninformative; that is, each of these 64 columns of the alignment has two or more types of bases, but at most one of these types of bases has two or more individuals. 35 of these positions are parsimony-informative; that is, each of these 35 columns of the alignment has two or more types of bases with at least two individuals each. The remaining 1468 positions are constant and uninformative. The percentage of parsimony-informative sites, calculated as the ratio of the number of parsimony-informative positions to the total number of positions, is 2.23% ($= 35/1567$).

Our maximum parsimony analysis produces two most parsimonious trees of 123 steps. The strict consensus of the two trees is depicted in Figure 1. Saururaceae (98%) (bootstrap value, the same thereafter), *Gymnotheca* (99%) and *Saururus* (70%) are monophyly; that is, each of these groups have an immediate common ancestor. *A. californica* is the sister group of *H. cordata* (75%), and they formed the basal of Saururaceae. *Saururus* is the sister group of *Gymnotheca* (99%).

Bootstrap is a method to estimate the confidence levels of inferred relationships. The process of bootstrapping is to iteratively create a new data matrix, that
has the same size as the original data matrix, by randomly resampling individual columns (*i.e.*, characters) of the original data matrix. In this process, some
characters of the original data matrix may be sampled more than once and some
may not be sampled at all. This process is repeated many times and phylogenies
are reconstructed each time. After all these processes of bootstrapping are finished, a majority-rule consensus tree is constructed from the optimal tree from
each bootstrap process. The bootstrap support value—*i.e.*, the bootstrap value—
for any internal branch is the number of times that it was recovered during these
processes of bootstrapping. Generally, the bootstrap process should be repeated
over 500 times. If the bootstrap value of a branch is above 70%, the branch can be
regarded as a reliable one because, according to the simulation study of Hillis and
Bull,[352] this bootstrap value corresponds to a probability of 95% that the branch
is real.

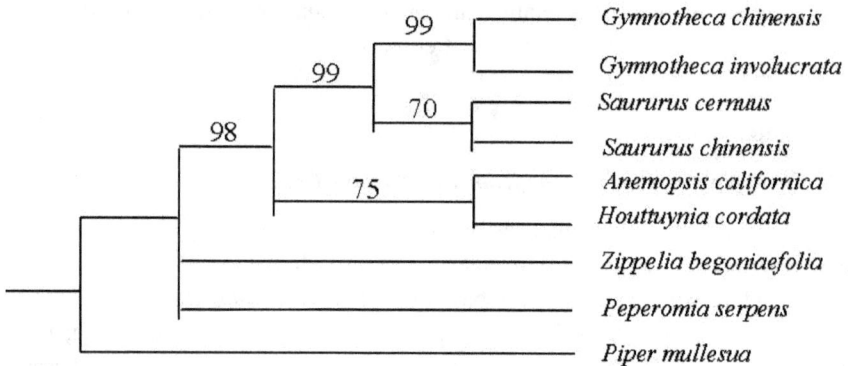

Fig. 1.   The strict consensus of the two most parsimonious trees of Saururaceae based on 18S nuclear
genes. Length = 123, CI =0.8943, RI = 0.7833, RC = 0.7005. Number of bootstrap replicates = 1000.
Bootstrap values (%) are above branches. The Consistency Index (CI) is a measure of how well an
individual character fits on a phylogenetic tree. It is calculated by dividing the minimum possible
number of steps by the observed number of steps. If the minimum number of steps is the same as
the observed number of steps, then the character has a CI of 1.0 (perfect fit). If a character is not
completely compatible with a tree, then it has a CI value less than 1.0 or even approaching zero (poor
fit). The CI value of a tree is the average CI value over all of the characters. There is a problem with
this value: It is always 1.0 for autapomorphies, which are character states that are seen in a single
sequence and no other. The Retention Index (RI) is similar to CI, but is also more resistant to bias due
to autopomorphies. The Re-scaled Consistency Index (RC) is also used to assess the congruency and
fit of characters to a tree (range from 0 to 1). It is computed as $RC = CI * RI$. A higher value of RC
indicates that the characters in the data set are more congruent with each other and with the given tree.

4.2. Phylogeny of Saururaceae from trnL-F Chloroplast DNA Sequences

Alignment of *trn*L-F DNA sequences produces a matrix of 1198 positions. 81 of these positions are variable and uninformative. 92 of these positions are parsimonious and informative. The percentage of parsimony-informative sites is 7.679%.

Our maximum parsimony analysis produces the single most parsimonious tree of 203 steps depicted in Figure 2. Saururaceae (100%), *Gymnotheca* (100%), and *Saururus* (100%) are monophyly. *A. californica* is the sister group of *H. cordata* (100%), and they form the first clade of Saururaceae. *Saururus* is the sister group of *Gymnotheca* (71%).

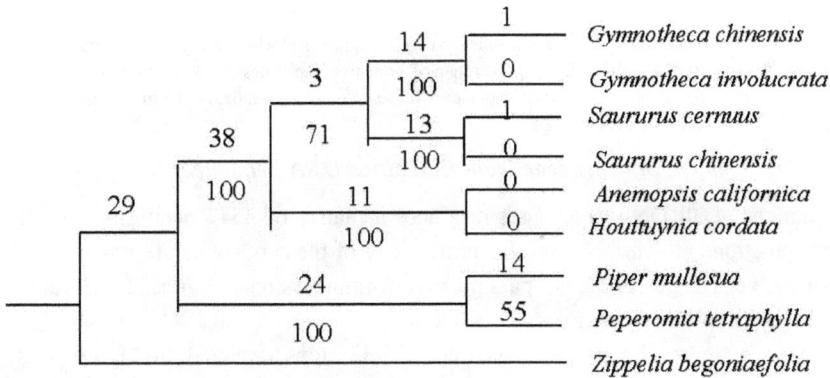

Fig. 2. The single most parsimony tree of Saururaceae based on *trn*L-F chloroplast DNA sequences. Length = 203, CI =0.9458, RI = 0.9231, RC = 0.8731. Number of bootstrap replicates = 1000. Base substitution values are shown above the branches, and bootstrap values (%) are shown below the branches.

4.3. Phylogeny of Saururaceae from matR Mitochondrial Genes

Alignment of *mat*R gene sequences produces a matrix of 1777 positions. 118 of these positions are variable. 43 of these positions are parsimonious and informative. The percentage of parsimony-informative sites is 2.42%.

Our maximum parsimony analysis yields the single most parsimonious tree of 136 steps depicted in Figure 3. Saururaceae (96%), *Gymnotheca* (100%), and *Saururus* (99%) are monophyly. *A. californica* is the sister group of *H. cordata* (83%), and they form the first clade of Saururaceae. *Saururus* is the sister group of *Gymnotheca* (95%).

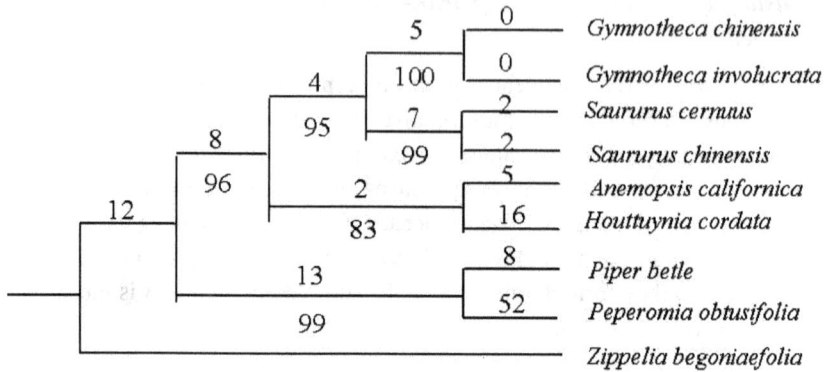

Fig. 3.   The single parsimony tree of Saururaceae based on *mat*R mitochondrial genes. Length = 136, CI =0.9118, RI = 0.8125, RC = 0.7408. Number of bootstrap replicates = 1000. Base substitution values are shown above the branches, and bootstrap values (%) are shown below the branches.

4.4. Phylogeny of Saururaceae from Combined DNA Sequences

Alignment of all DNA sequences produces a matrix of 4542 positions. 199 of these positions are variable-uninformative. 171 of these postions are parsimony-informative. The percentage of parsimony-informative sites is 3.76%. The P value of the partition-homogeneity test is 1.000000.

The single most parsimonious tree of 435 steps depicted in Figure 4 is produced by our maximum parsimony analysis. Saururaceae (100%), *Gymnotheca* (100%), and *Saururus* (100%) are monophyly. *A. californica* is the sister group of *H. cordata* (100%), and they form the first clade of Saururaceae. *Saururus* is the sister group of *Gymnotheca* (100%).

4.5. Phylogeny of Saururaceae from Morphological Data

1 character is constant. 16 characters are variable-uninformative. 41 characters are parsimony-informative. The percentage of parsimony-informative characters is 70.69%.

Our maximum parsimony analysis yields the single most parsimonious tree of 97 steps shown in Figure 5. Saururaceae (100%), *Gymnotheca* (92%), and *Saururus* (100%) are monophyly. *A. californica* is the sister group of *H. cordata* (65%), and they formed the first clade of Saururaceae. *Saururus* is the sister group of *Gymnotheca* (54%).

However, the bootstrap support values for *Gymnotheca-Saururus* and for *Anemopsis-Houttuynia* are somewhat weak. We think there is interference from

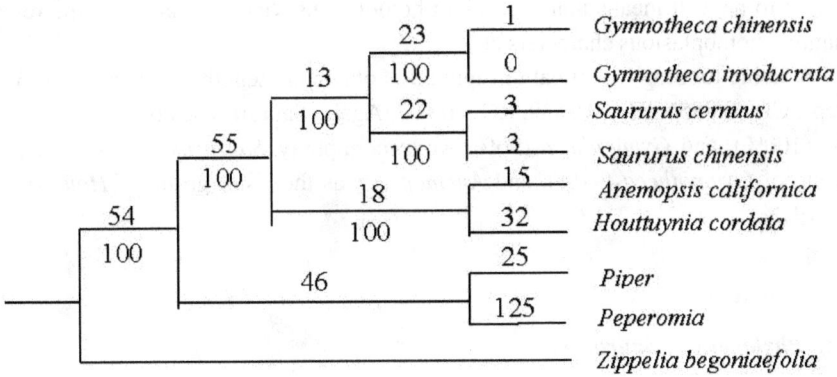

Fig. 4. The single parsimony tree of Saururaceae based on combined DNA sequences. Length = 435, CI =0.9264, RI = 0.8810, RC = 0.8162. Number of bootstrap replicates = 1000. Base substitution values are shown above the branches, and bootstrap values (%) shown below the branches.

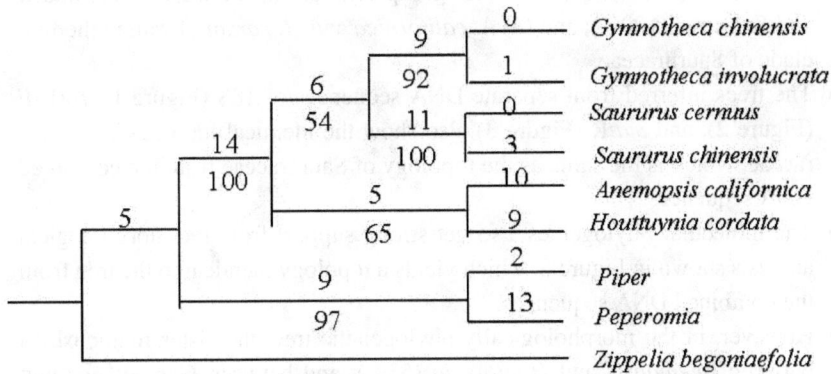

Fig. 5. The single parsimony tree of Saururaceae based on morphological data. Length = 97, CI =0.8351, RI = 0.7975, RC = 0.6659. Number of bootstrap replicates = 1000. Branch length values are shown above the branches, and bootstrap values (%) are shown below the branches.

the homoplasious characters. To reduce this interference, we should give more weight to the homologous characters in order to emphasize their effect. After weighting the characters according to re-scaled consistency indices (base weight = 2), we re-analyze the matrix once more with the same setting as before. "Base weight" is a degree of weighting. When analyzing a data matrix using equal weight, the base weight in all sites is equal to 1. For a phylogenetic program, it is the default status. In our study, we weight according to RC and set the base weight to be 2. It means that we weight homologous characters at 2 and the remaining homoplasious characters at 1.

A stable topology identical to Figure 5 is obtained. Length of the tree is 135 steps, CI = 0.9556, RI = 0.9439, RC = 0.902. Again, Saururaceae (100%), *Saururus* (100%), and *Gymnotheca* (96%) were monophyly. *Saururus* was the sister group of *Gymnotheca* (91%), and *Anemopsis* was the sister group of *Houttuynia* (92%).

5. Discussion

5.1. *Phylogeny of Saururaceae*

Let us first summarize the results from Section 4.

- We see from Figure 4 that the combined molecular data also strongly supports (1) the monophyly of Saururaceae (100%), *Saururus* (100%), and *Gymnotheca* (100%); (2) the sister group relationship between *Anemopsis* and *Houttuynia* (100%); (3) the sister group relationship between *Gymnotheca* and *Saururus* (100%); and (4) *A. californica* and *H. cordata* forming the first clade of Saururaceae.

- The trees inferred from separate DNA sequences of 18S (Figure 1), *trn*L-F (Figure 2), and *mat*R (Figure 3) also show the identical topology of Saururaceae, which is the same as the topology of Saururaceae from the combined DNA sequences.

- The molecular phylogenies also get strong support from the morphological analysis shown in Figure 5, which yields a topology identical to the tree from the combined DNA sequences.

- However, in the morphologically phylogenetic tree, the sister relationships between *Anemopsis* and *Houttuynia* (53%), and between *Gymnotheca* and *Saururus* (65%) are weak. Importantly, after weighting the characters according to RC indices, the sister relationships between *Anemopsis* and *Houttuynia* (92%), and between *Gymnotheca* and *Saururus* (91%) become strong.

- This result is surprising and differs from all the other phylogenetic opinions[484, 505, 631, 846, 908, 909] on Saururaceae.

Our results disagree with the systematic opinion of Wu and Wang[908, 909] on Saururaceae. However, our results partly agree with Wu,[906] who proposes that *Anemopsis* and *Houttuynia* are vicariant genera, and *S. chinensis* and *S. cernuus* are vicariant species. In a phylogenetic sense, vicariant genera or species may be interpreted as sister group. Our study well support the sister group relationships between *Anemopsis* and *Houttuynia*, and between *S. chinensis* and *S. cernuus*.

Our results are also not in consensus with Okada[631] and Lei *et al.*.[484] They suggest *Saururus* as the basal genus and *Anemopsis* and *Houttuynia* as "advanced" genera. Lei *et al.*[484] further suggest that *Gymnotheca* is the "most advanced." Okada and Lei *et al.* separately construct a phylogeny of Saururaceae only according to the basic chromosome number of Saururaceae. However, according to Figure 6, the basic chromosome numbers in Saururaceae (character 57) is homoplasious and is not a dominant characterisitic for reconstructing the phylogeny of Saururaceae. Moreover, it appears difficult to conclude that extant groups such as *Circaeocarpus*, *Anemopsis*, *Gymnotheca*, *Houttuynia* are derived from another extant group (*Saururus*).

In terms of the sister relationship of *Anemopsis* and *Houttuynia*, our results partly agree with Tucker, *et al.*[846] who has generated a tree identical to the combined DNA sequence tree in our study; see Figure 4 of this chapter and Figure 5 of Tucker *et al.*[846] Nevertheless, they treat *Saururus* as the first-derived genus in Saururaceae and believe that *Saururus* bear many plesiomorphies. The accepted tree in Tucker *et al.*[846] is supported with low bootstrap values.

We would like to address two points in Tucker *et al.*[846] One is the criterion for selecting out-group. *Cabomba*, *Magnolia*, *Chloranthus*, and even *Lactoris* and *Saruma* are not good out-groups for studying Saururaceae because they are too alien from Saururaceae according to the present understanding of angiosperm phylogeny.[27,688,786] Piperaceae is the best out-group of Saururaceae. The other is the interpretation of character 20 in Tucker *et al.*[846] on "whether a pair of stamens originated from separate primordia (0) or a common primordium (1)". The stamens of *Houttuynia* are from separate primordia (0).[505,841] However, this character is coded as 0 or 1 in Tucker *et al.*. When we correct it and re-analyze the same matrix using *Zippelia*, *Piper*, and *Peperomia* as out-groups, the topology is the same as Figure 5 in Section 4, and the bootstrap supports are high.

Liang[505] overweights a morphological character, "rhizomatous or not", and hence supports the monophyly of Saururaceae. She treats "stoloniferous" and "separate initiation of bract-flower" as synapomorphies, and hence supports the sister relationship of *Gymnotheca* and *Anemopsis*. She also treats "common primordium initiation of bract-flower" as the synapomorphies of *Saururus* and *Houttuynia*. According to Figure 6, "stoloniferous or erect stem" (character 0) and "the ontogeny of bract-flower" in Saururaceae (character 33) are homoplasious and are not suitable to reconstruct phylogeny of Saururaceae as dominant characters. Therefore, our study does not support an overweight on a few particular morphological or ontogenetic characters.

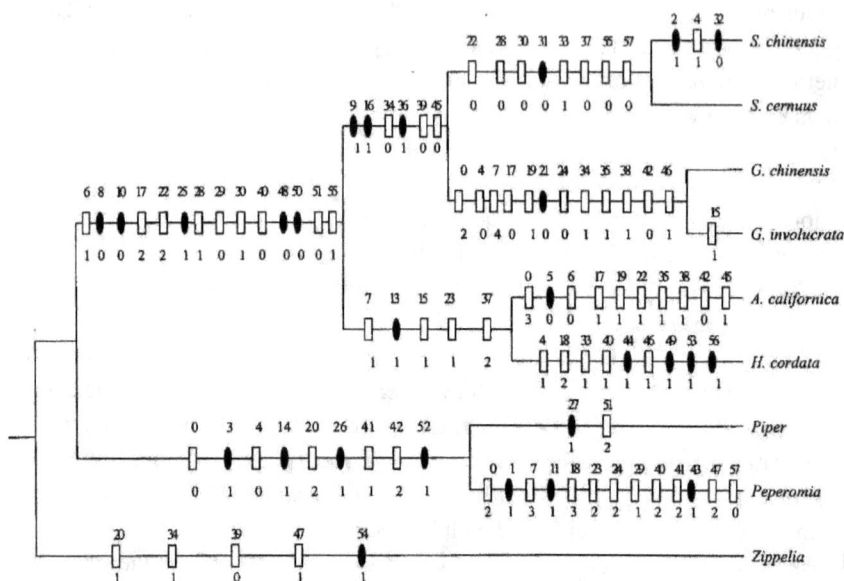

Fig. 6. The distribution of morphological characters. Character numbers are shown above the branches, and character states are shown below the branches. Black ovals represent homologous characters. Blank rectangles represent homoplasious or reversal characters. Homology is a similarity due to common evolutionary origin. Homoplasy is the existence of characters that have been subject to reversals, convergences, or parallelisms. Reversal is an evolutionary change whereby a character changes state and then changes back again. Convergence is an evolutionary event where the similarity between two or more characters is not inherited from a common ancestor. Parallelism is an evolutionary event where two identical changes occur independently. Convergences differ from parallelisms. In convergences, the ancestral characters are not the same. However, in parallelism, the ancestral characters are the same.

5.2. The Differences Among Topologies from 18S, trnL-F, and matR

The topologies from *trn*L-F (Figure 2) and *mat*R (Figure 3) are identical. In terms of arrangement of out-groups, the topology from 18S (Figure 1) slightly differs from the topology from *trn*L-F and *mat*R. Two points may cause the different arrangements of the out-groups. Firstly, we use different out-groups in separate analysis of different gene sequences. Besides *Z. begoniaefolia*, the out-groups are *Pe. serpens* and *P. mullesua* in the analysis of 18S sequences, *Pe. tetraphylla* and *P. mullesua* in the analysis of *trn*L-F sequences, and *Peperomia obtusifolia* and *Piper betle* in the analysis of *mat*R sequences. Secondly, Piperaceae is a large family, and so the arrangement of out-groups should be identical if we use more species of *Piper* and *Peperomia*.

Nevertheless, for the study on Saururaceae, the treatment for out-group in this chapter is suitable. Here, the main out-groups are *Z. begoniaefolia*, *Pe. tetraphylla* and *P. mullesua*. For the fast-mutating *trn*L-F, we use these three out-groups. For the slow-mutating functional 18S and *mat*R genes, we replace *Pe. tetraphylla* and *P. mullesua* with close species because the sequences of these close species are available in GenBank. Actually, the 18S sequences between *Pe. tetraphylla* and *Pe. serpens* are almost identical. So, in the phylogenetic analysis of 18S sequences of Saururaceae, it is suitable to replace *Pe. tetraphylla* with *Pe. serpens*. Similarly, in the analysis of *mat*R sequences, it is suitable to replace *P. mullesua* with *P. betle* and to replace *Pe. tetraphylla* with *Pe. obtusifolia*. Of course, support values should be higher if the identical out-groups are used.

The support values for each branch of Saururaceae based on 18S, *trn*L-F, and *mat*R are slightly different, and a few of them are quite low (Figures 1–3). The reasons are as follow. Firstly, since different genes have different characters, the phylogenetic trees based on different genes are different. Secondly, since sequences are from different sources, and thus have different systematic errors, the phylogenetic trees based on different sequences are different. In particular, the 18S sequence of *A. californica* and the *mat*R sequences of *A. californica*, *H. cordata*, and *S. cernuus* are from GenBank; but the other sequences are from the Laboratory for Plant Biodiversity and Biogeography, Kunming Institute of Botany, Chinese Academy of Sciences.

In conclusion, different out-groups, different genes, and different laboratory systems result in the slight difference among separate phylogenetic topologies. However, the difference is so small that it does not affect the topologies greatly. Moreover, the combination of DNA sequences reduces the differences among separate DNA sequence alignments (Figure 4).

5.3. Analysis of Important Morphological Characters

Some authors[484, 505, 631, 846, 908, 909] think that the ancestral Saururaceae is similar to the extant *Saururus*, which has free carpels, free stamens and superior ovaries. Moreover, they believe that six stamens (character 18), free stamens (character 19), hypogynous stamens (character 17), four carpels (character 23), superior carpels (character 22), free carpels (character 24), freedom of stamens and carpels (character 21), and marginal placenta (character 28) are primitive features of *Saururus*. However, these primitive characters can be interpreted as synplesiomorphies in most cases and are not important for phylogenetic reconstruction.[349] As shown in Figure 6, characters 18, 19, 23, 24 are homoplasious, and characters 17, 22, 28 are reverse in Saururaceae. These characters should not be used as dominant factors when reconstructing phylogeny of Saururaceae. In all analysis of this study, Figures 1–6, *Saururus* appears not to be the first derived genus in Saururaceae.

Free-carpel has long been regarded as a relictual feature in angiosperms.[79, 379] However, in the case of Saururaceae, according to the out-group comparison with *Zippelia*, *Piper*, and *Peperomia* in Figure 6, free-carpel (character 24) should be recognized as a homoplasious character. So should more-stamens (character 18), free-stamens (character 19), and more-carpels (character 23) be thought as homoplasious characters. Hypogynous-stamen (character 17), superior-ovary (character 22), and marginal-placenta (character 28) are regarded as primitive characters in phylogenetic reconstruction. However, they are reverse in Saururaceae. All of the above characters should not be dominant characters when reconstructing phylogeny of Saururaceae. Similar situation occurred in *Archidendron* (*Leguminosae*). Taubert[821] and following authors put *Archidendron* on a primitive position because *Archidendron* has several ovaries. But evidences from flowers, pollens, and wood anatomy, as well as the whole sequence of specialization from Ceaesalpinieae to Ingeae, indicate that *Archidendron* is highly advanced and its immediate ancestor has single ovary.[671]

6. Suggestions

Finally, let us note in the following subsections some of the skills that are important when reconstructing phylogeny.

6.1. *Sampling*

If a different set of sample is used in the study, we can expect the results to be different. For example, the tree produced from the 18S DNA sequences is slightly different from the tree produced from morphological data, mainly on the positions of out-groups and bootstrap supports. Concretely, *Piper* diverges earlier than *Peperomia* in the tree from the 18S DNA sequences in Figure 1, but *Peperomia* is the sister group of *Piper* in the tree from morphological data in Figure 5.

What is the reason? In the analysis of the 18S DNA sequences, the out-group *Piper* includes only *P. mullesua*; and *Peperomia* includes only *Pe. Serpens*. However, in the analysis of morphological data, the out-group *Piper* includes all species of the whole genus, and so does *Peperomia*. This difference in sampling causes the slight difference in the resulting topologies. So it is better to sample all species if the studied taxon is small, and to sample as many and as representative as possible if the studied taxon is big.

6.2. Selecting Out-Group

It is common that different out-group gives different result. What is the best out-group for a studied taxon? Hennig[349] points out that the sister group of a taxon is the best out-group, and one of the main tasks of phylogenetic analysis is to look for the sister group. In our case, identifying the sister group of Saururaceae becomes the critical procedure of the whole analysis.

In order to look for the sister group of Saururaceae, we have checked many classical works. Hutchinson[379] and Cronquist[179] both put Piperaceae, Saururaceae, and Chloranthaceae in Piperales. Melchior—see pages 764–796 of Brummitt's famous volume[113]—circumscribes Saururaceae, Piperaceae, Chloranthaceae, and Lactoridaceae in Piperales. In the systems of Dahlgren,Thorne, and Takhtajan—see respectively pages 777–783, 770–776, and 790–799 of Brummitt's famous volume[113]—Piperales only includes Saururaceae and Piperaceae although Takhtajan[815] separates Peperomiaceae from Piperaceae. Chase[139] backs the sister relationship between Piperaceae and Saururaceae in an analysis for *rbc*L sequences. So does Qiu *et al.*[688] for *rbc*L, *atp*B, 18S, *mat*R and *atp1* from three genomes, and Hoot[363] and Soltis *et al.*[786] for *atp*B, *rbc*L and 18S.

In conclusion, according to not only classical morphology systematics but also molecular systematics,[27, 688, 786, 846, 907] it is clear that Piperaceae is the sister group of Saururaceae. Thus Piperaceae is the best out-group when studying Saururaceae.

6.3. Gaining Sequences

Mistakes in sequences are not corrected even by the best analysis methods. So it is important to ensure that the sequences are accurate. How to guarantee accurate sequences? Firstly, the experimental skills and the experimental system should be good. Secondly, check each sequence with its electrophoretic map using SeqEd program (Applied Biosystems) or other softwares. Thirdly, determine the ends of sequences by comparing with similar sequences. *E.g.*, in our case, the ends of 18S gene are determined by comparing with the sequences of AF206929. It is more convenient to directly cite sequences from GenBank,[77] EMBL, or other databases. But some sequences and other data in these databases are rough. So better check the downloaded data before using them.

6.4. Aligning

When aligning sequences using Clustal-X, MEGA2b3, or other softwares, different input ordering of sequences can result in different alignment matrices, and consequently different phylogenetic trees. It is better to input high-quality and similar sequences first when using these softwares. When DNA are introns or gene spacers, alignment parameters should be small. When DNA are functional genes, alignment parameters should be big. Generally, the values should be bigger when genes are more conservative.

6.5. Analyzing

When reconstructing phylogeny, distance matrix methods, maximum parsimony methods, maximum likelihood methods, and methods of invariants are frequently used. No method can resolve all problems in all cases. So it is necessary to know which method should be used in different cases.

Different methods make different assumptions. Maximum parsimony methods assume that the evolutionary process is parsimonious. Hence it considers a tree with fewer substitutions or homoplasy as better than a tree with more substitutions or homoplasy. When the divergence between characters or sequences is small, maximum parsimony methods work well. In contrast, when the divergence is large, they work badly. Particularly, the parsimony methods are easily mislead when analyzing sequences that are evolving fast.

The unweighted pair-group with arithmetic mean method (UPGMA) assumes a constant rate in all branches. This assumption is often violated and thus the UPGMA tree often has errors in branching order.

Other distance methods assume that unequal rates among branches can be corrected by distances in a distance matrix. The performance of correction is affected by accuracy of the distances estimated. Normally, when the distances are small or the sequences of DNA or protein are long, the distance methods work well. However, when the sequences are short, distances are large or the differences of rates among sites are large, distance methods work badly.

Maximum likelihood methods make clear assumption on evolutionary rate or the base substitution model. Using maximum likelihood methods, we can develop a program with options for different evolutionary rates or different base substitution models. However, maximum likelihood methods need too much computation. So they are difficult to run fast in today's computer systems when the operational taxonomic units (OTU) are many or the sequences are long.[501]

The computational time is also different in different methods. In general, distance methods take the least time, and maximum likelihood methods take the most time. When reconstructing phylogeny, the common search techniques used are heuristic, branch-and-bound, and exhaustive. Heuristic search is rough but fast. Exhaustive search is accurate but slow, and its computational need is tremendous.

Branch-and-bound search is in the middle. Generally, use exhaustive search when the number of OTU is small, especially less than 11. Use heuristic search when the number of OTU is very large. Use branch-and-bound search when the number of OTU is not very large.

We can use many different programs to reconstruct phylogeny, such as PAUP, PHYLIP, MEGA, MacClade, WINCLADA, PAML and DAMBE. Although different programs have their own virtues, the most popular are PAUP and PHYLIP. Browse http://evolution.genetics.washington.edu to view some of these phylogenetic softwares.

For molecular phylogenetics, it is better to reconstruct phylogeny using as many different types of DNA sequences as possible. Analysis of separate DNA sequence matrices should be under the same setting. The separate DNA sequence matrices can also be combined into a larger matrix, and then analyze under the same setting. Generally, a combined matrix is better than separate matrices; because the combined matrix has more informative sites, and thus is more helpful to reconstruct a reliable phylogeny.

6.6. *Dealing with Morphological Data*

There are many articles on how to reconstruct phylogeny using morphological data,[255] so we only address two points here. Firstly, morphological characters should be polar and unordered. Typically, 0 is taken to represent the most primitive when numbering a character series, and the other positive integers are taken to represent different numbers and do not represent the evolutionary order. It means that, for example, the evolutionary step from 0 to 1 is one, and the evolutionary step from 0 to 4 is also one. Secondly, the values numbering out-groups' characters have not only 0, but also the other positive integers, such as 1, 2, 3, and 4.

6.7. *Comparing Phylogenies Separately from Molecular Data and Morphological Data*

A trend in phylogenetics is to combine matrices from DNA sequences and morphological data in order to make a more comprehensive analysis.[516] It is a reasonable method if a trade-off can be obtained between the two types of data. However, because a DNA sequence matrix is much longer and have much more informative sites than a morphological matrix, the information of the DNA sequences often overwhelms the information of the morphological data when analyzing a combination of the two types of data. Of course, we can enlarge the effect of the morphological matrix by weighting it. However, what is the criterion for the extra weight? How much extra should the weight be?

In our opinion, It is easier to compare the phylogenetic tree from molecular data with the phylogeny from morphological data. It is ideal if all of the phylogenies based on different data from variable subjects are identical. However, differences and discords among results from different data often occur. In our case, the phylogenies of Saururaceae from molecular data and morphological data are identical. However, they are almost opposite to the traditional opinions on the phylogeny of Saururaceae. Traditional opinions assert that *Saururus* is the most primitive genus in Saururaceae. However, we regard the *Saururus-Gymnotheca* clade as the sister group of the *Anemopsis-Houttuynia* in Saururaceae. What is the matter? We have to check. It costs much time because we have to check every original datum and every procedure of analysis. At last, we find out the reasons that cause the difference between our result and others' results.

6.8. *Doing Experiments*

Even though the conclusions of our case study are reasonable, they are based only on deduction. So, it is necessary to confirm our results by doing wet experiments. In order to check whether the concluded homoplasies of some morphological characters are true, we should design and perform experiments on floral development and histochemistry of Saururaceae and Piperaceae. Similarly, if it is possible, one should confirm one's conclusions by wet experiments after reconstructing phylogeny.

Acknowledgements

We would like to express our hearty thanks to De-Zhu Li, Han-Xing Liang, Shirley C. Tucker, Andrew W. Douglas, and Louxin Zhang for helpful suggestions; to Jun-Bo Yang, Zhen-Hua Guo, Yong-Yan Chen, Feng Wang, and Haiquan Li for their valuable assistance.

characters	character states
0	Growth habit	wood or liana(0), erect herb(1), stolon(2), lotiform plant(3).
1	Leaves position	alternate(0); alternate, lumbricine or whorled(1).
2	Terminal leaf of stem in reproductive period	green(0), white(1).
3	Stipule	adnation with a stipe of a leaf(0), not present(1).
4	Tomentum on lamina	no tomentum(0), tomentum on underside(1), tomentum on both sides(2).
5	Leaf venation	pinnate venation(0), hyphodromous(1), palmate venation(2).
6	Lateral leaf venation	no lateral leaf venation(0), dichotomous(1), not dichotomous(2).
7	Areoles	areolation lacking(0), incomplete(1), incomplete or imperfect(2), imperfect(3), imperfect or perfect(4).
8	Number of stem vascular cylinder	1(0), 2(1).
9	Fibre in stem	absent(0), discontinuous(1), continuous(2).
10	Perforation plate type in vessel members	scalariform perforation plate(0), simple perforation plate(1).
11	Number of inflorescence at one site	one(0), many(1).
12	Floral symmetry	radial symmetry(0), dorsiventral or zygomorphic symmetry(1).
13	Peloria	no abnormal regular flower(0), abnormal regular flower present(1).
14	Shape of floral bracts	lanceolate(0), peltate(1).
15	Color of inflorescence involucrum	green(0), showy(1).
16	Flower-bract stalk	no stalk(0), flower-bract stalk present(1).
17	Stamens position	hypogynous(0), perigynous(1), epigynous(2).
18	Number of stamens	6(0), 4(1), 3(2), 2(3).
19	Stamen fusion	free(0), connate(1).
20	Anther dehiscence	stomium along entire length of anther(0), stomium predominantly in proximal position(1), in distal position(2).

Fig. 7. Morphological characters and their states.

characters	character states
21 Adnation of stamens and carpels	free(0), fused partly(1).
22 Ovary position	superior ovary(0), perigynous ovary(1), inferior ovary(2).
23 Number of carpels	4(0), 3(1), 1(2).
24 Carpels adnation	free(0), fused(1), single carpel(2).
25 Style presence	no style(0), style present(1).
26 Stigma shape	stigmatic stylar cleft(0), capitate or tufted(1), divided stigma(2).
27 Ovule number per carpel	1(0), < 1(1).
28 Placenta	marginal placenta(0), parietal placenta(1), basal placenta(2).
29 Ovules per carpel	≥ 3(0), 1(1), < 1(2).
30 Number of carpel vascular bundle	2(0), coadnate(1), 1(2).
31 Vascular bundle fusion of stamens and carpels	free(0), fused partly(1).
32 Fusion of Adaxial and abaxial carpel bundle	free(0), fused partly(1).
33 Genesis of bract-flower	discrete bract and flower intition(0), common primordial inition(1).
34 Genesis order of carpels	middle primordium fi rst(0), bilateral primordium fi rst(1), appear simultaneous or single or common primordium(2).
35 Genesis of stamens	discrete primordium(0), common primordium(1).
36 Genesis order of stamens	bilateral stamens fi rst(0), middle stamens fi rst(1).
37 Genesis pattern of median sagittal stamens	in pair(0), adaxial axis fi rst(1), no adaxial or abaxial stamen(2).
38 Genesis pattern of bilateral stamen pair	discrete primordium(0), common primordium(1).
39 Median sagittal carpels	adaxial and abaxial carpels(0), adaxial(1).
40 Germinal aperture	anasulcate(0), anasulcate and anatrichotomosulcate(1), inaperturate(2).

Fig. 8. Morphological characters and their states (continued).

characters		character states
41	Ornamentation of pollen exine	foveolae(0), verruculose(1), large verruculose(2).
42	Small verruculose at the edge of foveolae of pollen tectate	absent(0), present(1), narrow belt of granule on tectate(2).
43	Number of microsporosac	4(0), 2(1).
44	Microspore gennesis	simultaneous(0), successive(1).
45	Type of minor tetrad	bilateral symmetry, T-shape and cross-shape(0), bilateral symmetry and cross-shape(1), bilateral symmetry(2).
46	Pollen abortion	no abortion(0), abortion(1).
47	Layers of integument	two layers(0), outer layer degradation(1), only inner layer(2).
48	Micropyle	consist of inner and outer integument(0), consist of inner integument(1).
49	Nucellus	crassinucellate ovule(0), tenuinucellate ovule(1).
50	Functional megaspore	from one of megaspore tetrad(0), from four of megaspore tetrad(1).
51	Type of embryo sac	polygonum type(0), drusa or peperomia type(1), fritillaria type(2).
52	Fusion of two central nuclei of embryo sac	before fertilization(0), after fertilization(1).
53	Apomixis	apomixis absent(0), apomixis present(1).
54	Perisperm type	cellular type perisperm(0), nuclear type perisperm(1).
55	Fruit type	folicule(0), capsule(1), berry(2).
56	Ploid	biploid(0), polyploid(1).
57	Basic number of chromosome	11(0), other number(1).

Fig. 9.   Morphological characters and their states (continued).

```
Taxon/Node                  111111111122222222223333333333444444444455555555
                  0123456789012345678901234567890123456789012345678901234567

S. chinensis      1010121201001000100000000100000001001000001000000000000000
S. cernuus        1000221201001000100000000100000011001000001000000000000000
G. chinensis      2000021401001000120101201100101110111110000000100000000101
G. involucrata    2000021401001001120101201100101110111110000000100000000101
A. californica    3000200?0?000101010101111100101110210211000001000000000101
                              1
H. cordata        1000121102000101022001211100101111200201101012100100010111
                              1
Z. begoniaefolia  1000222212101000000011001000222110100100101002011011?01201
                                                              1
Piper             0001000012101010000020001011222 1?020010111?002001012100201
                      1221          1  1 1  2                           1 1
                         2
Peperomia         2101012310111010003021022010212 1?0200?0?22210202101 1?00200
                      22   2                                              1
```

Fig. 10. The matrix of the morphological characters.

CHAPTER 12

FUNCTIONAL ANNOTATION AND PROTEIN FAMILIES: FROM THEORY TO PRACTICE

Noam Kaplan

The Hebrew University of Jerusalem
kaplann@pob.huji.ac.il

Ori Sasson

The Hebrew University of Jerusalem
ori@cs.huji.ac.il

Michal Linial

The Hebrew University of Jerusalem
michall@cc.huji.ac.il

We discuss two bioinformatics tools, ProtoNet and PANDORA, that deal with different aspects of protein annotations and functional predictions. ProtoNet uses an approach of protein sequence hierarchical clustering to detect remote protein relatives. PANDORA uses a graph-based method to interpret complex protein groups through their annotations.

ORGANIZATION.

Section 1. We introduce the objectives and challenges of computationally inferring function from sequence information. We discuss the shortcomings of some commonly used tools for this purpose. ProtoNet and PANDORA are two tools designed to complement these commonly used tools and alleviate their shortcomings.

Section 2. Then we present ProtoNet. ProtoNet is centered around the concept of homology transitivity. ProtoNet clusters protein sequences into a hierarchy based on sequence similarity and homology transitivity. Thus the hierarchy parallels the evolutionary history of these protein sequences. We then illustrate—using Histone proteins—the application of ProtoNet to detect remote protein relatives.

Section 3. We describe ProTarget, which is built on top of ProtoNet. ProTarget is a useful tool for selection of protein targets that have a high probability of exhibiting to a new fold.

Section 4. Next we present PANDORA. PANDORA starts from a binary protein-
annotation matrix and builds a PANDORA graph. Basically, each node of the graph
represents a set of proteins that share the same combination of annotations, and the
nodes are connected based on their inclusion-intersection relationships. We show how
to use PANDORA to assess functional information of protein families.

Section 5. Finally, we discuss the use of PANDORA in large-scale proteomic studies. We
mention PAGODA, an advanced option in PANDORA that detects outlier proteins in
the sense that they share some annotations but disagree on some other annotations.

1. Introduction

One of the major goals of bioinformatics is to gain biological insights about a
protein or gene from sequence information.[68, 93, 172, 489] The motivation for this
is that sequence information is relatively easy to obtain. The reason this goal
is realistic follows from the notion that two proteins with highly similar sequences are likely to be evolutionarily related and thus may share some biological
properties.[185, 723] Consequently, any knowledge gained with regard to one protein, may allow the inference of biological conclusions regarding any other protein
that exhibits high similarity to the former protein. Such biological information is
stored in protein databases and is generally referred to as annotations.

Inferring annotations based on sequence similarity opens the door to automatic
high-throughput annotation of whole genomes,[100, 489] and significantly reduces
the need for tedious and labor-intensive study of every single protein sequenced.
As attractive as annotation inference based on similarity is, there are inherent difficulties to be considered.[510] The main challenge is that protein function can often
vary significantly with the change of only a few amino acids. For example, changing a few amino acids at an enzyme's active site may drastically alter its function.
On the flip side, there are also instances of proteins that share the same function
despite having non-significant sequence similarity. But perhaps the most acute
difficulty encountered when attempting to transfer functional information among
proteins stems from the multi-domains nature of proteins. It is widely accepted
that the function of a protein is defined by the composition and organization of
its domains rather than from a local significant similarity.[401, 583] Consequently,
inference of biological characteristics from one protein to others requires not only
high global similarity but also validated biological knowledge about at least one of
the proteins. In other words, the protein database that is used must be rich enough
to contain a highly similar and well-annotated sequence for every new protein
sequence that we wish to learn about.

Naturally, the key drivers that determine the success of annotation inference
are the sequence comparison methods used and their sensitivity. The sensitivity of

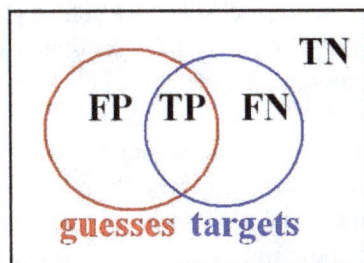

Fig. 1.   The 4 categories of guesses vs. targets.

a sequence comparison method determines its ability to detect proteins that belong to the same family and are thus homologous, even when their sequences exhibit low similarity.[23, 344, 487]

Before dealing with the practice of a bioinformatician in extracting functional information from a set of sequences at hand, let us review the basic notion of success in functional inference. Success can be easily translated to the balance between "precision" and "sensitivity" measures. Say we are looking at a certain set of specimens—*e.g.*, proteins—and some of them have a certain attribute. The set sharing this attribute shall be called the "target" set, and we choose using some prediction method a set of "guesses"—*i.e.*, a set consisting of the best guesses for specimens having the attribute. For example, we may look at a set of proteins and try to predict which proteins are receptors. In this case, the set of receptors that we predict are the guesses and the actual set of receptors are the targets. Given a guess, the specimen space can be divided into 4 categories depicted in Figure 1:

(1) True Positives (TP)—targets that are correctly predicted.
(2) True Negatives (TN)—non-targets that are correctly predicted.
(3) False Positives (FP)—non-targets that are predicted to be targets.
(4) False Negatives (FN)—targets that are predicted to be non-targets.

Sensitivity is defined as $TP/(TP + FN)$, or the percentage of targets that we guess successfully. Precision is defined as $TP/(TP + FP)$, or the percentage of our guesses that are targets. These two measures are closely linked, and in fact they trade off one for another. Often sensitivity is replaced by the term "coverage" and precision by "purity". At one end of the spectrum are the extreme cases of having a guess set which includes all specimens. Then the sensitivity is 1.0. However, such a guess set usually also contains a high number of false positives, and thus the precision is usually low. At the other end are cases of a very narrow guess set, say with only one element and that element is a correct guess. That provides a high precision of 1.0. However, assuming that there is more than a single element having the attribute in question, such a guess set implies many false negatives and thus low sensitivity.

Now that we have defined success, we can move to mention the set of tools that provide the "guess set" for any of the queries. The most widely used method for sequence-based functional annotation is a local alignment using BLAST.[23] While this method is shown to be powerful in terms of precision, BLAST often suffers from a low degree of sensitivity. It is thus unable to detect distant evolutionary relatives. PSI-BLAST[24] is a variation on BLAST that has gained popularity. PSI-BLAST increases the sensitivity dramatically, but often at the cost of low precision because of "drifting." Several other sequence-based functional prediction methods have been developed in an attempt to increase the sensitivity of searches without significant loss of precision.[647] Biological sequences such as proteins are not uniform in view of their evolutionary history, length, or information content. As such, no single method can be used optimally for all proteins. Instead, to retrieve maximal information on a query, it is adviceable to apply alternative tools that provide one or more functional prediction. The ultimate validation for any prediction obtained by computational tools is always in the laboratory.

In this chapter we discuss two bioinformatics tools, ProtoNet and PANDORA, that deal with different aspects of protein annotations and functional predictions. ProtoNet uses an approach of protein sequence hierarchical clustering in order to reconstruct an evolutionary tree of all proteins, thereby enabling highly sensitive detection of remote protein relatives.[746] PANDORA uses a graph-based method in order to provide means of interpreting complex protein groups through their annotations.[411]

2. ProtoNet — Tracing Protein Families

2.1. *The Concept*

One of the difficulties in designing a classification method using the information derived from simple pairwise alignment between proteins is incorporating biological knowledge into the method while keeping it fully automatic and unbiased by manual editing. ProtoNet[746] uses biological reasoning based on the notion of homology transitivity in order to deal with this problem. The definition of homology is simple. Two proteins are considered homologous if and only if they have evolved from a common ancestor protein. An interesting aspect of homology is that it may be considered to be transitive. The reason is that if proteins A and B are homologous and proteins B and C are homologous, then proteins A and C are homologous by definition. Homology transitivity is a powerful tool for overcoming the difficulty of noticing common evolutionary roots of proteins that do not exhibit high sequence similarity. When comparing two proteins that—by simple pairwise alignment—have a low degree of similarity, there might be a third protein that has a higher degree of similarity to both of them. This third protein can be used to establish a biological connection between these proteins and is often refer to as intermediate sequence.[647]

The ProtoNet system is designed around this concept. ProtoNet takes the protein universe that currently contains over one million sequences as an input. Utilizing an all-against-all comparison all using BLAST, ProtoNet builds a tree-like hierarchical organization of all proteins. This method enables a highly sensitive detection of distant relatives and is assume to capture the tale of protein family evolutionary history.[745]

2.2. The Method and Principle

ProtoNet uses a clustering technique called hierarchical agglomerative clustering. It comes down to a pretty simple idea, which is to iteratively merge together clusters which exhibit the most similarity. In our case, we start off with each protein being a cluster of its own—*i.e.*, "singleton"—and start a sequence of mergers based on similarities. For clusters that are non singletons, we use the average similarity between clusters as a measure of cluster similarity.

The use of hierarchical agglomerative clustering builds upon homology transitivity, since similar proteins are drawn together into the same cluster, even if they do not exhibit direct similarity. For example, if A and B show great similarity, and so do B and C, then all three sequences are most likely end up in the same cluster, thus putting A and C together. Such an association becomes tricky in the case of multi-domain proteins, which are quite common. In such a case, we might be drawing the wrong conclusion from the similarity information. Any way, with respect to ProtoNet, the process takes place very slowly, one merger at a time. This reduces the risk of less desirable mergers that put together a pair of proteins A and C with nothing in common other than each sharing a different domain with a protein B. Therefore it is more common to see a cluster having proteins with a single domain at the vicinity of other clusters that include such a domain albeit jointly with other non-related domain composition.

Due to the careful averaging rules along the merger process, the final tree is of gigantic proportions, as typically the number of merger steps measures in hundreds of thousands. This creates a problem of presentation of the most informative sub-tree of a homologous family. This difficulty is resolved by "condensing" the original tree. This condensing of the tree is done based on a criterion called "life-time." Intuitively, the lifetime measures the duration on time for which the cluster exists in the larger clustering process. The higher the lifetime, the more "stable" the cluster is in the dynamic process of mergers. Using this criteria for condensation allows us to ignore insignificant clusters, which are very quickly merged into other others. In other words, a set of merges are considered *en-bloc* as a single larger merger. This reduces the size of the tree, and makes it possible to navigate.

2.3. *In Practice*

In order to demonstrate the way ProtoNet can be used for browsing through the protein space, while keeping track of annotations, we describe a specific example. We look at Histone proteins that have been studied extensively. Histones facilitate the condensation of DNA by wrapping it around them to form a nucleosome. There are four conserved families of histone proteins that participate in this process—H2A, H2B, H3, and H4. Nucleosomes are essential for the compaction of the DNA in all eukaryotes. Histone H1 links the nucleosomes into a high order chromatin structure. Depending on the family, homology between members within each family can range from almost perfect (H4) to 50% (H1) sequence identity. A sixth histone, the poorly characterized H5, plays a role as a linker similar to H1 histone, and as such it is not a genuine part of the nucleosome structure. We start our browsing of the protein space with the protein marked in the SWISS-PROT database[86] as H1_ONCMY, which is a classical H1 histone protein.

As a biologist, you may find yourself with a sequence of an open reading frame (ORF) or a set of sequences for which you aim to retrieve maximal biological knowledge. The first step in studying a given protein sequence is to perform a sequence search using one of the various search engines. The most commonly used technique is BLAST[23] or its PSI-BLAST[24] variant. The result of such a search procedure is a hit list in which the top of the list indicates proteins with high score and associated with a statistical significance of that score in the appropriate database that is searched. The significance is measured by the expectation-value (E-value) and in case the hit list did not provide you any result with an E-value better than a predetermined threshold, let's say $E = 0.01$, you may find yourself going down a dead-end street.

ProtoNet can be used as the search engine for a specific protein in two distinct ways. The trivial case is when your protein is already included in the database of ProtoNet—*i.e.*, it is found in either SWISS-PROT or TrEMBL databases. In this case, the entire information gained by the construction of ProtoNet is available. In a more general setting, your protein is not part of the database, and in that case a

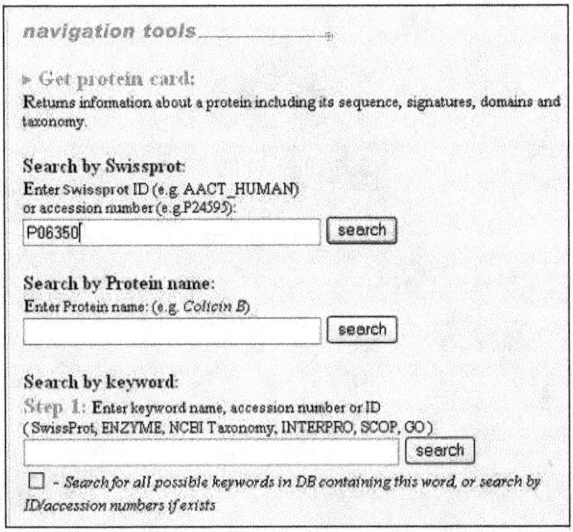

Fig. 2.   Protein Search Page.

local BLAST search is activated, followed by a computational procedure that emulates the presence of your protein in the database. In both cases, a connection of your protein to an already preformed cluster is expected. Note that in the case that your protein is remote from any other proteins, it may be reported as a singleton with no connection—this is quite a rare occurrence.

Searching for a specific protein by name or keyword is easy using ProtoNet. Once the ProtoNet main page at www.protonet.cs.huji.ac.il loads, we choose "Get Protein Card" from "Navigation Tools" in the main menu. This brings up the window shown in Figure 2. We type in the protein SWISS-PROT accession number (P06350) or the protein name. Alternatively, keyword search is possible. For our specific example, a search of the keywords "histone", "nucleosome", or "chromatin" leads you to a full list of proteins that are already annotated by these functionality terms. One of the proteins in this list is H1_ONCMY.

Clicking the search button brings up the data for this specific protein, as shown in Figure 3. The data includes among other things the protein name, sequence, accession number, length—measured in amino-acids—and any PDB[880] solved structures associated with it. An important control is the button allowing you to go to the cluster corresponding to this protein. This would be the lowest (significant) cluster in the hierarchy containing this protein.

Further down in the page additional information is provided as shown in

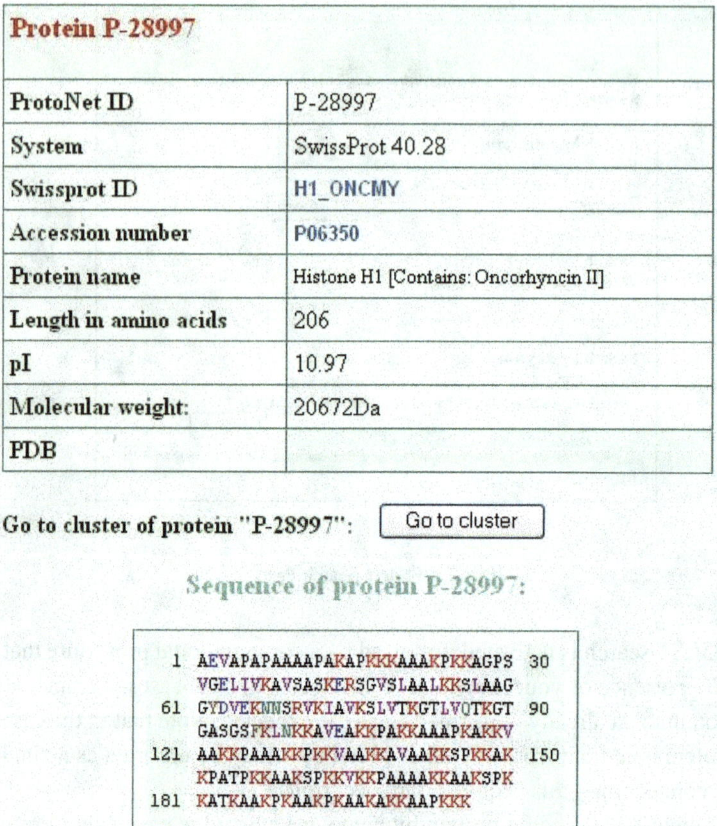

Protein P-28997	
ProtoNet ID	P-28997
System	SwissProt 40.28
Swissprot ID	H1_ONCMY
Accession number	P06350
Protein name	Histone H1 [Contains: Oncorhyncin II]
Length in amino acids	206
pI	10.97
Molecular weight:	20672Da
PDB	

Go to cluster of protein "P-28997": [Go to cluster]

Sequence of protein P-28997:

```
  1  AEVAPAPAAAAPAKAPKKKAAAKPKKAGPS  30
     VGELIVKAVSASKERSGVSLAALKKSLAAG
 61  GYDVEKNNSRVKIAVKSLVTKGTLVQTKGT  90
     GASGSFKLNKKAVEAKKPAKKAAAPKAKKV
121  AAKKPAAAKKPKKVAAKKAVAAKKSPKKAK  150
     KPATPKKAAKSPKKVKKPAAAAKKAAKSPK
181  KATKAAKPKAAKPKAAKAKKAAPKKK
```

Get motifs and domains of protein

Fig. 3.   Protein page for H1_ONCMY.

Figure 4. Relevant annotations are shown, based on SWISS-PROT[86] keywords, InterPro[29] classes, and GO[285] annotations. The taxonomy of the organism in which the protein is found is also shown in a concise form.

It is worth noting that direct links to relevant databases and primary sources is given, as well as a detailed graphical representation of the domain information within a sequence as combined by InterPro. The actual sequence on the protein that is defined as histone by several of the domain-based classification tools is

Keywords	
Swissprot	Acetylation, Chromosomal protein, Nuclear protein, Multigene family, DNA-binding
InterPro accession number	IPR005818, IPR005819
GO	**GO cellular component:** Cell, Cellular_component, Chromatin, Chromosome, Intracellular, Nucleosome, Nucleus **GO molecular function:** DNA binding, Ligand binding or carrier, Molecular_function, Nucleic acid binding **GO biological process:** DNA metabolism, DNA packaging, Biological_process, Cell growth and/or maintenance, Cell organization and biogenesis, Chromatin assembly/disassembly, Chromosome organization and biogenesis (sensu Eukarya), Establishment and/or maintenance of chromatin architecture, Metabolism, Nuclear organization and biogenesis, Nucleobase, nucleoside, nucleotide and nucleic acid metabolism, Nucleosome assembly

NCBI Taxonomy

```
SUPERKINGDOM - eukaryota
|_ KINGDOM - metazoa
   |_ PHYLUM - chordata
      |_ SUBPHYLUM - craniata
         |_ SUPERCLASS - gnathostomata
            |_ CLASS - actinopterygii
               |_ ORDER - salmoniformes
                  |_ SUBORDER - salmonoidei
                     |_ FAMILY - salmonidae
                        |_ GENUS - oncorhynchus
                           |_ SPECIES - oncorhynchus mykiss
```

Fig. 4. Keyword information for protein H1_ONCMY.

illustrated in "domain and motif" presentation window.

From the protein page we can go to a "cluster page" using the button shown in

General Info	
Size	32 proteins
Amount of Solved structures (in PDB)	1, total 1 entries
Amount of proteins without Prosite ID	32
Amount of Hypothetical proteins	0
Amount of Fragments	1
Average length and standard deviation of proteins	211.1 ± 20.3
Amount of clusters at ProtoLevel at creation	get data
Amount of singletons at ProtoLevel at creation	get data
ProtoLevel at creation	1.43707678224355e-06
ProtoLevel at termination	3.12758266305272
Lifetime	3.12758122597594
Fraction of "parent" cluster	97%

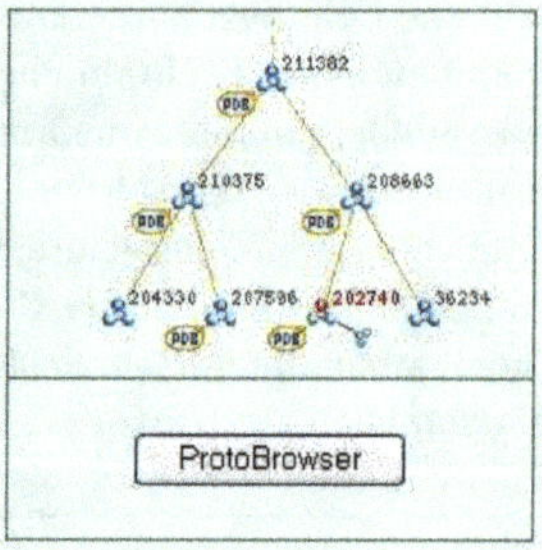

Fig. 5.   Cluster page for H1_ONCMY.

Figure 3. The cluster page corresponds to a grouping—using the ProtoNet algo-rithms explained earlier—of proteins together into a set of proteins based on their similarity. Clicking this button brings up a cluster description card as shown in Figure 5. The most important information on a cluster is its size and composition. In our case the cluster is numbered 202740 and consists of 32 proteins.

Other information available in the cluster page relates to the number of solved structures and how many PDB entries relate to them, the number of hypothetical proteins, and the number of fragments. Additional statistics relating to the Pro-toNet clustering process is given.

Below this information we have a localized view of the clustering tree. This allows us to go up and down the tree, browsing for the correct level of granularity that we are interested in. The reason we might need to do that relates directly to the

Keyword description	NumIN	Keyword Deviation from Expectation	Keyword frequency among proteins (%)		NumOUT	General keyword frequency (%)
			in this cluster	in children and appearances in merged proteins		
☐ 2 keywords of InterPro All						
Histone H1/H5 (IPR005818) / all clusters /	32	200.19	100%	▭	59	0.07%
Histone H5 (IPR005819) / all clusters /	29	197.23	90.62%	▭	48	0.06%

Fig. 6.   InterPro keyword breakdown for H1_ONCMY protein.

issue of sensitivity and precision discussed previously. Due to the different rates of evolutionary diversion and due to the characteristics of specific applications, there is no single "resolution"—or in terms of BLAST search, there is no single E-value threshold—that can be used universally. In the ProtoNet realm, relaxing your search threshold is equivalent to going up the tree and restricting it means going down the tree. The nice thing with this tree is that going one step up or one step down does not involve coming up with arbitrary thresholds—*e.g.*, going from E-value of 0.01 to 0.001 is quite arbitrary, and might not change the results in some cases, while changing them dramatically in other cases.

In our specific example, we have 32 proteins in the cluster, and we would like to know if this is the appropriate resolution for looking at Histone proteins. ProtoNet provides us a way to quickly break down the protein to groups based on keywords—as before, SWISS-PROT, InterPro, and GO keywords among others are supported. Looking at the InterPro keywords for this cluster, we get the result shown in the Figure 6.

What we see here is that the protein matches two families in InterPro—where one is actually a subfamily of the other—and indeed all cluster members are classified as H1/H5. However, we can notice that 59 proteins in this family are outside the cluster. That is, the sensitivity is rather low with a very high precision. This brings us to the conclusion we are better off go one step up the tree. We reach cluster 211832 with 87 proteins. This cluster captures the vast majority of protein annotated as H1/H5 by InterPro (87 out of 91), and all proteins in the cluster share this annotation. A statistical estimation for the deviation from expectation is provided at each of the steps. This measure is based on a distribution of the keywords and accounts for keyword abundance.

Keyword description	NumIN	Keyword Deviation from Expectation	Keyword frequency among proteins (%)		NumOUT	General keyword frequency (%)
			in this cluster	in children and appearances in merged proteins		
⊟ **13 keywords** of InterPro All						
Histone H1/H5 (IPR005818) / all clusters /	90	243.54	52.63%	🖵	1	0.07%
Histone H5 (IPR005819) / all clusters /	73	214.72	42.69%	🖵	4	0.06%
Histone H3 (IPR000164) / all clusters /	51	182.53	29.82%	🖵	1	0.04%
Linker histone, N-terminal (IPR003216) / all clusters /	39	161.17	22.8%	🖵	0	0.03%
High mobility group proteins HMG-I and HMG-Y (IPR000116) / all clusters /	8	72.98	4.67%	🖵	0	0%
Histone-fold/TFIID-TAF/NF-Y domain (IPR004822) / all clusters /	48	71.04	28.07%	🖵	252	0.26%

Fig. 7.   Keyword breakdown for cluster 223741.

Going further up in the tree does not improve the sensitivity level because we do not stumble upon a cluster with 91 proteins covering this exact family. The next significant merge brings us to cluster 223741, with 171 proteins. Part of the keyword breakdown for this cluster is shown in Figure 7. Interestingly, we can see that 90 out of the 91 H1/H5 proteins are detected here, and we also have 51 (out of 52) H3 proteins, and 39 (out of 39) Linker histone N-terminal proteins. Other proteins that are included share the properties of being DNA binding proteins that are instrumental in gene expression regulation. In addition, about 20% of the proteins marked "histone-fold TFIID-TAF-NF-Y domain" are included in the cluster. This cluster can be thus considered having a high level of functional abstraction.

To summarize, depending on the level of detail we are interested in, we can look at cluster 211832 if we are interested in H1/H5 proteins only, or we can go to cluster 223741 for a higher level view of a wider family. Similarly we can go

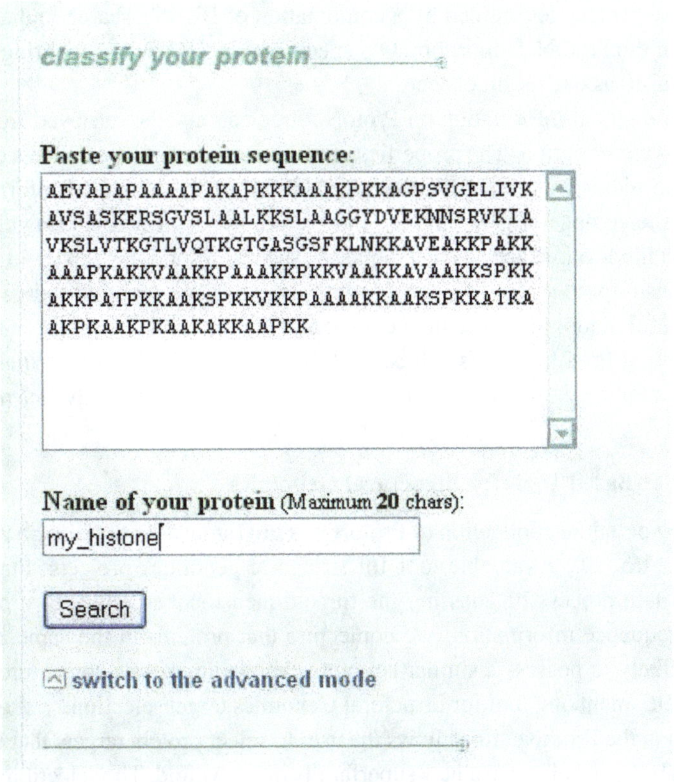

classify your protein

Paste your protein sequence:

```
AEVAPAPAAAAPAKAPKKKAAAKPKKAGPSVGELIVK
AVSASKERSGVSLAALKKSLAAGGYDVEKNNSRVKIA
VKSLVTKGTLVQTKGTGASGSFKLNKKAVEAKKPAKK
AAAPKAKKVAAKKPAAAKKPKKVAAKKAVAAKKSPKK
AKKPATPKKAAKSPKKVKKPAAAAKKAAKSPKKATKA
AKPKAAKPKAAKAKKAAPKK
```

Name of your protein (Maximum **20** chars):

```
my_histone
```

[Search]

⌃ switch to the advanced mode

Fig. 8. Classify-your-protein window.

higher in the tree to find larger clusters containing this cluster as well as other histone proteins.

The significant of clustering "histone-fold TFIID-TAF-NF-Y domain" with the major group of H1/H5 and H3 proteins cannot be fully appreciated at that stage. A use of PANDORA—to be described in the next section—can illuminate on the functional connectivity between the classical histones and of this group of proteins. A direct link is given from each cluster page to PANDORA.

As mentioned above, another entry point into this process is by showing a specific protein sequence. This is achieved by "Classify your protein" in the main menu of ProtoNet. In the example shown in Figure 8, we enter the sequence of the same protein studied above with the last amino-acid omitted—admittedly this is an artificial example, but it simplifies our presentation.

The output from this search is the most appropriate cluster in the hierarchy for the given sequence, determined by a combination of BLAST search and a slight variation of the ProtoNet algorithm. Our specific search is trivial, and brings us to cluster 202740 as one might expect.

Valuable global information on ProtoNet tree can also be retrieved from the "horizontal view" option. The properties of all clusters that are created at a certain level of the hierarchy are summarized in addition to the statistical information regarding the compactness of clusters and the properties of neighboring clusters in that specific level. More advanced queries such as to find the cluster in which two proteins—for example, histone H1 and histone H2A—are first merged may be very useful in looking for remote connectivity of proteins.

Navigating ProtoNet in its full capacity is aided by the detailed "Site Map" tour while additional options are only available by activating the "Advance mode".

3. ProtoNet-Based Tools for Structural Genomics

An exciting potential application of ProtoNet is ProTarget at `www.protarget.cs.huji.ac.il`, a valuable tool for structural genomics projects. Structure prediction is a process of guessing the three-dimensional structure of a protein given its sequence information. We conjecture that proteins in the same cluster are more likely to possess a similar structure. Validation of this conjecture leads to the development of a tool for Structural Genomics target selection.[675] Rational navigation in the ProtoNet tree allows the user to select protein targets that have a high probability to belong to a new superfamily or a new fold. The selection of targets can be done iteratively and changed dynamically by the user. Consequently, recently solved structures can be marked and ProTarget algorithm can then provide a new updated list of the best candidates for structural determination.[511]

Automatic detection of protein families is a challenging and unsolved problem. ProtoNet offers some automation but does not go all the way in determining what a family is—thus the need to go up and down the tree. One way to enhance ProtoNet toward automatic family detection is to study quantitative aspects of clusters and their annotations. A first step in this direction is done with PANDORA described in the following section.

4. PANDORA — Integration of Annotations

4.1. *The Concept*

A multitude of tools have been developed to assign annotations to proteins based on sequence properties, and these are constantly being improved. Such automatic methods are important for large-scale proteomic and genomic research, since they reduce bottlenecks associated with obtaining specific biological knowledge that is needed for each protein and each gene.

Large-scale proteomic and genomic research is often characterized by dealing with large sets of proteins or genes. Experimental methods such as DNA microarrays, 2D electrophoresis, and mass spectrometry provide means of dealing with complex biological processes and diseases by simultaneous inspection of hundreds of genes or proteins simultaneously. Computational proteomic family research can often deal with families of hundreds of proteins spanning several different proteomes—ProtoNet clusters, for example. Although high-throughput functional annotation eliminates the need to study each individual protein in these multi-protein sets, it shifts the bottleneck to the biological analysis of the results, a phase that requires manual inspection of the protein set, and often does not provide high-level biological insights about the proteins of the set.

PANDORA is a web-based tool to aid biologists in the interpretation of protein sets without the need of examining each individual protein. Furthermore, a major goal is to provide a global view of the protein set and of relevant biological subsets within it that are often hard to detect through normal manual inspection.

The general approach that PANDORA uses is based on annotations. In PANDORA, annotations are treated as binary properties that can be assigned to proteins. For example, for a given annotation "kinase", a protein may either have or not have the property "kinase", but cannot be half kinase. Each protein may have any amount of annotations assigned to it.

Annotations are often derived from different sources, each source possessing different characteristics. PANDORA uses annotation sources that cover a wide range of biological aspects: function, structure, taxonomy, cellular localization, biological pathways and more. Understanding the unique characteristics of each annotation source can greatly enhance the use of PANDORA.

4.2. The Method and Principle

The input to PANDORA is a set of P proteins. Based on the annotation sources chosen by the user, each protein may have any amount of annotations assigned to it by these sources. As mentioned, each annotation is treated as a binary property. Let the total amount of annotations used by this set of proteins be designated as K. Now, this data can be represented in a binary matrix, with P columns and K rows. Each cell in the matrix is binary, and can be occupied by a 1 or a 0, designating whether the protein represented by that column has the annotation represented by that row, as depicted in Part I of Figure 9.

Each column in the matrix represents all the annotations given to a single

(I) A binary protein-annotation matrix—indicating which protein possesses which property—underlies every PANDORA graph.

	Protein A	Protein B	Protein C	Protein D	Protein E
Kinase	1	1	1	0	0
Membrane	1	0	0	0	0
Transcription	0	0	1	1	0
Nuclear	0	0	1	1	0
Tyrosine Kinase	1	1	0	0	0
Viral Envelope	0	0	0	0	1

(II) A PANDORA graph derived from the protein-annotation matrix above. Each node of such a graph contains proteins that share a unique combination of annotations. An edge is then drawn between two nodes if one node is a superset of the other node. The edge is directed and is implicitly indicated by placing the first node above the second node.

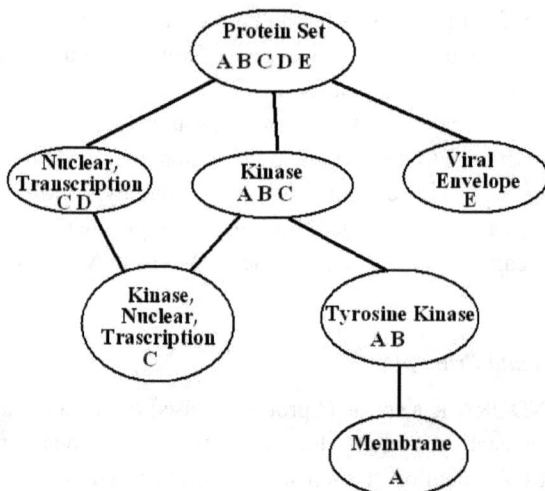

Fig. 9.   A PANDORA graph and its underlying protein-annotation matrix.

protein. Looking at the columns provide us with the "conventional" view of the annotations: For each protein, we look at the list of its annotations. Looking at the rows provide us with a more global view of the data. Each row represents a set of proteins that share an annotation that is represented by that row. These protein sets are the building blocks of PANDORA's graph. Each of these sets is checked for intersection with the other sets, and a graph is constructed. The graph constructed is an intersection-inclusion directed acyclic graph (DAG). This is a hierarchical graph representing intersection and inclusion relations between sets. In our case, each node in the graph represents a set of proteins that share a unique combination of annotations. The edges of the graph represent the hierarchy: An edge between two nodes shows that the upper node is a superset of the lower node—also referred to as the "parent" node. How is this graph constructed? The basic protein sets—represented by rows in the matrix—are checked for intersection amongst them. If two sets are equal, they are merged and become one set of proteins that has both annotations. If there is some partial degree of intersection between the nodes, a new "intersection node" is created. This node represents the set of proteins that belong to both nodes, and has both annotations. Also, this node is added to the graph hierarchy as a "daughter" of both nodes.

As shown in Part II of Figure 9, the PANDORA graph shows all the protein annotations in the matrix in a graphical manner, making it easy to detect relevant biological subsets of proteins that share a unique combination of annotations.

Consider what is visible in a PANDORA graph. All possible annotation combinations that have underlying proteins are shown. Each unique combination of annotations has its own node. Theoretically a graph with K annotations may have up to 2^K nodes—the amount of all the possible combinations. Thus even if there are only 20 annotations the graph can potentially have more than a million nodes! But this worst-case scenario never happens, mainly due to the fact that annotations are seldom randomly dispersed because they hold biological "meanings"—*e.g.*, the same annotations tend to appear on the same kinds of proteins. While cases of such extreme complexity never appear, the actual graphs associated with some annotations are very complex. At times so complex that you are probably be better off to abandon it and check the proteins one by one.

PANDORA offers a method to mitigate this complexity, by using variable resolution. Resolution is a parameter that can be used to simplify the graph. Recall that the graph shows the entire data that is in the binary matrix. Changing the graph would mean losing data. However, this is not necessarily a bad thing, because it is often not useful to see all the tiny details at once. Think about how you interpret complex data: first look at the data in low-detail to see the bigger picture showing main groups and relations, and then focus on specific parts of the data that are relevant to you and view them in high detail. This is exactly the concept behind varying resolution. Resolution is defined as the number of proteins that will be considered as an insignificant error when building the graph. Basically what this means is that if we set the resolution to 2 proteins, the graph will be simplified under the constraint that there are no errors in accuracy of more than 2 proteins. For example, one possible simplification would be if two nodes differ by only one protein. These nodes could be considered to be equal and merged. Although this is not entirely accurate, the error is relatively small—an error of 1 protein—and can be considered insignificant for most purposes. So, the higher the value of the resolution parameter is set, the graph becomes simpler but less accurate, thus providing a global view of the data. It is possible that there may be multiple ways to construct a graph with a specific resolution. In such a situation, we arbitrarily pick one of the possible ways—we believe that if a view in a certain resolution is meaningful, it is unlikely to depend much on which way is picked.

Once we have a simplification of the graph, PANDORA provides "zooming" in order to allow focusing on areas of interest. Zooming is a simple concept: you choose a node from a "low-detail" graph, and display it in a new window in higher detail. This allows you to focus on subsets that are relevant to your biological question one at a time, and study them separately without overloading the graph.

4.3. *In Practice*

A typical example of large protein sets whose study can gain from PANDORA is protein clusters such as the ones created by ProtoNet. In the previous section we considered the example of histone proteins by initiating a search in the ProtoNet tree starting by a H1 histone representative. For simplicity and consistency let us carry over the discussion of this example into this section. We use PANDORA to gain further functional understanding on the cluster with 171 proteins that was already discussed (cluster 223741). Inspecting Figure 7 and the summary of keyword appearances on the proteins indicate the presence of H1 and H1/H5 as well as H3 proteins but none of the keywords appear on a majority of the cluster's proteins. Now let's see the PANDORA graph of this cluster depicted in Figure 10:

To understand how to read PANDORA graph we should remember that the groups of proteins represented may have inclusion and intersection relations with other nodes. These relations are represented by the graph's hierarchy: If node A is connected to node B which is beneath it, A is a superset of B. This provides a simple yet important rule to follow: Each of the proteins of a node share not only the keywords of that node, but also those of all it's ancestors in the graph. The Basic Set (BS, all 171 proteins in the cluster) appears at the top of the graph. Clicking this node opens a window that lists the protein of this set as a list.

PANDORA captures also the biological "quality" or significance of the pro-

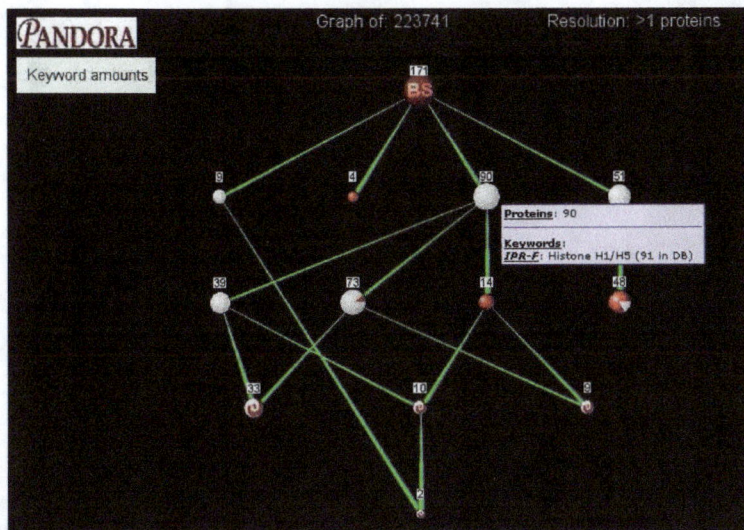

Fig. 10.   PANDORA view of cluster 223741 by InterPro keywords.

tein set. To understand this concept, consider a case where your protein set of 50 proteins share a common keyword. How significant is this biologically? Well, this depends on what is the keyword that they share. Obviously, if they share a keyword that appears only 50 times in the database, this should be considered significant. Conversely, if the annotation is highly abundant in the database—*e.g.*, "enzyme" or "membranous"—it may be less interesting biologically. To implement this concept we use sensitivity measure as explained in the Section 1. Sensitivity is defined here as the fraction of proteins in the node that have a common keyword out of the total amount of proteins in the database that have that keyword. The coloring of a node represents the sensitivity for that node's keyword: White represents the proteins in the node that have the keyword (TP); red represents proteins not in the node that have the keyword (FN). Therefore, a node that is completely white has a sensitivity of 1, because there are no other proteins in the database that have the keyword. Conversely, a node that contains a small fraction of the proteins having that keyword is coloured white only in a small portion and the rest of it is coloured red. The exact number of instances in the database of every keyword is visible via a "tool-tip" opened from the node, as shown in Figure 10. Sensitivity is more complex in the case of intersection nodes. To avoid this complexity, such nodes appear as a red-white swirl.

In addition to the graphical coloring method for quality assessment for a set

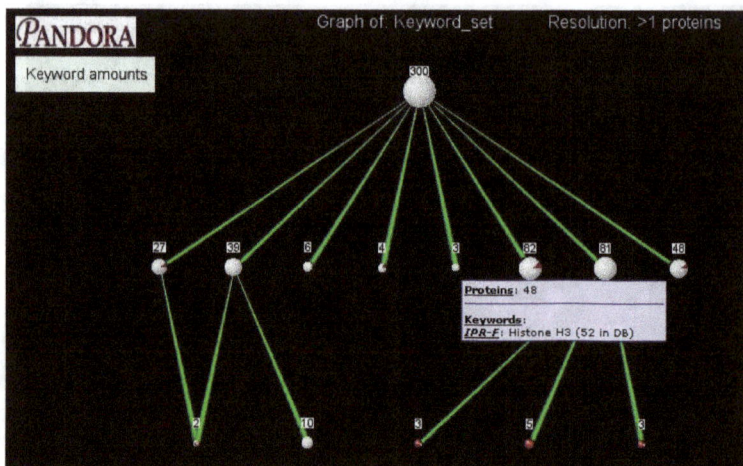

Fig. 11. PANDORA for InterPro annotation "Histone-fold/TFIID-TAF/NF-Y domain" (300 proteins).

of proteins, one can use the "show statistics" option. A table of the keywords that participate in the graph are sorted by the $\log(observed/expected)$ score, where the "expected" score is defined as the frequency of the keyword in the total population of protein sequences. This provides further means of assessing significance of keywords in your set.

Now that we know how to read a PANDORA graph we can come back to the tested example of cluster 224741 from ProtoNet. The relationships by the intersection and inclusion representation now becomes evident; see Figure 10. For example, the node in the graph that marks Histone H1/H5 (90 proteins) is the parent of 3 other nodes of Histone H5 (73 proteins); Linker histone, N-terminal (39 proteins); and Proline-rich extensin (14 proteins). The first two share 33 proteins that carry both terms of Histone 5 and Linker histone, N-terminal.

We can now turn to learn more on the basic set of the proteins in cluster 223741 through additional annotation sources. For example, annotation sources covering biochemical function, cellular localization and participation in biological processes are provided by Gene Ontology database. Other annotation sources provide information about the 3D structure, taxonomy, and more.

PANDORA is also useful to visually appreciate the connectivity of a specific node, as it reflects a keyword or a unification of keywords. For example, it can be seen that the Histone 3 set of proteins (51 proteins) are rather separated from the rest and the group; furthermore, 48 proteins that are marked "Histone-fold/TFIID-

Keyword type	Keyword	Amount	Expected	Log₂(obs/exp)
InterPro: Domain	Histone-fold/TFIID-TAF/NF-Y domain	300	0.7892	8.5704
InterPro: Domain	Transcription factor CBF/NF-Y/archaeal histone	39	0.1026	8.5703
Interpro: Family	Histone-like transcription factor CBF/NF-Y/archaeal histone, subunit A	9	0.0237	8.5689
InterPro: Domain	TATA box binding protein associated factor (TAF)	6	0.0158	8.5689
InterPro: Domain	Transcription initiation factor TFIID	3	0.0079	8.5689
Interpro: Family	Histone H2A	81	0.2210	8.5177
InterPro: Domain	Histone H2B	82	0.2289	8.4848
Interpro: Family	Histone H3	48	0.1368	8.4548
InterPro: Domain	Histone H4	27	0.0816	8.3702
InterPro: Domain	Transcription factor TAFII-31	4	0.0132	8.2433
Interpro: Family	Histone-like transcription factor/archaeal histone/topoisomerase	10	0.0816	6.9372

Fig. 12.   PANDORA statistical list for annotations associated with Histone-fold/TFIID-TAF/NF-Y domain.

TAF/NF-Y domain" are included within Figure 10. The red coloring of the node for the 48 proteins reects the low sensitivity of this group—only 48 out of total of 300 proteins in the database. A rational for the connection of the "Histone-fold/TFIID-TAF/NF-Y domain" to the cluster that is mostly composed of H1/H5 and H3 proteins can be sought by applying PANDORA to all proteins that are listed as "Histone-fold/TFIID-TAF/NF-Y domain" (300 proteins).

Figure 11 shows the PANDORA graph of annotation "Histone-fold/TFIID-TAF/NF-Y domain" (300 proteins). Most of the nodes are white, suggesting the high sensitivity of these nodes. The right most is the already discussed 48 proteins set that are combined with the term "H3 histone". The rest of the nodes global information is presented by clicking on the "statistical view" option; see Figure 12.

Now we can understand better the link of "Histone-fold/TFIID-TAF/NF-Y domain" to the cluster of histones. The 300 proteins that are included under this InterPro term combine the core of the nucleosomes—H2A, 2B, H3 and H4—and other histone-like proteins that are active in transcription by binding to promoter region, such as TAF and CBP. Our initial observation that this group of proteins is not tightly linked to the H1/H5 histones has corroborated itself.

Using PANDORA you can easily test potential functional connectivity. Continuing with our histone-fold example, one can ask whether bacterial DNA-binding proteins—recall that bacteria have no nucleosomes—share functional properties with the group of "Histone-fold/TFIID-TAF/NF-Y domain" that we have just discussed. We can collect a set of proteins that we would like in a combined PANDORA graph—for example, the 124 proteins that are annotated Histone-like bacterial DNA-binding protein, as well as the 31 proteins marked Histone-like transcription factor/archaeal histone/topoisomerase.

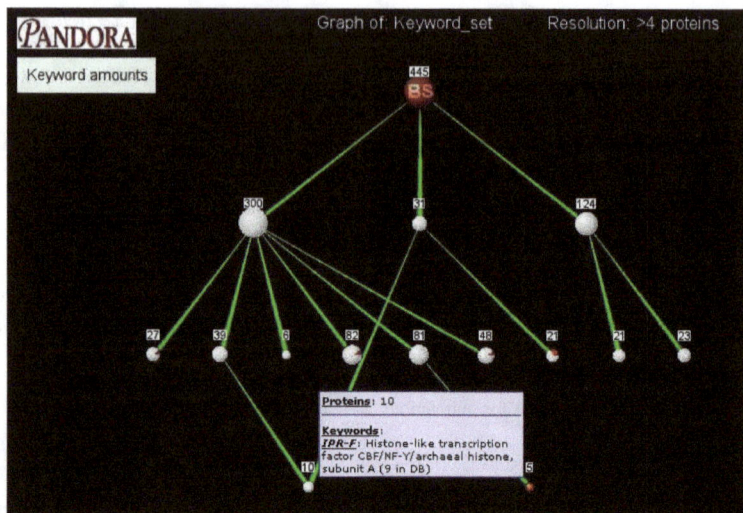

Fig. 13. PANDORA set of histone-like proteins from bacterial, archeal, and histone-fold/TFIID-TAF/NF-Y domain.

Figure 13 illustrates the result of such a query. In the Basic Set (BS is of 445 proteins), one can see that the bacterial—the rightmost note—shares no common keyword with "Histone-fold/TFIID-TAF/NF-Y domain". However, out of the 31 archaeal proteins 10 are shared with a new intersecting node of Histone-like transcription factor CBF/NF-Y/archaeal histone, subunit A. Note that in this case as in the previous examples, we gain biological understanding without the need to get the information through inspecting individual proteins.

5. PANDORA-Based Tools for Functional Genomics

We have illustrated the power of PANDORA for assessing valuable functional information in view of protein families. However, PANDORA is a generic tool that is suitable for any large-scale proteomic study that results in a large list of genes and proteins. Indeed, the input for PANDORA is either a ProtoNet cluster or any "User Set". The sets may be proteins that appear in our database—currently SWISS-PROT and TrEMBL—but also any of your sequence that is locally BLASTed against our database. The best matching protein above a certain threshold for each sequence is returned as input for PANDORA search.

Recall that PANDORA deals with binary properties that a protein may either have or not have. However, there are many biological properties that are naturally not binary, but quantitative. To be able to provide PANDORA capacity to quantitative measures a new advanced addition has been implemented. A classic example is taken from proteomic experiments that compare expression levels of proteins at different states or following a pharmacological treatment. In this case, it would be important to ask not only whether a protein's expression has changed or not, but also by how much. For this purpose we allow users to input values for quantitative properties on proteins.

A representative graph is shown in Figure 14. Looking at the graph, you notice the colorful bars beneath each node. These are color histograms indicating the distribution of the property on the proteins of the node. This provides a visual cue that simplifies the task of identifying nodes that are "interesting" in terms of the quantitative property. For example, if our quantitative property is "change in expression level", a node that shows increased expression means that there is a group of proteins that share some biological traits—the annotations of the node— and whose expression level is increased. Naturally this can be very helpful in obtaining biological understanding of such complex results and sets. Point the mouse over the color bars to see normal histograms of the distribution. On the upper left corner you see a color legend, with tool-tips showing the value range represented by each color.

It is important to mention that you are not limited to any specific kind of quantitative property. The values that are entered with the proteins can signify anything—disease linkage, toxicity, or even protein length. Any quantitative property that is interesting to look at in the context of biological sets can be used.

An additional tool that is essential for large-scale functional genomics is called PAGODA (Probing a Group of Disagreed Annotation). The basic principle is to apply the consistency of all annotations that are associated with proteins to automatically detect nodes that are outliers and are in disagreement with the rest of the annotations. This tool is still under development and will be available as part of PANDORA's advanced options.

Acknowledgements

This study is partially supported by the Sudarsky Center for Computational Biology. We thank all former and present members of the ProtoNet team for their enthusiasm and endless effort in developing, maintaining, and advancing our web-based systems.

Supporting Servers

ProtoNet (version 3.0)	www.protonet.cs.huji.ac.il
PANDORA (version 1.1)	www.pandora.cs.huji.ac.il

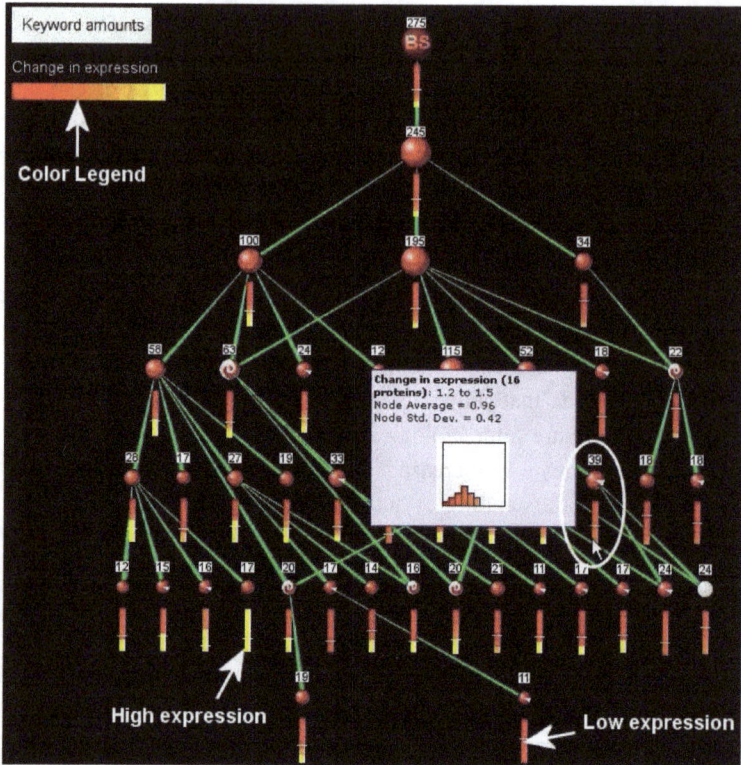

Fig. 14. PANDORA graph for the expression proteomic profile of 275 proteins analyzed by the quantitative option.

Related Web-sites and Resources

EBI GO Annotation	www.ebi.ac.uk/GOA
ENZYME	www.expasy.org/enzyme
Gene Ontology (GO)	www.ebi.ac.uk/go
InterPro	www.ebi.ac.uk/interpro
NCBI Taxonomy	www.ncbi.nlm.nih.gov
SCOP	scop.mrc-lmb.cam.ac.uk/scop
SWISS-PROT	www.exapsy.org/swissprot
ProTarget	www.protarget.cs.huji.ac.il

CHAPTER 13

DISCOVERING PROTEIN-PROTEIN INTERACTIONS

Soon-Heng Tan

Institute for Infocomm Research
soonheng@i2r.a-star.edu.sg

See-Kiong Ng

Institute for Infocomm Research
skng@i2r.a-star.edu.sg

The genomics and proteomics efforts have helped identify many new genes and proteins in living organisms. However, simply knowing the existence of genes and proteins does not tell us much about the biological processes in which they participate. Many major biological processes are controlled by protein interaction networks. A comprehensive description of protein-protein interactions is therefore necessary to understand the genetic program of life. In this chapter, we provide an overview of the various current methods for discovering protein-protein interactions experimentally and computationally.

ORGANIZATION.

Section 1. We introduce the term "protein interactome", which is the complete set of protein-protein interactions in the cell.

Section 2. Then we describe some common experimental approaches to detect protein interactions. The approaches described include traditional experimental methods such as co-immunoprecipitation and synthetic lethals. The approaches described also include high-throughput experimental methods such as yeast two-hybrid, phage display, affinity purification and mass spectrometry, and protein microarrays.

Section 3. Next we present various computational approaches to predict protein interactions. The approaches presented include structure-based predictions such as structural homology. The approaches presented also include sequence-based predictions such as interacting orthologs and interactiing domain pairs. Another class of approaches presented are the genome-based predictions such as gene neighborhood, gene fusion, phylogenetic profiles, phylogenetic tree similarity, and correlated mRNA expression.

Section 4. Lastly, we conclude with a caution on the need to counter check detected and/or predicted protein interactions using multiple approaches.

1. Introduction

Identifying and sequencing the genes is the monumental task that has been undertaken and completed by the Human Genome Project. However it does not provide sufficient information to develop new therapies. In the cells, genes are merely blueprints for the construction of proteins who are the actual workhorses for the different biological processes occurring in the cells. Protein-protein interactions— as elementary constituents of cellular protein complexes and pathways—are the key determinants of protein functions. For the discovery of new and better drugs for many diseases, a comprehensive protein-protein interaction map of the cell is needed to fully understand the biology of the diseases.

The term proteome, coined in 1994 as a linguistic equivalent to the concept of genome, is used to describe the complete set of proteins that is expressed by the entire genome in a cell. The term proteomics refers to the study of the proteome using technologies for large-scale protein separation and identification. The nomenclature has been catching on. The generation of messenger RNA expression profiles, which revolves around the process of transcription, has been referred to as transcriptomics, while the set of mRNAs transcribed from a cell's genome is called the transcriptome. In a similar vein, we can use the term "interactome" to describe the set of biomolecular interactions occurring in a cell. Since many of the key biological processes are controlled by protein interaction networks, we use the phrase "protein interactome" to refer to the complete set of protein-protein interactions in the cell. Other biomolecular interactions in a cell include protein-DNA and protein-small molecule interactions. This chapter is devoted to providing an overview of "protein interactomics"—the dissection of the protein interactome using technologies of large-scale protein interaction detection. Both experimental and computational methods are discussed, as computational approaches are rapidly becoming important tools of the trade in the molecular biology laboratories in the post-genome era.

2. Experimental Detection of Protein Interactions

In this section, we describe the various common experimental approaches to detect protein interactions. We classify the experimental methods into two classes: traditional experimental methods and high-throughput detection methods. The former contains methods for assaying protein interactions with limited throughput (sometimes only individually), while the latter describe technologies addressing the genome era's push for large-scale data generation. We cover the various computational means for predicting protein interactions in another section.

2.1. Traditional Experimental Methods

Traditionally, protein-protein interactions can be assayed biochemically by a variety of co-purification, gradient centrifugation, native gel, gel overlay, and column chromatography methods. Alternatively, protein-protein interactions can also be assessed indirectly by investigating their corresponding genetic interaction at the genome level. In this section, we highlight two representative experimental methods using biochemical and genetic approaches.

2.1.1. Co-Immunoprecipitation

A protein-protein interaction can be detected biochemically by selectively picking up one of the proteins from a mixture and then showing that the other protein, the interacting partner, is also picked up from the mixture.

In "co-immunoprecipitation",[296] the biochemical agent used to pick up selected proteins are specific antibodies. First, protein mixtures containing potential interacting protein partners are prepared in a cell lysate. An antibody designed to pick up—that is, immunoprecipitate—a specific protein is then applied. If the protein had been involved in a protein-protein interaction, its interacting protein partners would have also been picked up, or co-immunoprecipitated, along with it. The presence of this interacting protein partner can then be separated and identified using gel electrophoresis and mass spectrometry techniques.

The co-immunoprecipitation method is a laborious process. It is also restricted by the need of having specific antibodies against the proteins of interest. As a result, co-immunoprecipitation is not very amenable for the systematic large-scale detection of protein interactions which has become a necessary consideration in the post-genome era. As such, scientists have been working on ways to adapt it for large scale analysis of interactomes. One recent adaptation attempts to eliminate the limitation of having to have specific antibodies by using short protein sequences as "tags"[391, 615] to attach to the proteins of interest. In this way, tag-specific antibodies instead of protein-specific antibodies can be used for co-immunoprecipitating any proteins of interest. With this and other creative technological refinements, co-immunoprecipitation has the potential to develop into a high-throughput protein interaction detection method suitable for post-genome discoveries.

2.1.2. Synthetic Lethal Screening

Unlike co-immunoprecipitation which directly assays for the protein interactions, synthetic lethal screening is a genetic method that detects functional linkages between two proteins from which possibility of interactions can be suggested.[28, 74] Using mutations in the genes that encode the proteins of interest (e.g., gene deletions), we can observe the phenotypic effects of mutations in a pair of proteins from which functional linkages can be inferred. In synthetic lethal screening, the strategy is to screen for cases in which a mutation in a single protein is non-lethal but cell survival is destroyed when it is coupled with a mutation in another protein. Such a synthetic lethality occurrence can be explained by two scenarios:

(1) The two proteins are playing back-up or redundant roles in an essential pathway—therefore, loss of function only occurs when both are simultaneously disabled.
(2) The two proteins are performing discrete steps in an essential pathway. A mutation in either of the proteins only weakens the functioning of the pathway, but the combined detrimental effect of concurrent mutations in both proteins is sufficient to eliminate the essential pathway from the cell.

The second scenario can suggest the potential existence of physical interaction between the two proteins. Note that this method is only applicable for proteins that are involved in essential pathways. It is therefore not amenable for proteome-wide investigation of the interactome. While the method also does not provide direct information regarding the "biochemical distance" between the proteins—two proteins involved in a synthetic lethal interaction could be as close as interacting subunits of a protein complex or be dozens of steps away in a complex branching pathway—its results can still be useful for providing further "hints" or evidences for the exploration of the vastly uncharted protein interactome.

2.2. High Throughput Experimental Methods

The Human Genome Project—with its ambitious goal of assembling the entire sequence of the human genome—has ignited the now-prevalent emphasis on high-throughput data generation in molecular biology. It has catalyzed a major paradigm shift in modern biology: the scale of experimental investigations in biology has taken a great leap from studying single genes, proteins, and interactions to screening whole genomes, proteomes, and interactomes. Biologists can now study the living systems in both comprehensive systemic scope and exquisite molecular details. In this chapter, we describe several high-throughput experimental methods suitable for large-scale detection of the always formidable interactome. The practical bioinformatician is often tasked to analyze data generated from such high-throughput detection methods—it is useful to understand how these data are generated to understand better their various strengths as well as weaknesses.

2.2.1. *Yeast Two-Hybrid*

Yeast geneticists have developed a clever way of seeing whether two proteins can physically associate using the yeast as an *in vivo* platform. To do so, they enlist the service of a third protein—called a transcriptional activator—that has the ability to cause specific detectable "reporter genes" to be switched on. The scientists can experimentally separate an activator protein into two functional fragments, and then attach them separately to each of the candidate interacting proteins. If the two proteins—or rather, the two "hybrid" proteins, since they each has a part of an activator protein attached—interact, then the two fragments of the activator are reunited and switch on the associated "reporter gene" which produces a color change in the yeast cells. This is called the yeast two-hybrid (or Y2H) method,[250] the "two-hybrid" referring to the usage of the two hybrid candidate interacting proteins in the detection process.

The yeast scientists separate the transcriptional activator protein used in Y2H systems based on its two key functional parts: a DNA-binding domain and a trans-activation domain. The DNA-binding domain of a transcriptional activator is fused to a candidate protein known as the "bait", while its trans-activation domain is fused to the candidate protein's potential interacting protein partners known as the "prey". Since the yeast has two sexes, the "baits" and "prey" can easily be introduced into the same yeast cell by mating. If they physically interact, the DNA-binding and trans-activation domains are closely juxtaposed and the reconstituted transcriptional activator can mediate the switching-on of the associated reporter gene; see Figure 1.

Yeast two-hybrid was first described in 1989 by Fields and Song from State University of New York.[250] It has since become a routine method in biological labs to detect interaction between two proteins, albeit in a rather low-throughput manner. In recent years, Y2H has been successfully adapted for systematic high-throughput screening of protein-protein interaction. The first major high through-put genome-wide analysis of protein-protein interaction using yeast two-hybrid was applied to the yeast (or *Saccharomyces cerevisiae*) proteome itself.[386, 849] Of course, yeast is only an *in vivo* platform, detection of protein-protein interaction is not restricted to only the yeast proteome. In fact, large-scale identification of protein interaction using yeast two-hybrid has been carried out successfully on non-yeast proteomes, such as the proteomes of *Caenorhabditis elegans*[868] and

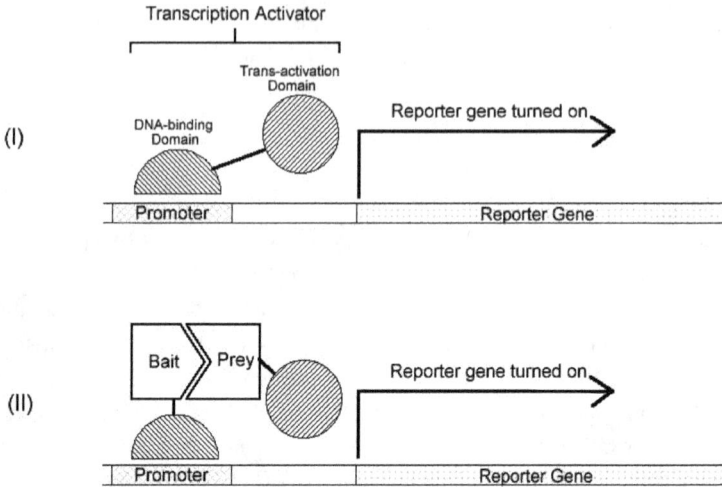

Fig. 1.   Interaction Detection by Yeast Two-Hybrid Assay. (I) Activation of reporter gene by transcriptional activator. (II) Activation of reporter gene by reconstituted transcriptional activator.

Helicobacter pylori.[695]

Large-scale studies of protein-protein interaction detection using the yeast two-hybrid method have revealed many interactions not detected previously by any genetic and biochemical studies. However, a prudent bioinformatician must not take all of the detected interactions at their face values. Several recent studies on the reliability of high-throughput detection of protein interaction using yeast-two hybrids have revealed high error rates,[483, 864] some reporting as high as 50% false positive rates.[793] There are inherent limitations in the Y2H method that can lead to false positives or biologically meaningless interactions:

- Some proteins exhibit transcriptional properties and can cause the reporter gene to be switched on by themselves, leading to artifactual interactions in which a positive signal is detected even though the two proteins do not interact with each other.
- Hybrid proteins can also adopt non-native interacting folds as a result of the fusion or in a foreign environment, giving rise to artificial interactions that may not occur naturally in the cellular environment.

As a choice method for large-scale genome-wide screening for protein-protein interactions, the yeast two-hybrid also suffers in terms of coverage. Neither of the two key comprehensive yeast interactome studies by Ito *et al.*[386] and Uetz *et al.*[849] using yeast-two-hybrid assays have recapitulated more than ~13% of the published interactions detected by the yeast biologist community using conventional single protein analyses.[337] The high false negative rate could be due to inherent experimental limitations of the Y2H method such as:

- In yeast two-hybrid systems, interactions are detected in the nucleus where transcription occurs. The method is therefore weak in detecting interactions for cytoplasmic proteins.
- Just as the non-native foldings of the hybrid proteins can give rise to artificial interactions, they can also prevent the interaction of two interacting proteins.
- Many proteins require post-translation modification for interaction, but this is not accommodated by the yeast two-hybrid approach.

In general, a prudent bioinformatician must be aware of the potential errors in experimental data, especially those generated by high-throughput methods. The detection of a protein-protein interaction—experimentally or computationally—must always be confirmed by at least two or more independent means. It is therefore important to develop other alternative methods for protein interaction detection and discovery. We describe a few other high-throughput experimental methods below, and leave the alternative computational methods for the next section.

2.2.2. *Phage Display*

Phages (or rather, bacteriophages) are viruses that infect bacterial cells and take over the hosts' cellular machinery for its reproduction. A phage is a very simple and efficient parasitic machine, made up of a genetic material (in form of either RNA or DNA) encapsulated by a protein coat assembled from viral proteins; see Part I of Figure 2. A key feature of phages is that they can accommodate segments of foreign DNA—a gene segment from another species, or stretches of chemically synthesized DNA—as "inserts" in their DNA. As the virus' DNA is replicated in the bacteria host, the foreign insert is also replicated along with it as a sort of passenger. This makes phages a choice vehicle in the laboratory for replicating other types of DNA.

In a phage display vector, we use the phages' DNA insertion templates to program the host bacteria cells to synthesize specific foreign peptides. By choosing one of the genes that make coat proteins for the phages to insert a foreign DNA into, hybrid fusion coat proteins that contain the foreign peptides are synthesized and used to construct the protein coats of the replicated phages. In this way, the expressed foreign peptides are "displayed" on the outer surface of the replicated phages for easy detection; see Part II of Figure 2.

Unlike yeast two-hybrid that detects interaction between two full length

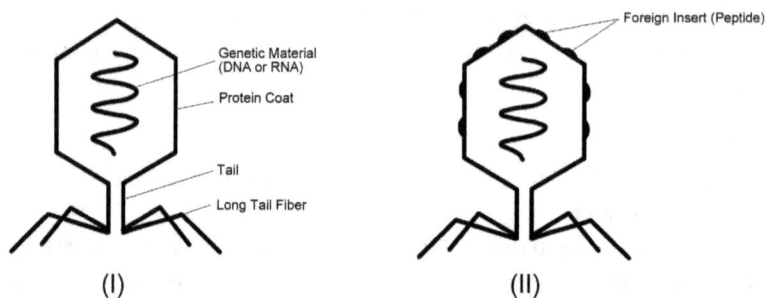

Fig. 2.   Schematic diagrams of (I) a phage; and (II) interaction detection by phage display.

proteins, phage display can be used to determine the binding of short protein sequences—for example, only around 10 amino acids in length—to proteins. To detect what peptide sequences bind to a protein, we immobilize the protein on a solid surface as a "bait", and then expose it to a large library of phages with different display peptide sequences. Phages with display peptides that bind to the "bait" protein can then be selected for infection on a bacterial host, to amplify the binding sequence to an appropriate amount to allow for identification by sequencing.

Phage Display was first reported by Smith in 1985.[778] To-date, phage display has been used effectively to determine short binding sequences for the SH3,[243] WW,[419] SH2,[169] and PDZ[477] protein domains. While phage display is best used for determining short binding sequences, it can also be used to detect—or rather, predict—protein-protein interaction based on the fact that most protein-protein interactions involve direct contact of very small numbers of amino acids. One example is the combined use of phage display and yeast two-hybrid technologies to respectively predict and validate a network of interactions between most of the SH3 domain containing proteins in yeast by Tong et al.[833] In this study, the SH3 domain binding motifs derived from phage display are used to predict a network of hypothetical interactions—between proteins with SH3 domain and those with sequences that matches the detected binding motifs—that are then experimentally validated using yeast two-hybrid screens.

Phage display is a powerful tool to identify partners of protein-protein interactions—it is one of the most established techniques to generate lead molecules in drug discovery. The method is easily amenable for rapid high-throughput, combinatorial detection. Phage display libraries containing 10^6 to 10^{10} independent clones can be readily constructed, with each clone carrying a different foreign DNA insert and therefore displaying a different peptide on its surface. However, detection of protein-protein interaction by phage display is an indirect method that predicts interaction between two proteins containing the detected short binding peptide sequences. Furthermore, phage display is most suitable for detecting interactions with short sequences instead of full-length proteins—as such, it may not work for all proteins. As in the other experimental approaches, phage display needs to work in combination with other complementary interaction detection methods—experimental and computational—in order to map out the vast and complicated interactome fully and accurately.

2.2.3. *Affinity Purification and Mass Spectrometry*

Interactions between proteins are not limited to pair-wise interactions such as those detected by the above methods—several proteins (sometimes as many as 20 or more) can come together to form a multimeric protein complex. Many functional pathways in the cell involve multi-protein complexes. The detection of protein interactions in the form of multi-protein complexes is therefore important for understanding the biochemical mechanisms of the living cell.

The so-called "affinity purification" process can be used to identify groups of proteins that interact together to form a complex.[67] To do so, a "bait" protein is first immobilized on a matrix or a solid surface such as the internal of a column. This can be done by attaching an affinity tag to the bait protein which helps stick it to the solid surface. Then, a mixture of candidate proteins passes through the column: proteins binding to the immobilized protein are thus retained and captured while non-interacting proteins are eluted away. The captured proteins in turn serve as additional baits to capture other proteins, leading to formation of protein complexes; see Figure 3. The bound proteins are subsequently collected from the column by washing it with a solution that decreases the binding affinity of the bound proteins, or using an enzyme to cleave the affinity tag to remove the bound proteins from the column. As with the genome-wide two-hybrid system, robotics made the assays high-throughput.

Traditional protein identification methods such as Edman sequencing and Western blots are tedious, time-consuming and not easily scalable for large-scale identification of proteins. For throughput, mass spectrometry provides a fast and accurate means for dissecting the protein complexes. To detect the mass fingerprint of a protein, the protein is first cleaved into many short-sequence peptides using proteases that cut proteins at specific sites. The masses of these cleaved peptide fragments are then determined in a mass spectrometry process known as Matrix-Assisted Laser Desorption/Ionization (MALDI) to generate a series of peaks, each describing the molecular mass of a single peptide in the mixture. Because the proteases cut the protein at specific sites, it is possible to know exactly

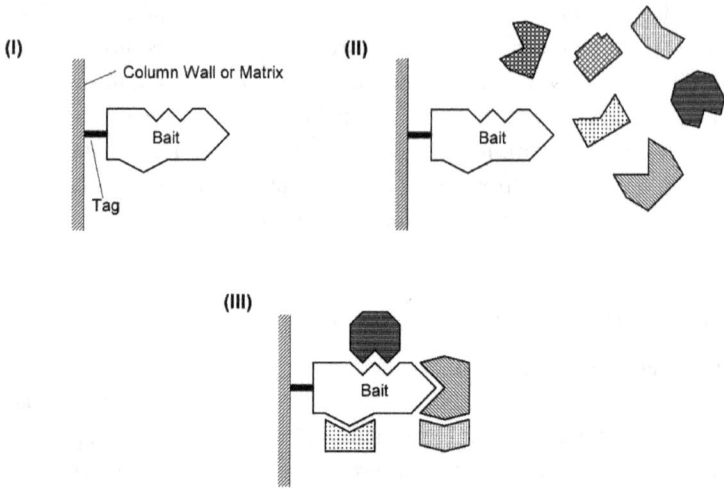

Fig. 3.   Interaction Detection by Affinity Purification.

which cleaved peptide fragments any given protein can generate. Each protein in the proteome can therefore be characterized with a "fingerprint" consisting of the set of peptide masses resulting from the mass spectrometry experiment. With the recent completion of many large-scale sequencing projects, huge sequence databases are now available to enable researchers to compare an observed peptide fingerprint with, for example, every possible human protein to identify an unknown protein. For identifying a mixture of proteins such as a protein complex from affinity purification studies, the proteins are first separated using gel electrophoresis and then individually identified using mass spectrometry.

Researchers have applied this approach on a proteome-wide scale. Gavin et al.[278] have found 232 yeast protein complexes using an affinity purification process. Subsequent protein identification using MADLI has revealed 1,440 distinct captured proteins, covering about 25% of the yeast's proteins. Ho et al.[358] have applied the same general approach to identify 3,617 interactions involving 1,578 yeast proteins. As in the other experimental methods, protein coverage is still a limitation. Furthermore, it has the caveat that a single bait protein may occur in more than one complex in a cell: it may therefore link with two or more proteins that never actually occur in the same complex, giving the illusion that the detected protein complex is bigger and more elaborate than it actually is. Both groups also report a significant number of false-positive interactions, while failing to identify many known associations.[462] So, as the results from most large-scale studies have illustrated, there are no perfect detection methods for mapping the interactome. It is essential to integrate data from many different sources in order to obtain an accurate and comprehensive understanding of protein networks.

2.2.4. *Protein Microarrays*

In the past, the study of molecular biology focused on studying a single gene or protein at a time. Today, with the many advancements in high-throughput technologies, it is now possible for biologists to perform global informational analyses in their discovery pursuits. For example, the DNA microarray technology has made possible the analysis of the expression levels of hundreds and thousands of genes simultaneously, allowing the biologists to analyze gene expression behaviors at the whole-genome level. As we will see in Section 3.3.5, scientists have even been able to use gene expression data to decipher the encoded protein networks that dictate cellular function. However, most cellular functions are manifested by the direct activities of the translated proteins and not by the genes themselves. In fact, protein expression levels often do not correlate with mRNA expression levels.[310] Expression analysis at the proteomic level is a more superior approach as proteins are one step closer to biochemical activities than genes are.

Researchers have recently begun to focus on developing protein microarray methods for the high-throughput analysis of proteins. Just like the gene microarrays, a protein microarray consists of tens to thousands of proteins, individually spotted at unique addresses in a micro- or even nano-scale matrix, so that interactions between the bait proteins and the test samples can easily be identified. The detection process in protein chip is very similar to the affinity purification technique described in the previous section. The bait proteins are purified and spotted separately onto a small solid surface such as a glass slide for capturing testing proteins in solution. The solid surface is then overlaid with a testing protein for interaction with the baits, washed and then assayed for protein binding at each microarray spot. Usually, the testing protein is attached to a suitable dye or enzyme that makes it easy for the bound proteins to be detected.

MacBeath and Schreiber[534] describe a proof-of-principle work in 2000 of spotting purified proteins onto glass slides using the existing DNA microarrayer and scanning tool, and showing that the purified proteins retained their activities when spotted onto chemically-treated glass slides. Since then, many researchers have worked on using protein microarray to detect protein-protein interaction on a massive scale. For example, Zhu *et al.*[939] construct a genome-wide protein chip and use it to assay interactions of proteins and phospholipids in yeast. A total of 5,800 predicted yeast's ORFs are cloned and 80% of these are purified to a detectable amount and then spotted on glass slides to construct a yeast proteome microarray to screen for their ability to interact with proteins and phospholipids. Their results illustrate that microarrays of an entire eukaryotic proteome can be prepared and screened for diverse biochemical activities.

Protein microarrays hold great promise for revolutionizing the analysis of entire proteomes, just as what DNA microarrays have done for functional genomics. However, developing protein microarrays is a much harder problem than making DNA microarrays. Proteins are heterogeneous, making it difficult to develop methods to attach them to biochips and have them remain functional. Proteins are also more difficult to synthesize than DNA and are more likely to lose structural or functional properties in different environments or when modified. Unlike DNA, where the sequence is all that matters, a protein's three-dimensional structure must be preserved. However, one can be confident that novel technologies will continue to expand the power of protein arrays so that it will soon play a major role—together with the other protein-protein interaction techniques described in this chapter–in deciphering the protein networks that dictate cellular functions.

3. Computational Prediction Protein Interaction

As we have seen in the previous sections, even the best experimental methods for detecting protein-protein interactions are not without their limitations. As such, the detection—or rather, prediction—of protein-protein interactions using computational approaches in a rapid, automatic, and reasonably accurate manner would complement the experimental approaches. Toward this end, bioinformaticians have developed many different computational approaches to screen entire genomes and predict protein-protein interactions from a variety of sources of information:

(1) *Structure-based predictions.* Interactions between proteins can be deemed as biophysical processes whereby the shapes of the molecules play a major role. Structural biologists have long been exploiting the structural information of the protein molecules to determine whether they interact. However, the determination of the three-dimensional structure of proteins is still a major bottleneck today, greatly limiting the use of structure-based prediction for unraveling protein-protein interactions at the proteome level as the structural data of most proteins are still unavailable.

(2) *Sequence-based predictions.* On the other hand, the genetic and amino acid sequences of most proteins are now available. This has prompted to resourceful bioinformaticians to find ways to predict protein-protein interactions based on sequence information of the proteins alone.

(3) *Genome-based predictions.* The complete sequences of many genomes are also available. To-date, according to the Entrez website at the US National Center for Biotechnology Information, more than 170 species already have their complete genetic sequences mapped. These entire genomes of multiple species can be used to screen for genome-level contextual information such as gene co-localizations, phylogenetic profiles, and even gene expression to infer interactions between proteins.

In this section, we describe a variety of computational methods that bioinformaticians have developed under each of the above categories. While it is clear that computational approaches will never be able to replace experimental methods, by combining the results from multiple approaches—*in silico* or otherwise—we can improve both the quantity and quality of protein interaction detected by leveraging on the complementary strengths of the different detection methods. *E.g.*, experimental methods typically suffer from limited coverage, whereas computational methods usually have broad coverage as they are less sensitive to the *in vivo* and *in vitro* biochemical intricacies. It is thus important for the bioinformaticians to continue to develop computational methods for the detection of protein-protein interactions. The combination of experimental and computational data will eventually lead to the complete set of information for us to understand the underlying interaction networks that govern the functioning of the living cell.

3.1. *Structure-Based Predictions*

Much of the focus in structure-based predictions related to protein-protein interactions is in the prediction of interaction sites, also known as the docking problem if the structures of the two interacting proteins are known. Docking is the process whereby two molecules fit together in three-dimensional space. However, knowing the induced fit based on the unbound, isolated structures of two protein molecules do not immediately imply that the two proteins will interact, because proteins undergo conformational changes upon binding. As such, most docking algorithms are used mainly to predict whether and how small molecules, such as drug candidates, interact with known protein targets.

However, even if we can solve the docking problem for protein-protein interactions, it is still hindered by the very small number of protein structures available. In order to handle genome-wide protein interaction prediction, structure-based methods must be able to infer from proteins whose structures are not yet known based on knowledge derived from limited number of known structures of protein-protein interactions—usually complexes—in an approach similar to sequence homology.

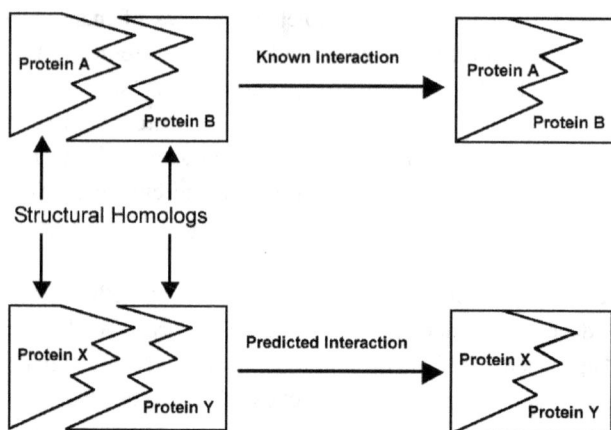

Fig. 4.   Structure-based Prediction of Protein-Protein Interactions.

3.1.1. *Structural Homology*

The premise behind protein-protein interaction prediction by structural homology is fairly straightforward: if protein A interacts with protein B, and two new proteins X and Y each looks structurally like proteins A and B respectively, then protein X might also interact with protein Y; see Figure 4. Given that most proteins do not have known structures, the first step is to predict the structure of a protein from its primary sequence. This can be done by a computational process known as "threading", in which we align the sequence of the protein of interest to a library of known folds and find the closest matching structure. If the structures are known to interact—from existing 3D structures of protein complexes, say—we can then compute the interfacial energy and electrostatic charge to further confirm whether the partners form a stable complex. Lu *et al.*[527] use such a threading-based algorithm to assign putative structures for predicting interaction between yeast proteins. They have predicted 2,865 interactions, with 1,138 interactions verified in the Database of Interacting Proteins (DIP).[910]

However, as we are only interested in the parts of the proteins' structures that are involved in the interactions, we can focus on the structures on key areas of proteins such as the protein domains that are most likely to be involved in the protein-protein interactions. Aloy and Russell[22] use pairs of interacting Pfam domains[64] from known three-dimensional complex structures for prediction. Pairs of proteins with sequences homologous to a known interacting domain pair can then be scored for how well they preserve the atomic contacts at the predicted interaction interface by using empirical potentials to confirm the predicted interactions.

The current lack of protein 3-D structures clearly limits the global application of structure-based approaches for genome-wide protein-protein interaction predictions, even with an approach that uses structural homology. With the increasing pace of structure determination and structural genomics efforts, we hope that the structures for many more protein complexes will be available in the future. In the meantime, the vast amount of information in protein and gene sequences can be used as an alternative source for inferring protein-protein interactions.

3.2. Sequence-Based Predictions

While protein structures may be the most informative source for protein-protein interactions, protein sequences can also be used, together with existing protein interaction data, for predicting new protein interactions. In this section, we describe two representative sequence-based approaches: one approach is based on conventional sequence homology across various species, while a second approach uses protein domains as an abstraction of proteins for their interactions, and then reduces protein-protein interactions into domain-domain interactions which can in turn be used for predicting new interactions.

3.2.1. Interacting Orthologs

A widely used approach of assigning function to newly sequenced genes is by comparing their sequences with that of annotated proteins in other species. If the new gene or protein's sequence bears significant similarity to the sequence of a gene or protein—namely, its ortholog—in an annotated database of another species, it can be assumed that the two proteins are either the same genetic instantiation, or at the very least, share very similar properties and functions. As such, if protein A interacts with protein B, then the orthologs of A and B in another species are also likely to interact.

A study by Matthews *et al.*[552] has investigated the extent to which a protein interaction map generated in one species can be used to predict interactions in another species under the interacting orthologs or "interologs" principle. In their study, Matthews *et al.* compare protein-protein interactions detected in *S. cerevisiae* to interactions detected in *C. elegans* using the same experimental method yeast two-hybrid. Although only 31% of the high-confidence interactions detected in *S. cerevisiae* are also detected in *C. elegans*, it confirmed that some interactions are conserved between organisms, and we should expect more interologs between more closely related species than *S. cerevisiae* and *C. elegans*.

3.2.2. *Interacting Domain Pairs*

The interolog method described above scans proteins full-length to look for co-evolved interactions. Since protein interactions usually involve only small regions of the interacting molecules, conservation of interactions theoretically only requires that these key subregions on the interacting proteins be conserved. One approach is to treat proteins as collections of conserved domains, where each domain is responsible for a specific interaction with another domain. Protein domains are modules of amino acid sequence on proteins with specific evolutionarily conserved motifs—these protein domains are therefore quite likely the structural or functional units that participate in intermolecular interactions. As such, the existence of certain domains in proteins can be used to suggest the possibility of two proteins to interact or form a stable complex. In fact, Wojcik and Schächter[892] have shown that the use of domain profile pairs can provide better prediction of protein interactions than the use of full-length protein sequences.

Researchers have begun to use domain-domain interactions to predict protein-protein interactions with promising results.[199, 298, 609] For example, Deng *et al.*[199] predict yeast protein-protein interactions using inferred domain-domain interactions, and they achieve 42.5% specificity and 77.6% sensitivity using the combined data of Uetz *et al.*[849] and Ito *et al.*[386] showing that interacting domain pairs can be useful for computational prediction of protein-protein interactions. Note that the relatively low specificity may be caused by the fact that the observed protein-protein interactions in the Uetz-Ito combined data represent only a small fraction of all of the real interactions. However, one major drawback of this approach is that there are currently no efficient experimental methods for detecting domain-domain interactions—the number of experimentally derived interacting domain pairs is highly limited. As such, researchers can only used inferred domain-domain interactions in the prediction of protein-protein interactions, the accuracy of which may be further thwarted by the inference errors associated with the inferred domain-domain interactions.

3.3. *Genome-Based Predictions*

Given that a rapidly increasing number of genomes have already been sequenced, we can transcend conventional homology-based methods such as those described in the previous sections, and take into account the genomic context of proteins and genes within complete genomes for the prediction of interactions. By mining entire genomes of different species, we can discover cross-genome contextual information that are useful for predicting protein-protein interactions—usually indirectly through functional linkages—such as:

(1) *Gene locality context.* We can track the localities of genes in different species and use such information to infer functional linkages and possible interactions. We can explore the idea of co-localization or gene neighborhood, which is based on notion that genes which interact or are at least functionally associated will be kept in physical proximity to each other on the genome.[184] Alternatively, we can search for gene fusion events, whereby the fusion of two genes in one species can indicate possible interactions.

(2) *Phylogenetic context.* Instead of tracking the spatial arrangements of genes on the genomes, we can also track the evolutionary patterns of the genes, using the notion that genes that are functionally related tend to be inherited together through evolution.[660]

A third source of genome-wide information for protein-protein interaction prediction can also be gleaned, albeit indirectly, from gene expression experimental data:

(3) *Gene expression context.* Microarray technologies has enabled quantitative measurement of genome-wide gene expression levels simultaneously. To reveal the various functions of the genes and proteins, the gene expression profiles of a series of experimental conditions can be analyzed so that the genes can be grouped into clusters based on the similarity in their patterns of expression. The co-expression clusters can then be interpreted as potential functional linkages from which we may infer protein interactions.

Below, we describe the use of these three categories of genome-based information for the *in silico* detection of protein-protein interaction in details.

3.3.1. *Gene Locality Context: Gene Neighborhood*

One of the earlier attempts at genome-based prediction of protein-protein interactions is based on the notion of conservation of gene neighborhood. We can predict functional linkage between a pair of genes if their orthologs tend to be in close physical proximity in many genomes, as shown in Figure 5. In fact, studies have revealed that genes that participate in the same biological pathway tend to be neighbors or be clustered into discrete region along the genomic DNA. The most well-known example occurs in the bacterial and archael genomes, which are organized into regions such as operons that code for functionally-related proteins.[85]

As functionally-related proteins are clearly more likely to interact than unrelated ones, genes conserved as neighbors across genomes indicate possible interactions between their protein products. In a study by Dendekar *et al.*,[184] ~300 genes were identified to be conserved in neighboring clusters across different bac-

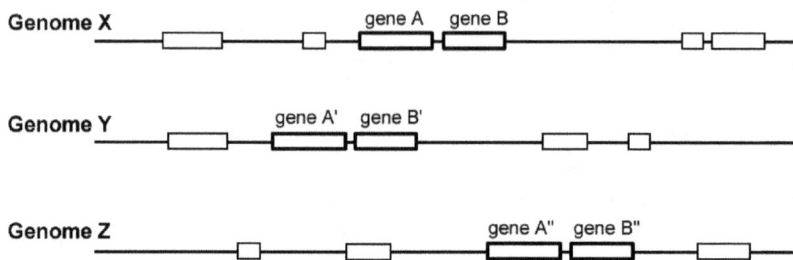

Fig. 5.   Interaction Prediction by Gene Neighborhood.

terial genomes. Out of these ∼300 genes, 75% have been previously described to be physically interacting, while another 80% of the remaining conserved neighbors have functions that are highly indicative of interactions between them. Their results show that gene neighborhood can be a powerful method for inferring protein-protein interactions in bacteria. In fact, Overbeek et al.[641] has successfully used this method to detect missing members of metabolic pathways in a number of prokaryotic species. While the gene neighborhood method has worked well with bacteria, this method may not be directly applicable to the higher eukaryotic species, in which the correlation between genome order and biological functions is less pronounced since the co-regulation of genes is not imposed at the genome structure level. For these other species, alternative genome-based methods must be used instead.

3.3.2. Gene Locality Context: Gene Fusion

One alternative method that is quite similar to the gene neighborhood approach is the so-called Rosetta Stone[541] or gene fusion[234] method. In fact, the complete fusion of two genes into one single unit can be deemed the ultimate form of gene proximity. It has been observed that many genes become fused through the course of evolution due to selective pressure—for example, fusion of two genes may allow the metabolic channeling of substrates or decrease the regulatory load in the cell. Gene fusion events have been observed frequently in evolution; some well-known examples include the fusion of tryptophan synthetase α and β subunits from bacteria to fungi,[126] and that of TrpC and TrpF genes in E. coli and

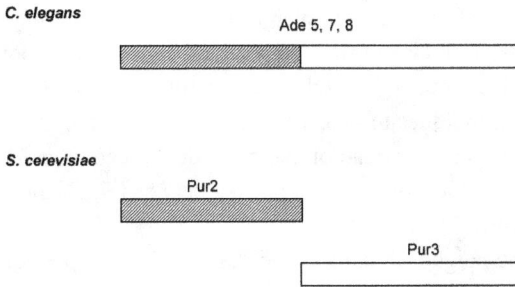

Fig. 6.   Interaction Prediction by Gene Fusion.

H. influenzae.[720] Figure 6 depicts another example gene fusion event: the proteins *Pur2* and *Pur3* are two separate interacting proteins in yeast (*S. cerevisiae*), but their orthologs are found fused into one protein in *C. elegans*.

The numerous observed examples suggest that the protein products of the fused genes either physically interact or are at least closely functionally associated. As such, computational detection of gene-fusion events in complete genomes can be used to infer functional linkage or even physical interaction between proteins. In a study by Marcotte *et al.*,[541] they detected ∼7,000 putative protein-protein interactions in *E. coli*, and ∼45,500 putative protein-protein interactions in yeast by gene fusion analysis, demonstrating that the gene fusion phenomenon is quite widespread.

Recently, a similar approach based on protein domains has been proposed. As we have explained in Section 3.2.2, protein domains are evolutionarily conserved modules of amino acid sequence on proteins that can be deemed the structural or functional units that participate in intermolecular interactions. To exploit the gene fusion concept in protein interaction prediction, we can treat a protein as a set of conserved domains, where each domain is responsible for a specific interaction with another one.[417, 541] In this domain fusion method, we computationally detect fused composite proteins in a reference genome with protein domains that correspond to individual full-length component proteins in other genomes.

Marcotte *et al.*[541] use predefined protein domains as a basis for searching fused—*i.e.*, multi-domain—proteins to detect gene fusion from a database of protein sequences. Using the SWISS-PROT database[86] annotated with domain information from ProDom,[765] they have detected ∼7,842 so-called Rosetta Stone domain fusion links in yeast and ∼750 high-confidence ones in *E. coli*, indicating that the domain fusion phenomenon—even using only ProDom domains—is widely observed and suitable as a basis for predicting protein interactions. However, the use of pre-defined domains such as those in ProDom may limit the coverage of the approach, since a portion of proteins may not have pre-assigned domains. Enright and Ouzounis[233] used an alternative approach that employed sequence alignment techniques to detect regions of local similarities between proteins from different species instead of using pre-defined domains. They have successfully detected 39,730 domain fusion links between 7,224 proteins from the genomes of 24 species.[233]

Unlike the gene neighborhood method described in the previous section, the gene or domain fusion method does not require the individual genes to be proximal along the chromosomes. As such, the method can be applied to eukaryotic genomes.[234] The occurrence of shared domains in distinct proteins is a phenomenon whose true extent in prokaryotic organisms is still unclear,[853] limiting the use of the domain fusion method for protein-protein interaction predictions in the prokaryotes. This shows that just as it is in the case for experimental approaches, the coverage of various computational detection methods can also differ—it is therefore necessary to explore multiple complementary approaches such that complete information about interaction networks can be obtained.

3.3.3. Phylogenetic Context: Phylogenetic Profiles

During evolution, functionally-linked proteins tend to be either preserved or eliminated in a new species.[660] This means that if two proteins are functionally associated, their corresponding orthologs will tend to occur together in another genome. We can exploit such evolutionary patterns to predict if proteins interact.

One approach, called phylogenetic profiling, is to detect the presence or absence of genes in related species for suggesting possible interaction. The phylogenetic profiling method is based on the notion that interacting or functionally linked proteins must be jointly present or jointly absent in different organisms. A phylogenetic profile describes an occurrence of a certain protein in a set of organisms. Proteins whose genes have highly correlated phylogenetic profiles can then be inferred as physically interacting or at least functionally linked.

The phylogenetic profile of a protein is typically represented as a string that encodes the presence or absence—in form of 1 or 0—of a protein in a given number of genomes; see Figure 7. Using this binary vector representation, the phylogenetic profiles of proteins are computationally constructed across different genomes. Then, proteins that share similar profiles are clustered together, and functional linkage or even physical interaction can be predicted for proteins that are clustered together, as shown in the figure.

In a study by Pellegrini et al.,[660] they apply this method to detect possible

org W
prot A
prot D
prot C

org X
prot B
prot D
prot E

org Y
prot A
prot B
prot C

org Z
prot E
prot D

	organism			
	W	X	Y	Z
prot A	1	0	1	0
prot B	0	1	1	0
prot C	1	0	1	0
prot D	1	1	0	1
prot E	0	1	0	1

prot A	1	0	1	0
prot C	1	0	1	0

prot A ←→ prot C

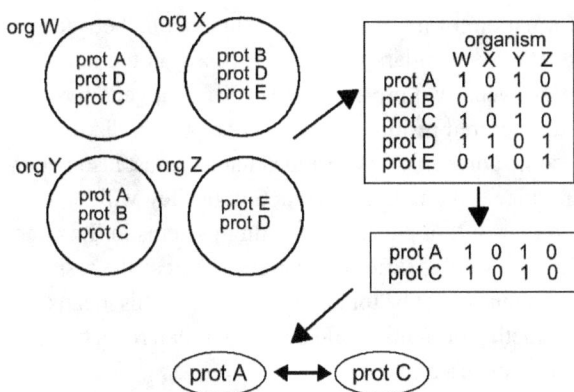

Fig. 7. Interaction Prediction by Phylogenetic Profiling.

functional linkages between 4,290 *E. coli* proteins using 16 different genomes. They demonstrate that comparing with random groups of proteins, the clusters of proteins formed by similar phylogenetic profiles tend to share the same functional annotation under SWISS-PROT. Since this method requires the detection of the absence of a protein in a genome, it can only be applied to complete genomes. However, this limitation should not be a key concern as an increasing number of complete genome sequences are becoming available. In fact, the method is expected to become more powerful since more completely-sequenced genomes will allow for larger and more accurate profiles to be constructed for each protein.

One limitation with phylogenetic profiling is its inability to detect linkages for proteins that are essential and common to most species. This group of proteins constitute a major portion of entire gene set in a genome. Also, as with most of the other computational methods, phylogenetic profiling can only be used to suggest possible functional linkages—a direct physical interaction between the proteins is not necessarily implied. It is prudent for the practical bioinformatician to be mindful when using these information for further discoveries.

3.3.4. *Phylogenetic Context: Phylogenetic Tree Similarity*

The co-evolution of interacting protein pairs has long been observed in such well-known interacting protein pairs as dockerins and cohexins,[644] as well as insulin and its receptors[266]—the corresponding phylogenetic trees of these proteins show a significantly greater degree of similarity than non-interacting proteins are expected to show. As such, phylogenetic tree similarity is another suitable form of evolutionary information for inferring possible interaction between two proteins.

The phylogenetic tree similarity method is based on the notion of co-related residues changes between two proteins across different genomes. The reasoning is as follows: if an incurred residue change in one protein disrupts its interaction with its partner, some compensatory residue changes must also occur in its interacting partner in order to sustain the interaction or they will be selected against and eliminated. As a result, a pair of interacting proteins in the course of evolution would go through similar series of changes, whereas the residue changes for non-interacting proteins would be totally uncorrelated. This means that the phylogenetic tree of interacting proteins would be very similar, reflecting their similarity in their evolutionary histories.

While phylogenetic profiling looks for proteins that co-exist (or otherwise) in different genomes, the phylogenetic tree similarity method looks for co-related residues changes between two proteins across different genomes. Although the name of the method may imply a direct comparison of the phylogenetic tree structures, we can measure tree similarity by comparing the correlation between the distance matrices of protein orthologs from different species. Distance matrices are typically used in the construction of phylogenetics trees—a distance matrix is an $n \times n$ matrix that contains the pairwise distances between n sequences in a set. The distance between two sequences are measured by their sequence alignment scores, which could simply be the number of mismatchs in the alignment. In this way, we can account for and compare the structure of the underlying phylogenetic trees between orthologous proteins. Note that the phylogenetic profiling method described in the previous section can be considered as a simplification of the phylogenetic tree similarity method, where the "distance matrix" for each protein is merely a binary vector indicating the presence or absence of the protein ortholog in a particular species.

In a study by Goh *et al.*,[294] they apply this procedure—also known as mirrortree—to the two interacting domains of phosphoglycerate kinase. They found a high correlation coefficient of 0.8 between two corresponding distance matrices between the two interacting protein domains. This value was later confirmed by Pazos *et al.*[650] in a larger scale experiment for predicting protein interactions in *E. coli*. In a control set of 13 known interactions, they found that the interacting protein pairs have high correlation values in their distance matrices—in fact, 9 out of 13 have correlation coefficient values higher than 0.77. A total of 67,000 unknown pairs of proteins are then compared across 14 genomes; 2,742 pairs have correlation coefficient values greater than 0.8—they can therefore be inferred as interacting protein pairs.

The basic steps in this method are shown in Figure 8. To determine if a pair of

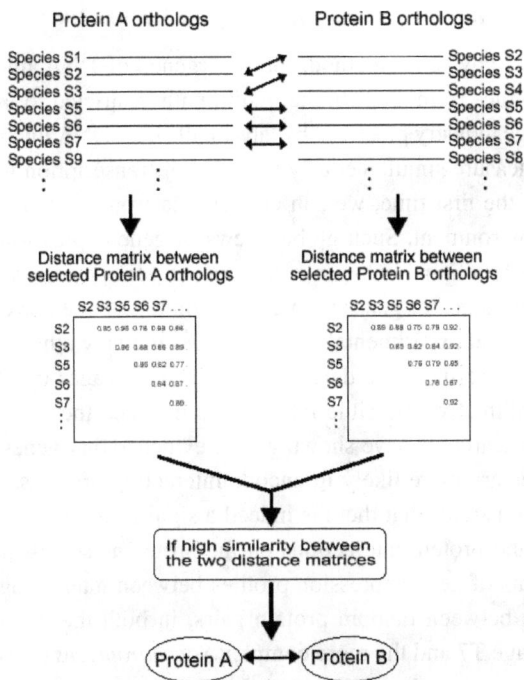

Fig. 8. Interaction Prediction by Phylogenetic Tree Similarity.

candidate proteins interact, we search for their orthologs across different genomes to form as many ortholog pairs from coincident species as possible. For each protein, we construct its distance matrix by pairwise sequence comparison between its various orthologs found. If the two proteins have a high correlation coefficient value between their two distance matrices, they can be predicted to be a possibly interacting pair.

A clear advantage of this method over the phylogenetic profiling method is that it does not require the presence of fully-sequenced genomes. Furthermore, while many of the other genome-based methods detect functional linkages and infer interactions indirectly, the phylogenetic tree similarity method predicts interactions that are more likely to be physically interacting. However, a main limitation of this method is that we must find a sufficient number orthologs pairs in order to make a reasonable postulation about interaction. In the study by Pazos *et al.*,[650] 11 was the minimum number of species required. As the sequence of many other species become available, this requirement will soon not be a constraint of the phylogenetic tree similarity method.

3.3.5. *Gene Expression: Correlated mRNA Expression*

Perhaps more than any other method, the emergence of DNA microarrays has transformed genomics from a discipline traditionally restricted to large sequencing labs to a cottage industry practiced by labs of all sizes. DNA microarrays allow investigators to measure simultaneously the level of transcription for every gene in a genome. For the first time, we can collect data from a whole genome as it responds to its environment. Such global views of gene expression can be used for elucidating the functional linkages of the various genes by applying clustering techniques to group together genes with expression levels that correlate with one another under different experimental conditions. As in many other genome-based methods we have described, the detected functional linkages can then be used to infer—albeit indirectly—possible interactions between the proteins that they encode. Several researchers have shown global evidence that genes with similar expression profiles are more likely to encode interacting proteins. For example, Grigoriev[301] demonstrates that there is indeed a significant relationship between gene expression and protein interactions on the proteome scale—the mean correlation coefficients of gene expression profiles between interacting proteins are higher than those between random protein pairs, in both the genomically simplistic bacteriophage T7 and the more complex *Saccharomyces cerevisiae* (yeast) genomes. In a separate study by Ge *et al.*[280] on yeast, they compare the interactions between proteins encoded by genes that belong to common expression-profiling clusters with those between proteins encoded by genes that belong to different clusters, and found that proteins from the intra-group genes are more than 5 times likely to interact with each other than proteins from the inter-group genes.

In another work to relate whole-genome expression data with protein-protein interactions, Jansen *et al.*,[390] find that while the subunits of the permanent protein complexes do indeed share significant correlation in their RNA expression, the correlation expression method is understandably relatively weak in detecting transient interactions. However, they have also observed weak correlated RNA expression patterns between interacting proteins determined by genome-wide yeast two-hybrid studies, indicating potential limitations in using this approach for protein-protein interaction prediction. On the other hand, while this method by itself is relatively weak for accurate interaction detection, it can serve as an excellent complementary method to validate interaction generated from other experimental methods. In a comprehensive study conducted by Kemmeren *et al.*,[426] up to 71% of biologically-verified interactions can be validated with the gene co-expression approach. Integration of expression and interaction data is thus a way to improve the confidence of protein-protein interaction data generated by high-throughput technologies.

4. Conclusion

Before the advent of high-throughput experimental and computational methods, protein-protein interactions have always been studied in the molecular biology laboratories in a relatively small scale using conventional experimental techniques. However, in order to understand, model, and predict the many unfathomable rules that govern protein-protein interactions inside the cell on the genomic level, large scale protein interaction maps must be generated. In this chapter, we have provided an overview of the various current high-throughput protein-protein interaction detection methods. In particular, we have shown that both the conventional experimental approaches and the new computational approaches can be useful for mapping the vast interactomes. We have also shown that there is no single best method for large-scale protein-protein interaction detection—each method, experimental or otherwise, has its own advantages and disadvantages.

The advent of the various high-throughput detection and prediction technologies has brought about a major paradigm shift in modern molecular biology research from single-molecule experiments to genome and proteome-level investigations. With the current high throughput approaches powerful enough to generate more data than those accumulated over many decades from small scale experiments, predictive research has become a mainstay of knowledge discovery in modern molecular biology. This has led to the tendency for experiments to be technology-driven rather than hypothesis-driven, with datasets routinely generated without much specific knowledge about the functions of genes being investigated. This can be problematic because the high-throughput data have been shown to exhibit high error rates. For example, a recent rigorous study by Sprinzak *et al.*[793] has revealed that the reliability of the popular high-throughput yeast-two-hybrid assay is only about 50%. Another comprehensive survey on current protein-protein interaction detection technologies done by von Mering *et al.*[864] showed that different experimental methods cover rather different classes of protein interactions. This indicates the possibility of high false negative rates in the interaction data in addition to the many false positive detections. As practicing bioinformaticians, we should always be mindful about how the data that we are analyzing are generated in order to have a good grasp of the data quality. In this way, we can then be sufficiently vigilant in detecting the inherent data artifacts to avoid making spurious conclusions.

Interaction data from traditional small-scale experiments are generally more reliable because their biological relevance is often very thoroughly investigated by the researchers. In fact, the published results are oftentimes based on repeated observations by multiple research groups. The current explosive rate of data generation fueled by the powerful high-throughput interaction detection technologies has made it impractical for their verification by traditional methods in small scale experiments. Nevertheless, we can still generate high-quality interaction data by using an integrative approach. In the von Mering study,[864] interactions confirmed by two or more detection methods are found to have a higher percentage of true positives than those that are detected by only individual methods. Interactions confirmed by three or more detection methods have an even higher degree of accuracy. This means that in order to generate an accurate map of the interactomes, each experiment indicating a particular protein-protein interaction must be confirmed by at least two or more independent means, computationally and/or experimentally. Fortunately, as we have shown in this chapter, the concerted efforts by the industrious biologists and bioinformaticians have already resulted in a wide array of methods for discovering protein-protein interactions in high throughput, each with its own strengths and specialties. These methods, together with the continuing efforts by investigators in developing and refining further innovative interaction discovery techniques—for example, automatic text mining for discovering annotated protein interactions from the literature,[536, 543, 608] and the formulation of mathematical measures for assessing the reliability of protein-protein interactions in terms of the underlying interaction network topology,[734, 735] to name just a couple—will, in the near future, lead us to a complete and accurate map of the interactome.

CHAPTER 14

TECHNIQUES FOR ANALYSIS OF GENE EXPRESSION DATA

Jinyan Li

Institute for Infocomm Research
jinyan@i2r.a-star.edu.sg

Limsoon Wong

Institute for Infocomm Research
limsoon@i2r.a-star.edu.sg

The development of microarray technology in the last decade has made possible the simultaneous monitoring of the expression of thousands of genes. This development offers great opportunities in advancing the diagnosis of dieases, the treatment of diseases, and the understanding of gene functions. This chapter provides an in-depth survey of several approaches to some of the gene expression analysis challenges that accompany these opportunities.

ORGANIZATION.

Section 1. We begin with a brief introduction to microarrays in terms of how they are made and how they are used for measuring the expression level of thousands of genes simultaneosly.

Section 2. Then we discuss how to diagnose disease subtypes and states by microarray gene expression analysis. We present a standard approach that combines gene selection and subsequent machine learning. Besides applying gene selection and machine learning methods from Chapter 3, we also present the shrunken centroid method of Tibshirani *et al.* in some detail. The subtype diagnosis of childhood acute lymphoblastic leukaemia is used as an example.

Section 3. Next we consider the problem of discovering new disease subtypes by means of microarray gene expression analysis. We relate the discovery of a novel subtype of childhood acute lymphoblastic leukaemia via a hierarchical clustering approach. We also describe the discovery of novel transcription factor binding sites and novel gene funtional groupings via a fuzzy k-means clustering approach.

Section 4. Then we look at the problem of infering how the expression of one gene influences the expression of another gene. We present two approaches; one is based on mining of association rules, the other is based on an ingenious use of normal classifiers. We also describe the concept of interaction generality in the related problem of detecting false positives from high-throughput protein-protein interaction experiments.

Section 5. Lastly, we present a highly speculative use of gene expression patterns to formulate treatment plans.

1. Microarray and Gene Expression

Microarrays or DNA chips are powerful tools for analyzing the expression profiles of gene transcripts under various conditions.[518, 890] These microarrays contain thousands of spots of either cDNA fragments corresponding to each gene or short synthetic oligonucleotide sequences. By hybridizing labeled mRNA or cDNA from a sample to a microarray, transcripts from all expressed genes can be assayed simultaneously. Thus one microarray experiment can yield as much information as thousands of Northern blots. It is hopeful that better diagnosis methods, better understanding of disease mechanisms, and better understanding of biological processes, can be derived from a careful analysis of microarray measurements of gene expression profiles.

There are two main types of microarray. The first type is based on the scheme of Fodor *et al.*[253] that uses lithographic production techniques to synthesize an array of short DNA fragments called oligos. Here is a brief outline of their scheme. First a silicon surface is coated with linker molecules that bind the four DNA building blocks, adenine (A), cytosine (C), guanine (G), and thymine (T). These linkers are initially capped by a "blocking" compound that can be removed by exposure to light. By shining light through a mask, those areas of the silicon surface that correspond to holes in the mask become exposed. The chip is then incubated with one of the four DNA building blocks, say adenine, which then binds to those exposed areas. After that, the blocking compound is reapplied. By repeating this process with different masks and different DNA building blocks, an array of different oligos can be built up easily, as shown in Figure 1.

Each oligo can bind stretches of DNA that have complementary sequences to the oligo in the usual Crick-Watson way. Then the following procedure is followed to use the microarray to monitor the expression of multiple genes in a sample. RNAs are isolated from samples, converted into cDNAs, and conjugated to biotin. These biotin-conjugated cDNAs are then fragmented by heat, and hybridized with the oligos on the microarray. A washing step then follows to get rid of unbound cDNAs. The strands that are bound to the microarray can then be stained by a streptavidin-linked fluorescent dye, and detected by exciting the fluorescent tags with a laser. Since the sequence of each oligo on the microarray is known by construction, it is easy to know the sequence of the cDNA that is bound to a

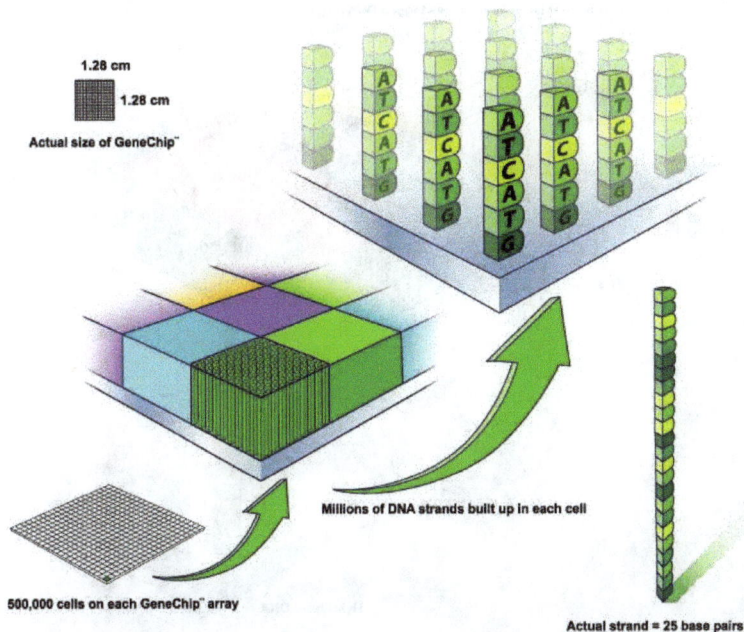

Fig. 1. A cartoon of the oligos on a microarray. Notice that the sequence of each oligo and their position on the microarray are known by construction. The oligo-based microarrays made by Affymetrix are called GeneChip® microarrays, and the oligos are 25 nucleotides in length. (*Image credit: Affymetrix.*)

particular position of the microarray, as shown in Figure 2.

The second popular type of microarrays is based on the scheme developed at Stanford.[752,770] Here, cDNAs are directly spotted onto a glass slide, which is treated with chemicals and heat to attach the DNA sequences to the glass surface and denature them. This type of microarray is primarily used for determining the relative level of expression of genes in two contrasting samples. The procedure is as follows. The fluorescent probes are prepared from two different mRNA sources with the use of reverse transcriptase in the presence of two different fluorophores. The two set of probes are then mixed together in equal proportions, hybridized to a single array, and scanned to detect fluorescent color emissions corresponding to the two fluorophores after independent excitation of the two fluorophores. The differential gene expression is then typically calculated as a ratio of these two fluorescent color emissions.

Shining a laser light at GeneChip causes tagged DNA fragments that hybridized to glow

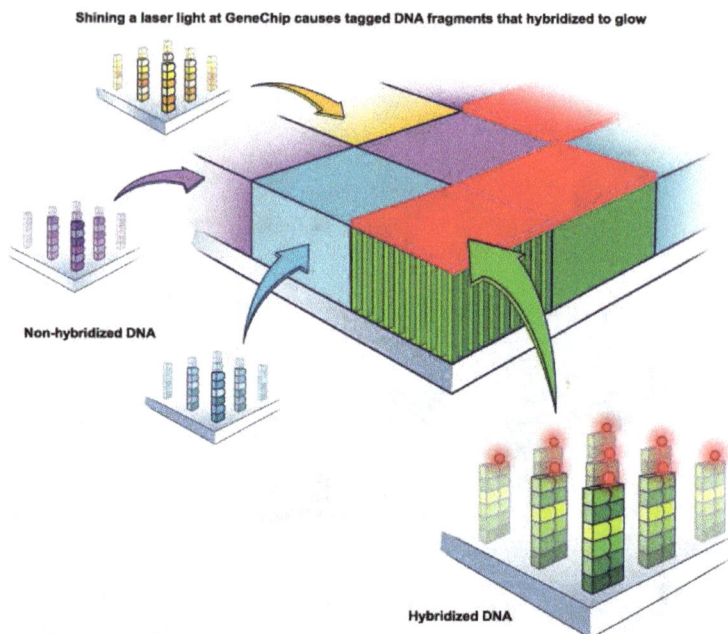

Non-hybridized DNA

Hybridized DNA

Fig. 2.   A cartoon depicting scanning of tagged and un-tagged probes on an Affymetrix GeneChip®️ microarray. (*Image credit: Affymetrix.*)

2. Diagnosis by Gene Expression

A single microarray experiment can measure the expression level of tens of thou-sands of genes simultaneously.[518, 698] In other words, the microarray experiment record of a patient sample—see Figure 3 for an example—is a record having tens of thousands of features or dimensions. A major excitement due to microarrays in the biomedical world is the possibility of using microarrays to diagnose disease states or disease subtypes in a way that is more efficient and more effective than conventional techniques.[20, 269, 297, 663, 918]

Let us consider the diagnosis of childhood leukaemia subtypes as an il-lustration. Childhood leukaemia is a heterogeneous disease comprising more than 10 subtypes, including T-ALL, E2A-PBX1, TEL-AML1, BCR-ABL, MLL, Hyperdiploid>50, and so on. The response of each subtype to chemotherapy is different. Thus the optimal treatment plan for childhood leukaemia depends criti-cally on the subtype.[686] Conventional childhood leukaemia subtype diagnosis is a difficult and expensive process.[918] It requires intensive laboratory studies com-prising cytogenetics, immunophenotyping, and molecular diagnostics. Usually,

probe	pos	neg	pairs in avg	avg diff	abs call	Description
...	...	...	...	...	...	...
107_at	4	4	15	3723.3	A	Z95624 Human DNA ...
108_g_at	5	2	15	1392.4	A	Z95624 Human DNA ...
109_at	6	2	16	2274.7	M	Z97074 Human mRNA ...
...	...	...	...	...	...	...

Fig. 3. A partial example of a processed microarray measurement record of a patient sample using the Affymetrix® GeneChip® U95A array set. Each row represents a probe. Typically each probe represents a gene. The U95A array set contains more than 12,000 probes. The 5th column contains the gene expression measured by the corresponding probe. The 2nd, 3rd, 4th, and 6th columns are quality control data. The 1st and last columns are the probe identifier and a description of the corresponding gene.

these diagnostic approaches require the collective expertise of a number of professionals comprising hematologists, oncologists, pathologists, and cytogeneticists. Although such combined expertise is available in major medical centers in developed countries, it is generally unavailable in less developed countries. It is thus very exciting if microarrays and associated automatic gene expression profile analysis can serve as a single easy-to-use platform for subtyping of childhood ALL.

2.1. *The Two-Step Approach*

The analysis of gene expression profiles for the diagnosis of disease subtypes or states generally follows a two-step procedure first advocated by Golub *et al.*,[297] *viz.*

(1) selecting relevant genes and
(2) training a decision model using these genes.

The step of selecting relevant genes can be performed using any good feature selection methods such as those presented in Chapter 3—signal-to-noise measure,[297] t-test statistical measure,[133] entropy measure,[242] χ^2 measure,[514] information gain measure,[692] information gain ratio,[693] Fisher criterion score,[251] Wilcoxon rank sum test,[742] principal component analysis,[399] and so on. The step of decision model construction can be performed using any good ma-

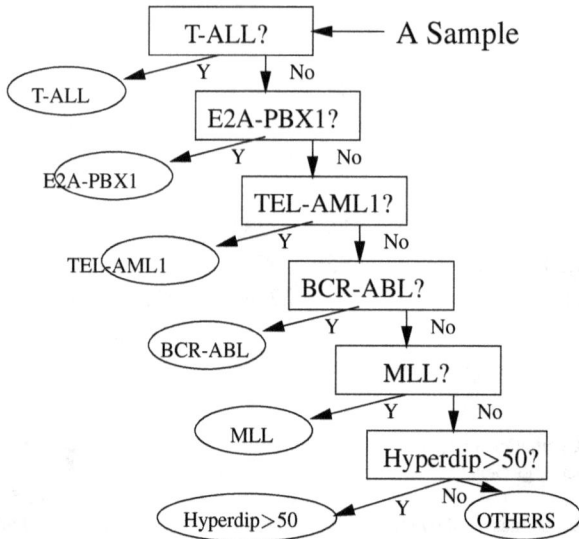

Fig. 4. The classification of the ALL subtypes is organized in a tree. Given a new sample, we first check if it is T-ALL. If it is not classified as T-ALL, we go to the next level and check if it is an E2A-PBX1. If it is not classified as E2A-PBX1, we go to the third level and so on.

chine learning methods such as those presented in Chapter 3—decision tree induction methods,[693] Bayesian methods,[214] support vector machines (SVM),[855] PCL,[492, 497] and so on.

We illustrate this two-step procedure using the childhood acute lymphoblastic leakaemia (ALL) dataset reported in Yeoh et al.[918] The whole dataset consists of gene expression profiles of 327 ALL samples. These profiles were obtained by hybridization on the Affymetrix® GeneChip® U95A array set containing probes for 12558 genes. The data contain all the known acute lymphoblastic leukaemia subtypes, including T-ALL, E2A-PBX1, TEL-AML1, BCR-ABL, MLL, and Hyperdiploid>50. The data are divided by Yeoh et al. into a training set of 215 instances and an independent test set of 112 samples. The original training and test data are layered in a tree-structure, as shown in Figure 4. Given a new sample, we first check if it is T-ALL. If it is not classified as T-ALL, we go to the next level and check if it is an E2A-PBX1. If it is not classified as E2A-PBX1, we go to the third level and so on.

Li et al.[492] are the first to study this dataset. At each level of the tree, they first use the entropy measure[242] and the χ^2 measure[514] to select the 20 genes that are most discriminative in that level's training data. Then they apply the PCL

Testing Data	Error rate of different models			
	C4.5	SVM	NB	PCL
T-ALL vs OTHERS1	0:1	0:0	0:0	0:0
E2A-PBX1 vs OTHERS2	0:0	0:0	0:0	0:0
TEL-AML1 vs OTHERS3	1:1	0:1	0:1	1:0
BCR-ABL vs OTHERS4	2:0	3:0	1:4	2:0
MLL vs OTHERS5	0:1	0:0	0:0	0:0
Hyperdiploid>50 vs OTHERS	2:6	0:2	0:2	0:1
Total Errors	14	6	8	4

Fig. 5.   The error counts of various classification methods on the blinded ALL test samples are given in this figure. PCL is shown to make considerably less misclassifications. The OTHERSi class contains all those subtypes of ALL below the ith level of the tree depicted in Figure 4.

classifier[492] on the training data using those 20 genes to construct a decision model to predict the subtypes of test instances of that level. The entropy measure, the χ^2 measure, and the PCL classifier are described in Chapter 3.

For comparison, Li *et al.* have also applied several popular classification methods described in Chapter 3—C4.5,[693] SVM,[855] and Naive Bayes (NB)[214]—to the same datasets after filtering using the same selected genes. In each of these comparison methods, the default settings of the weka package are used. The weka package can be obtained at http://www.cs.waikato.ac.nz/ml/weka.

The number of false predictions on the test instances, after filtering by selecting relevant genes as described above, at each level of the tree by PCL, as well as those by C4.5, SVM, and NB, are given in Figure 5. The results of the same algorithms but without filtering by selecting relevant genes beforehand are given in Figure 6, which clearly shows the beneficial impact of the step of gene selection.

2.2. *Shrunken Centroid Approach*

Tibshirani *et al.*[827] also use the same two-step approach to diagnose cancer type based on microarray data. However, the details of their approach are different from those basic methods already described in Chapter 3. Their approach performs well, is easy to understand, and is suitable for the situation where there are more than two classes.

Testing Data	Error rate of different models		
	C4.5	SVM	NB
T-ALL vs OTHERS1	0:1	0:0	13:0
E2A-PBX1 vs OTHERS2	0:0	0:0	9:0
TEL-AML1 vs OTHERS3	2:4	0:9	20:0
BCR-ABL vs OTHERS4	1:3	2:0	6:0
MLL vs OTHERS5	0:1	0:0	6:0
Hyperdiploid>50 vs OTHERS	4:10	12:0	7:2
Total Errors	26	23	63

Fig. 6. The error counts of various classification methods on the blinded ALL test samples without filtering by selecting relevant genes are given in this figure. The OTHERSi class contains all those subtypes of ALL below the ith level of the tree depicted in Figure 4.

The gist of the approach of Tibshirani et al. is as follows. Let X be a matrix of gene expression values for p genes and n samples. Let us write $X[i, j]$ for the expression of gene i in sample j. Suppose we have k classes and we write C_h for the indices of the n_h samples in class h. We create a "prototype" gene expression profile vector Y_h, also called a "shrunken" centroid, for each class h. Then given a new test sample t, we simply assign to it the label of the class whose prototype gene expression profile is closest to this test sample.

Let us use the notations $\langle e_i \mid i = 1 \ldots n \rangle$ to mean the vector $\langle e_1, \ldots, e_n \rangle$. Then the usual centroid for a class h is the vector

$$Z_h = \left\langle \frac{\sum_{j \in C_h} X[i, j]}{n_h} \,\middle|\, i = 1 \ldots p \right\rangle$$

where the ith component $Z_h[i]$ is the mean expression value in class h for gene i. And the overall centroid is a vector O where the ith component $O[i] = \sum_{j=1}^{n} X[i, j]/n$ is the mean expression value of gene i over samples in all classes.

In order to give higher weight to genes whose expression is stable within samples of the same class, let us standardize the centroid of each gene i by the within-class standard deviation in the usual way, viz.

$$d_{ih} = \frac{Z_h[i] - O[i]}{\sqrt{\dfrac{1}{n_h} + \dfrac{1}{n}} \times (s_i + s_0)}$$

where s_i is the pooled within-class standard deviation for gene i, *viz.*

$$s_i^2 = \frac{1}{n-k} \times \sum_h \sum_{j \in C_h} (X[i,j] - Z_h[i])^2$$

and the value s_0 is a positive constant—with the same value for all genes—included to guard against the possibility of large d_{ih} values arising by chance from genes with low expression levels. Tibshirani *et al.* suggest to set s_0 to the median value of s_i over the set of genes.

Thus d_{ih} is a t-statistics for gene i, comparing class h to the overall centroid. We can rearrange the equation as

$$Z_h[i] = O[i] + \sqrt{\frac{1}{n_h} + \frac{1}{n}} \times (s_i + s_0) \times d_{ih}$$

Tibshirani *et al.* shrink d_{ih} toward zero, giving d'_{ih} and yielding the shrunken centroid or prototype Y_h for class h, where

$$Y_h[i] = O[i] + \sqrt{\frac{1}{n_h} + \frac{1}{n}} \times (s_i + s_0) \times d'_{ih}$$

The shrinkage they use is a soft thresholding: each d_{ih} is reduced by an amount Δ in absolute value and is set to 0 if it becomes less than 0. That is,

$$d'_{ih} = \begin{cases} sign(d_{ih})(|d_{ih}| - \Delta) & \text{if } |d_{ih}| - \Delta > 0 \\ 0 & \text{otherwise} \end{cases}$$

Because many of the $Z_h[i]$ values are noisy and close to the overall mean $O[i]$, soft thresholding usually produces more reliable estimates of the true means. This method has the advantage that many of the genes are eliminated from the class prediction as the shrinkage parameter Δ is increased. To see this, suppose Δ is such that $d'_{ih} = 0$. Then the shrunken centroid $Y_h[i]$ for gene i for any class h is $O[i]$, which is independent of h. Thus gene i does not contribute to the nearest shrunken centroid computation. By the way, Δ is normally chosen by cross-validation.

Let t be a new sample to be classified. Let $t[i]$ be the expression of gene i in this sample. We classify t to the nearest shrunken centroid, standardizing by $s_i + s_0$ and correcting for class population biases. That is, for each class h, we first compute

$$\delta_h(t) = \sum_{i=1}^{p} \left(\frac{t[i] - Y_h[i]}{s_i + s_0} \right)^2 - 2 \times \log(\pi_h)$$

The first term here is simply the standardized squared distance of t to the hth shrunken centroid. The second term here is a correction based on the class prior probability π_h, which gives the overall frequency of class h in the population, and is usually estimated by $\pi_h = n_h/n$. Then we assign t the label h that minimizes $\delta_h(t)$. In other words, the classification rule is

$$C(t) = \alpha, \text{ where } \delta_\alpha(t) = \min_h \delta_h(t)$$

This approach appears to be very effective. On the acute lymphoblastic leukaemia dataset of Golub et al.,[297] Tibshirani et al.[827] report that at $\Delta = 4.06$— the point at which cross-validation error starts to rise quickly—yields 21 genes as relevant. Furthermore, these 21 genes produce a test error rate of 2/34. In comparison, Golub et al.[297] use 50 genes to obtain a test error rate of 4/34. On the childhood small round blue cell tumours dataset of Khan et al.,[430] Tibshirani et al. report that at $\Delta = 4.34$ there are 43 genes that are relevant. Furthermore, these 43 genes produce a test error rate of 0. This result is superior to that of Khan et al., who need 96 genes to achieve the same test error rate.

3. Co-Regulation of Gene Expression

In the preceding section we see that it is possible to diagnose disease subtypes and states from gene expression data. In those studies, we assume that all the disease subtypes are known. However, in real life, it is possible for a heterogeneous disease to have or to evolve new subtypes that are not previously known. Can computational analysis of gene expression data help uncover such new disease subtypes? Similarly, there are still many genes and their products whose functions are unknown. Can computational analysis of gene expression data help uncover functionally related gene groups? and can we infer the functions and regulation of such gene groups? Unsupervised machine learning methods, especially clustering algorithms, are useful for these problems. This section present two examples.

3.1. Hierarchical Clustering Approach

Let us use the childhood ALL dataset from Yeoh et al.[918] from Subsection 2.1 for illustration. As mentioned earlier, childhood ALL is a heterogeneous disease comprising 6 known major subtypes, viz. T-ALL, hyperdiploid with > 50 chromosomes, BCR-ABL, E2A-PBX1, TEL-AML1, and MLL gene rearrangements. However, the dataset from Yeoh et al. also contain some samples that are not assigned to any of these subtypes—these are the group marked as "OTHERS" in Figure 4.

This "OTHERS" group presents an opportunity for identifying new subtypes of childhood ALL. To do so, Yeoh et al.[918] perform a hierarchical clustering on their 327 childhood ALL samples using all the 12558 genes measured on their Affymetrix® GeneChip® U95A array set and using Pearson correlation as the

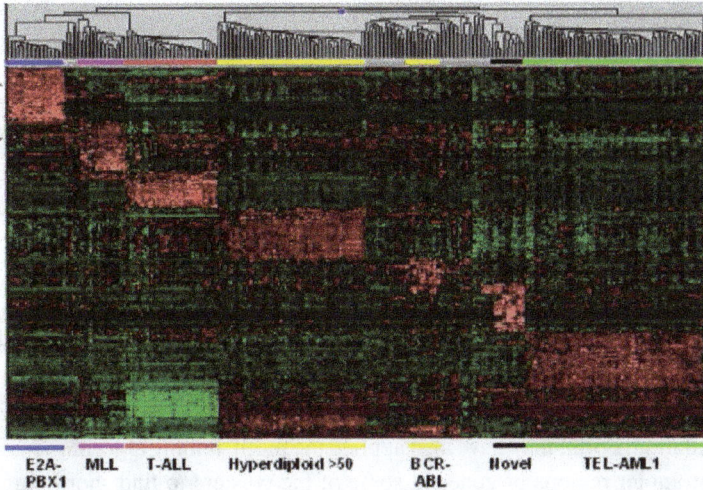

Fig. 7. Hierarchical clustering of 327 childhood ALL samples and genes chosen by χ^2 measure. Each column represents a sample, each row represents a gene. Note the 14 cases of the novel subtype.

distance between samples. Remarkably, this analysis clearly identifies the 6 major childhood ALL subtypes mentioned above. Moreover, within the "OTHERS" group, a novel subgroup of 14 cases are identified that have a distinct gene expression profile. These 14 cases have normal, pseudodiploid, or hyperdiploid karyotypes, and lack any consistent cytogenetic abnormality.

Figure 7 depicts the result of a hierarchical clustering of the 327 childhood ALL samples. To improve visualization clarity, instead of presenting a clustering involving all 12558 genes, only the top 40 genes selected using the χ^2 measure for each of the 6 major groups and the novel group are retained in this figure. The 14 cases of the novel subtype is clearly visible.

Thus, clustering algorithms can be used to discover new disease subtypes and states. As an introduction to hierarchical cluster algorithms and to the χ^2 measure can be found in Chapter 3, we omit them in this chapter. The definition of Pearson correlation is given in the next subsection—however, for the current subsection, the G and H in that formula should be interpreted as vectors representing the expression values of genes in sample g and sample h, and thus $G[i]$ is the expression of gene i in sample g and $H[i]$ is the expression of gene i in sample h.

3.2. Fuzzy K-Means Approach

Gasch and Eisen[277] use a technique called fuzzy k-means[80] to cluster a large collection of gene expression data obtained under a variety of experimental conditions. The dataset comprises 6153 genes in 93 microarray experiments taken from genomic expression data of wild-type *S. cerevisiae* responding to zinc starvation,[532] phosphate limitation,[625] DNA damaging agents,[275] and stressful environmental changes.[276]

They have obtained several very interesting results from analysing the resulting clusters. First, they have identified some meaningful clusters of genes that hierarchical and standard k-means clustering methods are unable to identify. Second, many of their clusters that correspond to previously recognized groups of functionally-related genes are more comprehensive than those clusters produced by hierarchical and standard k-means clustering methods. Third, they are able to assign many genes to multiple clusters, revealing distinct aspects of their function and regulation. Forth, they have also applied the motif-finding algorithm MEME[39] to the promoter regions of genes in some of the clusters to find short patterns of 6 nucleotides that are over represented and thus identified a few potentially novel transcription factor binding sites.

Before we proceed to describe the fuzzy k-means clustering method, let us first fix some notations. Let X be a matrix of gene expression values of $|X|^r$ genes under a variety of $|X|^c$ conditions. We write $X[g,i]$ for the expression of gene g in condition i in X. We write $X[g,\_]$ for the vector $\langle X[g,i] \mid i = 1 \ldots |X|^c \rangle$ representing the expression pattern of gene g in all the conditions in X. We write $X[\_,i]$ for the vector $\langle X[g,i] \mid g = 1 \ldots |X|^r \rangle$ representing the expression of genes in condition i in X. Similar notations are used for other two dimensional matrices. Also, for a vector G, we write $|G|$ for its size and $G[j]$ for its jth element.

For any two vectors G and H of gene expression patterns of genes g and h over the same conditions, so that $G[i]$ is the expression of gene g in condition i and $H[i]$ is the expression of gene h in condition i, the Pearson correlation coefficient of the observations of G and H is defined as:

$$S(G,H) = \frac{1}{|G|} \times \sum_{i=1}^{|G|} \frac{G[i] - \mu_G}{\sigma_G} \times \frac{H[i] - \mu_H}{\sigma_H}$$

where μ_G and μ_H are the mean of observations on G and H, and σ_G and σ_H are the standard deviation of G and H:

$$\mu_G = \sum_{i=1}^{|G|} \frac{G[i]}{|G|} \quad \text{and} \quad \sigma_G = \sqrt{\sum_{i=1}^{|G|} \frac{(G[i] - \mu_G)^2}{|G|}}$$

and similarly for μ_H and σ_H. The corresponding Pearson distance $D(G, H)$ is defined as $1 - S(G, H)$.

Let V be a matrix representing $|V|^r$ cluster centroids of averaged gene expression values in $|V|^c$ conditions. The fuzzy k-means algorithm[80] is based on the minimization of the objective function below, for a given fuzzy partition of the dataset X into $|V|^r$ clusters having centroids V:

$$J(X, V) = \sum_{g=1}^{|X|^r} \sum_{j=1}^{|V|^r} M(X[g, \_], V[j, \_], V)^2 \times D(X[g, \_], V[j, \_])^2$$

where $M(X[g, \_], V[j, \_], V)$ is the membership of gene g in cluster j.

The cluster membership function is a continuous variable from 0 to 1. Its value is to be interpreted as the strength of a gene's membership in a particular cluster. That is, under fuzzy k-means, a gene can belong to several clusters. The cluster membership function is defined as:

$$M(X[g, \_], V[j, \_], V) = \frac{1}{D(X[g, \_], V[j, \_])^2} \Bigg/ \sum_{j=1}^{|V|^r} \frac{1}{D(X[g, \_], V[j, \_])^2}$$

During a cycle of fuzzy k-means clustering, the centroids are refined repeatedly. A centroid $V[j, \_]$ is refined to $V[j, \_]'$ on the basis of the weighted means of all the gene expression patterns in the dataset X according to

$$V[j, \_]' = \left\langle \frac{\sum_{g=1}^{|X|^r} M(X[g, \_], V[j, \_], V)^2 \times W(X, g) \times X[g, i]}{\sum_{g=1}^{|X|^r} M(X[g, \_], V[j, \_], V)^2 \times W(X, g)} \;\middle|\; i = 1 \ldots |V|^c \right\rangle$$

where the gene weight $W(X, g)$ is defined empirically as

$$W(X, g) = \left(\sum_{h=1}^{|X|^r} \frac{S(X[g, \_], X[h, \_]) - C}{1 - C} \right)^2$$

and C is a correlation cutoff threshold. In the work of Gasch and Eisen,[277] they set $C = 0.6$.

In each clustering cycle, the centroids are iteratively refined until the average change in gene memberships between interations is < 0.001. After each clustering cycle, the centroids are combined with those identified in previous cycles, and replicate centroids are averaged as follows. Each centroid is compared to all other centroids in the set, and centroid pairs that are Pearson correlated at > 0.9 are replaced by the average of the two vectors. The new vector is compared to the remaining centroids in the set and is again averaged with those to which it is Pearson correlated at > 0.9. This process continues until each centroid is compared to all other existing centroids in the set.

At the end of a clustering cycle, those genes with a Pearson correlation at > 0.7 to any of the identified centroids are taken as belonging to the respective clusters. These genes are then removed from further consideration. The next cycle of fuzzy k-means clustering are carried out on the remaining genes—*i.e.*, those with Pearson correlation at ≤ 0.7 to all the centroids. Incidentally, by considering a gene whose Pearson correlation to a centroid is > 0.7 as belong to the cluster of that centroid, it is therefore possible for a gene to belong simultaneously to multiple clusters. This is a great advantage of the fuzzy k-means method over other clustering methods that do not allow a gene to belong to more than one cluster. The reason is that many genes in real life do have multiple functional roles and thus naturally should belong to multiple clusters.

Gasch and Eisen[277] perform 3 successive cycles of fuzzy k-means clustering. Since k clusters are desired at the end of the 3 cycles, they aim to produce $k/3$ clusters in each cycle. The first cycle of clustering is initialized by using the top $k/3$ eigen vectors from a principle component analysis[399] on their dataset as prototype centroids for that clustering cycle. Subsequent cycles of clustering are initialized similarly, except that principle component analysis is performed on the respective data subset used in that clustering cycle. As details of principle component analysis have already been described in Chapter 3, we do not repeat here.

4. Inference of Gene Networks

A large number of genes can be differentially expressed in a microarray experiment. Such genes can serve as markers of the different classes—such as tumour vs. normal—of samples in the experiment. Some of these genes can even be the primary cause of a sample being tumour. In order to decide which gene is part of the primary cause and which gene is merely a down-stream effect, the underlying molecular network has to be assembled and considered. After the causal genes are identified, we may want to further develop drug substances to target them. The two major causes of treatment failure by drug substances are side effects and compensation effects. Side effects arise because genes and their protein products other than the intended ones are also modulated by the drug substances in unexpected ways. Compensation effects arise due to existence of parallel pathways that perform similar functions of the genes and proteins targeted by the drug substances and these parallel pathways are not affected by those drug substances. An understanding of the underlying molecular network is also useful for suggesting how best to target the causal genes. Motivated by these reasons, construction of a database of molecular network on the basis of microarray gene expression experiments has been attempted.

Let us recall that in analysing microarray gene expression output in the last two sections, we first identify a number of candidate genes by feature selection. Do we know which ones of these are causal genes and which are mere surrogates? Genes are "connected" in a "circuit" or network. The expression of a gene in a network depends on the expression of some other genes in the network. Can we reconstruct the gene network from gene expression data? For each gene in the network, can we determine which genes affect it? and how they affect it—positively, negatively, or in more complicated ways? There are several techniques to reconstructing and modeling molecular networks from gene expression experiments. Some techniques that have been tried are Bayesian networks,[263] Boolean networks,[16, 17] differential equations,[153] association rule discovery,[643] classification-based methods,[783] and several other approaches to related problems.[380, 734, 735]

We devote the rest of this section to describe the classification-based method of Soinov *et al.*,[783] the association rules method of Creighton and Hanash,[176] and the interaction generality method of Saito *et al.*[734] The last method—interaction generality[734, 735]—is actually concerned more with assessing the reliability of protein-protein interaction networks than with gene networks. However, it has been shown[280, 301, 390] that the average correlation coefficient of gene expression profiles that correspond to interacting gene products is higher than that of random pairs of gene products. Therefore, one might conceivably apply it in the context of gene networks.

4.1. *Classification-Based Approach*

In this subsection, we describe the classification-based method of Soinov *et al.*[783] for inferring molecular networks. Let a collection of n microarray gene expression output be given. For convenience, this collection can be organized into a gene expression matrix X. Each row of the matrix is a gene, each column is a sample, and each element $X[i, j]$ is the expression of gene i in sample j. Then the basic idea of the method of Soinov *et al.*[783] is as follows.

First determine the average value a_i of each gene i as $(\sum_j X[i, j])/n$. Next, denote s_{ij} as the state of gene i in sample j, where $s_{ij} = up$ if $X[i, j] \geq a_i$, and $s_{ij} = down$ if $X[i, j] < a_i$. Then, according to Soinov *et al.*,[783] to see whether the state of a gene g is determined by the state of other genes G, we check whether $\langle s_{ij} |\ i \in G \rangle$ can predict s_{gj}. If it can predict s_{gj} with high accuracy, then we conclude that the state of the gene g is determined by the states of other genes G.

Any classifier can be used to see if such predictions can be made reliably, such as C4.5,[693] PCL,[497] SVM,[855] and other classifiers described in Chapter 3.

Naturally, we can also apply feature selection methods described in Chapter 3—such as Fisher criterion score[251] or entropy-based methods[242]—to select a subset of genes from G before applying the classifiers to the selected subset of genes. Furthermore, to see how the state of a gene g is determined by the state of other genes, we apply C4.5, PCL, or other rule-based classifiers described in Chapter 3 to predict s_{gj} from $\langle s_{ij} |\ i \in G \rangle$ and extract the decision tree or rules used.

This interesting method has a few advantages: It can identify genes affecting a target genes in an explicit manner, it does not need a discretization threshold, each data sample is treated as an example, and explicit rules can be extracted from a rule-based classifier like C4.5 or PCL. For example, we generate from the gene expression matrix a set of n vectors $\langle s_{ij} |\ i \neq g \rangle \rightarrow s_{gj}$. Then C4.5 (or PCL) can be applied to see if $\langle s_{ij} |\ i \neq g \rangle$ predicts s_{gj}. The decision tree (or emerging patterns, respectively) induced would involve a small number of s_{ij}. Then we can suggest that those genes corresponding to these small number of s_{ij} affect gene g.

One other advantage of the Soinov method[783] is that it is generalizable to time series. Suppose the matrices X^t and X^{t+1} correspond to microarray gene expression measurements taken at time t and $t + 1$. Suppose s_{ij}^t and s_{ij}^{t+1} correspond to the state of gene i in sample j at time t and $t + 1$. Then to find out whether the state of a gene g is affected by other genes G in a time-lagged manner, we check whether $\langle s_{ij}^t |\ i \in G \rangle$ can predict s_{gj}^{t+1}. The rest of the procedure is as before.

Of course, there is a major caveat that this method as described assumes that a gene g can be in only two states, *viz.* $s_{gj} = up$ or $s_{gj} = down$. As cautioned by Soinov et al.,[783] it is possible for a gene to have more than two states and thus this assumption may not infer the complete network of gene interactions. Another caution is that if the states of two genes g and h are strongly co-related, the rules $s_{hj} \rightarrow s_{gj}$ and $s_{gj} \rightarrow s_{hj}$ saying that h depends on g and g depends on h are likely to be both inferred, even though only one of them may be true and the other false. Hence, further confirmation by experiments is advisable.

We do not have independent results on this approach to reconstructing molecular networks. However, we refer the curious reader to Soinov et al.[783] for a discussion on experiments they have performed to verify the relevance of this method. In particular, Soinov et al. have applied this method to the microarray datasets of Spellman and Cho for the *Saccharomyces cerevisiae* cell cycle.[157, 792] They consider a set of well-defined genes that encode proteins important for cell-cycle regulation and examine all extracted relations with respect to the known roles of the selected genes in the cell cycle. They have shown that in most cases the rules confirm the *a priori* knowledge.

4.2. *Association Rules Approach*

Recall from Chapter 3 that an association rule generally has the form $\alpha \to^{\mathcal{D}} \beta$, where α and β are disjoint sets of items, and the β set is likely to occur whenever the α set occurs in the context of a dataset $\mathcal{D}$. Note that we often drop the superscript $\mathcal{D}$ if the dataset $\mathcal{D}$ is understood or unimportant. As mentioned in Chapter 3, the support of an association rule $\alpha \to^{\mathcal{D}} \beta$ is the percentage of transactions in $\mathcal{D}$ that contains $\alpha \cup \beta$; and its confidence is the percentage of transactions in $\mathcal{D}$ containing α that also contain β.

In this subsection, we concentrate on the approach of Creighton and Hanash[176] for inferring associations between gene expression that is based on association rules. Let a collection of n microarray gene expression output be given as a gene expression matrix X so that each element $X[i, j]$ is the expression of gene i in sample j. Then the basic idea of the method of Creighton and Hanash is as follows.

Each element $X[i, j]$ is discretized into a state s_{ij} that indicates whether the gene i in sample j is considered up ($s_{ij} = up$), down ($s_{ij} = down$), or neither up nor down ($s_{ij} = neither$). This discretization to 3 states—*up, down*, and *neither*—is important because there is a good deal of noise in the data[373, 823] and binning whole ranges of gene expression values into a few states is a good way to alleviate problems with noise. Creighton and Hanash[176] decide on the assignment of *up, down, neither* by setting $s_{ij} = up$ if the expression value of gene i in sample j is greater than 0.2 for the log base 10 of the fold change, $s_{ij} = down$ if the expression value of gene i in sample j is less than –0.2 for the log base 10 of the fold change, and $s_{ij} = neither$ if the expression value of gene i in sample j is between –0.2 and 0.2 for the log base 10 of the fold change.

Then a dataset $\mathcal{D} = \{T_1, ..., T_n\}$ of n transactions is formed, where each sample j is treated as a transaction $T_j = \{gene_1 = s_{ij}, ..., gene_k = s_{kj}\}$. Then association rule mining algorithms described in Chapter 3 such as the Apriori algorithm[12] and the Max-Miner algorithm[70] can be used to mine for useful association rules. As many association rules can potentially be produced, Creighton and Hanash[176] adopt three measures for restricting the association rules to the most interesting ones, *viz.*

(1) they consider only those association rules that have support $\geq 10\%$ and confidence $\geq 80\%$;
(2) they consider only rules of the form $\alpha \to^{\mathcal{D}} \beta$ where α is a singleton; and
(3) they consider only the so-called "closed" rules, where a rule $\alpha \to^{\mathcal{D}} \beta$ is closed in the sense that there is no other rule $\alpha \to^{\mathcal{D}} \beta'$ such that $\beta \subset \beta'$ and has support $\geq 10\%$ and confidence $\geq 80\%$.

Creighton and Hanash[176] have applied this method to mine association rules from the gene expression profiles of 6316 transcripts corresponding to 300 diverse mutations and chemical treatment in yeast produced by Hughes *et al.*[373] They have obtained about 40 rules that contain ≥ 7 genes such as $\{YHM1 = up\} \rightarrow \{ARG1 = up, ARG4 = up, ARO3 = up, CTF13 = up, HIS5 = up, LYS1 = up, RIB5 = up, SNO1 = up, SNZ1 = up, YHR029C = up, YOL118C = up\}$. To see that these rules are significant, Creighton and Hanash also construct a randomized dataset and carry out association rule mining on this randomized dataset. On the randomized dataset, Creighton and Hanash is able to find only one rule. Hence, it is very likely that all the rules that are found by Creighton and Hanash from the dataset of Hughes *et al.* are not likely to have existed by chance.

This interesting method has two advantages. First, while we have made each transaction T_j to take the form $\{gene_1 = s_{1l}, ..., gene_k = s_{kj}\}$, it is possible to generalize it to include additional information such as environment and effects. As an example, consider $T_j = \{heatshock = 1, gene_1 = s_{1j}, ..., gene_k = s_{kj}\}$, where we use the item $heatshock = 1$ to indicate that a heat shock treatment has been first given to a sample j before profiling, and $heatshock = -1$ otherwise. Then we would be able to mine rules such as $\{heatshock = 1\} \rightarrow \{gene_h = up, gene_i = down\}$. That is, association rules may be helpful in relating the expression of genes to their cellular environment.

Second, the same gene is allowed to appear in several rules, in contrast to the clustering situation where each gene is normally required to appear in one cluster. A typical gene can participate in more than one gene network. Therefore, the association rule approach may be more useful in helping to uncover gene networks than the clustering approach. Furthermore, association rules also describe how the expression of one gene may be associated with the expression of a set of other genes.

Of course, there is a similar major caveat to that of the Soinov method.[783] This method as described above also assumes that a gene g can be in only three states, viz. $s_{gj} = up$, $s_{gj} = down$, or $s_{gj} = neither$. As cautioned by Soinov *et al.*,[783] it is possible for a gene to have more than three states and thus this assumption may not infer the complete network of gene interactions.

4.3. *Interaction Generality Approach*

In the two previous subsections, we have presented two techniques for inferring gene networks from microarray data. Both of these techniques can be said to work from a "positive" perspective in the sense that they assume there are no relationship between the genes by default and attempt to directly infer rules that connect the state of one or more genes to the state of another gene.

Is it possible to work from a "negative" perspective in the sense of assuming every pair of genes affect each other by default and attempt to eliminate those that have no effect on each other? It turns out that this approach has been used in the related problem of eliminating false positive interactions from certain type of high-throughput protein-protein interaction experiments by Saito *et al.*[734, 735]

A network of protein-protein interactions can be represented as an undirected graph $\mathcal{G}$, where each node represents a protein and each edge connecting two nodes represent an interaction between the two proteins corresponding to the two nodes. Given an edge $X \leftrightarrow Y$ connecting two proteins, X and Y, the "interaction generality" measure $ig^{\mathcal{G}}(X \leftrightarrow Y)$ of this edge as defined by Saito *et al.*[734] is equivalent to

$$ig^{\mathcal{G}}(X \leftrightarrow Y) = 1 + |\{X' \leftrightarrow Y' \in \mathcal{G} \mid X' \in \{X,Y\}, \ deg^{\mathcal{G}}(Y') = 1\}|$$

where $deg^{\mathcal{G}}(U) = |\{V \mid U \leftrightarrow V \in \mathcal{G}\}|$ is the degree of the node U in the undirected graph $\mathcal{G}$. Note that in an undirected graph, an edge $X \leftrightarrow Y$ is the same one as the edge $Y \leftrightarrow X$. This measure is based on the idea that interacting proteins that appear to have many other interacting partners that have no further interactions are likely to be false positives.

Uetz *et al.*[849] and Ito *et al.*[386] independently screen yeast protein-protein interactions. Saito *et al.*[734] determine the interaction generality of all the interactions detected by the screens of Uetz *et al.* and Ito *et al.*. While only 72.8% of interactions that are detected exclusively by the screen of Ito *et al.* have interaction generalities ranging from 1 to 5, as many as 94.7% of interactions that are detected by both screens have interaction generalities ranging from 1 to 5. As the portion of protein-protein interactions that are detected in both screens are considered to be reliable—whereas those that are detected in one screen are considered very likely to be false positive interactions—this indicates that true positive interactions tend to be associated with low interaction generalities.

It is also widely accepted that interacting proteins are likely to share a common cellular role,[633] to be co-localized,[759] or to have similar gene expression profiles.[280, 301, 390] If interaction generality is indeed inversely related to true positive protein-protein interactions, then the proportion of protein-protein interaction pairs that share a common cellular role, that are co-localized or have similar gene expression profiles, must be increasing as we look at protein-protein interaction pairs of decreasing interaction generality. This is confirmed by Saito *et al.*[734] in the datasets of Ito *et al.* and Uetz *et al.*

The interaction generality measure of Saito *et al.*[734] does not take into consideration the local topological properties of the interaction network surrounding the candidate interacting pair. Saito et al.[735] have also developed an improved interaction generality measure $ig_2^{\mathcal{G}}(X \leftrightarrow Y)$ that incorporates the local topological properties of interactions beyond the candidate interacting pair. They consider 5 local topological relationships between the candidate interacting pair and a third protein. The improved interaction generality measure is then computed as a weighted sum of the 5 topological relationships with respect to the third protein.

Most recently, our colleagues—Jin Chen, Wynne Hsu, Mong Li Lee, and See-Kiong Ng (private communication)—have proposed an "interaction pathway believeability" measure $ipb^{\mathcal{G}}(X \leftrightarrow Y)$ for assessing the reliability of protein-protein interactions obtained in large-scale biological experiments. It is defined as

$$ipb^{\mathcal{G}}(X \leftrightarrow Y) = \max_{\phi \in \Phi^{\mathcal{G}}(X,Y)} \prod_{(U \leftrightarrow V) \in \phi} \left(1 - \frac{ig^{\mathcal{G}}(U \leftrightarrow V)}{ig^{\mathcal{G}}_{\max}}\right)$$

where $ig^{\mathcal{G}}_{\max} = \max\{ig^{\mathcal{G}}(X \leftrightarrow Y) \mid (X \leftrightarrow Y) \in \mathcal{G}\}$ is the maximum interaction generality value in $\mathcal{G}$; and $\Phi^{\mathcal{G}}(X,Y)$ is the set of all possible non-reducible paths between X and Y, but excluding the direct path $X \leftrightarrow Y$. This measure can be seen as a measure on the global topological properties of the network involving X and Y in the sense that it evaluates the "credibility" of the non-reducible alternative path connecting X and Y, where the "probability" of each edge $U \leftrightarrow V$ in that path is $1 - ig^{\mathcal{G}}(U \leftrightarrow V)/ig^{\mathcal{G}}_{\max}$. Here, a path ϕ connecting X and Y is non-reducible if there is no shorter path ϕ' connecting X and Y that shares some common intermediate nodes with the path ϕ.

Jin Chen, Wynne Hsu, Mong Li Lee, and See-Kiong Ng further show that $ipb^{\mathcal{G}}(X \leftrightarrow Y)$ is better at separating true positive interactions from false positive interactions than $ig^{\mathcal{G}}(X \leftrightarrow Y)$ and $ig_2^{\mathcal{G}}(X \leftrightarrow Y)$. E.g., on a large dataset of protein-protein interactions—comprising that of Uetz et al.[849], Ito et al.[386], and Mewes et al.[564]—the difference between the average value of $ig^{\mathcal{G}}(X \leftrightarrow Y)$ and $ig_2^{\mathcal{G}}(X \leftrightarrow Y)$ on true positive and false positive interactions are 7.37% and 7.83% respectively; but that of $ipb^{\mathcal{G}}(X \leftrightarrow Y)$ is 29.96%.

As mentioned earlier, Saito et al.[735] have identified 5 local topological relationships, between a candidate pair of interacting proteins and a third protein, that are particularly useful in distinguishing true positive protein-protein interactions from false positive interactions. Actually, Milo et al.[567] have also studied similar kind of local topological relationships in complex networks, including gene networks. They call these topological relationships network motifs. In particular, they[567] have reported two such network motifs for gene regulation networks of E. coli and S. cerevisiae. However, they have not explored using these network motifs to distinguish true positive interactions in gene networks from false positives.

5. Derivation of Treatment Plan

In Section 2, we see that the entropy measure can be used to identify genes that are relevant to the diagnosis of disease states and subtypes. Let us now end this chapter with a provocative idea of Li and Wong[496, 498] of the possibility of a personalized "treatment plan" that converts tumor cells into normal cells by modulating the expression levels of a few genes.

Let us use the colon tumour dataset of Alon *et al.*[21] to demonstrate this highly speculative idea. This dataset consists of 22 normal tissues and 40 colon tumor tissues. We begin with finding out which intervals of the expression levels of a group of genes occur only in cancer tissues but not in the normal tissues and vice versa. Then we attempt an explanation of the results and suggest a plan for treating the disease.

We use the entropy measure[242] described in Chapter 3 to induce a partition of the expression range of each gene into suitable intervals. This method partitions a range of real values into a number of disjoint intervals such that the entropy of the partition is minimal. For the colon cancer dataset, of its 2000 genes, only 135 genes can be partitioned into 2 intervals of low entropy.[496, 498] The remaining 1865 genes are ignored by the method. Thus most of the genes are viewed as irrelevant by the method.

For the purpose of this chapter we further concentrate on the 35 genes with the lowest entropy measure amongst the 135 genes. These 35 genes are shown in Figure 8. This gives us an easy platform where a small number of good diagnostic indicators are concentrated. For simplicity of reference, the index numbers in the first column of Figure 8 are used to refer to the two expression intervals of the corresponding genes. For example, the index 1 means $M26338 < 59.83$ and the index 2 means $M26383 \geq 59.83$.

An emerging pattern, as explained in Chapter 3, is a pattern that occurs frequently in one class of samples but never in other classes of samples. An efficient border-based algorithm[207, 495] is used to discover emerging patterns based on the selected 35 genes and the partitioning of their expression intervals induced by the entropy measure. Thus, the emerging patterns here are combinations of intervals of gene expression levels of these relevant genes.

A total of 10548 emerging patterns are found, 9540 emerging patterns for the normal class and 1008 emerging patterns for the tumour class. The top several tens of the normal class emerging patterns contain about 8 genes each and can reach a frequency of 77.27%, while many tumour class emerging patterns can reach a frequency of around 65%.

These top emerging patterns are presented in Figure 9 and Figure 10. Note

Our list	accession number	cutting points	Name
1,2	M26383	59.83	monocyte-derived neutrophil-activating ...
3,4	M63391	1696.22	Human desmin gene
5,6	R87126	379.38	myosin heavy chain, nonmuscle (Gallus gallus)
7,8	M76378	842.30	Human cysteine-rich protein (CRP) gene ...
9,10	H08393	84.87	COLLAGEN ALPHA 2(XI) CHAIN ...
11,12	X12671	229.99	heterogeneous nuclear ribonucleoprotein core ...
13,14	R36977	274.96	P03001 TRANSCRIPTION FACTOR IIIA
15,16	J02854	735.80	Myosin regulatory light chain 2 ...
17,18	M22382	447.04	Mitochondrial matrix protein P1 ...
19,20	J05032	88.90	Human aspartyl-tRNA synthetase alpha-2 ...
21,22	M76378	1048.37	Human cysteine-rich protein (CRP) gene ...
23,24	M76378	1136.74	Human cysteine-rich protein (CRP) gene ...
25,26	M16937	390.44	Human homeo box c1 protein mRNA
27,28	H40095	400.03	Macrophage migration inhibitory factor
29,30	U30825	288.99	Human splicing factor SRp30c mRNA
31,32	H43887	334.01	Complement Factor D Precursor
33,34	H51015	84.19	Proto-oncogene DBL Precursor
35,36	X57206	417.30	1D-myo-inositol-trisphosphate 3-kinase B ...
37,38	R10066	494.17	PROHIBITIN (Homo sapiens)
39,40	T96873	75.42	Hypothetical protein in TRPE 3'region ...
41,42	T57619	2597.85	40S ribosomal protein S6 ...
43,44	R84411	735.57	Small nuclear ribonucleoprotein assoc. ...
45,46	U21090	232.74	Human DNA polymerase delta small subunit
47,48	U32519	87.58	Human GAP SH3 binding protein mRNA
49,50	T71025	1695.98	Human (HUMAN)
51,52	T92451	845.7	Tropomyosin, fibroblast and epithelial ...
53,54	U09564	120.38	Human serine kinase mRNA
55,56	H40560	913.77	THIOREDOXIN (HUMAN)
57,58	T47377	629.44	S-100P PROTEIN (HUMAN)
59,60	X53586	121.91	Human mRNA for integrin alpha 6
61,62	U25138	186.19	Human MaxiK potassium channel beta subunit
63,64	T60155	1798.65	Actin, aortic smooth muscle (human)
65,66	H55758	1453.15	ALPHA ENOLASE (HUMAN)
67,68	Z50753	196.12	H.sapiens mRNA for GCAP-II/uroguanylin ...
69,70	U09587	486.17	Human glycyl-tRNA synthetase mRNA

Fig. 8.   The 35 top-ranked genes by the entropy measure. The index numbers in the first column are used to refer to the two expression intervals of the corresponding genes. For example, the index 1 means M26338 < 59.83 and the index 2 means M26383 ≥ 59.83.

Emerging patterns	Count & Freq. (%) in normal tissues	Count & Freq. (%) in cancer tissues
$\{25, 33, 37, 41, 43, 57, 59, 69\}$	17(77.27%)	0
$\{25, 33, 37, 41, 43, 47, 57, 69\}$	17(77.27%)	0
$\{29, 33, 35, 37, 41, 43, 57, 69\}$	17(77.27%)	0
$\{29, 33, 37, 41, 43, 47, 57, 69\}$	17(77.27%)	0
$\{29, 33, 37, 41, 43, 57, 59, 69\}$	17(77.27%)	0
$\{25, 33, 35, 37, 41, 43, 57, 69\}$	17(77.27%)	0
$\{33, 35, 37, 41, 43, 57, 65, 69\}$	17(77.27%)	0
$\{33, 37, 41, 43, 47, 57, 65, 69\}$	17(77.27%)	0
$\{33, 37, 41, 43, 57, 59, 65, 69\}$	17(77.27%)	0
$\{33, 35, 37, 41, 43, 45, 57, 69\}$	17(77.27%)	0
$\{33, 37, 41, 43, 45, 47, 57, 69\}$	17(77.27%)	0
$\{33, 37, 41, 43, 45, 57, 59, 69\}$	17(77.27%)	0
$\{13, 33, 35, 37, 43, 57, 69\}$	17(77.27%)	0
$\{13, 33, 37, 43, 47, 57, 69\}$	17(77.27%)	0
$\{13, 33, 37, 43, 57, 59, 69\}$	17(77.27%)	0
$\{13, 32, 37, 57, 69\}$	17(77.27%)	0
$\{33, 35, 37, 57, 68\}$	17(77.27%)	0
$\{33, 37, 47, 57, 68\}$	17(77.27%)	0
$\{33, 37, 57, 59, 68\}$	17(77.27%)	0
$\{32, 37, 41, 57, 69\}$	17(77.27%)	0

Fig. 9.   The top 20 emerging patterns, in descending frequency order, in the 22 normal tissues. The numbers in the emerging patterns above refer to the index numbers in Figure 8.

that the numbers in the emerging patterns in these figures, such as $\{2, 10\}$ in Figure 10, refer to the index numbers in Figure 8. Hence, $\{2, 10\}$ denotes the pattern $\{M26383 \geq 59.83, H08393 \geq 84.87\}$.

The emerging patterns that are discovered are the most general ones. They occur in one class of data but do not occur in the other class. The discovered emerging patterns always contain only a small number of the relevant genes. This result reveals interesting conditions on the expression of these genes that differentiate between two classes of data.

Each emerging pattern with high frequency is considered as a common prop-

Emerging patterns	Count & Freq. (%) in normal tissues	Count & Freq. (%) in cancer tissues
{2, 10}	0	28 (70.00%)
{10, 61}	0	27 (67.50%)
{10, 20}	0	27 (67.50%)
{3, 10}	0	27 (67.50%)
{10, 21}	0	27 (67.50%)
{10, 23}	0	27 (67.50%)
{7, 40, 56}	0	26 (65.00%)
{2, 56}	0	26 (65.00%)
{12, 56}	0	26 (65.00%)
{10, 63}	0	26 (65.00%)
{3, 58}	0	26 (65.00%)
{7, 58}	0	26 (65.00%)
{15, 58}	0	26 (65.00%)
{23, 58}	0	26 (65.00%)
{58, 61}	0	26 (65.00%)
{2, 58}	0	26 (65.00%)
{20, 56}	0	26 (65.00%)
{21, 58}	0	26 (65.00%)
{15, 40, 56}	0	25 (62.50%)
{21, 40, 56}	0	25 (62.50%)

Fig. 10. The top 20 emerging patterns, in descending frequency order, in the 40 cancer tissues. The numbers in the emerging patterns refer to the index numbers in Figure 8.

erty of a class of cells. Based on this idea, Li and Wong[496, 498] propose a strategy for treating colon tumors by adjusting the expression level of some improperly expressed genes. That is, to increase or decrease the expression levels of some particular genes in a cancer cell, so that it has the common properties of normal cells and no properties of cancer cells. As a result, instead of killing the cancer cell, it is "converted" into a normal one. We show later that almost all "adjusted" cells are predicted as normal cells by a number of good classifiers that are trained to distinguish normal from colon tumor cells.

As shown in Figure 9, the frequency of emerging patterns can reach a very high level such as 77.27%. The conditions implied by a highly frequent emerging pattern form a common property of one class of cells. Using the emerging pattern $\{25, 33, 37, 41, 43, 57, 59, 69\}$ from Figure 9, we see that each of the 77.27% of the normal cells simultaneously expresses the eight genes—M16937, H51015, R10066, T57619, R84411, T47377, X53586, and U09587 referenced in this emerging pattern—in such a way that each of the eight expression levels is contained in the corresponding interval—the 25th, 33th, 37th, 41st, 43rd, 57th, 59th, and 69th—as indexed in Figure 8.

Although a cancer cell may express some of the eight genes in a similar manner as normal cells do, according to the dataset, a cancer cell can never express all of the eight genes in the same way as normal cells do. So, if the expression levels of those improperly expressed genes can be adjusted, then the cancer cell can be made to have one more common property that normal cells exhibit. Conversely, a cancer cell may exhibit an emerging pattern that is a common property of a large percentage of cancer cells and is not exhibited in any of the normal cells. Adjustments should also be made to some genes involved in this pattern so that the cancer cell can be made to have one less common property that cancer cells exhibit. A cancer cell can then be iteratively converted into a normal one as described above.

As there usually exist some genes of a cancer cell which express in a similar way as their counterparts in normal cells, less than 35 genes' expression levels are required to be changed. The most important issue is to determine which genes need an adjustment. The emerging patterns can be used to address this issue as follows. Given a cancer cell, first determine which top emerging pattern of normal cells has the closest Hamming distance to it in the sense that the least number of genes need to be adjusted to make this emerging pattern appear in the adjusted cancer cell. Then proceed to adjust these genes. This process is repeated several times until the adjusted cancer cell exhibits as many common properties of normal cells as a normal cell does. The next step is to look at which top emerging pattern of cancer cells that is still present in the adjusted cancer cell has the closest Hamming distance to a pattern in a normal cell. Then we also proceed to adjust some genes involved in this emerging pattern so that this emerging pattern would vanish from the adjusted cancer cell. This process is repeated until all top emerging patterns of cancer cells disappear from our adjusted cancer cell.

We use a cancer cell (T1) of the colon tumor dataset as an example to show how a tumor cell is converted into a normal one. Recall the emerging pattern $\{25, 33, 37, 41, 43, 57, 59, 69\}$ is a common property of normal cells. The eight genes involved in this emerging pattern are M16937, H51015, R10066, T57619, R84411, T47377, X53586, and U09587. Let us list the expression profile of these eight genes in T1:

genes	expression levels in T1
M16937	369.92
H51015	137.39
R10066	354.97
T57619	1926.39
R84411	798.28
T47377	662.06
X53586	136.09
U09587	672.20

However, 77.27%—17 out of 22 cases—of the normal cells have the following expression intervals for these 8 genes:

genes	expression interval
M16937	<390.44
H51015	<84.19
R10066	<494.17
T57619	<2597.85
R84411	<735.57
T47377	<629.44
X53586	<121.91
U09587	<486.17

Comparing T1's gene expression levels with the intervals of normal cells, we see that 5 of the 8 genes—H51015, R84411, T47377, X53586, and U09587—of the cancer cell T1 behave in a different way from those the 22 normal cells commonly express. However, the remaining 3 genes of T1 are in the same expression range as most of the normal cells. So, if the 5 genes of T1 can be down regulated to scale below those cutting points, then this adjusted cancer cell will have a common property of normal cells. This is because $\{25, 33, 37, 41, 43, 57, 59, 69\}$ is an emerging pattern which does not occur in the cancer cells. This idea is at the core of Li and Wong[496, 498]'s suggestion for this treatment plan.

Interestingly, the expression change of the 5 genes in T1 leads to a chain of other changes. These include the change that 9 extra top-ten EPs of normal cells are contained in the adjusted T1. So all top-ten EPs of normal cells are contained in T1 if the 5 genes' expression levels are adjusted. As the average number of top-ten EPs contained in normal cells is 7, the changed T1 cell will now be considered as a cell that has the most important features of normal cells. Note that we have adjusted only 5 genes' expression level so far.

It is also necessary to eliminate those common properties of cancer cells that are contained in T1. By adjusting the expression level of 2 other genes, M26383 and H08393, the top-ten EPs of cancer cells all disappear from T1. According to the colon tumor dataset, the average number of top-ten EPs of cancer cells contained in a cancer cell is 6. Therefore, T1 is converted into a normal cell as it now holds the common properties of normal cells and does not hold the common properties of cancer cells.

By this method, all the other 39 cancer cells can be converted into normal ones after adjusting the expression levels of 10 genes or so, possibly different genes from person to person. Li and Wong[496, 498] conjecture that this personalized treatment plan is effective if the expression of some particular genes can be modulated by suitable means.

Lastly, we discuss a validation of this idea. The "adjustments" made to the 40 colon tumour cells are based on the emerging patterns in the manner described above. If these adjustments have indeed converted the colon tumour cells into normal cells, then any good classifier that can distinguish normal vs colon tumour cells on the basis of gene expression profiles is going to classify our adjusted cells as normal cells. So, Li and Wong[496, 498] establish a SVM model using the original entire 22 normal plus 40 cancer cells as training data. The code for constructing this SVM model is available at http://www.cs.waikato.ac.nz/ml/weka. The prediction result is that all of the adjusted cells are predicted as normal cells. Although Li and Wong's "therapy" is not applied to the real treatment of a patient, the prediction result by the SVM model partially demonstrates the potential biological significance of this highly speculative and provocative proposal.

CHAPTER 15

GENOME-WIDE CDNA OLIGO PROBE DESIGN AND ITS APPLICATIONS IN *SCHIZOSACCHAROMYCES POMBE*

Kui Lin

Beijing Normal University
linkui@bnu.edu.cn

Jianhua Liu

Genome Institute of Singapore
liujh@gis.a-star.edu.sg

Lance Miller

Genome Institute of Singapore
millerl@gis.a-star.edu.sg

Limsoon Wong

Institute for Infocomm Research
limsoon@i2r.a-star.edu.sg

Microarrays are glass surfaces bearing arrays of DNA fragments—also known as "probes"—at discrete addresses. These DNA fragments on the microarray are hybridized to a complex sample of fluorescently labeled DNA or RNA in solution. After a washing and staining process, the addresses at which hybridization has taken place can be determined and the expression level of the corresponding genes derived. Today, a single microarray can contain several tens of thousands of DNA fragments. Thus, microarrays are a technology for simultaneously profiling the expression levels of tens of thousands of genes in a sample.[518, 698]

In this chapter, we present a method for selecting probes to profile genome-wide gene expression of a given genome. We demonstrate our method on the genome of *Schizosaccharomyces pombe*. *S. pombe* or fission yeast is a single-celled free living Ascomycete fungus with many of the features found in the cells of more complicated eukaryotes. *S. pombe* is the second yeast, after *S. cerevisiae*, whose genome has been completely sequenced. Due to the fact that *S. cerevisiae* has undergone genome duplication and gene lineage loss and divergence, *S. pombe* can be a better model organism for the study of gene expression, especially for those genes whose products are not present in *S. cerevisiae*.

ORGANIZATION.

Section 1. The biological background of *Schizosaccharomyces pombe* is summarized.

Section 2. The problem of designing oligo probes for genome-wide gene expression profiling by microarrays is formalized.

Section 3. An overview of our approach to solving this problem is then given, and its advantages briefly discussed. The approach has three modules. The first module extracts the coding regions from the given genome. the second module produces candidates oligos from the coding regions satisfying certain constraints such as G+C content, cross homology, *etc.* The third module selects from the amongst the candidates an optimal probe set satisfying some additional criteria such as minimizing distance to the 3' end of genes.

Section 4. The detailed implementation is then presented, such as aspects of data schema, object creation, criteria for probe production, probe production, and optimal probe set selection.

Section 5. The program is then run on the *S. pombe* genome to design a set of oligo probes for genome-wide gene expression profiling of *S. pombe*. The quality statistics of the resulting probe set is reported.

1. Biological Background

Schizosaccharomyces pombe was first isolated from an East African millet beer, called Pombe. It lives mainly as a haploid, divides by cell fission, and responds to nutrient starvation by mating to a partner with opposite mating type and forming four-spored asci. Its diploid cells can be maintained in the laboratory if zygotes are transferred to a rich medium. Like its distant relative, *S. cerevisiae*, *S. pombe* is amenable to genetic, biochemical, cellular, molecular, and functional genomic studies. *S. pombe* has served as an excellent model organism for the study of cell-cycle control, mitosis and meiosis, DNA repair and recombination, and checkpoint controls important for genome stability.

    S. pombe, normally a haploid, spends most of its time in G2 and controls its cell cycle by regulating the G2-M phase transition. By contrast, *S. cerevisiae*, normally a diploid in the wild, has a long G1 phase and the major decision point for its cell-cycle entry occurs at the G1-S phase transition. Thus, both yeast species have provided important contributions to the discovery of the basic mechanisms of cell division.

    Both *S. pombe* and *S. cerevisiae* have the same total DNA content, but *S. cerevisiae* divides its genome amongst 16 chromosomes, while *S. pombe* has just three.

That may explain why *S. cerevisiae* has simple structures of DNA-replication origins and centromeres, whereas *S. pombe* contains relatively complex organization of DNA-replication origins and centromeres, although not as complex as that found in higher eukaryotes.

S. pombe is the sixth eukaryotic genome to be sequenced,[898] following *Saccharomyces cerevisiae*,[292] *Caenorhabditis elegans*,[128] *Drosophila melanogaster*,[7] *Arabidopsis thaliana*,[31] and *Homo sapiens*.[467] The comparison of *S. pombe* and *S. cerevisiae* revealed that the genome of *S. cerevisiae* was more redundant and underwent lineage-specific gene loss. This makes *S. pombe* a more attractive model organism for functional genomic studies.

Some gene sequences are as equally diverged between the two yeasts as they are from their human homologs, probably reflecting a more rapid evolution within fungal lineages compared with those of the Metazoa. Due to the fact that *S. cerevisiae* has undergone genome duplication and gene lineage loss and divergence, *S. pombe* can be a better model organism for the study of gene expression, especially for those genes whose products are not present in *S. cerevisiae*.

2. Problem Formulation

The design of oligonucleotide probes for genome-wide microarray gene expression analysis can be formulated as follows. Assume that there are m protein-coding genes annotated in a completely sequenced genome, denoted here as $G = \{g_i : |g_i| \geq L, 1 \leq i \leq m\}$ where L is the minimum length of a gene considered and is set to 50 nucleotides. The goal of the design of oligonucleotide probes for genome-wide microarray gene expression analysis is to identify, for each gene g_i of length $|g_i|$ nucleotide bases, at least one sequence segment $g_i^{u,v}$ between positions u and v of g_i such that

- $1 \leq u, v \leq |g_i|$,
- $v - u + 1 = L$, and
- $g_i^{u,v}$ is specific to g_i only.

The specificity (*i.e.*, uniqueness) of one segment of DNA sequence is defined as the similarity between that segment and any portion of any g_j of the same length L, for $1 \leq j \leq m$ and $j \neq i$. The degree of specificity must be kept below some threshold specified *a priori* under some predefined similarity measure method. Typically, different oligo design softwares rely on different similarity measures. For instance, PRIMEGENS[911] uses the minimal length and expectation value of the local alignment algorithm BLASTN[23] to measure the similarity between two DNA sequences.

K. Lin, J. Liu, L. Miller, & L. Wong

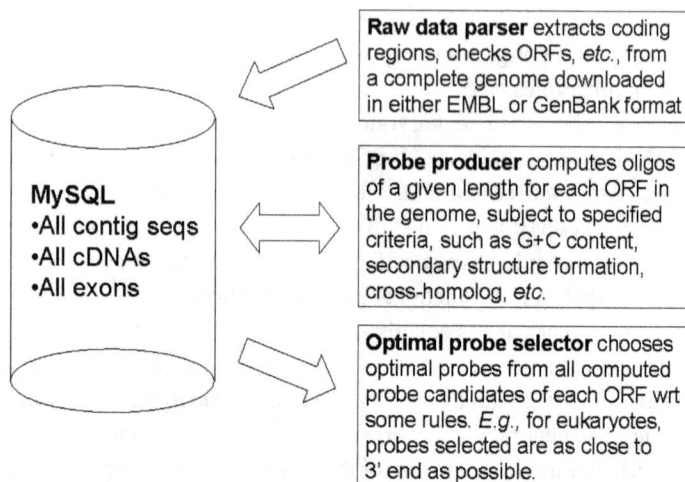

Fig. 1.   The main modules of our oligo probe design program are the raw data parser module, the probe producer module, and the optimal probe selector module.

In our algorithm to be described shortly, the similarity measure used is called Hamming distance,[317] which is a measure of the specificity between two equal length DNA sequences. The Hamming distance $D_H(s_1, s_2)$ between two sequences s_1 and s_2 is defined as the number of mismatched nucleotides in the alignment of s_1 and s_2. For each probe candidate computed, our algorithm keeps the minimum Hamming distance that indicates how specific the probe is. The larger the minimum Hamming distance, the better the probe candidate is.

3. Algorithm Overview

We develop our oligo probes design program using the C programming language and using the MySQL relational database.[917] MySQL is used as a backend data warehouse for the storage of all raw genomic data, some intermediate data, and all final result. Figure 1 presents the main modules—raw data parser, probe producer, and optimal probe selector—of our program for the design of DNA oligo probes of *S. pombe*.

While our C-plus-MySQL framework increases IO operation time, we find that it also has many advantages. Firstly, we can integrate a large amount of related information that is important for computing a more optimal set of oligo probes for the complete genome. For example, we can use exon information to avoid those probes whose locations span two contiguous exons in one ORF. This increases hybridization sensitivity in expression experiments.

Secondly, we are interested to parallelize the computation of oligo probe design for complete genome that may consist of more than several thousands of ORFs. We think it is easier to implement parallel computation on top of this C-plus-MySQL framework while maintaining consistency. For example, one way to accomplish this is to lock records or tables when we produce a new putative probe and insert it into specific table(s) for each individual process/thread running concurrently. See Figure 2.

Thirdly, this framework allows us to evaluate the quality of the selected set of probes. Having kept many intermediate results in the database, we can easily trace the hybridization result of each probe to the computed and logged information to see which factors are most related to the hybridization results, especially for those with "bad" hybridization signals. Are these "bad" results due to the "bad" selection of these probes? Or do they have a biological or experimental origin in for example, sample preparation, mRNA labeling, hybridization control, *etc.*? This analysis allows us to improve on the quality of our probe design in future experiments.

Lastly, as the probe selection computation is a time-consuming job in terms of CPU, it is advantageous if the program can be interrupted at any time, and be resumed at a later time if necessary. This functionality is very useful in situations such as when the server needs to be shutdown for system maintenance, or when the heavy load our of program is interfering with programs of other users, and so on. Under our C-plus-MySQL framework, the program can be easily interrupted and resumed according to the requirements of computational environment during the course of computing a microarray probe design.

4. Implementation Details

In this section, we present the details of our implementation, including aspects of data schema, object creation, criteria for probe production, probe production, and optimal probe set selection.

4.1. *Data Schema*

In our program, there are at least 5 core tables that must be created on the MySQL relational database before the probe computation can be carried out. We describe them in details as follows:

- `Contig` table stores all chromosomes, contigs, cosmids, and plasmids in the given complete genome.
- CDS table stores all ORF and CDS information, including location, strand, cDNA sequence, title, and number of probes computed so far, and so on.
- Exon table stores all exon information for each ORF.
- `Region` table holds specific information of each ORF and is used to help pick the candidate segment for the next round of computation with respect to pre-defined rules.
- `Probe` table stores all computed oligo probes. Each record contains useful information related to the probe, including G+C content, minimum Hamming distance, maximum number of contiguous nucleotide, location in the ORF, *etc.*

4.2. *Data Objects Creation*

A global array of `Contig` objects is designed to hold all contigs parsed from a given annotation file, which serves as the starting point of the design. The following segment is excerpted from the feature table of the annotated file of *Schizosaccharomyces pombe* chromosome I from EMBL database. It describes that a gene, sod2, consists of two exons which are located from nucleotide 62341 to 62464 and 62542 to 63824 in chromosome I. It encodes a protein called na(+)/h(+) antiporter in *S. pombe*.

```
FT   CDS          join(62341..62464,62542..63824)
FT                /gene="sod2"
FT                /note="SPAC977.10, len:468"
FT                /codon_start=1
FT                /label=sod2
FT                /product="na(+)/h(+) antiporter"
FT                /protein_id="NP_592782.1"
FT                /db_xref="GI:19113694"
FT                /db_xref="SWISS-PROT:P36606"
```

Note that, in our program, a contig represents an individual genomic sequence that is assembled in a given genome. It could be a chromosome, a contig, a cosmid, or a plasmid depending on the types of genomic sequences. Each `Contig` object in C is defined as follows:

```
struct Contig               /* Contig object */
{
   char id[15];             /* Identifier of the contig */
   unsigned long len;       /* DNA sequence length */
   unsigned int cdsNum;     /* number of CDS the contig has */
   char * title;            /* contig description */
   char * seq;              /* DNA sequence */
   struct CDS * cds; /* pointer to CDS list */
};
```

Each `Contig` may contain one or many CDS/ORFs, which are organized as a single linked list in the `Contig` object, and each CDS/ORF represents a coding region in the Contig. We define ORF and CDS objects in C as follows:

```
struct CDS                        /* CDS object */
{
    char id[21];                  /* Identifier of the CDS */
    char * title;                 /* gene/ORF's name */
    int complement;               /* is it complement? 1=yes, 0-no */
    struct Exon * exons;          /* pointer to the exon list */
    unsigned long start, end, len; /* ORF location */
    unsigned char * cDNA;         /* pointer to cDNA sequence */
    unsigned int probeNum;        /* number of probes computed so far */
    char isDuplicated;            /* Is it duplicated ORF? */
    unsigned long codons[62];     /* 61 codons and their sum */
    struct CDS * next;            /* pointer to the next CDS */
};
```

Exon objects are defined so that they can be used to select more robust or specific oligo probes for each ORF. For example, select a probe that do not span two contiguous exons of the ORF. We define Exon objects as follows.

```
struct Exon                       /* exon object */
{
    unsigned long start, end;     /* location */
    int phase, endPhase;          /* for splicing site */
    struct Exon * next;
};
```

A Probe object is defined to keep as much information pertaining to its computation as possible. Such information is important in selecting an optimal set of probes for each ORF. We define Probe objects as follows.

```
struct Probe                      /* oligo probe object */
{
    unsigned char * nt;           /* pointer to the probe size */
    int maxC;                     /* maximum contiguous number */
    int Tm;                       /* melting temperature */
    int GC;                       /* G+C content */
    int minHD;                    /* minimum Hamming distance */
    unsigned long start;          /* start position at the cDNA sequence */
};
```

When the input annotation file is parsed by the raw data parser module, an array of Contig objects is created in memory and all CDS objects—corresponding to ORFs in a contig—are also created as a single linked list that is linked to its corresponding Contig object. At the same time, all Exon objects belonging to each CDS are also created and organized as a single linked list which is attached to the CDS automatically. After having done some necessary checking work—such as start codon, stop codon, and so on—all types of objects are automatically imported into their specified MySQL tables whose schema are identical to the corresponding objects in C. Then the program terminates normally from the raw data parser module and the oligo probe production module can be launched.

4.3. *Criteria for Probe Production*

Before discussing probe production, we need to spend some time to characterize the criteria that the program uses in the oligo probe production module. Different strategies can be used in the computation of putative probes—*e.g.*, starting from the 3' end of each ORF first, or from some specific regions in the ORF at first. Regardless of the strategiy used, the selected segment needs to be checked against predefined constraints before it can be compared to other equal size segments from all the other ORFs.

Different programs typically use different constraints. However, the following constraints are usually taken into account in many probe and primer design programs:[409, 490, 518, 694]

- Each probe should have minimal secondary structure.
- Each probe should have no contiguous complementary stretches > 15bp.
- Each probe should have minimal distance from the 3' end, taking into account poly-A prediction if possible.
- The combination of probes should be a maximal representation of alternative splice variants if possible.
- The combination of probes should avoid cross-hybridization.
- The combination of probes should have homogeneity in G+C content and melting temperature

It has been reported that, for good gene specificity, non-target cDNAs ought to be less than 75% in sequence similarity compared to the target region (50bp in size) to prevent significant cross-hybridization.[409] In our program, we define the following global variables to characterize the constraints on the oligo probes that we want:

```
MAX_Tm = 65;                        /* 1. max melting temperature */
MIN_Tm = 45;                        /* 2. min melting temperature */
MAX_GC = 0.65;                      /* 3. max G+C content */
MIN_GC = 0.45;                      /* 4. min G+C content */
MAX_SINGLE_NT_CONTENT = 0.50;   /* 5. max ratio of single nt */
MAX_CONTIGUOUS_SINGLET_NT = 12;/* 6. max contiguous singlet nt */
MAX_3PRIME_NUM = 4;               /* 7. max contiguous identity at 3' end */
MAX_COMP_CONTIGUOUS_NUM = 5;    /* 8. max contiguous complementarity */
MAX_PAIR_CONTIGUOUS_NUM = 12;   /* 9. max contiguous pair comparison */
MAX_PAIR_SIMILARITY = 0.75; /*10. max allowed similarity for pair probes */
MAX_COMP_SIMILARITY = 0.75; /*11. max allowed similarity for self-complement */
```

In the above, MAX_SINGLE_NT_CONTENT is the maximum ratio of the occurrence of any single nucleotide letter—*i.e.*, A, C, G, or T—to the length of the probe. MAX_CONTIGUOUS_SINGLET_NT is the maximum length of the consecutive occurrence of any single nucleotide letter in the probe. MAX_3PRIME_NUM is the maximum length of complementary base pairs consecutively between a given probe's 3' end and some part within the probe. MAX_COMP_CONTIGUOUS_NUM

1. Create objects from the database

2. Select one CDS, *e.g.*, pick one segment from 3' end

n   3. Does segment satisfy constraints?

y

4. Compare to all the other segments from all the other CDSs

n   5. Is minimum Hamming distance > threshold?

y

6. Insert into Probes Table in the database

n   7. Sufficient number of probes found or no more candidate?

y

8. Stop

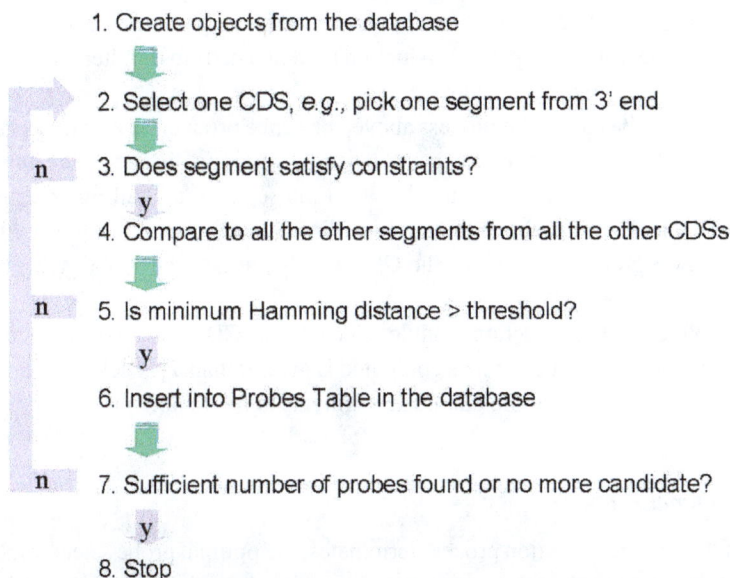

Fig. 2.   The main steps of the probe production module of our oligo probe design program.

is the maximum length of complementary base pairs consecutively between any two parts of the probe. MAX_PAIR_CONTIGUOUS_NUM is the maximum length of complementary base pais consecutively between a probe and its non-target cDNA. MAX_PAIR_SIMILARITY is the maximum total identity, based on Hamming distance, between a probe and its non-target cDNAs.

Note that Constraints 7, 8, and 11 are used to eliminate as many probe candidates that would form internal secondary structures as possible. Although the 3' end checking is not as important in probe design as in primer design, we include it in our program and hope to produce probes with better quality than without this constraint.

4.4. *Probe Production*

Figure 2 describes the probe production process. At the start of the computation and production of putative oligo probes, the probe production module creates all objects described above in the memory from the database. Simply speaking, an array of *Contig* objects is dynamically created from the database first.

Next, the probe production module selects one CDS and tries to pick one segment along the CDS according to a pre-specified picking strategy. Having picked the putative segment, the probe production module needs to do filtering under the given constraints mentioned above.

After a probe passes the process above, the probe production module performs a whole cDNA sequence set comparison using Hamming distance to evaluate the segment. If the Hamming distance between the segment and all other segments from all other CDSs is above the minimal dissimilarity constraint threshold, it is selected as a probe candidate of the CDS and is put into the probes table in the database.

Then the probe production module selects a new CDS and repeats the process above until the number of probes computed is greater than a predefined number— *e.g.*, 10—or there is no more candidate segments to be found.

4.5. *Optimal Probe Set*

When the probe production process terminates, the optimal probe selector module allows the user to select the most optimal set of probes of each ORF according to a definition of what an optimal probe is. In theory, an oligo probe is optimal if the following 3 conditions are satisfied:

- It is as close to the 3' end as possible for optimal cDNA synthesis and labeling of eukaryotic mRNA templates, which utilizes oligo dT primers that bind the 3' polyA tail for transcription initiation.
- It maximizes the Hamming distance between it and all the other CDS to minimize cross-hybridization.
- It has minimal secondary structure so that the sensitivity of hybridization is maximized.

For maximum detection sensitivity of gene expression, the G+C content should be normalized among probes so that the melting temperature is uniform across the probe set. The optimal range is from 45% to 65% in the *S. pombe* genome.

In practice, we use the following rules to select the optimal set of probes for each ORF in the *S. pombe* genome:

- For probes whose distance from the 3' end are less than 500 nucleotide bases, we select those candidates that maximize the Hamming distance;
- For probes whose distances from 3' end are greater than 500 nucleotide bases, we select those candidates that minimize the distance from 3' end.

We see in the next section that the resulting probes selected using these two rules have comparable statistics to the probes designed by some companies we have consulted.

5. Results and Discussions

There are 3 chromosomes in *S. pombe*. We download all 11 contigs in EMBL format (November 2001) from http://www.sanger.ac.uk/Projects/S_pombe of the genome sequencing group at the Sanger center. We parse and store all data into our MySQL database at the Genome Institute of Singapore using our program mentioned earlier. Our oligo probe size is 50mer in length.[409] Due to limitations of computing resources, for each ORF excluding duplicates, we produce only about 10 oligo probe candidates for the optimal selection of probes. In order to guarantee the sensitivity of the hybridization and the reliability of experiments, the top two probe candidates are picked for each ORF to construct the whole set of 50mer oligo probes of the *S. pombe* genome.

In our fission yeast functional genomic studies, the initial stage is to design about 10,000 oligos that depict about 5,000 ORFs in total. ORFs under 100 amino acids are excluded by the Genefinder program (C. Wilson, L. Hilyer, and P. Green; unpublished) used for gene prediction. Nevertheless, there are 147 ORFs less than 100 amino acids in length that are included because they either are confirmed experimentally or are reliably predicted with strong significance scores. Currently, there are 4987 annotated ORFs from Sanger Centre (November 2001). From these ORFs, we design about 10,000 oligo probes as described earlier for the Genome Institute of Singapore (January 2002). The probe set is subsequently manufactured by Genset.

To explore whether there are more small ORFs in the genome, we are taking the Matrix-Assisted Laser Desorption Ionization (MALDI) Mass spectrometry approach. The next stage is to include oligos that represent those small polypeptides whose ORFs may not be included in the current annotation.

Some main features of the oligo probes we have produced from the *S. pombe* ORFs described above are as follow:

- A total of 9859 50mer probes are selected for 4929 ORFs (98.84%), after excluding 58 duplicated ORFs.
- The average distance of the probes from the 3' end is 191 nucleotides and the median distance is 141 nucleotides.
- The average cross-homology between the probes is $\leq 61\%$.
- 97.9% of the probes have a distance from the 3' end of ≤ 500 nucleotides.
- 99.9% of the probes have a distance from the 3' end of ≤ 1000 nucleotides.

- 88.1% of the probes have $\leq$ 60% cross-homology with other probes.
- 98.9% of the probes have $\leq$ 68% cross-homology with other probes.
- 98.4% of the probes have 45% $\leq$ G+C content $\leq$ 65%.

The availability of genetic sequence information in both public and private databases over the world has shifted genome-base research away from pure sequencing towards functional genomics and genotype-phenotype studies. A powerful and versatile tool for functional genomics is DNA microarray technology which has been vastly applied in monitoring thousands or tens of thousands of genes in parallel and has provided biological insights into gene function and the relevance of the genetic loci for phenotypic traits.

As described earlier, we have developed a program containing several modules that can be used to compute and select optimal oligo probes for an entire genome. It has been successfully applied to the *S. pombe* genome which consists of about 5000 ORFs (in November 2001). The optimal set of oligo probes is composed of two probes from about 10 candidates of each ORF in the genome. From the *in silico* point of view, the quality of the set of probes is very reasonable. However, many microarray experiments across different biological contexts must be analyzed to thoroughly validate the efficacy of the probe designs.

For the improvement of the program that we have developed, the speed of computation of probe candidates should be increased before the program can be used for selecting oligo probes for larger complete genomes.

Acknowledgements

We are grateful to Yuyu Kuang for her useful SQL help; and to Phil Long, Prasanna Kolatkar, Edison Liu, and Mohan Balasubramanian for their invaluable suggestions, comments, and encouragements.

CHAPTER 16

MINING NEW MOTIFS FROM CDNA SEQUENCE DATA

Christian Schönbach

RIKEN Genomic Sciences Center
schoen@gsc.riken.jp

Hideo Matsuda

Osaka University
matsuda@ist.osaka-u.ac.jp

General biological databases that store basic information on genome, transcriptome, and proteome are indispensable sequence discovery resources. However, they are not necessarily useful for inferring functions of proteins. To see this, we observe that SWISS-PROT[41]—a protein knowledgebase containing curated protein sequences and functional information on domains and diseases—has grown a mere 26-fold in 15 years, from 3,939 entries in 1986 to 126,147 entries in 2003. Similarly, despite the human draft genome and the mouse draft genome and transcriptome, the number of human and mouse protein sequences with some functional information has remained low—7,471 (7.4%) for man and 4,816 (4.7%) for mouse—compared to an estimated proteome of 0.5–1.0×10^6 sequences.[60]

The majority of sequences in the TrEMBL database of SWISS-PROT/TrEMBL, FANTOM,[714] and other similar databases are hypothetical proteins, or are uninformative sequences described as "similar to DKFZ ..." or "weakly similar to KIAA ...." These sequences have no informative homolog that had diverged from a common ancestor, and have matched to a non-informative homolog. Algorithms for identification of motifs are commonly used to classify these sequences, and to provide functional clues on binding sites, catalytic sites, and active sites, or structure/functions relations. For example 5,873 of 21,050 predicted FANTOM1 protein sequences contain InterPro motifs or domains. In fact, the InterPro name is the only functional description of 900 sequences.

Extrapolations from current mouse cDNA data indicate that the proteome is significantly larger than the genome. This underlines the importance of exploring protein sequences, motifs, and modules, to derive potential functions and interactions for these sequences. Strictly defined new protein sequence motifs are either conserved sequences of common ancestry, or are convergence (functional motifs)

within several proteins that group together for the first time by similarity search
and show statistical significance.

However, some motifs often do not reflect common ancestry. Examples include motifs occurring in paralogs, motifs occurring in mosaic protein sequences,
and structural motifs based on secondary structure or active site conservation.
Therefore, biological interpretation of motif findings requires additional efforts—
for example literature, structural, and phylogenetic analysis.

This chapter presents a case study of mining new motifs from the FANTOM1
mouse cDNA clone collection by a linkage-clustering method, with an all-to-all
sequence comparison, followed by visual inspection, sequence, topological, and
literature analysis of the motif candidates. Initially the ratio of true positive to
false positive new motifs turned out to be about 1:7 due to sequence redundancy
and retrotransposons inserted in to the coding sequence. After filtering out those
false conserved region, the ratio improves to 1:3.

ORGANIZATION.

Section 1. We briefly introduce the concept of motifs. Then we mention several broad
categories of approaches to recognize and discover motifs. We also discuss some of
the difficulties involved in discovering new motifs from cDNA sequences.

Section 2. We have designed a pipeline to discover new motifs in the FANTOM[714] data
set. This section gives an overview of this pipeline.

Sections 3–9. The next seven sections dive into the details of the key steps of our pipeline.
These key steps are: prepare a non-redundant translated data set from the FANTOM
data set for motif discovery, cluster the non-redundant sequences into groups of homologous sequences, extract blocks from these groups, form a block graph from
these blocks, detect homologous regions using a maximum density subgraph algorithm, eliminate those detected regions that contain known motifs, enrich the remaining blocks with additional sequences that match HMM built from these blocks.
These blocks give us our candidate new motifs. Visual inspection are then carried out
on these candidate motifs considering issues such as chromosomal localization, secondary structures, cellular localization, phylogenetic relations, and literature to assess
which candidate motifs are true and novel. A discussion of the various categories of
false positives is also given to illustrate this final manual assessment step.

Sections 10–11. Finally, we close the chapter with an extensive discussion on the true
positives and their biological interpretations. We also offer a speculation on the number
of new motifs that remain to be discovered.

1. What is a Motif?

Motifs are traditionally defined as conserved sequence patterns within a larger set
of protein sequences that share common ancestry. Conserved motifs may be used
to predict the functions of novel proteins if the relationship among the encoding
genes is orthologous.[818] However, the increasing number of paralogs and mosaic
proteins evolved from gene duplications and genomic rearrangement mechanisms

has led to enlarged interpretations of the term motif and to the concept of modules as conserved building blocks of proteins that have a distinct function.[91,345]

A module can consist of one motif or multiple adjacent motifs. Many mammalian extracellular proteins, proteins involved in signaling cascades, and disease related positionally cloned genes, contain modules with multiple adjacent motifs or domains that are involved in different functions, such as catalytic, adaptor, effector, and/or stimulator functions. For example, C2H2 zinc fingers, leucine zippers, and POU domains are DNA-binding modules.

The variety of domains in multi-domain proteins, structural motifs or active site conservation in a short stretch of sequences seldom reflects common ancestry. Therefore, the biological interpretation of motifs—particularly new motifs—requires additional efforts, for example, literature searching and reading, structural and phylogenetic analysis.

There are so many motif discovery methodologies. References and URLs to some of these are listed in Figure 1. Which one should we use? The methods can be broadly divided into the following five categories:

(1) manual, as in PROSITE;[238] automated, as in PRODOM;[175] and mixed approaches, as in PRINTS[37] and MDS;[424]
(2) regular expressions and profiles, as in PROSITE;
(3) hierarchical clustering-based sequence similarity and derivatives with position weight matrices, as in BLOCKS[343] and PRINTS;
(4) non-linear approaches, as in the hidden Markov models (HMM) of Pfam[65], ProtFam[563], and TIGRFAMs,[313] or the neural networks of ProClass;[370] and
(5) graph-based linkage clustering, as in MDS.

Manual approaches tend to be highly specific but lack broad coverage, whereas results of completely automated methods need careful post-processing and curation to avoid misclassification of sequences. The threshold settings of algorithms in categories 1–3 determine the coverage and specificity of the motifs. Category 4 methods are dependent on the initial seed alignments and number of training set sequences. Method 5 is robust towards cut-off thresholds and data size but may cause biological false positives if conserved regions of paralogs are detected as motif members.

Each method has its strength, weakness, and the potential to miss out novel motifs or motif members. None of the above methods is "the best method" for identifying a known motif or discovering new motifs because the cut-off thresholds are either predefined or not comparable. Thus, the success of any motif analysis depends on applying and comparing multiple existing methodologies. InterPro[30] integrates various motif information and increases the confidence if the results are overlapping.

Several of the mentioned pattern discovery methods—InterProScan[927] and Pfam hmmsearch—are applied during the FANTOM sequence annotation to classify sequences, to assign gene names, and to perform indirect functional assignment by motif-gene ontology mapping. Yet existing motif discovery methods do not yield previously unknown motifs nor refined functional classifications of submotifs that indicate potential new functions of known and new genes. None of the existing methods is designed to detect biologically relevant conserved regions of distantly related proteins without including segments that look similar by chance.

A major problem with existing methods is that cut-off scores are either predetermined by users or are empirically determined by the developers of the motif detection algorithms. In biology, the notion of a conserved region is fuzzy and depends on the hierarchical context of the protein sequences. Therefore, the cut-off thresholds of conserved regions among superfamilies, familes, and subfamilies are different.

2. Motif Discovery

We have designed a pipeline to discover and characterize new motifs in the FANTOM (Functional Annotation of Mouse for RIKEN full-length cDNA clones)[714] dataset. FANTOM is part of a systematic approach to determine the full coding potential of the mouse genome and assign functional annotations to uncharacterized cDNAs. The FANTOM1 data set that is analyzed by us consists of 21,076 full-length clones (see also http://fantom.gsc.riken.go.jp). This pipeline is generic and is applicable to other large-scale sequence collections. The pipeline is depicted in Figure 2. It comprises an automated part for discovery of motif sequences using a maximum density subgraph method, and a semi-automated part for exploration of motif sequences. The latter consists of visual inspection, sequence analysis, topological analysis, and literature analysis, of motif candidate members. The thorough case-by-case exploration minimizes the effect of misannotations and error propagation.

The maximum density subgraph method (MDS)[550] is a graph-based maximum-linkage clustering method. It avoids the problems of single-linkage and hierarchical clustering, such as similarity by chance and under-clustering—too many small clusters and a few large clusters—caused by a single threshold. The MDS method applies a very low cut-off threshold to detect all related sequence pairs. Irrelevant sequence pairs are filtered out by calculating the density of blocks—the ratio of the sum of similarity scores between ungapped subsequences to the number of the subsequences—in sequence pairs. The blocks are ordered by density and blocks of the highest density cluster are selected first. The process is repeated until no high density cluster blocks are found.

Sections 3 to 9 are devoted to a more detailed exposition of the main steps of our pipeline for motif discovery.

3. Preparations for Computational Motif Detection

Implementation of our motif discovery pipeline requires that we have a UNIX or LINUX operating system and several locally installed programs and databases as listed in Figure 3. The starting point of our motif discovery is the preparation of a non-redundant set of translated sequences. We have chosen DECODER[268] to predict the open reading frame (ORF) because of its ability to correct frame shifts. In retrospect, we recommend the application of multiple programs—*e.g.*, ESTSCAN[384] and OrfFinder[882]—because the positions of the ORF can differ significantly depending on the algorithm used.

DECODER prediction yields 21,050 potential coding sequences. Since the clone set shows redundancies that can lead to false positive motifs, we cluster the putative translations using DDS[371] and ClustalW.[826] From each cluster we select the longest sequence as the representative of the cluster. As a result, we obtain 15,631 non-redundant sequences.

4. Extraction of Homologous Sequences and Clustering

The non-redundant sequences are compared against each other by BLASTP[24] using a E-value of 0.1, the BLOSUM62 matrix, and the SEG filter option[900] to remove low-complexity regions. The E-value is set low to detect all possible sequence pairs.

Each sequence pair is then analyzed using a clustering algorithm[551] based on graphy theory to extract homologous groups of sequences. Each sequence is considered as a vertex. If the similarity between any pair of sequences exceeds the user-defined E-value threshold of 0.1, an arc is drawn between the two vertices corresponding to the sequence pair. The algorithm repeatedly extracts subgraphs whose vertices are connected with at least a fraction P—a user-defined ratio, here $P = 40\%$—of the other vertices until the subgraphs cover the whole graph or no further subgraphs can satisfy the conditions. The groups of subgraphs may overlap with each other if some sequences, such as the multi-domain containing sequences, share two or more homologous regions with a different set of sequences.

The method is equivalent to complete-linkage clustering if P is set to 100%. In contrast, single-linkage clustering requires only one arc to any member in a group and P becomes virtually 0% when the number of members is large. The linkage-clustering method with all-to-all sequence comparison[551] results in 2,196 homologous groups of non-redundant sequences.

5. Detection of Homologous Regions with Maximum-Density Subgraphs

Next, we extract all subsequences of at least 20 amino acid residues length in a sequence. Then we perform ungapped pairwise alignments among all subsequence pairs to obtain blocks. A block must contain at least four subsequences. Subsequence pairs may overlap with each other if some sequences share two or more homologous regions with a different set of sequences. In this case, the overlapping pairs are merged to the same blocks step by step in descending order of their similarity scores. However, the merge is not performed if it cause the accidental join of two independent pairs (non-overlapping or partialy-overlapping pairs in the previous merge step).[550]

The alignments are scored using the BLOSUM50 score matrix. A block graph is constructed by regarding blocks as vertices. Two vertices are connected by weighted arcs if the corresponding blocks show at least the user-defined level of similarity according to their BLOSUM50 score.

Highly connected components in the block graph are detected using a maximum-density subgraph algorithm (MDS).[550] Here, "density" is a graph-theoretic term that is defined as the ratio of the sum of the similarity scores between blocks to the number of blocks. Homologous regions longer than 20 amino acids are obtained by combining overlapping blocks. The MDS algorithm yields 465 blocks that contain at least 4 sequences, and a total of 1,531 motif candidates (i.e., sequences which share similar regions over more than 20 amino acid residues). The 465 blocks occur in the 3,202 conceptually translated mouse cDNA sequences and the blocks overlap 12,251 conserved regions. Conserverd regions are defined as those regions detected by HMMER in Pfam, BLASTP in ProDom and InterPro Scan in InterPro databases.

6. Visualization of Graph-Based Clustering

The original publication[424] of the MDS method does not have room to visualize the graph-based clustering. So we take the opportunity of this chapter to illustrate the visualization. For this purpose, we conduct a small experiment with 35 known members of the Inhibitor of Growth (ING) subfamilies and two control sequences from yeast, YNJ7_YEAST and YHP0_YEAST, that share a PHD domain with ING members but are otherwise unrelated.

The BLAST scores of the sequences are computed from SWISS-PROT/TrEMBL NRDB (SWISS-PROT 40.14, 03-Apr-2002). The thresholds for drawing an arc between the sequence pairs and extracting subgraphs whose vertices are connected are set to $E < 10^{-20}$ and $P = 80\%$. The results are shown and explained in Figure 4.

7. Filtering Out Known Motifs

The detected blocks are then searched for already reported conserved regions with HMMER in Pfam (Release 5.5), BLASTP in ProDom (Release 2000.1), and InterProScan in InterPro (Release 2.0) databases. Blocks that overlap with one or more residues of known conserved regions (motifs or domains reported in Pfam, InterPro, or ProDom) are discarded. The remaining 49 blocks, containing 139 sequences and 216 conserved regions, are labeled as new motif candidates with the original discovery date.

8. Extension of New Motif Candidates

In order to expand the number of motif members, we construct new candidates from the conserved blocks using the HMMER hmmbuild program, with the –f option for a local alignment of multiple domains HMM. The HMM profile is searched with hmmsearch, with E-value < 0.1, against the SPTR-NRDB database, the 10,603 DECODER predicted FANTOM1 translations, and the 1,908 DECODER predicted translation of the EST assemblies that are not included in GenBank nor SPTR.

The hmmsearch for the 49 motif candidates increases the number of sequence from 139 to 277 sequences. The HMM expanded candidate motif sequences are aligned and displayed together with their HMM score, E-value, start position, end position, and chromosomal localization information if available to facilitate visual inspections.

9. Motif Exploration and Extended Sequence Analysis

The interpretation of 49 motif candidates is a manual process that requires biological expertise. Visual inspections of all conserved regions are carried out under consideration of species distribution, chromosomal localization, secondary structures, cellular localization, phylogenetic relations, and publications.

On the basis of the inspections, 7 of the 49 motif candidates are assessed as true and new motifs (MDS00105, MDS00113, MDS00132, and MDS00145–MDS00148). These 7 motifs are present in 28 FANTOM and 108 SPTR derived sequences. The remaining 42 of the 49 motif candidates are assessed as false positive motifs in the sense that they are either not true or not new.

The 42 false positives fall into the following 7 categories:

(1) Two motifs overlap with a published domain or motif that has not yet been incorporated into InterPro, Pfam, and ProDom releases at the time of the analysis.

(2) One motif turns out to be a low complexity region that is missed by the SEG filter.

(3) Twenty-four motifs are generated by sequence redundancy. The detection of redundant sequences after applying a clustering program shows that one should not rely on a single program. In retrospect, we should have applied two different clustering algorithms.

(4) Alternative splicing or the presence of unspliced introns cause another three false positives.

(5) Eight motifs are detected only in mouse sequences of the FANTOM clone set. Since mouse-specific motifs are unlikely to occur, the motif members may be derived from paralogs.

(6) The last category of false positives comprises repeat elements because the cDNA sequences have not been masked before predicting open reading frames. Depending on the data sources, it is recommended to check at the beginning of the pipeline for computational translated repeat elements using the RepeatMasker, which can be obtained from `http://ftp.genome.washington.edu/RM/RepeatMasker.html`.

When we to scanned the candidate motif sequences retrospectively for repeat elements and compared the positions of the repeats within coding regions (CDS) to the motif candidate positions four (8%) of the motif candidates contained B1, B2, intracisternal A-particle LTR, and mammalian apparent LTR-retrotransposon repeat elements, respectively. Details and sequence alignments are shown at `http://motif.ics.es.osaka-u.ac.jp/MDS/falsepositives.html`.

For the 7 true positive motifs, we search PubMed with the informative gene or protein names of their member sequences to collect articles that contain biochemical, structural, or disease information. In addition, we carry out for all sequences additional sequence analyses. Secondary structure analyses of sequences are performed with locally installed ANTHEPPROT V5.0 rel.1.0.5 software,[195] DSC package,[432] and on the external PredictProtein server.[722] Functional sites— for example, phosphorylation and N-glycosylation sites—are predicted using ANTHEPROT from the PROSITE database. The locally installed PSORTII program is also used to predict the cellular localization of proteins. Chromosomal localization information is retrieved from the FANTOM map of RIKEN clones to human and mouse chromosomes as well as LocusLink[684]; it is used to judge orthology and paralogy. Multiple sequence alignments of motif member sequences are performed using locally installed ClustalW 1.8. The alignments motif sequences and motif member sequences are post-processed with the coloring software MView 1.41[112]. In some cases, the alignments are also edited by hand to improve the alignment quality. The colored alignments by amino acid properties are helpful in inferring possible functions. Phylogenetic trees are constructed from the motif member sequences by the maximum-likelihood method using MOLPHY ProtML[6] to re-assess common ancestry. The tree is obtained by the "Quick Add Search," using the Jones-Taylor-Thornton model[400] of amino acid substitution, and 300 top ranking trees are retained (options -jf -q -n 300). Bootstrap values of the tree are calculated by analyzing 1,000 replicates using the resampling of the estimated log-likelihood (RELL) method.[433]

10. Biological Interpretation of Motifs

In general the impact of bioinformatics-aided functional predictions depends on a close collaboration with biologists who put the new findings into the functional context of existing data. Since the scope of the book is on Bioinformatics, we give only an abridged version of the biological findings that have been published by Kawaji *et al.*[424] The sequence alignments of all motifs described in this section are available at http://motif.ics.es.osaka-u.ac.jp/MDS.

Three of the 7 new motifs given in Section 9 have been found in hypothetical proteins. Since we lack experimental information on these proteins, we briefly summarize the predicted functions. MDS00132 members are encoded by mouse 2210414H16Rik and 330001H21Rik, and human DKFZP586A0522 loci (SPTR accessions Q9H8H3 Q9H7R3, Q9Y422, AAH08180). The human proteins belong to the generic methyltransferase family (InterPro IPR001601) and contain, adjacent to the N-terminal located MDS00132, a SAM (S-adenosyl-L-methionine) binding motif (IPR000051). Considering the 80% sequence identity and 90% similarity to DKFZP586A0522 over 146 residues (data not shown), it is likely that hypothetical proteins 2210414H16Rik and 3300001H21Rik belong to the methyltransferase family.

Motif MDS00146 comprises 21 members of hypothetical proteins or fragment derived from human, mouse, rat, fruitfly, and worm. Three members, mouse 1200017A24Rik (SPTR accession Q9DB92) and human DOCK8 (Q8NF50) previously represented by BA165F24.1 and FLJ00026, carry at their C-terminus an aminoacyl-transfer RNA synthetases class-II signature (IPR002106), indicating possible involvement in the protein synthesis.

Motif MDS00147 is located at the N-terminus of four mouse and two human hypothetical proteins. No other motifs have been detected in the sequences. This motif is an example where even motif analysis fails to add functional information. However, from the perspective of experimental biologists, the non-informative motifs are most interesting as they provide new discovery targets for protein interactions and biochemical reactions.

Motif MDS00105 is specific for the ING family, comprising three subfamilies: the ING1/ING1L subfamily,[308, 733, 929] the ING3 subfamily, and the ING1-homolog subfamily including distant homologues in *D. melanogaster*, *A. thaliana*, and *S. pombe*. The tree submotifs allow classification of sequences into the subfamilies which represent binding sites for distinct subfamily-specific protein-protein interaction candidates with HAT, HDAC, MYC,[340] and other cell cycle related proteins, while the unique regions of each subfamily member may modulate interactions. Motif MDS00105 is a candidate for protein-protein interaction experiments to define the physiological roles of the three ING subfamily members.

Motif MDS00145 is specific for mammalian 1-acyl-SN-glycerol-3-phosphate acyltransferases AGPAT3 and AGPAT4. RIKEN clones 4930526L14 and 2210417G15 represent Agpat3 (Chr 10 41.8 cM), which is the ortholog of human AGPAT3 on Chr 21q22.3. The FANTOM1 mapping of RIKEN clones 4930526L14 and 2210417G15 to Chr 16 69.90–71.20 appears to be caused by an Agpat3 related sequence on Chr 16. Agpat4 (clone 1500003P24) has been mapped to mouse Chr 17, 7.3–8.2 cM and a syntenic region on human Chr 6 that contains AGPAT4 and is close to MAP3K3. AGPAT4 is a paralog of AGPAT1 (Chr 6p21) located in major histocompatibility complex class III region. According to PSORT predictions,[597] AGPAT3 and AGPAT4 are endoplasmic reticulum (ER) membrane proteins. The latter is in concordance with previous findings for human AGPAT1.[13] The Pfam-defined acyltransferase domain is located between second and third transmembrane helix and shared by all AGPAT members. The divergence of AGPATs in the ER-sided region around MDS00145 motif of AP-GAT3 and AGPAT4 suggests a regulatory function of MDS00145 for transacylation specificity or activity.

Motif MDS00148 spans a 35–37 amino acids long extracellular oriented loop region between two transmembrane domains that is conserved among members of the mammalian solute carrier family 21 (organic anion transporters), organic anion transporter polypeptide-related (OATPRP) and related *Drosophila* and *C. elegans* organic anion transporters.[838] MDS00148 represents a novel module with some structural similarities to the kazal-type domain. MDS00148 might have evolved from a kazal-like protease inhibitor domain but acquired different functionality related to substrate binding in the Na^+-independent transport of organic anions, conjugated and unconjugated bile acids when transferred into an ancestral transmembrane SLC21 family protein.

Motif MDS00113 includes 20 members with conserved sequences of 43 amino acids length that carry either a leucine zipper signature characteristic for Fos related antigen 1 (FRA1) or a leucine zipper-like motif (16 members). We analyze 13 representative members in detail. Ten out of the 13 contain a leucine heptad repeat in their sequences. Since the leucine repeat could have occurred by chance, it would be risky to infer from it the leucine zipper function of DNA binding. We therefore re-analyze the sequences for features of known and characterized leucine zippers occurring in transcription factors:

(1) alpha helical coiled-coil region[639] with mostly 3,4-hydrophobic repeat of apolar amino acids at positions a and d of the helix,
(2) overlap of the coiled-coil region with the leucine heptad repeat,
(3) a coiled-coil trigger sequences[257] or the 13-residue trigger motif,[407, 905]
(4) a basic DNA binding region preceding the heptad repeat and
(5) a nuclear localization signals.[351]

When these these criteria are applied, only the FRA1 sequences qualify as basic leucine zipper with DNA binding function.[468, 469] Eleven sequences bear a coiled-coil trigger sequence or trigger motif. Given the sequence conservation with FRA1, it is conceivable that an ancestral functional basic leucine zipper region was subjected to recombination and mutation events degenerated the basic leucine zipper. We therefore suggest that the tandem coiled-coil containing proteins bind to proteins, rather than DNA, in a similar fashion as the group D basic helix-loop-helix (HLH) proteins.[36]

11. How Many New Motifs are Waiting to be Discovered?

Let us conclude this chapter with a speculation on the number of new motifs that are waiting to be discovered. To answer this question, we estimate the motif coverage and do some extrapolations.

The coverage of the 7 MDS motifs in Section 9 is 0.224% (28 sequences out of 12,511 sequences comprising 10,603 DECODER predicted FANTOM1 translations and 1,908 DECODER predicted translation of EST assemblies thatare not included in GenBank nor SPTR). If we extrapolate from the 136 hits of the 7 motifs to 707,571 sequences of the non-redundant SPTR database, excluding the 10,465 FANTOM1 sequences, the estimated number of new MDS motif containing sequences would be 927 (0.133% of 697,106 sequences). Since the number of sequences per MDS motif varies from 4—the minimum number of motif containing sequences that our method detects—to 57 (MDS00148), the estimated number of not-yet-discovered MDS motifs in the current release of SPTR should range from 16 to 231.

The low number of new motifs may reflect a constraint on the number of possible functions and interactions for a given protein in the proteome. In addition, some of the new motifs may be lineage-specific due to species-specific expansion of regulatory genes.[578]

Any way, each motif discovery strategy provides different results and views. We have presented an alternative strategy that has potential to extract many new motifs from existing data. Motifs can provide a rich data source of functional clues that support initial steps of protein-protein interaction, and regulatory and active site target selections in a drug discovery process.

However, their discovery on transcriptome or genome-scale requires careful data preparation. Otherwise, too much time is spent on filtering out false positive motifs. Before embarking on a motif discovery and exploration journey, one should keep mind that proteins function in a cellular context. One or multiple functions are the results of protein structure, which is dependent on the protein sequence, transcription, translation, post-translational modifications, and cellular localization at a given time within a complex network. Motif discovery can give at best some answers for one layer of complexity—at the level of protein sequence—that can enable us to ask new questions.

Database	URL	Motifs/ Domains	Comment
GRAPH-BASED LINKAGE CLUSTERING			
MDS	`motif.ics.es.osaka-u.` `ac.jp/MDS`	7	curated
REGULAR EXPRESSIONS			
PROSITE	`www.expasy.ch/prosite`	1,517	curated
PROFILES & HIDDEN MARKOV MODELS			
PFAM	`pfam.wustl.edu`	3,621	
ProtFam	`mips.gsf.de/proj/protfam`		
TIGRFAMs	`www.tigr.org/TIGRFAMs`	1,415	
BLOCKS	`blocks.fhcrc.org/blocks`	2,101	curated
PRINTS	`www.bioinf.man.ac.uk/` `dbbrowser/PRINTS`	1,650	curated
SBASE	`www3.icgeb.trieste.it/` `~sbasesrv/main.html`		
	SBASE-A (consolidated domains)	2,425	
	SBASE-B (unconsolidated domains)	739	
CONSENSUS SEQUENCES			
ProDom	`prodes.toulouse.inra.fr/` `prodom/doc/prodom.html`	108,076	fully automated
DOMO	`www.infobiogen.fr/` `services/domo`	8,877	fully automated
NEURAL NET			
PROCLASS	`pir.georgetown.edu/` `gfserver/proclass.html`	6,171	curated
INTEGRATED DATABASE			
Interpro	`www.ebi.ac.uk/interpro`	4,691	curated
HITS	`hits.isb-sib.ch`	4,547	fully automated
SMART	`smart.embl-heidelberg.` `de`	631	curated
eMOTIF	`dna.stanford.edu/` `identify`	170,294	
MetaFAM	`metafam.ahc.umn.edu`	2,793	fully automated

Number of motifs/domains as of 7 April 2002.

Fig. 1.   Motif Databases

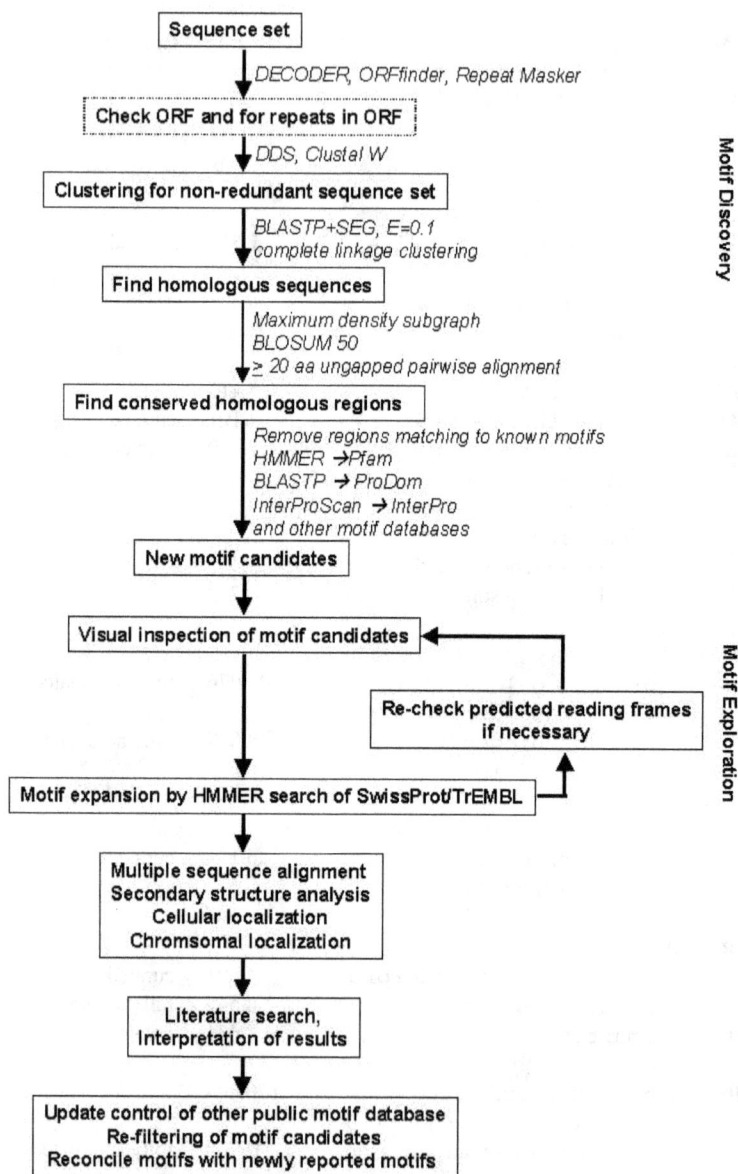

Fig. 2.   A flowchart of the maximum-density subgraph motif discovery and exploration process.

Components	URL/Contact
GENERAL TOOLS	
BioPERL	www.bioperl.org
Apache HTTP server	www.apache.org
PostgresSQL	www.postgresql.org
	Results are stored in a relational database. Note that the added value is low and can increase maintenance costs compared to semi-structured flat file format.
SPECIFIC TOOLS	
DECODER	Available from rgscerg@gsc.riken.go.jp
DDS	Available as part of the AAT package at ftp://ftp.tigr.org/pub/software/AAT
BLASTP	ftp://ncbi.nlm.nih.gov/toolbox/ncbi_tools
InterProScan	ftp://ftp.ebi.ac.uk/pub/databases/interpro/iprscan
HMMER	hmmer.wustl.edu
MDS	Available from matsuda@ist.osaka-u.ac.jp
SEView	www.isrec.isb-sib.ch/ftp-server/SEView
ClustalW	ftp://ftp.ebi.ac.uk/pub/software/unix/clustalw
MView	mathbio.nimr.mrc.ac.uk/~nbrown/mview
PSORTII	psort.ims.u-tokyo.ac.jp
DCS	ftp://ftp.dcs.aber.ac.uk/pub/users/rdk/dsc
PredictProtein	maple.bioc.columbia.edu/pp
MINIMUM DATABASE SETS	
InterPro	ftp://ftp.ebi.ac.uk/pub/databases/interpro
ProDoM	ftp://ftp.toulouse.inra.fr/pub/prodom/current_release
SPTR	ftp://ftp.ebi.ac.uk/pub/databases/sp_tr_nrdb

Fig. 3. Tools for motif discovery

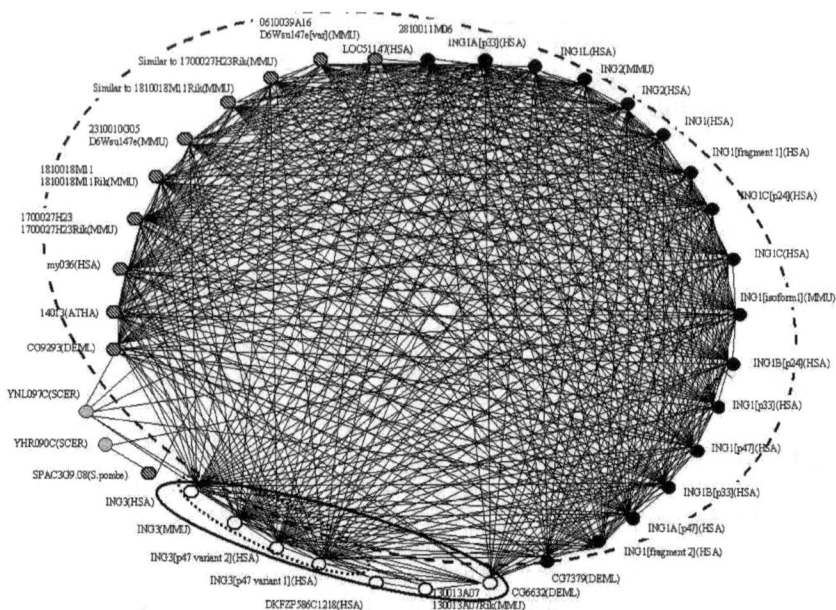

Fig. 4. The maximum density subgraph of ING subfamily members. ING members split into three clusters which are denoted by ovals. The fi rst cluster (broken line) contains 27 members ING1/ING1L and ING1-homolog subfamilies. The second cluster (solid line) is composed of 7 ING3 subfamily members. The third cluster (dotted line) contains 4 ING3 members that belong to both cluster 1 and 2. White circles indicate ING3 members. ING1-homolog members are symbolized by hatched circles. The black circles denote ING1/ING1 members. The two yeast sequences, YNL097C (YNJ7_YEAST) and YHR090C (YHP0_YEAST), that contain a PHD domain but are otherwise unrelated to ING1 family are shown in grey circles. SPAC3G9.08 (O42871) belongs to the ING1-homolog subfamily but was not included into cluster 1 under the threshold setting of this experiment ($P = 80\%$ and E-value $< 10^{-20}$).

CHAPTER 17

TECHNOLOGIES FOR BIOLOGICAL DATA INTEGRATION

Limsoon Wong

Institute for Infocomm Research
limsoon@i2r.a-star.edu.sg

The process of building a new database relevant to some field of study in biomedicine involves transforming, integrating, and cleansing multiple data sources, as well as adding new material and annotations. We review in this chapter some of the requirements and technologies relevant to this data integration problem.

ORGANIZATION.

Section 1. We begin with a detailed account of the motivations and requirements for a general integration system for biological data. We also discuss the motivations and requirements for locally warehousing such data.

Section 2. Then we review some representative technologies for data integration in biological and medical research. The technologies surveyed include EnsEMBL, GenoMax, SRS, DiscoveryLink, OPM, Kleisli, and XML.

Section 3. Following that, we highlight some of the features that distinguish the more general data integration technologies from the more specialized ones. The features considered include data model, data exchange format, query capability, warehousing capability, and application programming interface.

Section 4. Lastly, we compare the surveyed technologies and comment on selecting such technologies. We also briefly discuss TAMBIS and the semantics aspect of data integration.

1. Requirements of Integration Systems for Biological Data

In a dynamic heterogeneous environment such as that of bioinformatics, many different databases and software systems are used. A large proportion of these databases were designed and built by biologists. When these databases were first created, the amount of data was small and it was important that the database entries were human readable. Database entries were therefore often created as flat files. As new types of data were captured, new databases were created using a variety of flat file formats. We ended up with a large number of different databases in different formats, typically using non-standard query softwares.[52] Many questions that a biologist is interested in could not be answered using any single data source. However, some of these queries can be satisfactorily solved by using information from several sources. Unfortunately, this has proved to be quite difficult in practice.

These databases and systems often do not have anything that can be thought of as an explicit database schema, which is a formalized queriable catalogue of all the tables in the database, the attributes of each of these tables, and the meaning of and indices on each of these attributes. Further compounding the problem is that research biologists demand flexible access and queries in ad-hoc combinations. Simple retrieval of data is not sufficient for modern bioinformatics. The challenge is how to manipulate the retrieved data derived from various databases and restructure the data in such a way to investigate specific biomedical problems.[1]

As observed by Baker and Brass,[52] many existing biology data retrieval systems are not fully up to the demand of painless and flexible data integration. These systems rely on low-level direct manipulation by the user, where she uses a keyword to extract summary records, then clicks on each resulting record to view its contents or to perform operations. This works well for simple actions. However, as the number of actions or records increases, such direct manipulations rapidly becomes a repetitive drudgery. Also when the questions become more complex and involve many databases, assembly of the data needed is likely to exceed the skill and patience of the biologist. Merely providing a library package that interfaces to a large number of databases and analysis softwares is also not useful if it requires long-winded and tedious programming to make use of and adding to the package.

The systems provided by bioinformaticians in answer to the challenge above can roughly be divided into "point" and "general" solutions. A point solution is a highly specialized system: the data sources to be considered are small and fixed; the biomedical research questions to be addressed are small and fixed; and the point solution is a specific software that provides the expected answers and nothing else.[1] Hence, there is little database design and consideration for extensibility nor for flexibility. In contrast, a general solution is not designed with a specific set of biomedical research questions in mind nor with a specific set of data sources in mind. It must be designed with extensibility and flexibility in mind. A general solution can serve as the platform upon which to shorten the time needed for constructing various point solutions, just as a relational database management system can serve as the platform upon which to build specific accounting systems.

A system that aims to be a general integration mechanism in the bioinformatics environment described earlier must satisfy at least the following four conditions, which were identified previously by Wong.[894]

(1) It must not count on the availability of schemas. It must be able to compile any query submitted based solely on the structure of that query. If it needs a schema before it can compile a query, then it would be hard to use for our purpose because biomedical databases often do not have usable schemas.

(2) It must have a data model that the external database and software systems can easily translate to, without doing a lot of type declarations. If it does not have such a data model, then there would be a significant impedance in moving external data into the system, in moving internal data into external databases, and in manipulating the data when they are brought into the system.

(3) It must shield existing queries from evolution of the external sources as much as possible. For example, an extra field appearing in an external database table must not necessitate the recompilation or rewriting of existing queries over that data source. The external data sources used by a bioinformatician are typically owned by different organizations who have autonomous right to evolve their databases. It is therefore important for a general data integration solution to be robust when the data sources evolve.

(4) It must have a data exchange format that is straightforward to use, so that it does not demand too much programming effort or contortion to capture the variety of structures of output from external databases and softwares. The data exchange format is the standard by which the system exchange data with the external data sources. If it is not straightforward to use, then great effort would be needed for connecting the system to the external data sources.

Besides the ability to query, assemble, and transform data from remote heterogeneous sources, it is also important to be able to conveniently warehouse the data locally. The reasons to create local warehouses are given below, some of which were identified previously by Davidson *et al.*[188]

(1) It increases efficiency. It is clear that we do not want to be choked by the slowest external data source nor by communication latency in the execution of our queries, especially if we own a fast computer. Warehousing gives us as much efficiency as we can afford to pay for.

(2) It increases availabilty. It is clear that we do not want to be unable to run our queries at a time we wish because a needed external source is unavailable. Warehousing guarantees that the data we need for our queries are always available whenever we need them.

(3) It reduces risk of unintended "denial of service" attacks on the original sources. Some data sources, such as the Entrez website at the National Center for Bioinformatic Information, impose a strict limit on the number of times or the amount of data that we can access within a single day. If we exceed that limit, then we risk being banned from the site. Unfortunately, some of our queries may require very intensive access to data held in such sites. Warehousing protects us from this risk by rendering it unnecessary for us to access the remote site.

(4) It allows more careful data cleansing that cannot be done on the fly. It is widely acknowledged that many of the biomedical sources contain a large number of errors.[104] For example, Schönbach et al.[756] reported that up to 30% of the database records that they accessed when constructing their warehouse on swine major histocompatibility complexes contained errors. Some of these errors can be detected and corrected on the fly, some of these errors cannot. It therefore makes sense that if our queries are sensitive to certain errors that cannot be detected nor corrected on-the-fly, then we should warehouse the data after careful cleansing.

Creating warehouses leads to other requirements on a general data integration solution. Specifically the general data integration solution must provide for the construction of warehouses that have the following properties.

(1) The warehouse should be efficient to query.
(2) The warehouse should be easy to update. There are two aspects to this issue of ease of update. The first aspect is of making an individual change to the warehouse, such as modifying an existing record, deleting an existing record, or adding a new record. This aspect is a fundamental characteristic of the data integration tools that are used for maintaining the warehouse. The second aspect is that of the number of such individual changes that need to be made to bring the warehouse up to date. The second aspect is more a consideration for the strategy for maintaining the warehouse and is dictated by the interval between updates to the warehouse and the amount of changes that the underlying data sources can accummulate during the interval. A data integration tool that offers greater ease on the first aspect obviously also allows a greater range of strategies on the second aspect.
(3) Equally important in the biology arena is that the warehouse should model the data in a conceptually natural form. Although a relational database system is efficient for querying and easy to update, its native data model of flat tables forces us to unnaturally and unnecessarily fragment our data in order to fit our data into the third normal form.[850] For example, a record in the popular

SWISS-PROT database[41] would be fragmented into almost 30 tables in order to be stored in accordance to the third normal form. This unnatural fragmentation brings forth two problems. Firstly, it increases the mental load of the programmer and the possibility of programming errors in answering a queries for several reasons: (i) the implementer of a query at a later date may not be the same person who did the third normal form conversion and (ii) the implementer of a query may not be the biologist who asks the query. Secondly, it increases the cost of certain queries significantly. For example, if the query needs to reconstruct a large portion of a SWISS-PROT record, we would be required to perform 10-20 joins on the tables.

It is also important to realise that no single system is complete for all possible uses. A data integration system is rightly focused on

(1) reading data from multiple sources for integration,
(2) simple database-style transformation of data to facilitate data being passed from one application to the next, and
(3) writing data to warehouses.

There are certain types of analysis and manipulations of data that a data integration system is not expected to perform but is merely expected to facilitate. These analysis and manipulations include bioinformatics specific operations such as multiple alignment and visualization-specific operations such as display data in a graphical user interface. These operations are best implemented either in a specialized scripting language designed for those purpose or in a full strength common programming language. In order to facilitate the programming of these operations, the general data integration system must provide a means for these scripting and programming languages to interface to it, via a language embedding or via an application programming interface for these languages.

Lastly, the semantics issue may also be important.[188] This issue concerns the equivalence and consistency between parts of records in different data sources, as well as the mappings between these parts. A data integration technology that understands which parts of two data sources have the same meanings and should be consistent with each other is desirable. However, it must be recognized that the same record in a database can sometimes be interpreted in different ways depending upon the purpose and requirement of the user. Consequently, this issue is sometimes considered as a part of building a specific application or integrated database, as opposed to as a part of the tools used for building that integrated database or application.

2. Some Data Integration Solutions

We survey here a few alternative solutions to the data integration and warehousing problem in biomedicine. The surveyed solutions include EnsEMBL,[372] Geno-Max, SRS,[235] DiscoveryLink,[311] OPM,[145] Kleisli,[894] and XML.[2, 5] These examples are chosen to span specialized point solutions to increasingly general solutions. For each of these systems, we provide an overview and a discussion of their strong and weak points.

2.1. *EnsEMBL*

EnsEMBL is a software system jointly developed by the European Bioinformatics Institute and the Sanger Institute.[372] It provides easy access to eukaryotic genomic sequence data. It also performs automatic prediction of genes in these sequence data and assembles supporting annotations for these predictions. It is not so much an integration technology. However, it is an excellent example of a very successful integration of data and tools for the highly specific purpose of genome browsing.

EnsEMBL organizes raw sequence data from public databases into its internal database. It then assembles these sequences into their proper place in the genome. After that, it runs GenScan[123] to predict the location of genes and applies various analysis programs to annotate these predicted genes. Finally, the results of the process described above are presented for public access.

The main "entry points" to these results on the EnsEMBL Genome Browser are by

(1) searching by sequence similarity via the built-in BLAST component of the EnsEMBL Genome Browser;
(2) browsing from the chromosome level all the way down to the DNA sequence level;
(3) searching using special EnsEMBL identifiers; and
(4) free-text matching using annotation of databases linked to EnsEMBL, including OMIM,[318] SWISS-PROT,[41] and InterPro.[29]

It can also dump its data into Excel spreadsheets for use by external datamining softwares. Alternatively, the EnsMart data retrieval tool can also be used to access these results. EnsMart has a good query builder interface that allows a user to conveniently specify certain types of genomic regions and filters on these results. As a last resort, EnsEMBL provides a Perl-based programmatic interface for the most flexible access to its stored results.

Its strengths lie in its highly tailored functionalities for genome browsing. Once the sequences are imported into the system, assembly and annotation are automatically performed, the results are automatically prepared for browsing in a nice graphical user interface.

Its weaknesses lie also in its highly tailored point solution nature. It is not possible to ask EnsEMBL to perform an *ad hoc* query in general, unless that particular type of query has been anticipated by the designer of the EnsEMBL and its associated access tools. For example, while it is possible to ask a query such as "extract 500 bases flanking the translation initiation site of each confirmed gene in the database" using EnsMart, it does not seem possible to ask a query such as "extract the first exon of each confirmed gene in the database" using EnsMart at this moment. For the latter query, the user can resort to accessing EnsEMBL and extracting the required information by Perl programming. EnsEMBL also does not have a flexible data model nor exchange format, other than the structure of its highly specialized internal database. Thus, it is not straightforward to add new kinds of data sources, and it is also not straightforward to output or export data from EnsEMBL other than in the fixed export formats.

The weaknesses mentioned above are viewed from the perspective of the requirements of a general data integration system. However, one has to remember that EnsEMBL is intended as a point solution for the specific purpose of genome browsing. Within the context of this specific purpose, EnsEMBL works much better than virtually any other alternatives, as its design has anticipated the common queries a biologist may want to ask and makes it possible for her to ask them without requiring the help of a programmer.

2.2. GenoMax

GenoMax is an enterprise-level integration of bioinformatics tools and data sources developed by InforMax; see http://www.informaxinc.com/ solutions/genomax. It is a good illustration of an almagamation of a few point-solutions, including a sequence analysis module and a gene expression module, developed on top of a data warehouse of fixed design.[1] The warehouse is an ORACLE database designed to hold sequence data, gene expression data, 3D protein structures, and protein-protein interaction information. Load routines are built in for standard data sources such as GenBank and SWISS-PROT. The specialized point-solution modules provide capabilities such as performing BLAST[24] and GenScan[123] runs on sequences and computing differentially expressed genes from microarray experiments. A special scripting language of limited expressive power is also supported for building analytical pipelines.

Its strengths are twofold. Firstly, each of GenoMax's component point-solution modules is a very well designed application for a specific purpose. For example, its gene expression module provides self-organizing map clustering, principal component analysis, and so forth on microarray data via simple-to-use graphical user interfaces. Secondly, these components are integrated in a tight way via the specially designed data warehouse.

Its weakness is its tight point-solution-like application integration. While GenoMax has a broader scope than EnsEMBL, it does cover less data types and products than products such as SRS, DiscoveryLink, and Kleisli. For example, these latter systems can easily incorporate chemical assay data which are beyond the current data warehouse design of GenoMax. In addition, GenoMax's scripting language is not designed for large-scale database style manipulations and hence this type of *ad hoc* queries are not always straightfoward nor optimized in Geno-Max. There are also difficulties in adding new kinds of data sources and analysis tools. For example, it is probably impossible to express in the GenoMax scripting language the "rosetta stone" method for extracting protein interactions.[541]

2.3. SRS

SRS[235] is marketed by LION Bioscience and is arguably the most widely used database query and navigation system for the Life Science community. It provides easy-to-use graphical user interface access to a broad range of scientific databases, including biological sequences, metabolic pathways, and literature abstracts. SRS provides some functionalities to search across public, in-house and in- licensed databases.

In order to add a new data source into SRS, this data source is generally required to be available as a flat file and a description of the schema or structure of the data source must be available as an Icarus script, which is the special built-in wrapper programming language of SRS. The notable exception to this flat file requirement on the data source is when the data source is a relational database. SRS then indexes this data source on various fields parsed and described by the Icarus script. A biologist then accesses the data by supplying some keywords and constraints on them in the SRS Query Language. Then all records matching those keywords and constraints are returned.

The SRS Query language is primarily a navigational language. This query language has limited data joining capabilities based on indexed fields and has limited data restructuring capabilities. The results are returned as a simple aggregation of records that matched the search constraints. In short, in terms of querying power, SRS is essentially an information retrieval system. It brings back records matching specified keywords and constraints. These records can contain embedded links that a user can follow individually to obtain deeper information. However, it does not offer much help in organizing or transforming the retrieved results in a way that might be needed for setting up an analytical pipeline.

There is also a browser-based interface for formulating SRS queries and viewing results. In fact, this interface of SRS is often used by biologists as a unified front end to independently access multiple data sources, rather than learning the idiosyncrasies of the original search interfaces of these data sources. For this reason, SRS is sometimes considered[1] to serve "more of a user interface integration role rather than as a true data integration tool."

In summary, SRS has two main strengths. Firstly, it is very straightforward to add new data sources into the system, due to the use of the Icarus scripting language and due to the simplicity of flat file indexing. In fact, several hundred data sources have been incorporated into SRS to date. Secondly, it has a nice user interface that greatly simplifies query formulation, making the system usable by a biologist without the assistance of a programmer. In addition, SRS has an extension known as PRISMA that is designed for automating the process of maintaining a SRS warehouse. PRISMA integrates the tasks of monitoring remote data sources for new data sets, and downloading and indexing such data sets.

On the other hand, SRS also has some weaknesses. Firstly, it is basically a retrieval system that simply returns entries in a simple aggregation. If the biologist wishes to perform further operations or transformations on the results, she has to do that by hand or writes a separate postprocessing program using some external scripting languages like C or Perl, which is cumbersome. Secondly, its principally flat-file based indexing mechanism rules out the use of certain remote data sources—in particular, those that are not relational databases—and does not provide for straightforward integration with dynamic analysis tools. However, this latter shortcoming is mitigated by the SCOUT suite of applications marketed by LION Bioscience that are specifically designed to interact with SRS.

2.4. *DiscoveryLink*

DiscoveryLink[311] is an IBM product and, in principle, it goes one step beyond SRS as a general data integration system for biomedical data. The first thing that stands out—when DiscoveryLink is compared to SRS, EnsEMBL, and GenoMax—is the presence of an explicit data model. This data model dictates the way a DiscoveryLink user views the underlying data, the way she views results, as well as the way she queries the data.

The data model is the relational data model.[170] The relational data model is the de facto data model of most commercial database management systems, including the IBM's DB2 database management system upon which DiscoveryLink is based.

As a result, DiscoveryLink comes with a high-level query language, SQL, that is a standard feature of all such database management systems.

This gives DiscoveryLink several advantages over SRS. Firstly, not only can a user easily express SQL queries that go across multiple data sources—which a SRS user is able to do, but she can also perform further manipulations on the results—which a SRS user is unable to do. Secondly, not only are the SQL queries more powerful and expressive than those of SRS, the SQL queries are also automatically optimized by DB2. The use of query optimization allows a user to concentrate on getting her query right without worrying about getting it fast.

However, DiscoveryLink still has a some way to go in practice. The reason is twofold. The first reason is that DiscoveryLink is tied to the relational data model. This implies every piece of data that it handles must be a table of atomic objects like strings and numbers. Unfortunately, most of the data sources in biology are not that simple and are deeply nested. Therefore, there is severe impedance mismatch between these sources and DiscoveryLink. Consequently, it is not straightforward to add new data sources or analysis tools into the system. For example, to put the SWISS-PROT database into a relational database in the third normal form would require us to break every SWISS-PROT record into nearly 30 pieces in a normalization process! Such a normalization process requires a certain amount of skill. Similarly, to query the normalized data in DiscoveryLink requires some mental and performance overhead, as we need to figure out which part of SWISS-PROT has gone to which of the 30 pieces and we need to join some of the pieces back again.

The second reason is that DiscoveryLink supports only wrappers written in C++, which is not the most suitable programming language for writing wrappers. In short, it is difficult to extend DiscoveryLink with new sources. In addition, DiscoveryLink does not store nested objects in a natural way and is limited in its capability for handling long documents. It also has limitations as a tool for creating and managing data warehouses for biology.

In spite of these weaknesses, in theory, DiscoveryLink has greater generality than point solutions like EnsEMBL, specialized application integration like Geno-Max, and user interface integration solutions like SRS. Unfortunately, this greater generality is achieved at the price of requiring that SQL be used for expressing queries. While writing queries in SQL is generally simpler than writing in Perl, it is probably still beyond the skill of an average biologist. This is a disadvantage in comparison to EnsEMBL, GenoMax, and SRS, which have good user interfaces for a biologist to build the simpler queries.

2.5. *OPM*

OPM[145] was developed at Lawrence-Berkeley National Labs. OPM is a general data integration system. OPM was marketed by Gene Logic, but it sales was discontinued some time ago. It goes one step beyond DiscoveryLink in the sense that it has a more powerful data model, which is an enriched form of the entity-relationship data model.[149]

This data model can deal with the deeply nested structure of biomedical data in a natural way. Thus it removes the impedance mismatch. This data model is also supported by a SQL-like query language that allows data to be seen in terms of entities and relationships. Queries across multiple data sources, as well as transformation of results, can be easily and naturally expressed in this query language. Queries are also optimized. Furthermore, OPM comes with a number of data management tools that are useful for designing an integrated data warehouse on top of OPM.

However, OPM has several weaknesses. Firstly, OPM requires the use of a global integrated schema. It requires significant skill and effort to design a global integrated schema well. If a new data source needs to be added, the effort needed to re-design the global integrated schema potentially goes up quadratically with respect to the number of data sources already integrated. If an underlying source evolves, the global integrated schema tends to be affected and significant re-design effort is potentially needed. Therefore, it may be costly to extend OPM with new sources.

Secondly, OPM stores entities and relationships internally using a relational database management system. It achieves this by automatically converting the entities and relationships into a set of relational tables in the third normal form. This conversion process leads to an entity being broken up into many pieces when stored. This process is transparent to the OPM user. So she can continue to think and query in terms of entities and relationships. Nevertheless, the underlying fragmentation often causes performance problems, as many queries that required no join—when viewed at the conceptual level of entities and relations—are mapped to queries that required many joins on the physical pieces that entities are broken into.

Thirdly, OPM does not have a simple format to exchange data with external systems. At one stage, it interfaces to external sources using CORBA. The effort required for developing CORBA-compliance wrappers is generally significant.[764] Furthermore, CORBA is not designed for data intensive applications.

Although OPM's query language is at a higher level and is simpler to use than the SQL of DiscoveryLink, it shares the same disadvantage as DiscoveryLink from the perspective of an average biologist. The programming of queries other than the simplest kind is probably still beyond her expertise.

2.6. Kleisli

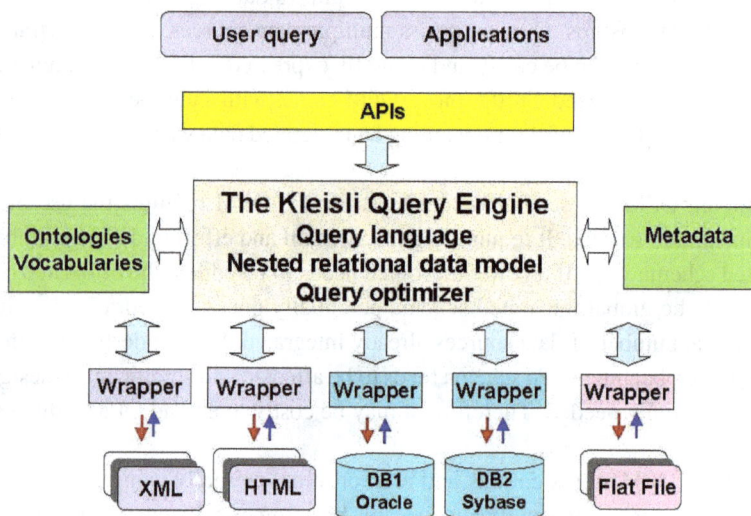

Fig. 1.   Kleisli, positioned as a mediator

Kleisli[162, 189, 894] is marketed by geneticXchange Inc of Menlo Park. It is one of the earliest systems that have been successfully applied to some of the earliest data integration problem in the human genome project, including the so-called US Department of Energy's "impossible" queries in early 1994.

The approach taken by the Kleisli system is illustrated by the diagram in Figure 1. It is positioned as a mediator system encompassing a nested relational data model, a high-level query language, and a powerful query optimizer. It runs on top of a large number of light-weight wrappers for accessing various data sources. There are also a number of application programming interfaces that allow Kleisli to be accessed in a ODBC- or JDBC-like fashion in various programming languages for a various applications.

The Kleisli system is highly extensible. It can be used to support several different high-level query languages by replacing its high-level query language module. Currently, Kleisli supports a "comprehension syntax"-based language called CPL[894] and a "nested relationalized" version of SQL called sSQL. The Kleisli system can also be used to support many different types of external data sources by adding new wrappers, which forward Kleisli's requests to these sources and translate their replies into Kleisli's exchange format. These wrappers are light weight and new wrappers are generally easy to develop and insert into the Kleisli system. The optimizer of the Kleisli system can also be customized by different rules and strategies.[894]

Besides the ability to query, assemble, and transform data from remote heterogeneous sources, it is also important to be able to conveniently warehouse the data locally. Kleisli does not have its own native database management system. Instead, Kleisli has the ability to turn many kinds of database systems into an updatable store conforming to its nested relational data model. In particular, Kleisli can use flat relational database management systems such as Sybase, Oracle, MySQL, *etc.* to be its updatable store. It can even use all of these systems simultaneously. It is also worth noting that Kleisli stores nested relations into flat relational database management systems using an encoding scheme that does not require these nested relations to be fragmented over several tables.

Kleisli possesses the following strengths.[894] It does not require data schemas to be available. It has a nested relational data model and a data exchange format that external databases and software systems can easily translate into. It shields existing queries, via a type inference mechanism, from certain kinds of structural changes in the external data sources. Kleisli also has the ability to store, update, and manage complex nested data. It has a good query optimizer. Finally, Kleisli is also equiped with two application programming interfaces so that it can be accessed in a JDBC-like manner from Perl and Java.[895]

However, Kleisli shares a common weakness with DiscoveryLink and OPM. Even though CPL and sSQL are both high-level query languages and protect the user from many low level details—such as communication protocols, memory management, thread scheduling, and so on—the programming of queries using CPL or sSQL other than the simplest kind is probably still beyond the expertise of an average biologist.

2.7. *XML*

XML is a standard for formatting document. As such, XML is not a data integration system by itself. However, there is a growing suite of tools based on XML that, taken as a whole, can be used as a data integration system. We therefore believe it is pertinent to include a discussion on XML and its associated tools in the context of this paper.

XML allows for a hierarchical nesting of tags and the set of tags can be defined flexibly. Thus XML can be viewed as a powerful data model[2] and a useful data exchange format, providing directly for two of the important ingredients of a general data integration solution for biomedicine. As a result, an increasing number of tools and sources in biomedicine such as PIR,[903] Entrez, and so on are becoming XML compatible.[5]

The intense interest in the development of query languages for semi-structured data[2] in the database community has also resulted in a number of powerful XML query languages such as XQL[716] and XQuery,[202] which provide the means for querying across multiple data sources and for transforming the results into more suitable form for subsequent analysis steps. Research and development works are also in progress on XML query optimization[246] and on XML data stores.[252]

A robust and stable XML-based general data integrating and warehousing system does not yet exist for biomedicine. However, once high-performance XML data stores become available, we can also expect the database research community to begin more research and development on data warehousing using these stores.

Consequently, we believe that given sufficient time, XML and the growing suite of XML-based tools can mature into an alternative data integration system in biomedicine that is comparable to Kleisli in generality and sophistication.

3. Highlight of Selected Features

This section highlights some features that distinguish the more general data integration technologies from the more specialized data integration solutions surveyed earlier.

3.1. *Data Model and Data Exchange Format*

A key feature that separates the more general data integration technologies—DiscoveryLink, OPM, Kleisli—from the more specialized technologies—EnsEMBL, GenoMax, SRS—is the explicit presence of a data model. From the point of view of the traditional database world,[3] a data model provides the means for specifying particular data structures, for constraining the data associated with these structures, and for manipulating the data *within* a database system. In order to handle data outside of the database system, this traditional concept of a data model is extended to include a data exchange format, which is a means for bringing data outside the database system into it and also for bringing data inside the database system outside. We use Kleisli's data model to illustrate this concept.

The data model underlying the Kleisli system is a complex object type system that goes beyond the "sets of records" or "flat relations" type system of relational databases.[170] It allows arbitrarily nested records, sets, and a few other data types.[894] Having such a "nested relational" data model is useful and matches the structure of biomedical data sources well. For example, if we are restricted to the flat relational data model, the GenPept report in Example 1 must necessarily be split into many separate tables in order to be losslessly stored in a relational database. The resulting multi-table representation of the GenPept report is conceptually unnatural and operationally inefficient.

Example 1: The GenPept report is the format chosen by the US National Center for Biotechnology Information to present amino acid sequence information. The feature table is the part of the GenPept report that documents the positions and annotations of regions of special biological interest. The following type represents the feature table of a GenPept report from Entrez.[758] Here we use $\{-$ and $\}-$brackets for sets, $(-$ and $)-$brackets for records, $[-$ and $]-$brackets for lists, and $\#l$: to label the field l of a record. In fact, the same bracketing scheme is used as the data exchange format of Kleisli.[895]

```
(#uid:num, #title:string, #accession:string, #feature:{(
    #name:string, #start:num, #end:num, #anno:[(
        #anno_name:string, #descr:string)])})
```

The feature table of GenPept report 131470, a tyrosine phosphatase 1C sequence, is shown partially below. The particular feature displayed goes from amino acid 0 to amino acid 594, which is actually the entire sequence, and has two annotations: The first annotation indicates that this amino acid sequence is derived from mouse DNA sequence. The second is a cross reference to the US National Center for Biotechnology Information taxonomy database.

```
(#uid:131470, #accession:"131470", #title:"... (PTP-1C)...",
    #feature:{(
        #name:"source", #start:0, #end:594, #anno:[
            (#anno_name:"organism", #descr:"Mus musculus"),
            (#anno_name:"db_xref", #descr:"taxon:10090")]),
        ...})
```

□

It is generally easy to develop a wrapper for a new data source, or modifying an existing one, and insert it into Kleisli. The main reason is that there is no impedance mismatch between the data model supported by Kleisli and the data model that is necessary to capture the data source. The wrapper is often a very

light-weight parser that simply parses records in the data source and prints it out
in Kleisli's very simple data exchange format.

Example 2: Suppose we want to implement a function `webomim-get-`
`-detail` that uses an OMIM[318] identifier to access the OMIM database and
returns a set of objects matching the identifier. Suppose the output is of type

```
{(#uid: num, #title: string, #gene_map_locus: {string},
  #alternative_titles:{string}, #allelic_variants:{string})}
```

Note that is this a nested relation: it is a set of records, and each
record has three fields that are also of set types, viz. `#gene_map_locus`,
`alternative_titles`, and `allelic_variants`. This type of output
would definitely present a problem if we had to give it to a system based on the
flat relational model, as we would need to arrange for the information in these
three fields to be sent into separate tables. Fortunately, such a nested structure can
be mapped directly into Kleisli's exchange format. So the wrapper implementor
would only need to parse each matching OMIM records and to write it out in a
format like this:

```
{(#uid: 189965,
  #title: "CCAAT/ENHANCER-BINDING PROTEIN, BETA; CEBPB",
  #gene_map_locus: "20q13.1",
  #alternative_titles: {
    "C/EBP-BETA",
    "INTERLEUKIN 6-DEPENDENT DNA-BINDING PROTEIN; IL6DBP",
    "LIVER ACTIVATOR PROTEIN; LAP",
    "LIVER-ENRICHED TRANSCRIPTIONAL ACTIVATOR PROTEIN",
    "TRANSCRIPTION FACTOR 5; TCF5"},
  #allelic_variants: {})}
```

Here, instead of needing to create separate tables to keep the sets nested inside
each record, the wrapper would simply print the appropriate set brackets { and
} to enclose these sets. Kleisli would automatically deal with them as they were
handed over by the wrapper. This kind of parsing and printing is extremely easy
to implement.   □

OPM shares with Kleisli a nested relational data model, except that the former
lacks a data exchange format. Hence the mapping of the examples to OPM's data
model is conceptually just as straightforward, but the practical implementation in
OPM demands considerably more effort. It is worth pointing out that while SRS
does not have an explicit data model, it does have an implicit one supported by its
Icarus language for scripting parsers. In the case of SRS, this implicit data model
in Icarus also greatly facilitates the rapid scripting of wrappers.

3.2. *Query Capability*

Another feature that separates the more general data integration technologies from the more specialized ones is the presence of a flexible high level query language for manipulating data conforming to the data model. We use sSQL, the primary query language of Kleisli to illustrate this feature. sSQL is based on the *de facto* commercial database query language SQL, except for extensions made to cater for the nested relational model and for the federated heterogeneous data sources.

Example 3: The feature table of a GenBank report has the type below. The field #position of a feature entry is a list indicating the start and stop positions of that feature. If the feature entry is a CDS, this list corresponds to the list of exons of the CDS. The field #anno is a list of annotations associated with the feature entry.

```
(#uid: num, #title: string, #accession: string, #seq: string,
 #feature: {(
    #name: string,
    #position: [(#start:num,#end:num,#negative:bool, ...)],
    #anno: [(#anno_name:string,#descr:string)], ...)}, ...)
```

Given a set DB of feature tables of GenBank chromosome sequences, we can extract the 500 bases up stream of the translation initiation sites of all disease genes—in the sense that these genes have a cross reference to OMIM—on the positive strand in DB as below. Here, 12s is a function that converts a list into a set.

```
select
  uid: x.uid,
  protein: r.descr,
  flank: string-span(x.seq, p.start - 500, p.start)
from
  DB x, x.feature f,
  {f.position.list-head} p,
  f.anno.12s a, f.anno.12s r
where not (p.#negative)
and    a.descr like "MIM:%" and a.anno_name = "db_xref"
and    r.anno_name = "protein_id"
```

Similarly, we can extract the first exons of these same genes as follows:

```
select
  uid: x.uid,
  protein: r.descr,
  exon1: string-span(x.seq, p.start, p.end)
from
  DB x, x.feature f,
  {f.position.list-head} p,
```

```
   f.anno.l2s a, f.anno.l2s r
where not (p.#negative)
and    a.descr like "MIM:%" and a.anno_name = "db_xref"
and    r.anno_name = "protein_id"
```

These two example queries illustrate the how a high level query language makes it possible to extract very specific output in a relatively straightforward manner.   □

We illustrate how to combine multiple sources using high level query languages. An *in silico* discovery kit (ISDK) prescribes experimental steps carried out in computers very much like the experimental protocol carried out in wet laboratories for specific scientific investigation. From the perspective of Kleisli, an *in silico* discovery kit is just a script written in sSQL and performs a defined information integration task. It takes an input data set and parameters from the user, executes and integrates the necessary computational steps of database queries and applications of analysis programs or algorithms, and outputs a set of results for specific scientific inquiry.

Example 4: The simple *in silico* discovery kit in Figure 2 demonstrates how to use an available ontology data source to get around the problem of inconsistent naming in genes and proteins, and to integrate information across multiple data sources. It is implemented in the sSQL script below.

```
create function get-info-by-genename (G) as
Select
   hugo: w, omim: y, pmid1-abstract: z,
   num-medline-entries: list-sum(
        lselect ml-get-count-general(n)
        from x.Aliases.s21 n)
from
   hugo-get-by-symbol(G) w,
   webomim-get-id(
        searchtime:0, maxhits:0,
        searchfields:{}, searchterms:G) x,
   webomim-get-detail(x.uid) y,
   ml-get-abstract-by-uid(w.PMID1) z
where
   x.title like ("%" ^ G ^ "%");
```

With the user input of a gene name G, the ISDK performs the following task: First, it retrieves a list of aliases for G from the Gene Nomenclature database provided by the Human Genome Organization (HUGO). Then it retrieves information for diseases associated with this particular protein in the Online Mendelian Inher-

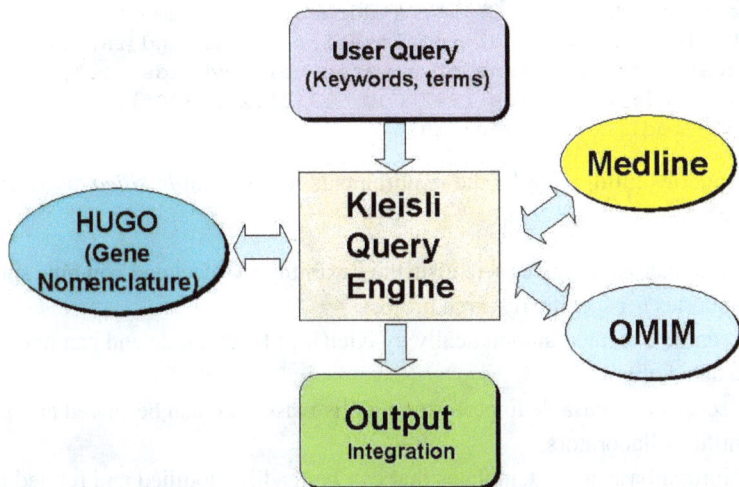

Fig. 2. An "*in silico* discovery kit" that uses an available ontology data source to get around the problem of inconsistent naming in genes and proteins, and integrates information across multiple data sources.

itance of Man Database (OMIM),[318] and finally it retrieves all relevant references from MEDLINE.

Here, s21 is a function that converts a set into a list; list-sum is a function to sum a list of numbers; ml-get-count-general is a function that accesses the MEDLINE database in Bethesda and computes the number of MEDLINE reports matching a given keyword; ml-get-abstract-by-uid is a function that accesses MEDLINE for report given a unique identifier; webomim-get-id is a function that accesses the OMIM database in Bethesda to obtain unique identifiers of OMIM reports matching a keyword; webomim-get-detail is a function that accesses OMIM for report given a unique identifier; and hugo-get-by-symbol is a function that accesses the HUGO database and return HUGO reports matching a given gene name.

For instance, this query get-info-by-genename can be invoked with the transcription factor CEBPB as input to obtain the following result.

```
{ (#hugo: (#HGNC: "1834",
      #Symbol: "CEBPB", #PMID1: "1535333", ...
```

```
     #Name: "CCAAT/enhancer binding protein (C/EBP), beta",
     #Aliases: {"LAP", "CRP2", "NFIL6", "IL6DBP", "TCF5"}),
   #omim: (#uid: 189965, #gene_map_locus: "20q13.1",
         #allelic_variants: {}, ...),
   #pmid1-abstract: (#muid: 1535333,
     #authors:"Szpirer C...", #address:"Departement ...",
     #title: "Chromosomal localization in man and rat ...",
     #abstract: "By means of somatic cell hybrids ...",
     #journal: "Genomics 1992 Jun; 13(2):292-300"),
     #num-medline-entries: 1936)}
```

Such queries fulfill many of the requirements for efficient *in silico* discovery processes:

(1) their modular nature gives scientists the flexibility to select and combine specific queries for specific research project;
(2) they can be executed automatically by Kleisli in batch mode and can handle large data volume;
(3) their scripts are reusable to perform repetitive tasks and can be shared among scientific collaborators;
(4) they form a base set of templates that can be readily modified and refined to meet different specifications and to make new queries; and
(5) new databases and new computational tools can be readily incorporated to existing scripts. ☐

The flexiblity and power shown in these sSQL examples can also be experienced in OPM, and to a lesser extent in DiscoveryLink. With good planning, a specialised data integration system can also achieve great flexibility and power within a narrower context. For example, the EnsMart tool of EnsEMBL is a very well designed interface that helps a non-programmer build complex queries in a simple way. In fact, an equivalent query to the first sSQL query in Example 3 can be also be specified using EnsMart with a few clicks of the mouse. Nevertheless, there are some unanticipated cases that cannot be expressed, such as the second sSQL query in Example 3.

3.3. *Warehousing Capability*

Besides the ability to query, assemble, and transform data from remote heterogeneous sources, a general data integration technology is also expected to be able to conveniently store data locally. Kleisli does not have its own native database management system. Instead, Kleisli has the ability to turn many kinds of database systems into an updatable store conforming to its complex object data model. In particular, Kleisli can use flat relational database management systems such as Sybase, Oracle, MySQL, *etc.* to be its updatable complex object store. It can even use all of these systems simultaneously. We illustrate this feature using the example of GenPept reports.

Example 5: Create a warehouse of GenPept reports. Initialize it to reports on protein tyrosine phosphatases. Note that Kleisli provides several functions to access GenPept reports remotely from Entrez.[758] One of them is the function `aa-get-seqfeat-general` used below, which retrieves GenPept reports matching a search string.

```
! connect to our Oracle database system
oracle-cplobj-add (name: "db", ...);
! create a table to store GenPept reports
create table genpept(uid:"NUMBER", detail: "LONG") using db;
! initialize it with PTP data
select (uid: x.uid, detail: x) into genpept
from aa-get-seqfeat-general("PTP") x using db;
! index the uid field for fast access
db-mkindex(table:"genpept",index:"genpeptindex",schema:"uid");
! let's use it now to see the title of report 131470
create view GenPept from genpept using db;
select x.detail.title from GenPept x where x.uid = 131470;
```

In this example, a table `genpept` is created in our Oracle database system. This table has two columns, `uid` for recording the unique identifier and `detail` for recording the GenPept report. A LONG data type is used for the `detail` column of this table. However, recall from Example 1 that each GenPept report is a highly nested complex objects. There is therefore a "mismatch" between LONG and the complex structure of a GenPept report. This mismatch is resolved by the Kleisli system which automatically performs the appropriate encoding and decoding.

Thus, as far as the Kleisli user is concerned, `x.detail` has the type of GenPept report as given in Example 1. So he can ask for the title of a report as straightforwardly as `x.detail.title`. Note that encoding and decoding are performed to map the complex object transparently into the space provided in the `detail` column; that is, the Kleisli system does not fragment the complex object to force it into the third normal form.   □

What distinguishes the ability to store or warehouse data in a system like Kleisli is the ease with which the store or warehouse can be specified and populated in theory. However, in practice, one does need to anticipate failures in a large update and to provide recovery mechanisms. OPM also enjoys this capability if one is not concerned with how the store or warehouse is supposed to be related to the other existing stores or warehouses already in the system. DiscoveryLink enjoys this capability to a lesser extent, as its data model does not permit nesting of data. The more specialized systems can also have warehousing capability, but with greater limitations on what can be warehoused by them.

3.4. *Application Programming Interfaces*

The high-level query languages of the more general data integration systems surveyed are all SQL-like and are thus designed to express traditional (nested relational) database-style queries. Not every query in bioinformatics falls into this class. For these non-database-style queries, some other programming languages can some time be a more convenient or more efficient means of implementation. Therefore, it is useful to develop some application programming interfaces to these more general data integration systems for various popular programming languages.

In the case of Kleisli, there is the Pizzkell suite[895] of interfaces to the Kleisli Exchange Format for various popular programming languages. Each of these interfaces in the Pizzkell suite is a library package for parsing data in Kleisli's exchange format into an internal object of the corresponding programming language. It also serves as a means for embedding the Kleisli system into that programming language, so that the full power of Kleisli is available within that programming language in a manner similar to that achieved by JDBC and ODBC for relational databases. The Pizzkell suite currently include CPL2Perl and CPL2Java, for Perl and Java.

The presence of such application programming interfaces may be even more crucial for the more specialized integration solutions. While a point solution like EnsEMBL is typically designed with a specific aim in mind, it is not unusual to subsequently discover that a user wants to use the integrated data in an unanticipated way. In such a situation, it would be convenient if an application programming interface is available on the integrated data. For example, in the case of EnsEMBL, as EnsEMBL is implemented in Perl using Bioperl[797] as the backbone, the same library of routines that have been accumulated in the course of implementing EnsEMBL would be the perfect application programming interface to EnsEMBL.

Bioperl[797] itself can also be thought of as a low level integration toolkit for biological data. Such a toolkit typically contains a set of library routines, for accessing some commonly used bioinformatics data sources and tools, that can be invoked as procedure or function calls from the underlying programming language. No high-level query support and no optimization are provided. There are also similar toolkits in other popular "open source" programming languages, such as BioPython and Biojava.[538]

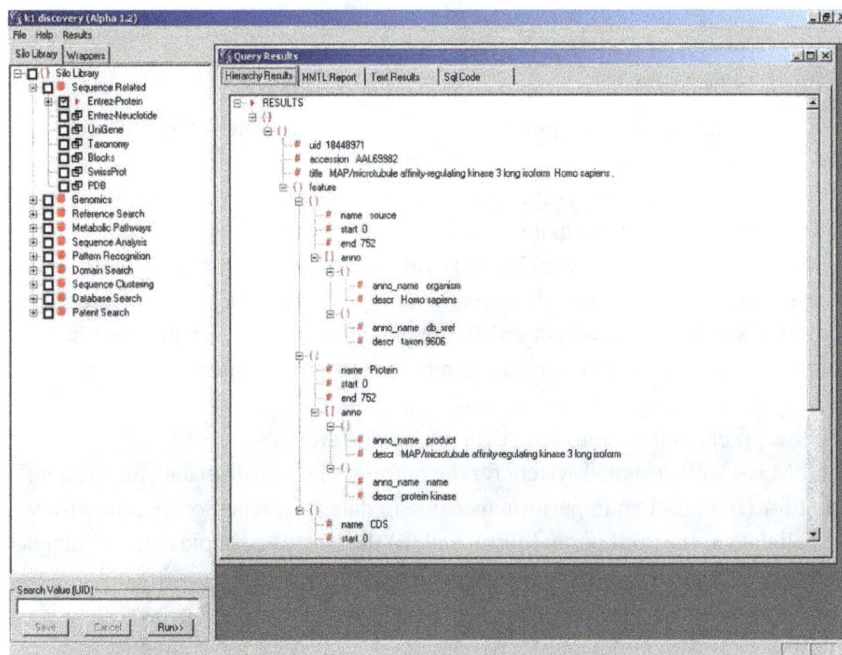

Fig. 3.   A screenshot of the Discovery Builder, a graphical interface to Kleisli.

There is also a graphical interface to the Kleisli system that is designed for non-programmers. It is called the Discovery Builder and is developed by folks at geneticXchange Inc. This graphical interface makes it far easier to visualize the source data required to formulate the queries and generates the necessary sSQL codes. It allows a user to see all available data sources and their associated meta-data and assists the user to navigate and to specify his query on these sources. A screenshot of Discovery Builder is presented in Figure 3. It provides all of the following key functions:

(1) a graphical interface that can "see" all of the relevant biological data sources, including metadata—tables, columns, descriptions, *etc.*—and then construct a query "as if" the data were local;

(2) add new wrappers for any public or proprietary data sources, typically within

hours, and then have them enjoined in any series of ad-hoc queries that can be created;

(3) execute the queries, which may join many data sources that can be scattered all over the globe, and get fresh result data quickly.

4. Concluding Remarks

Let us first summarize our opinion on how well each of the surveyed systems satisfy the requirements of a general data integration system for biomedicine. EnsEMBL and GenoMax are point solutions and thus naturally do not satisfy the requirements of a general data integration system well.

SRS and DiscoveryLink are claimed by their inventors as general data integration systems for biomedicine. However, in reality, SRS is a form of user interface integration and hence it does not satisfy the requirements well. On the other hand, while DiscoveryLink has most of the components required, these components are probably in the wrong flavour—the adoption of the flat relational model causes it to be less potent in the biomedical data integration arena.

OPM is a well designed system for the purpose of biomedical data integration, except for (i) a problem in performance due to data fragmentation as it unwisely maps all data to the third normal form, and (ii) the lack of a simple data exchange format, and (iii) the need of a global schema.

XML and Kleisli have all the qualities required for a good general data integration. However, compared to Kleisli, XML still need more time to mature especially in terms of query optimization and data warehousing capabilities.

Let us next look at these surveyed systems from the perspective of an average biologist. While general data integration systems such as DiscoveryLink, OPM, and Kleisli simplify the programming of *ad hoc* queries, it must also be acknowledged that the programming skills required are still significant. In contrast, data integration systems that are nearer to the point-solution end of the spectrum—such as EnsEMBL, GenoMax, and SRS—have considerably better user interfaces that help a biologist to build the simpler type of queries.

Of course, a biologist may find it frustrating that the graphical user interfaces of EnsEMBL, GenoMax, and SRS cannot let her express a particular *ad hoc* query such as the one that asks for the sequence of the first exon of all genes in a database. However, it is very likely that the same biologist may also find it equally frustrating that she does not know how to express that query in DiscoveryLink, OPM, and Kleisli, even though she knows that that query is expressible in these systems. In other words, the more general data integration systems can directly increase the productivity of a bioinformatics programmer, but they probably cannot directly increase the productivity of an average biologist.

Drawing from the remarks above, we see a dichotomy between expressiveness and simplicity. Therefore, which type of data integration system is preferred necessarily depends on the trade-off between these two factors. Many problems in biomedical research on drug targets and candidates require access to many data sources that are voluminous, heterogeneous, complex, and geographically dispersed. If these data sources are successfully integrated into a new database, researchers can then uncover relationships that enable them to make better decisions on understanding and selecting targets and leads. Therefore, a successful integration of data is crucial to improving productivity in this research.

It is important to stress that a successful data integration must be in support of a specific research problem, and different research problems are likely to need different ways of integrating and analysing data. Even though point solutions such as EnsEMBL does not fair well as a general data integration system, it works much better than any general data integration system in the specific context of genome browsing.

If one's data integration needs are of a more *ad hoc* nature, a general data integration system can often ease the implementation significantly as such a system provides greater adaptability. It is also worth remarking that the more specialized solutions may themselves be implemented on top of a more general data integration solution. One such example is TAMBIS,[291] which is built on top of Kleisli.

The systems surveyed so far generally do not consider the semantics aspect of the underlying data sources. Let us end this chapter with a brief mention of TAMBIS. TAMBIS[291] is a data integration solution that specifically addresses the semantics aspect. The central distinguishing feature of TAMBIS is the presence of an ontology and a reasoning system over this ontology. The TAMBIS ontology contains nearly 2000 concepts that describe both molecular biology and bioinformatics tasks. TAMBIS provides a user interface for browsing the ontology and for constructing queries.

A TAMBIS query is formulated by starting from one concept, browsing the connected concepts and applicable bioinformatics operations in the ontology, selecting one such connected concepts or applicable bioinformatics operation, and browing and selecting for further connected concepts and applicable bioinformatics operations. The ontology and the associated reasoning component thereby guide the formulation of the query, ensuring that only a query that is logically meaningful can be formulated. The query is then translated by TAMBIS and passed to an underlying Kleisli system for execution.

From the point of view of TAMBIS, Kleisli significantly simplifies the task of implementing TAMBIS, as the TAMBIS implementors can concentrate on the

ontology and reasoning components and leave the details of handling the underlying data sources to Kleisli. From the point of view of Kleisli, TAMBIS makes
it possible for a biologist to ask more complicated *ad hoc* queries on the data
sources integrated by Kleisli without the assistance of a programmer. The ontology of TAMBIS is currently being enriched by its inventors in the University of
Manchester to allow an even larger range of complicated queries to be expressed.

CHAPTER 18

CONSTRUCTION OF BIOLOGICAL DATABASES: A CASE STUDY ON THE PROTEIN PHOSPHATASE DATABASE (PPDB)

Prasanna R. Kolatkar

Genome Institute of Singapore
kolatkarp@gis.a-star.edu.sg

Kui Lin

Beijing Normal University
linkui@bnu.edu.cn

Biological data is being created at ever-increasing rates as different high-throughput technologies are implemented for a wide variety of discovery platforms. It is crucial for researchers to be able to not only access this information but also to integrate it well and synthesize new holistic ideas about various topics. A key ingredient in this process of data-driven knowledge-based discovery is the availability of databases that are user-friendly, that contain integrated information, and that are efficient at storage and retrieval of data.

Implementations of integrated databases include GenBank,[77] SWISS-PROT,[41] InterPro,[29] PIR,[903] *etc.* No single one of these databases contains all the information one might need to understand a specific topic. So unique databases are required to provide researchers better access to specific information. Databases can be built using a variety of tools, techniques, and approaches;[416, 756, 897] but there are some methods that are extremely powerful for management of large amounts of data and that can also integrate this information.

We have created several purpose-built integrated databases. We describe one of them—the Protein Phosphatase DataBase (PPDB)—in this chapter. PPDB has been constructed using a data integration and analysis system called Kleisli. Kleisli can model complex biological data and their relationships, and integrate information from distributed and heterogeneous data resources.[162, 189, 894]

ORGANIZATION.

Section 1. The importance of protein phosphatases in signal transduction in eukaryotic cells is briefly explained. The lack of a well-integrated protein phosphatase database is noted, motivating the construction of PPDB.

Section 2. An overview of the architecture of PPDB and its underlying workflow is then given.

Section 3. This is followed by a brief introduction to the data integration tool and the object representation used in constructing PPDB.

Sections 4–9. The next several sections present in greater detail the types of information provided in PPDB for protein phosphatases. The types of information include protein and DNA sequences, structure, biological function related information, disease related information, and so on, as well as the classification tree of protein phosphatases.

Sections 10–15. After that, the details of data collection, integration, and update are described. The techniques and rules used in PPDB for the automatic classification of protein phosphatases into one of 8 categories are also discussed.

1. Biological Background

Most cellular processes are extremely complex and require tightly controlled signal transduction events. One of the fundamental mechanisms that a cell utilizes to control its biological processes is via protein phosphorylation and dephosphorylation reactions that are catalyzed by protein kinases and protein phosphatases.[834, 936] This type of reversible covalent modification of proteins is of crucial importance for modulation of the cell's biochemical pathways.

Protein kinases catalyze the transfer of g-phosphate of a nucleotide triphosphate—usually ATP—to an acceptor residue in the substrate protein, while protein phosphatases do the reverse job of removing the phosphate. It was originally thought that protein kinases were the key enzymes controlling the phosphorylation. In fact, the protein phosphatases are also composed of a large family of enzymes that parallel protein kinases in terms of structural diversity and complexity. It is now clear that the protein phosphatases and kinases play equally important roles in signal transduction in eukaryotic cells as the Yin and Yang of protein phosphorylation and cell signaling function.

Papers published over the last several years document crucial physiological roles for protein phosphatase in a variety of mammalian tissues and cells.[32, 200, 323] However, to date, there are no unique, well-integrated protein phosphatase databases available that a researcher can access to locate a large amount of related information for his favorite phosphatase. We thus have implemented the Protein Phosphatase DataBase (PPDB) not only to integrate related information concerning protein phosphatases but also to classify and organize the data as hierarchical classes based on the specified biochemical, genetic, and biological characteristics. We believe that such data organization is helpful for researchers.

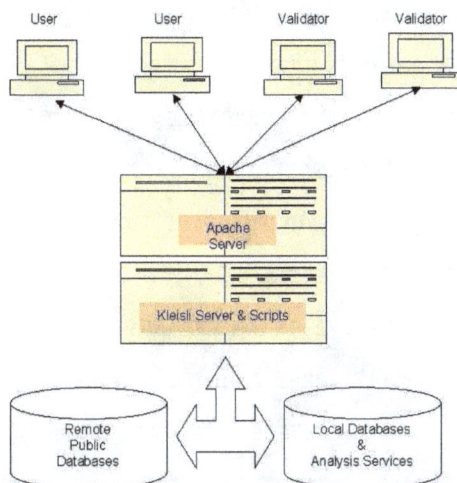

Fig. 1.   The 3-tier architecture of PPDB.

2. Overview of Architecture and Work ow of PPDB

In PPDB ver. 1.0, we focus only on human phosphatases. An overview of the 3-tier architecture of the database is shown in Figure 1. Users can access the database via the Internet by using their browsers. Figure 8 gives the general work- ow of how the PPDB is created using the Kleisli data integration and analysis system.[162, 189, 894]

At the time of writing (October 2001), there are 70 phosphatases, which are clustered into the 8 categories shown in Figure 2. Currently, we mostly focus on the classification of tyrosine phosphatases, which function Of this type of phosphatases, 4 subcategories—tyrosine specific phosphatase, dual specific phosphatase, low molecular weight phosphatase, and anti phosphatase—are devised to classify them. The tyrosine specific phopahatase subclass is in turn divided into 2 subclasses, termed receptor-like group[458] and cytosolic phosphatase.[200, 871] At the same time, we also provide two additional subcategories, "non-protein" and "not-sure" for storing those phosphatases that are not protein phosphatases or currently have unknown function.

In the following sections, we describe individual data types, methods for data collection and integration, automatic phosphatase classification schema, expertise

```
o phosphatase  ----------------------------  [    0 ]
|
|---o tyrosine  -------------------------  [    1 ]
|   |
|   |   ---o tyrosine-specific  ---------  [    0 ]
|   |   |
|   |   |   ---o receptor-like  -------  [   14 ]
|   |   |   |
|   |   |   |---o non-receptor  --------  [   11 ]
|   |   |
|   |   |---o dual specific  -----------  [   10 ]
|   |   |
|   |   |---o low molecular weight  -----  [    2 ]
|   |   |
|   |   |---o anti phosphatase/STYX-like  [    0 ]
|   |
|   |---o serine/threonine  ------------  [   18 ]
|
|---o other  --------------------------  [    9 ]
|   |
|   |---o non-protein  -----------------  [    3 ]
|   |
|   |---o not sure  --------------------  [    2 ]
```

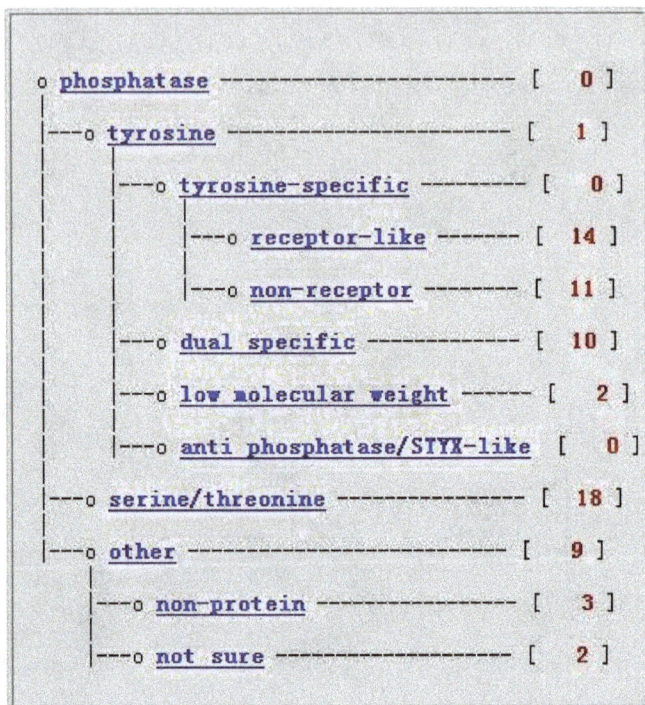

Fig. 2.   The classification and organization tree for phosphatases in PPDB. The number of entries in various subcategories is given in brackets.

for validation, data storage, content update, and visualization in detail.

3. Data Integration Tool and Object Representation

PPDB has been implemented using a powerful data integration and analysis system called Kleisli. Kleisli can facilitate efficient building of a prototype for complex biological data objects and their relationships, as well as integrating information from distributed and heterogeneous data resources.[162, 189, 894]

There are 3 types of object containers in Kleisli,[894] *viz.* set, bag, and list. Objects in a container can be either homogeneous or heterogeneous. In the latter, objects are tagged by different labels using variant data type. Record objects are the main data type to describe entities in the real world. Similar to the general records in relational databases, the records consist of several labeled fields as well. However, each field of a record can contain any type of data within Kleisli—for instance, it can be a bag, a list, a set, a record, *etc.* Hence, the type of data modeling facilitated by Kleisli is more powerful than other methods.[897]

Kleisli is equiped with many relevant built-in operators and functions. In particular, many existing bioinformatics tools, algorithms, and databases are integrated into Kleisli in a transparent way.[162, 894] Kleisli is hence not only a powerful data integration system but also a data analysis system. It is highly suitable for addressing bioinformatics problems.[146−148, 162, 189, 509, 895, 896]

In PPDB, each phosphatase object is presented as a complex data object—*e.g.*, Figures 3 and 4—in the data exchange format of Kleisli.[895] All the same type of objects are usually organized into a set or a list of Kleisli complex objects. Most of the objects are presented as Kleisli records in a set or a list. However, the fields of a record can also be any of types of Kleisli objects—*i.e.*, basic types such as string and number; or complex types such as record, set, or list. Thus, data objects are allowed to be deeply nested to reflect the natural complex structure of real-life biological data.

For example, the field #organism of a protein sequence in Figure 3, which is a complex record, includes the field #lineage, which is a list of taxonomy. Another object is the classification tree object in Figure 4, which is a recursive record to represent a hierarchical category data structure. In PPDB, the core data object is the tree object by which all of the other data objects are connected together, such as protein and DNA sequences, literature abstracts, related information from OMIM,[318] ENZYME,[40] and PDB[880] databases.

```
(#uid: 4096844,
 #title: "protein tyrosine phosphatase Cr1PTPase precursor",
 #accession: "4096844",
 #common: "human",
 #organism: (#genus: "Homo", #species: "sapiens",
             #lineage: ["Eukaryota", "Metazoa", ...]),
 #feature: {
   (#name: "Protein", #start: 0, #end: 656,
    #anno: [ (#anno_name: "product",
              #descr: "protein tyrosine phosphatase ...")]),
   (#name: "CDS", #start: 0, #end: 656,
    #anno: [ (#anno_name: "gene",
              #descr: "Ch-1PTPase alpha"),
             (#anno_name: "note",
              #descr: "receptor-type protein tyrosine ..."),
             ...]), ...}, ...)
```

Fig. 3.   A snapshot of phosphatase Cr1PTPase curated in PPDB in the Kleisli data exchange format. This figure only shows part of the whole protein information.

```
(#name: "phosphatase",
 #classes: [
    (#name: "tyrosine",
     #classes: [
        (#name: "tyrosine-specific",
         #classes: [
            (#name: "receptor-like", #classes:  []),
            (#name: "non-receptor", #classes:  []) ]),
        (#name: "dual specific", #classes:  []),
        (#name: "low molecular weight", #classes: []),
        (#name: "anti phosphatase/STYX-like",#classes:[])]),
    (#name: "serine/threonine", #classes: []),
    (#name: "other",
     #classes: [
        (#name: "non-protein", #classes: []),
        (#name: "not sure", #classes: []) ]) ])
```

Fig. 4.   The classification tree object in Kleisli data exchange format.

4. Protein and DNA Sequences

Figure 3 shows a Kleisli complex data object of the phosphatase protein called
"Cr1PTPase precursor" within the PPDB. The object is transformed from NCBI's
GenPept[78] database of translated protein sequences from GenBank. We also have
curated related DNA sequence information from GenBank,[77] and each DNA sequence object has a similar data structure like the protein object. A sequence
object is a record, and consists of several fields. Some of them are simple and
basic data types, such as unique identifier (number), title (string), and the amino
acid sequence (long string). Some are complex data types, for example, the field
#feature is a set of records of annotated information which is the most im-
portant part in any type of sequence database. Researchers usually extract useful
information from this field for each protein, DNA, EST, SNP, *etc.* For example,
we use the annotated information in this field to extract the gene name for each
phosphatase-related DNA sequence; see Figure 5.

5. Structure

When a new phosphatase is validated and curated into PPDB, the system automatically parses the potentially related Protein Data Bank (PDB)[475, 880] identifier
for the phosphatase. If there is a related entry found, the system downloads and

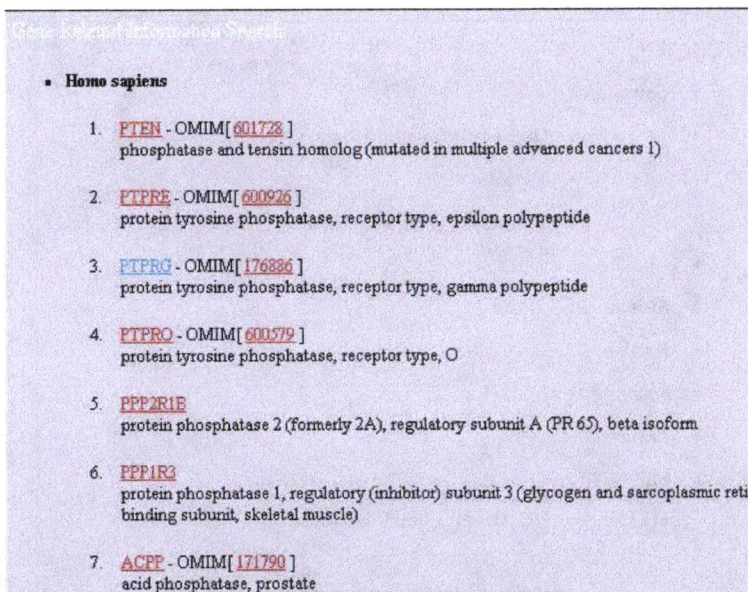

Fig. 5.   A view of the list of gene names in PPDB.

integrates that part of the information from PDB for the respective phosphatase. The information is represented and visualized in HTML format. Figure 6 shows the visualization of 3D structure information that is extracted from PDB for the phosphatase PTPmu/RPTPmu[360] in PPDB.

6. Biological Function and Related Information

In addition to the classification tree, the most interesting and important information in PPDB is the related biological function information linked to each phosphatase. This includes structural features, gene expression properties, catalytic properties, substrates, interacting proteins, cell effects, disease linkage, and knockout/transgenics properties, *etc.* Figure 7 lists the related biological, biochemical, and genetic information of the phosphatase PTPmu/RPTPmu in PPDB. This information is retrieved by the phosphatase expert and integrated into PPDB by the subsystem called InfoCollection which is validated by the expert via a user-friendly web interface.

PDB: 1RPM

- **Compound:**

    1. **Molecule:** Receptor Protein Tyrosine Phosphatase Mu

        - **Domain:**

        - **Fragment:** Cytosolic Membrane Proximal Catalytic Domain

- **Classification:** Receptor

- **Resolution:** 2.3[Å]

- **R value:** 0.195

- **Space group:** P 1 21 1

- **Atoms:** 4838

- **Unit cell:**

a[Å]	b[Å]	c[Å]	alpha[ ]	beta[ ]	gamma[ ]
95.54	36.29	95.38	90.0	95.66	90.0

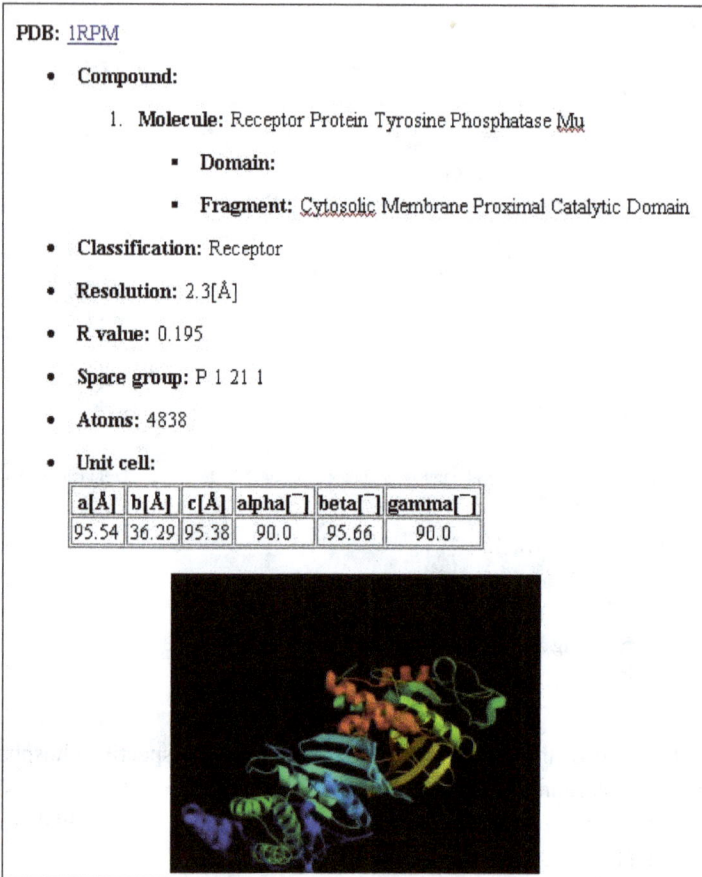

Fig. 6.   The 3D information of phosphatase PTPmu/RPTPmu.

7. Other Data Objects

There are other data objects integrated into PPDB, such as phosphatase related gene information, associated MEDLINE abstracts, and relevant disease information with links to the OMIM database.[318] These diverse types of information are used to classify phosphatases automatically according to pre-defined classification rules. The hotlinks to the corresponding databases are given in the presentation of query results, *e.g.*, Figure 5.

Class: [receptor-like] **Infosheet:** [1113] **Protein:** [1113] **3 D:** [1RPM]

- **Name:** PTPmu/RPTPmu
- **Structural Features:**
  - ○ **ECD**
    - ▪ Precursor with signal peptide
    - ▪ 1 MAM domain
    - ▪ 1 Ig-like domain
    - ▪ 4 FN type-III repeats
    - ▪ Potential N-linked glycosylation (13)
    - ▪ Post-translational proteolytic processing at site in ECD, non-covalent association of cleavage products[7961788][7559782]
  - ○ **ICD**
    - ▪ Juxtamembrane region homologous to conserved ICD of cadherins
    - ▪ 2 catalytic domains; D1 and D2
- **Expression:**
  - ○ Lung (most abundant), brain, heart[165529]
  - ○ Upregulated by increasing cell density[7559782]
- **Catalytic Properties:**
  - ○ D1 active, D2 inactive[7504951]
- **Substrates:**
- **Interacting Proteins:**
  - ○ **ECD**
    - ▪ Homophilic interactions[8394372][8393854]

Fig. 7.   The biological function information of phosphatase PTPmu/RPTPmu.

8. Classification Tree

The most important Kleisli data object in PPDB is the classification tree depicted in Figure 12. It integrates all the related information in PPDB together. It is also used to explore and visualize the whole database by allowing access to all relevant information. In Kleisli, the classification tree is a hierarchical object, as shown in Figure 4. We curate the data by automatic classification of newly deposited and unclassified phosphatases by integrating different types of information linked to the specified phosphatase, and displaying the related web links. The tree is used to visualize the query results as well.

9. Data Integration and Classification

We employ Kleisli to collect phosphatase information from several public protein databases, such as NCBI Entrez GenPept,[78, 758] GenBank,[77] PubMed, SWISS-PROT,[41] PDB,[880] *etc.* Each newly collected phosphatase is classified automatically into an existing category of the classification tree according to its related evidence. The evidence considered include sequence, annotation features, related PubMed abstracts, and specific domain knowledge.

The automatically classified result and the supporting evidence are then displayed to allow phosphatase experts to validate the findings. All consensus phosphatases are integrated into PPDB with their related data, which includes genomic information, pathway information, related disease information along with associated mutation data, and protein function information.

10. Data Collection

Although more complex searching mehcanisms can be applied, we decide to implement simple keyword searching using, for example, "phosphatase" to search the NCBI GenPept database via the "Title" field. Subsequent to the search, information pertaining to all the hits is downloaded and saved as a local dataset. This local dataset can be updated automatically based on the classification results and the phosphatase entries in our database can be curated and validated.

In addition to protein sequence data, we also need to collect those DNA and PubMed items that are directly linked to the phosphatase proteins in our database. For example, for PPDB ver. 1.0, we download about 2,600 protein records from GenPept and save them as RawPhosphatases, Figure 8. We filter out those entries whose titles include words like "similar" or "predict." We therefore obtain a relatively smaller data set as the input for the automatic classification subsystem.

11. Phosphatase Classification Strategy

Phosphatases that utilize phosphoproteins as substrates have been divided into two major categories based on their substrate specificity:[834] protein-tyrosine phosphatases (PTPs) and protein serine/threonine phosphatases. PTPs can then be further subdivided into four subclasses. The PPDB categorizes the PTPs according to their biological functions. Figure 2 shows the current classification tree of phosphatases used in PPDB.

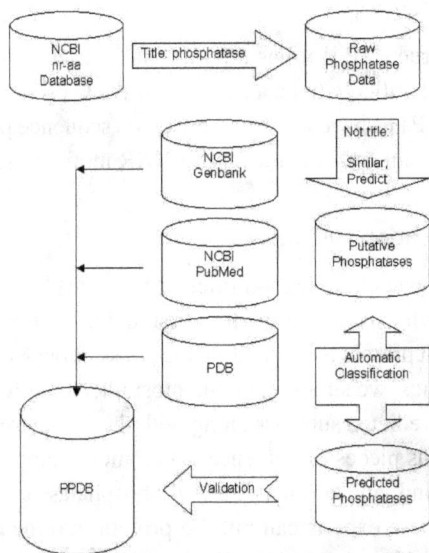

Fig. 8.   The processes of data collection and integration of PPDB.

Currently, we use 3 simple approaches for automatically classifying the input entries into their specific subgroups in the classification tree. These approaches are:

- keyword searching,
- specified sequence motif matching, and
- prosite database searching

For example, we utilize the following methods to distinguish the PTPs:

(1) Most (if not all) of the Y-specific PTPs have the active site motif [VI]HCXAGXXR[TS].[201, 806] Here "X" stands for any amino acid, and "[VI]" ("[TS]") stands for either the amino acid "V" ("T") or the amino acid "I" ("S").

(2) The Y-specific PTPs tend to have other regions of conservation in the catalytic domains near the N-terminus, which often has the sequence KNRY;[396] and further along the sequence are the residue stretches of DYINAS or XYINAX; and finally the WPD motif.

(3) If the protein sequence has a transmembrane region, it is classified as a receptor-like PTP (RPTP). To date, RPTPs are all Y-specific.

(4) If the protein sequence has two tandem catalytic domains, it is a RPTP (Y-specific).
(5) It is likely to be a true PTP if it has the minimal conserved active site motif CXXXXXR;[163] and unlikely to be a true PTP if the C or the R is substituted.
(6) The dual specificity PTPs have less well conserved sequence patterns throughout the catalytic domain and only the CXXXXXR motif is usually conserved.

12. Automatic Classification and Validation

According to the predefined classification rules of Section 11, we set the classification conditions for each subclass in the tree first and use the whole tree to discriminate each new input protein by evaluating the possibilities of all subclasses to which it may belong. Thus, we simply pick the most likely result as the predicted result, and publish the predicted subclass along with the whole tree in HTML format together with various pieces of evidence—*e.g.*, motifs along the amino acid sequence, keywords found, and so on—to allow phosphatase experts to validate the choice manually. These experts can put the protein into the correct class by just clicking a button on the web interface. The system then collects the validated result returned through the web interface and saves it as the trusted classification result.

13. Data Integration and Updates

After facilitating the validation, the system prompts the expert to supply specific biological information such as structural features, expression information, catalytic properties, substrates, interacting proteins, regulation, cell effects, disease linkages, crystallographic information, knockout/transgenics; see Figure 7. It subsequently combines all collected information in the database. In addition to this information, we also collect the related 3D structure information (Figure 10) from PDB,[880] disease linkage information (Figure 5) from OMIM,[318] and pathway information from ENZYME[40] (Figure 9 and 10).

To keep PPDB updated, we adopt three different strategies. The simplest strategy is to keep as many symbolic links to the specified databases as possible. For example, we use hotlinks for OMIM,[318] GenBank,[77] *etc.* The second strategy is to only re-create the webpages for the newly collected information of relevant phosphatases. That is, to only update existing phosphatases' with additional information, such as new literature links, instead of re-creating the whole database. The third strategy is the integration of new phosphates into PPDB in the future. This step is not trivial and we need to keep the internal unique identifiers for each phosphatase to preserve data integrity.

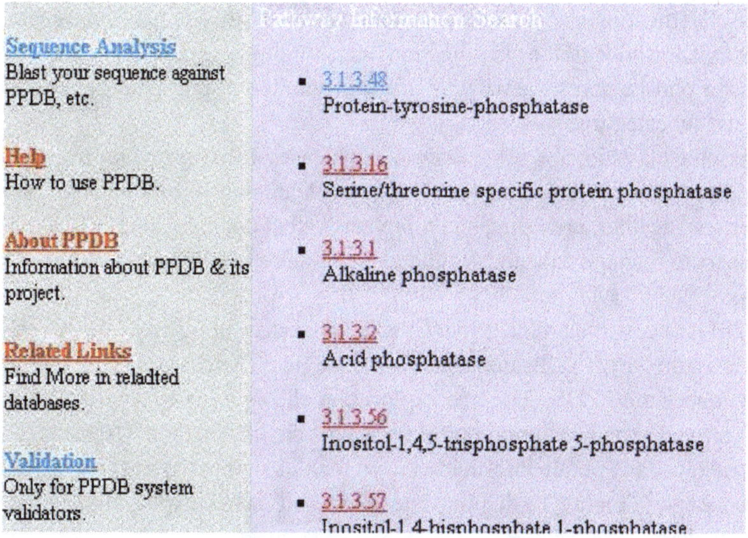

Fig. 9. A display of the list of EC numbers of enzymes in PPDB.

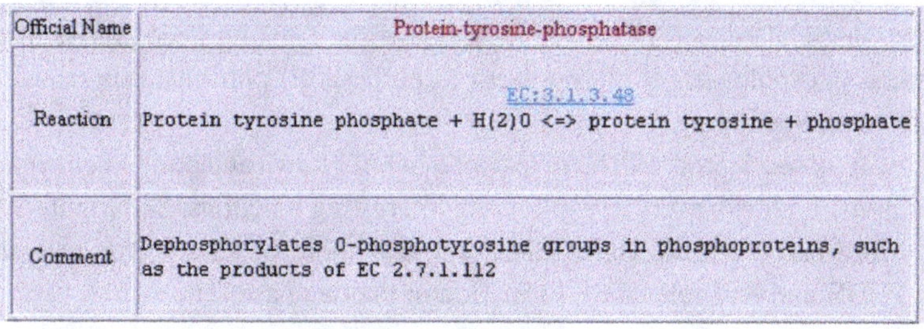

Official Name	Protein-tyrosine-phosphatase
Reaction	EC:3.1.3.48 Protein tyrosine phosphate + H(2)O <=> protein tyrosine + phosphate
Comment	Dephosphorylates O-phosphotyrosine groups in phosphoproteins, such as the products of EC 2.7.1.112

Fig. 10.   The visualization of the protein tyrosine phosphatase pathway.

14. Data Publication and Internet Access

In PPDB ver. 1.0, there are 70 phosphatases, which are clustered into 8 categories, as shown in Figure 2. One phosphatase, however, belongs to two separate categories corresponding to different functions when it is found at different locations within a cell. We provide additional subcategories for storing phosphatases that currently have unknown function. As new phosphatase information is discovered and their biochemical features become available, we will classify them with into the correct groups in the classification tree.

Our classification system is flexible and we can easily add new subcategories into the existing node of the classification tree. Although we currently concentrate on tyrosine phosphatases, we will classify serine/threonine phosphatases into their functional subcategories as well.

Using the classification scheme for phosphatases, PPDB provides researchers more powerful and systematic searches. A vast amount of related information is displayed together, providing a comprehensive but compact interface allowing researchers to manage data and facilitate use of more detailed and specific information.

PPDB is accessible online through the web interface at `http://www.gis.a-star.edu.sg/PPDB`, depicted in Figure 11. There are several different ways to query the PPDB. The tree exploration allows users to step through the classification tree and easily find interesting specific information. Users can click on any category and obtain information from for all members of that specific phosphatase category. Figure 12 demonstrates the result of tree exploration of the tyrosine subclass.

For general search functions, PPDB ver. 1.0 provides 3 different methods, *viz.* gene-based search, pathway-based search, and keyword-based search. These methods are described briefly below.

- Gene-based search. PPDB provides a list of the existing gene names of the curated phosphatases. It allows users to pull out all phosphatases related to each gene. See Figure 5.
- Pathway-based search. PPDB gives out a list of known and curated enzymatic reactions. It allows users to explore the interesting reactions. See Figure 9.
- Keyword-based search. Users can input any interesting keywords to search the PPDB and find relevant information in the database. The search function of PPDB currently simply implements the title content of related data objects.

Like many genomic databases, PPDB also provides additional tools to allow users to do sequence search and analysis as well. For example, a user can submit a protein sequence to BLAST search[24] against the PPDB protein dataset. The system returns to the user the phosphatase homologs in their classified categories. PPDB also provides a localized ProfileScan 2.0 tool[361] for domain hunting to perform sequence domain analysis.

15. Future Development

PPDB curate phosphatases and their related information from diverse public resources, especially for structural, pathway, and functional information. We plan to

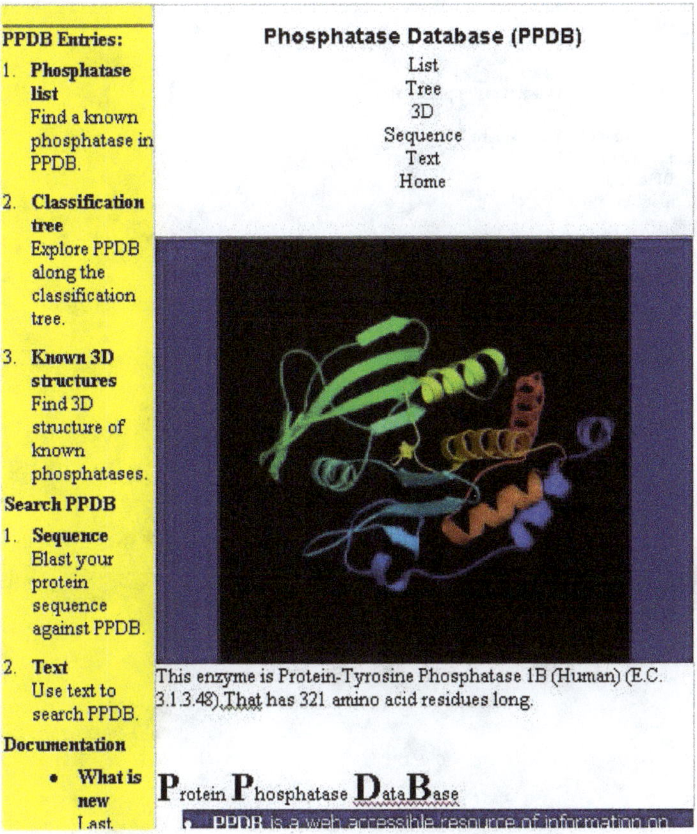

Fig. 11. A snapshot of the home page of PPDB.

include phosphatases from other important model species, such as *S. cerevisiae*, *E. coli*, *etc*. As the content increases, we also plan to do comprehensive domain comparisons within one species and cross species to better understand their biochemical functions in the cell.

Acknowledgement

We thanks Catherine J. Pallen and Kah-Leong Lim for helpful discussions and suggestions.

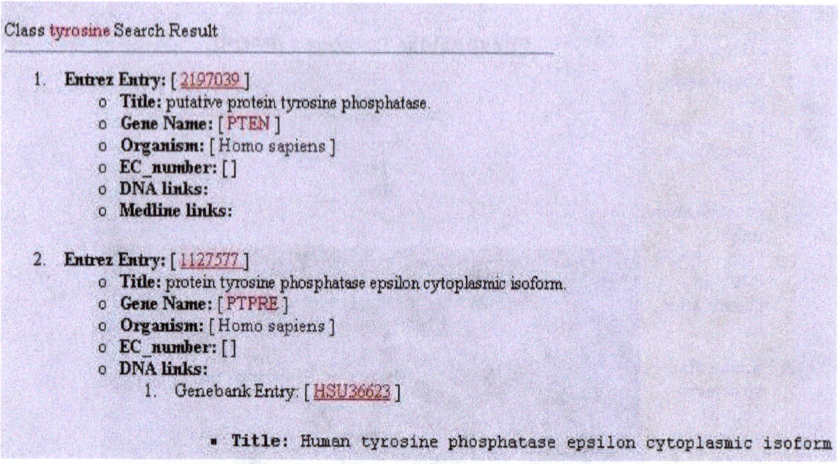

Class tyrosine Search Result

1. **Entrez Entry:** [2197039]
   o **Title:** putative protein tyrosine phosphatase.
   o **Gene Name:** [PTEN]
   o **Organism:** [Homo sapiens]
   o **EC_number:** [ ]
   o **DNA links:**
   o **Medline links:**

2. **Entrez Entry:** [1127577]
   o **Title:** protein tyrosine phosphatase epsilon cytoplasmic isoform.
   o **Gene Name:** [PTPRE]
   o **Organism:** [Homo sapiens]
   o **EC_number:** [ ]
   o **DNA links:**
       1. Genebank Entry: [HSU36623]

   - `Title: Human tyrosine phosphatase epsilon cytoplasmic isoform`

Fig. 12.   The result of tree exploration of tyrosine subclass.

CHAPTER 19

A FAMILY CLASSIFICATION APPROACH TO FUNCTIONAL ANNOTATION OF PROTEINS

Cathy H. Wu

Georgetown University Medical Center
wuc@georgetown.edu

Winona C. Barker

Georgetown University Medical Center
wb8@georgetown.edu

The high-throughput genome projects have resulted in a rapid accumulation of genome sequences for a large number of organisms. To fully realize the value of the data, scientists need to identify proteins encoded by these genomes and understand how these proteins function in making up a living cell. With experimentally verified information on protein function lagging far behind, computational methods are needed for reliable and large-scale functional annotation of proteins.

A general approach for functional characterization of unknown proteins is to infer protein functions based on sequence similarity to annotated proteins in sequence databases. While this is a powerful approach that has led to many scientific discoveries, accurate annotation often requires the use of a variety of algorithms and databases, coupled with manual curation. This complex and ambiguous process is inevitably error prone.[92] Indeed, numerous genome annotation errors have been detected,[104, 203] many of which have been propagated throughout other molecular databases. There are several sources of errors. Since many proteins are multifunctional, the assignment of a single function, which is still common in genome projects, results in incomplete or incorrect information. Errors also often occur when the best hit in pairwise sequence similarity searches is an uncharacterized or poorly annotated protein, is itself incorrectly predicted, or simply has a different function.

The Protein Information Resource (PIR)[903] provides an integrated public resource of protein informatics to support genomic and proteomic research and scientific discovery. PIR produces the Protein Sequence Database (PSD) of functionally annotated protein sequences, which grew out of the Atlas of Protein Sequence and Structure edited by Margaret Dayhoff.[191] The annotation problems are addressed by a classification-driven and rule-based method with evidence attribution, coupled with an integrated knowledge base system being developed. The knowledge base consists of two new databases to provide a comprehensive protein sequence collection and extensive value-added protein information, as well as sequence analysis tools and graphical interfaces. This chapter describes and illustrates how to use PIR databases and tools for functional annotation of proteins with case studies.

ORGANIZATION.

Section 1. We present a detail description of the classification-driver rule-based approach in PIR to the functional annotation of proteins.

Section 2. Then we illustrate the approach by two case studies. The first case study looks at the issue of error propagation to secondary databases using the example of IMP Dehydrogenase. The second case study looks at the issue of transitive identification error using the example of His-I bifunctional proteins.

Section 3. After that, we provide a careful discussion on the common identification errors and their causes.

Sections 4–5. Finally, we describe two new protein databases (NREF and iProClass) of PIR in details. We also discuss how this integrated knowledge base can facilitate protein function annotation in a way that goes beyond sequence homology.

1. Classification-Driven and Rule-Based Annotation with Evidence Attribution

1.1. *Protein Family Classification*

Classification of proteins is widely accepted to provide valuable clues to structure, activity, and metabolic role. This is increasingly important in this era of complete genome sequencing. Protein family classification has several advantages as a basic approach for large-scale annotation:

(1) it improves the identification of proteins that are difficult to characterize based on pairwise alignments;
(2) it assists database maintenance by promoting family-based propagation of annotation and making annotation errors apparent;
(3) it provides an effective means to retrieve relevant biological information from vast amounts of data; and
(4) it reflects the underlying gene families, the analysis of which is essential for comparative genomics and phylogenetics.

In recent years a number of different classification systems have been developed to organize proteins. Scientists recognize the value of these independent approaches, some highly automated and others curated. Among the variety of classification schemes are:

(1) hierarchical families of proteins, such as the superfamilies/families[62] in the PIR-PSD, and protein groups in ProtoMap;[919]
(2) families of protein domains, such as those in Pfam[64] and ProDom;[175]
(3) sequence motifs or conserved regions, such as in PROSITE[238] and PRINTS;[37]
(4) structural classes, such as in SCOP[517] and CATH;[651] as well as
(5) integrations of various family classifications, like ProClass/iProClass[370, 904] and InterPro.[29]

While each of these databases is useful for particular needs, no classification scheme is by itself adequate for addressing all genomic annotation needs.

The PIR superfamily/family concept,[193] the original such classification based on sequence similarity, is unique in providing comprehensive and non-overlapping clustering of sequences into a hierarchical ordering of proteins to reflect their evolutionary origins and relationships. Proteins are assigned to the same super-family/family only if they share end-to-end sequence similarity, including similar domain architecture (*i.e.*, the same number, order, and types of domains), and do not differ excessively in overall length (unless they are fragments or result from alternate splicing or initiators).

Other major family databases are organized based on similarities of domain or motif regions alone, as in Pfam and PRINTS. There are also databases that consist of mixtures of domain families and families of whole proteins, such as SCOP and TIGRFAMs.[313] However, in all of these, the protein-to-family relationship is not necessarily one-to-one, as in PIR superfamily/family, but can also be one-to-many. The PIR superfamily classification is the only one that explicitly includes this aspect, which can serve to discriminate between multidomain proteins where functional differences are associated with presence or absence of one or more domains.

Family and superfamily classification frequently allow identification or prob-able function assignment for uncharacterized ("hypothetical") sequences. To as-sure correct functional assignments, protein identifications must be based on both global (whole protein, *e.g.*, PIR superfamily) and local (domain and motif) se-quence similarities, as illustrated in the case studies.

1.2. *Rule-Based Annotation and Evidence Attribution*

Family and superfamily classification also serves as the basis for rule-based pro-cedures that provide rich automatic functional annotation among homologous se-quences and perform integrity checks. Combining the classification system and se-quence patterns or profiles, numerous rules have been defined to predict position-specific sequence features such as active sites, binding sites, modification sites, and sequence motifs. For example, when a new sequence is classified into a superfamily containing a "ferredoxin [2Fe-2S] homology domain," that sequence is automatically searched for the pattern for the 2Fe-2S cluster, and the feature "Binding site: 2Fe-2S cluster (Cys) (covalent)" is added if the pattern is found.

Such sequence features are most accurately predicted if based on patterns or profiles derived from sequences most closely related to those that are experimentally verified. For example, within the cytochrome c domain (PF00034), the "CXXCH" pattern, containing three annotatable residues, is easily identified and the ligands (heme and heme iron) are invariant. However, there is no single pattern derivable for identifying the Met that is the second axial ligand of the heme iron.

In contrast, within the many superfamilies containing the calcineurin-like phosphoesterase domain (PF00149), the metal chelating residues, the identity of the bound metal ion, and the catalytic activity are variable. In such a case, automated annotation must be superfamily-specific in order to be accurate. Integrity checks are based on PIR controlled vocabulary, standard nomenclature, and other ontologies. For example, the IUBMB Enzyme Nomenclature is used to detect obsolete EC numbers, misspelt enzyme names, or inconsistent EC number and enzyme name.

Attribution of protein annotations to validated experimental sources provides effective means to avoid propagation of errors that may have resulted from large-scale genome annotation. To distinguish experimentally verified from computationally predicted data, PIR entries are labeled with status tags of "validated," "similarity," or "imported" in protein Title, Function, and Complex annotations. The entries are also tagged with "experimental," "predicted," "absent," or "atypical" in Feature annotations.

The validated Function or Complex annotation includes hypertext-linked PubMed unique identifiers for the articles in which the experimental determinations are reported. The amount of experimentally verified annotation available in sequence databases, however, is rather limited due to the laborious nature of knowledge extraction from the literature.

Linking protein data to more bibliographic data that describes or characterizes the proteins is crucial for increasing the amount of experimental information and improving the quality of protein annotation. We have developed a bibliography system that provides literature data mining, displays composite bibliographic data compiled from multiple sources, and allows scientists/curators to submit, categorize, and retrieve bibliographic data for protein entries.

2. Case Studies

2.1. *IMP Dehydrogenase: Error Propagation to Secondary Databases*

During the PIR superfamily classification and curation process, at least 18 proteins were found to be mis-annotated as inosine-5'-monophosphate dehydrogenase (IMPDH) or related in various complete genomes. These "misnomers," all of which have been corrected in the PIR-PSD and some corrected in Swiss-Prot/TrEMBL,[41] still exist in GenPept (annotated GenBank translations) and RefSeq[684], see Figure 1.

The mis-annotation apparently resulted from local sequence similarity to the CBS domain. As illustrated in Figure 2, most IMPDH sequences (*e.g.*, PIR: A31997 in superfamily SF000130) have four annotated Pfam domains, the N-terminal IMPDH/GMP reductase domain (PF01574), the C-terminal IMPDH/GMP reductase domain (PF00478) associated with a PROSITE signature pattern (PS00487), and two CBS domains (PF00571). [Note added in press: PF01574 and PF00478 are now represented by one single Pfam domain PF00478.] Structurally, the N- and C-terminal domains form the core catalytic domain and the two CBS regions form a flanking CBS dimer domain.[932] There is also a well-characterized IMPDH (PIR: E70218 in SF000131)[938] that contains the N- and C-terminal catalytic domains but lacks the CBS domains, showing that CBS domains are not necessary for enzymatic activity.

The four misnomers shown in Figure 2, one from the *Methanococcus jannaschii* genome and three from *Archaeoglobus fulgidus*, all lack the functional region of an IMPDH but contain the two repeating CBS domains. Two of them also possess other domains, and have been classified into different superfamilies.

Many of the genome annotation errors still remain in sequence databases and have been propagated to secondary, curated databases. IMPDH occurs in most species, as the enzyme (EC 1.1.1.205) is the rate-limiting step in the de novo synthesis of guanine nucleotides. It is depicted in the Purine Metabolism pathway for *Archaeoglobus fulgidus* (afu00230) in the KEGG pathway database[410] based on the three mis-annotated IMPDH proteins shown above. However, there is no evidence that a homologous IMPDH protein actually exists in the *Archaeoglobus fulgidus* genome to substantiate its placement on the pathway. Indeed, the only three proteins annotated by the genome center as IMPDH are all misnomers; and no IMPDH can be detected after genome-wide search using either sequence similarity searches (BLAST[24] and/or FASTA[655]) against all known IMPDH proteins, or hidden Markov model search (HMMER[221]) against the N- and C-terminal IMPDH domains.

18 entries were found

ID	Organism	PIR	Swiss-Prot/TrEMBL	RefSeq/GenPept
NF00181857	Methanococcus jannaschii	E64381 conserved hypothetical protein MJ0653	Y653_METJA Hypothetical protein MJ0653	g1592300 inosine-5'-monophosphate dehydrogenase (guaB) / NP_247637 inosine-5'-monophosphate dehydrogenase (guaB)
NF00187788	Archaeoglobus fulgidus	C69355 MJ0653 homolog AF0847 ALT_NAMES: inosine-monophosphate dehydrogenase (guaB-1) homolog [misnomer]	O29441 INOSINE MONOPHOSPHATE DEHYDROGENASE (GUAB-1)	g2649754 inosine monophosphate dehydrogenase (guaB-1) / NP_069681 inosine monophosphate dehydrogenase (guaB-1)
NF00188267	Archaeoglobus fulgidus	E69514 yhcV homolog 2 ALT_NAMES: inosine-5'-monophosphate dehydrogenase (guaB-2) homolog [misnomer]	O28162 INOSINE MONOPHOSPHATE DEHYDROGENASE (GUAB-2)	g2648410 inosine monophosphate dehydrogenase (guaB-2) / NP_070943 inosine monophosphate dehydrogenase (guaB-2)
NF00188697	Archaeoglobus fulgidus	B69402 MJ0188 homolog ALT_NAMES: inosine monophosphate dehydrogenase homolog [misnomer]	O29009 Hypothetical protein AF1259	g2649320 inosine monophosphate dehydrogenase, putative / NP_070087 inosine monophosphate dehydrogenase, putative
NF00197776	Thermotoga maritima	B72265 hypothetical protein TM1354 ALT_NAMES: inosine-5'-monophosphate dehydrogenase-related protein [misnomer]	Q9X175 INOSINE-5-MONOPHOSPHATE DEHYDROGENASE-RELATED PROTEIN	g4981914 inosine-5-monophosphate dehydrogenase-related protein / NP_229155 inosine-5-monophosphate dehydrogenase-related protein
NF00414709	Methanothermobacter thermautotrophicus	C69030 MJ0653 homolog MTH1226 ALT_NAMES: inosine-monophosphate dehydrogenase related protein V [misnomer]	O27294 INOSINE-5-MONOPHOSPHATE DEHYDROGENASE RELATED PROTEIN V	g2622337 inosine-5'-monophosphate dehydrogenase related protein V / NP_276134 inosine-5'-monophosphate dehydrogenase related protein V
NF00414811	Methanothermobacter thermautotrophicus	D69035 MJ1232 protein homolog MTH126 ALT_NAMES: inosine-5'-monophosphate dehydrogenase related protein VII [misnomer]	O26729 INOSINE-5-MONOPHOSPHATE DEHYDROGENASE RELATED PROTEIN VII	g2621166 inosine-5'-monophosphate dehydrogenase related protein VII / NP_275209 inosine-5'-monophosphate dehydrogenase related protein VII
NF00414837	Methanothermobacter thermautotrophicus	H69232 MJ1225-related protein MTH992 ALT_NAMES: inosine-5'-monophosphate dehydrogenase related protein IX [misnomer]	O27073 INOSINE-5-MONOPHOSPHATE DEHYDROGENASE RELATED PROTEIN IX	g2622093 inosine-5'-monophosphate dehydrogenase related protein IX / NP_276127 inosine-5'-monophosphate dehydrogenase related protein IX
NF00414969	Methanothermobacter thermautotrophicus	B69077 yhcV homolog 2 ALT_NAMES: inosine-monophosphate dehydrogenase related protein X [misnomer]	O27616 INOSINE-5-MONOPHOSPHATE DEHYDROGENASE RELATED PROTEIN X	g2622697 inosine-5'-monophosphate dehydrogenase related protein X / NP_276687 inosine-5'-monophosphate dehydrogenase related protein X

Fig. 1.   A partial list of the 18 IMP dehydrogenase misnomers in complete genomes remaining in some protein databases.

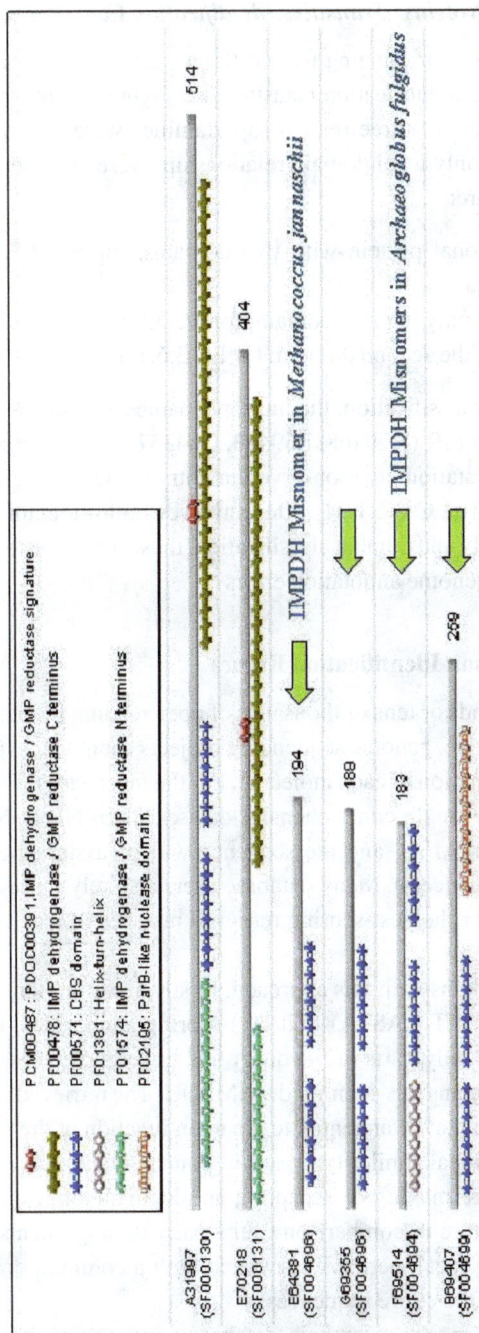

Fig. 2. Domain architectures of IMP dehydrogenase (IMPDH) and misnomers. A typical IMPDH (A31997) has two IMPDH domains that form the catalytic core and two CBS domains. A less common but functional IMPDH (E70218) lacks the CBS domains. All four misnomers show strong similarity to the CBS domains.

2.2. His-I Bifunctional Proteins: Transitive Identification Catastrophe

Several annotation errors originating from different genome centers have led to the so-called "transitive identification catastrophe." Figures 3 and 4 illustrate an example where members of three related superfamilies were originally mis-annotated, likely because only local domain relationships were considered. Here, the related superfamilies are:

- SF001258, a bifunctional protein with two domains, for EC 3.5.4.19 and 3.6.1.31, respectively;
- SF029243, containing only the first domain, for EC 3.5.4.19; and
- SF006833, containing the second domain, for EC 3.6.1.31.

Based on the superfamily classification, the improper names assigned to three sequence entries imported to PIR (H70468, E69493, G64337) were later corrected. The type of transitive annotation error observed in entry G64337 (named as EC 3.5.4.19 when it is actually EC 3.6.1.31) often involves multi-domain proteins. Comprehensive superfamily and domain classification, thus, allows systematic detection and correction of genome annotation errors.

3. Analysis of the Common Identification Errors

Faced with several thousands or tens of thousands of open reading frames to identify and functionally annotate, genome sequencing projects cannot be expected to perform a thorough examination of each molecule. For the most part, the sequence will be searched against a single comprehensive dataset, often NR at NCBI[882], PIR-PSD, or SwissProt/TrEMBL, and the sequence will be assigned the name of the highest-scoring sequence(s). Many database users also rely on searching a comprehensive database for the best-scoring retrieved matches in making identifications of unknown proteins.

There are several problems with this approach. Firstly, the common sequence searching algorithms (BLAST, FASTA) find best-scoring similarities; however, the similarity may involve only parts of the query and target molecules, as illustrated by the numerous proteins mis-identified as IMPDH. The retrieved similarity may be to a known domain that is tangential to the main function of the protein or to a region with compositional similarity, e.g., a region containing several trans-membrane domains. Before making or accepting an identification, users should examine the domain structure in comparison to the pairwise alignments and determine if the similarity is local, perhaps associated with a common domain, or extends convincingly over the entire sequences.

Secondly, annotation in the searched databases is at best inconsistent

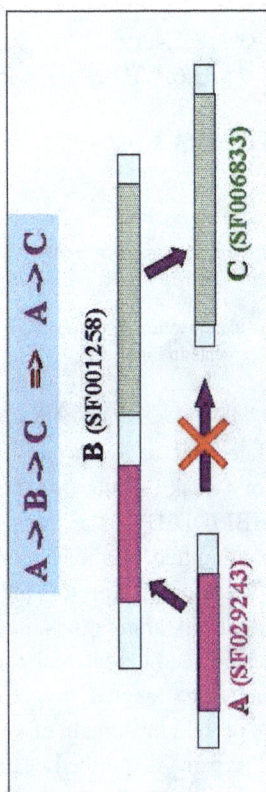

Fig. 3. FASTA neighbors of H70468 are in three superfamilies. G64337 is an example of mis-annotation by transitive identification error. It was named as EC 3.5.4.19 when it is actually EC 3.6.1.31.

PIR ID	Imported	Corrected	Superfamily
H70468	3.6.1.31	3.5.4.19/ 3.6.1.31	SF001258 hisI-bifunctional en- zyme
E69493	3.5.4.19/ 3.6.1.31	3.5.4.19	SF029243 phosphoribosyl- AMP cyclohydrolase
G64337	3.5.4.19	3.6.1.31	SF006833 phosphoribosyl-ATP pyeophosphatase

Fig. 4. The mis-identification of three proteins by genome centers was later corrected based on the superfamily assignments in Figure 3.

and incomplete and at worst misleading or erroneous, having been based on partial or weak similarity. The major nucleotide sequence database GenBank/EMBL/DDBJ[77] is an "archival" database, recording the original identifications as submitted by the sequencers unless a revision is submitted by the same group. Therefore, the protein identifications in GenPept, which are taken directly from GenBank annotations, may never be updated in light of more recent knowledge. Users need to realize that entries in a comprehensive database may be under-identified, e.g., labeled "hypothetical protein" when there is a convincing similarity to a protein or domain of known function; over-identified, e.g., the specific activity "trypsin" is ascribed when the less specific "serine proteinase" would be more appropriate; or mis-identified, as in the case studies discussed above.

Over-identification can be suspected when the similarity is not strong over the entire lengths of the query and target sequences. PIR defines "closely related" as at least 50% identity and assigns such sequences to the same "family." A PIR superfamily is a collection of families. Sequences in different families in the same superfamily may have as little as 18–20% sequence identity and their activities, while often falling within the same general class, may be different. For example, the long-chain alcohol dehydrogenase superfamily contains alcohol dehydrogenase (EC 1.1.1.1), L-threonine 3-dehydrogenase (EC 1.1.1.103), L-iditol 2-dehydrogenase (EC 1.1.1.14), D-xylulose reductase (EC 1.1.1.9), galactitol-1-phosphate 5-dehydrogenase (EC 1.1.1.251), and others. Of five sequences from the recently sequenced genome of *Brucella melitensis* that were identified specifically as alcohol dehydrogenase (EC 1.1.1.1), only two are closely related (60% identity) to well-characterized alcohol dehydrogenases. For the others, the functional assignment may be overly specific, as they are more distantly related (less than 40% identity). For the most part, users will need to inspect database entries and read at least the abstracts of published reports to ascertain whether a functional assignment is based on experimental evidence or only on sequence similarity. Users should also ascertain that any residues critical for the ascribed activity (*e.g.*, active site residues) are conserved.

Thirdly, in many cases a more thorough and time-consuming analysis is needed to reveal the most probable functional assignments. Factors that may be relevant, in addition to presence or absence of domains, motifs, or functional residues, include similarity or potential similarity of three-dimensional structures (when known), proximity of genes (may indicate that their products are involved in the same pathway), metabolic capacities of the organisms, and evolutionary history of the protein as deduced from aligned sequences. Bork and Koonin[92] discuss additional effective strategies. Iyer *et al.*[387] analyze several additional examples of mis-identifications and their subsequent correction.

4. Integrated Knowledge Base System to Facilitate Functional Annotation

To facilitate protein identification and functional annotation, two new protein databases (NREF and iProClass) have been developed and are being integrated into a knowledge base system with sequence analysis tools and graphical user interfaces.

4.1. *PIR-NREF Non-Redundant Reference Database*

The PIR-NREF database was designed to provide all the identifications in major databases for any given sequence, identified by source organisms. It is a timely and comprehensive collection of all protein sequence data containing source attribution and minimal redundancy. The database has three major features:

(1) comprehensiveness and timeliness: it currently consists of more than 920,000 sequences from PIR Protein Sequence Database, Swiss-Prot/TrEMBL, RefSeq, GenPept, and PDB,[880] and is updated biweekly;
(2) non-redundancy: it is clustered by sequence identity and taxonomy at the species level; and
(3) source attribution: it contains protein IDs, accession numbers, and protein names from source databases in addition to amino acid sequence, taxonomy,

(I)

● NREF Entry: NF00136823			Submit Bibliography	XML View	Last Updated: 10-Oct-2001

(Part I contains a PIR-NREF sequence entry report with protein name, taxonomy, binding entry, composite database, protein sequence, sequence length, and neighbor NREF sequences sections.)

Neighbor NREF sequences:

Seq. ID	Organism	Seq. Length	% identity	Overlap	Matched region	
NF00101962	Homo sapiens	146	99.275	138	21-158:9-146	
NF00100518	Homo sapiens	60	96.667	60	99-158:1-60	
NF00124041	Homo sapiens	73	100.000	49	1-49:1-49	
NF00087533	Homo sapiens	70	100.000	49	1-49:1-49	

(II)

Database	Protein ID	Accession	Taxon ID	Protein Name
PIR	T40073	T40073	4896	phosphoribosyl-AMP cyclohydrolase (EC 3.5.4.19) / phosphoribosyl-ATP pyrophosphatase (EC 3.6.1.31) [similarity]
TrEMBL	O59667	O59667	4896	PROBABLE PHOSPHORIBOSYL-AMP CYCLOHYDROLASE (EC 3.5.4.19) / PHOSPHORIBOSYL-ATP PYROPHOSPHOHYDROLASE (EC 3.6.1.31) / HISTIDINOL DEHYDROGENASE (EC 1.1.1.23) (HDH)
GenPept	g3006138	CAA18379.1	4896	probable phosphoribosyl-amp cyclohydrolase

Fig. 5. (I) PIR-NREF sequence entry report. Each entry presents an identical sequence from the same source organism in one or more underlying protein databases. (II) Discrepant protein names assigned by different databases reveal annotation errors.

and composite bibliographic data (Figure 5, Part I). Related sequences, including identical sequences from different organisms, as well as identical subsequences and highly similar sequences ($\geq 95\%$ sequence identity) are also listed. The NCBI taxonomy[882] is used for matching source organism names at the species or strain (if known) levels.

The PIR-NREF database can be used to assist functional identification of proteins, to develop an ontology of protein names, and to detect annotation errors. It is ideal for sequence analysis tasks because it is comprehensive, non-redundant, and contains composite annotations from source databases. The clustering at the species level aids analysis of evolutionary relationships of proteins. It also allows sequence searches against a subset of data consisting of sequences from one or more species. The composite protein names, including synonyms and alternate names, and the bibliographic information from all underlying databases provide an invaluable knowledge base for application of natural language processing or computational linguistics techniques to develop a protein name ontology.[355, 920]

The different protein names assigned by different databases may also reflect annotation discrepancies. As an example (Figure 5, Part II), the protein (PIR: T40073) is variously named as a monofunctional (EC 3.5.4.19), bifunctional (EC 3.5.4.19, 3.6.1.31), or trifunctional (EC 3.5.4.19, 3.6.1.31, 1.1.1.23) protein in three different databases. Thus, the source name attribution provides clues to incorrectly annotated proteins.

4.2. *iProClass Integrated Protein Classification Database*

A few sentences describing the properties of a protein may not be adequate annotation. What is required is as much reliable information as possible about properties like function(s) of the protein, domains and sites, catalytic activity, pathways, subcellular location, processes in which the protein may be involved, similarities to other proteins, *etc.* Thus, an ideally annotated protein database should

(1) include domain structure, motif identification, and classifications,
(2) distinguish experimentally determined from predicted information, with citations for the former and method for the latter, and
(3) include annotations of gene location, expression, protein interactions, and structure determinations.

However, in practice, it is unrealistic to expect that protein sequence databases can keep all (or even a substantial minority) of entries up-to-date with regard to all of the above. Nevertheless, much of this information is available in specialty databases.

The iProClass database was designed to include up-to-date information from many sources, thereby, providing much richer annotation than can be found in any single database. It contains value-added descriptions of all proteins and serves as a framework for data integration in a distributed networking environment. The protein information in iProClass includes family relationships at both global (superfamily/family) and local (domain, motif, site) levels, as well as structural and functional classifications and features of proteins. The database is extended from ProClass,[370, 902] a protein family database that organizes proteins based on PIR superfamilies and PROSITE motifs.

The version at the time of writing (May 2002) consists of more than 735,000 non-redundant PIR-PSD, SwissProt, and TrEMBL proteins organized with more than 36,000 PIR superfamilies, 145,000 families, 3700 Pfam and PIR homology domains, 1300 PROSITE/ProClass motifs, 280 RESID[270] post-translational modification sites, 550,000 FASTA similarity clusters, and links to over 45 molecular biology databases. iProClass cross-references include databases for protein families (e.g., COG[819] and InterPro), functions and pathways (e.g., KEGG[410] and WIT[642]), interactions (e.g., DIP[910]), structures and structural classifications (e.g., PDB, SCOP, CATH, and PDBSum[475]), genes and genomes (e.g., TIGR[666] and OMIM[318]), ontologies (e.g., Gene Ontology[285]), literature (e.g., NCBI PubMed), and taxonomy (e.g., NCBI Taxonomy).

The iProClass presents comprehensive protein and superfamily views as sequence and superfamily summary reports. The protein sequence report (Figure 6) covers information on family, structure, function, gene, genetics, disease, ontology, taxonomy, and literature, with cross-references to relevant molecular databases and executive summary lines, as well as a graphical display of domain and motif regions. The superfamily report (Figure 7) provides PIR superfamily membership information with length, taxonomy, and keyword statistics, complete member listing separated into major kingdoms, family relationships at the whole protein and domain and motif levels with direct mapping to other classifications, structure and function cross-references, and domain and motif graphical display.

4.3. *Analytical Tools and Graphical Interfaces*

Integrated with the protein databases are many search and analysis tools that are freely accessible from the PIR website (http://pir.georgetown.edu)[555]. These tools assist the exploration of protein structure and function for knowledge discovery. For reliable protein identification, search results should display more detailed information, including lengths of the query and target sequences and of the "overlap" (the local regions of each that are matched), and the percentage identity of the overlap region, as in the interface that displays FASTA neighbors (Figure 3).

Other useful features available from the PIR website are graphical displays, such as the domain and motif display in iProClass reports (Figures 6 and 7); ability to make multiple alignments and display domain structures; ability to sort search output by criteria (e.g., species) other than similarity score; and easy retrieval of full entries, citation abstracts, and classification information with multiple text search options.

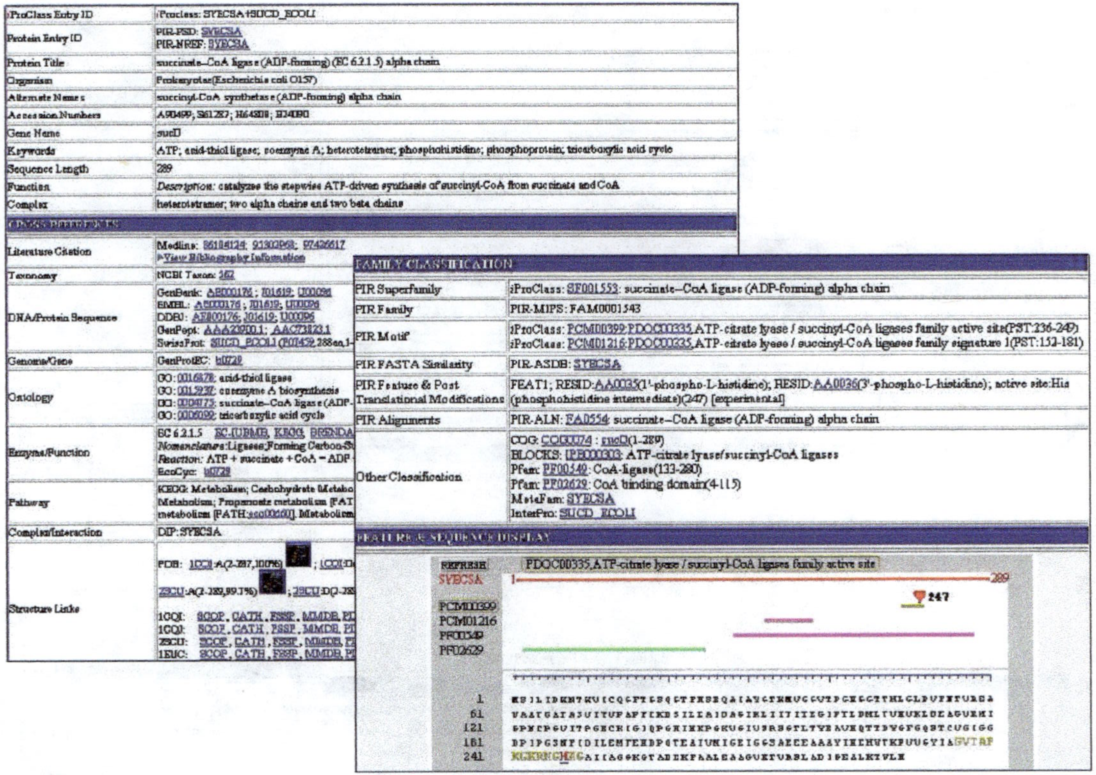

Fig. 6.   The iProClass sequence report for comprehensive value-added protein information.

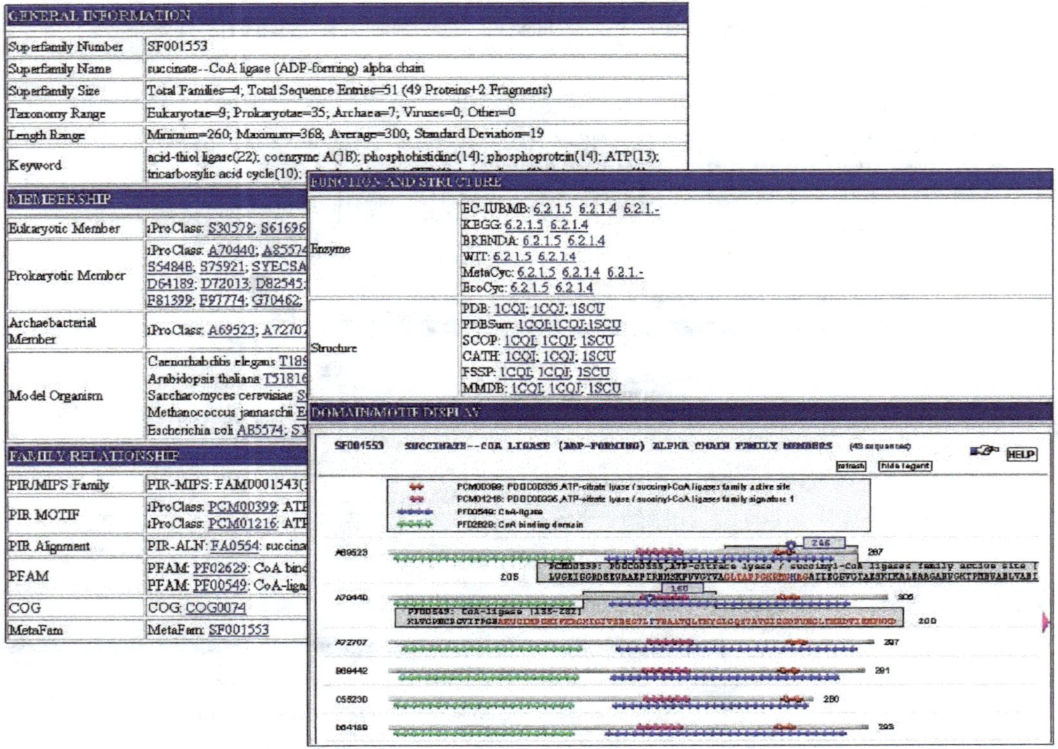

C. H. Wu & W. C. Barker

Fig. 7. The iProClass superfamily report with family relationship information.

5. Functional Associations Beyond Sequence Homology

The PIR serves as a primary resource for exploration of proteins, allowing users to answer complex biological questions that may typically involve querying multiple sources. In particular, interesting relationships between database objects, such as relationships among protein sequences, families, structures, and functions, can be revealed readily. Functional annotation of proteins requires association of proteins based on properties beyond sequence homology—proteins sharing common domains connected via related multi-domain proteins (grouped by superfamilies); proteins in the same pathways, networks, or complexes; proteins correlated in their expression patterns; and proteins correlated in their phylogenetic profiles with similar evolutionary patterns.[542]

The data integration in iProClass is important in revealing protein functional associations beyond sequence homology, as illustrated in the following example. As shown in Part I of Figure 8, the Adenylylsulfate kinase (EC 2.7.1.25) domain (PF01583) appears in four different superfamilies (*i.e.*, SF000544, SF001612, SF015480, SF003009), all having different overall domain arrangements. Except for SF000544, proteins in the other three superfamilies are bifunctional, all also containing sulfate adenylyltransferase (SAT) (EC 2.7.7.4) activity. However, the SAT enzymatic activity is found in two distinct sequence types, the ATP-sulfurylase (PF01747) domain and CYSN homology (PF00009+PF03144), which share no detectable sequence similarity. Furthermore, both EC 2.7.1.25 and EC 2.7.7.4 are in adjacent steps of the same metabolic pathway (Figure 8, Part II). This example demonstrates that protein function may be revealed based on domain and/or pathway association, even without obvious sequence homology. The iProClass database design presents such complex superfamily-domain-function relationships to assist functional identification or characterization of proteins.

The PIR, with its integrated databases and analysis tools, thus constitutes a fundamental bioinformatics resource for biologists who contemplate using bioinformatics as an integral approach to their genomic/proteomic research and scientific inquiries.

Acknowledgements

The PIR is supported by grant P41 LM05978 from the National Library of Medicine, National Institutes of Health. The iProClass database is supported by DBI-9974855 from the National Science Foundation.

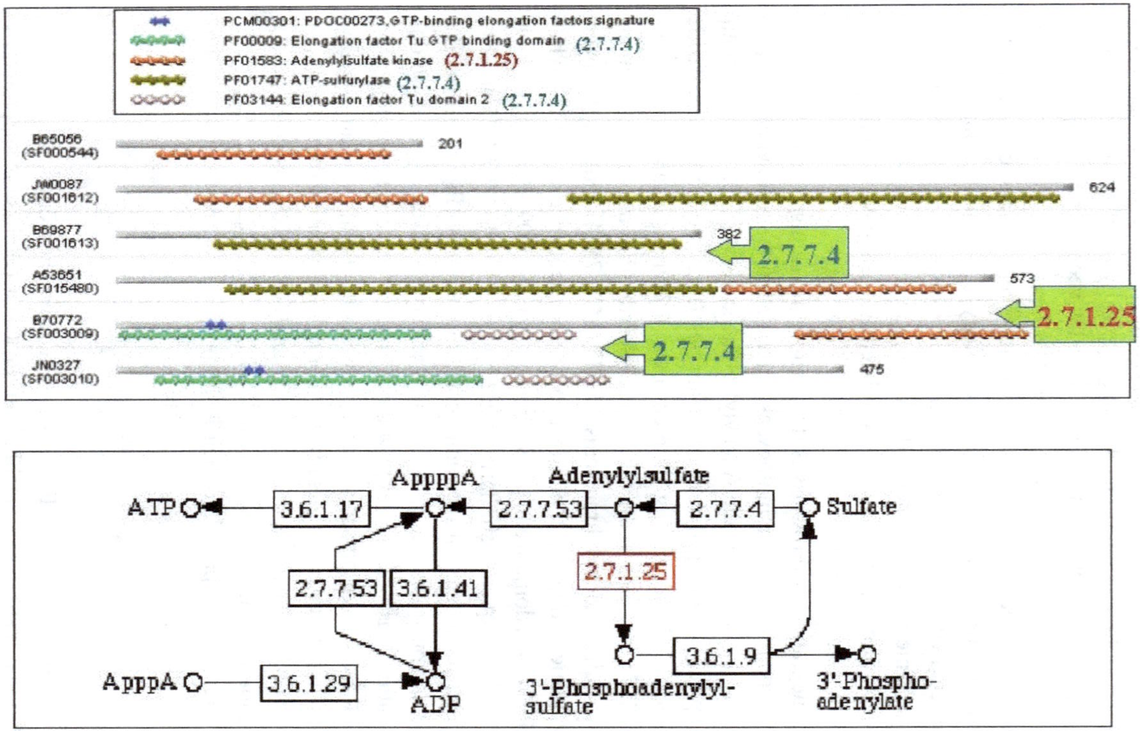

Fig. 8. (I) Superfamily-domain-function relationship for functional inference beyond sequence homology. Association of EC 2.7.1.25 and two distinct sequence types of EC 2.7.7.4 in multi-domain proteins. (II) Association of EC 2.7.1.25 and EC 2.7.7.4 in the same metabolic pathway.

CHAPTER 20

INFORMATICS FOR EFFICIENT EST-BASED GENE DISCOVERY IN NORMALIZED AND SUBTRACTED CDNA LIBRARIES

Todd E. Scheetz

University of Iowa
tscheetz@eng.uiowa.edu

Thomas L. Casavant

University of Iowa
tomc@eng.uiowa.edu

Beginning in 1997, large-scale efforts in EST-based gene discovery in Rat, Human, Mouse, and other species have been conducted at the University of Iowa. To date, these efforts have led to the sequencing of more than 400,000 ESTs accounting for more than 60,000 novel ESTs, including 40,000 previously undescribed genes.

This effort has been augmented by a set of custom informatics tools to gather, analyze, store, and retrieve the sequence data generated. The high rate of gene discovery associated with this work was primarily due to novel normalization and subtraction methodologies.[90] A critical aspect of this work is to periodically perform subtractive hybridizations of cDNA libraries against a set of previously sequenced clones from that library.

The informatics necessary for this effort consists first of examination of individual sequences to verify the presence of a 3' end, and to confirm the clone identity. Second, it is necessary to cluster the sequences to determine the current novelty of the library and to allow feedback to the cDNA library subtraction process to remove redundant clones and improve overall discovery efficiency. Finally, informatics is required for the submission of the EST sequences and associated annotation to both public and local databases.

In this chapter, the overall system of software is described, statistics concerning performance and throughput are given, and detailed methods of certain aspects of the pipeline are provided. In particular, our sequence editing, clustering, and clone verification methods are described in some detail. All originally produced softwares described here are available from our web server at http://genome.uiowa.edu, or by contacting genome@eng.uiowa.edu.

ORGANIZATION.

Section 1. We first provide an abridged biological experimentation background on EST-based gene discovery with cDNA libraries.

Section 2. Then we give an overview of a pipeline for EST sequence processing and annotation pipeline that we have developed for this purpose.

Section 3. The pipeline has five major components, *viz.* raw data gathering and archival, quality assessment and sequence editing, sequence annotation, novelty assessment, and submission to databases. These five components are described in detail in this section.

Section 4. Finally, we close with a discussion on the computational and storage resources required by various components of our pipeline.

1. EST-Based Gene Discovery

EST-based gene discovery is an efficient strategy in defining a gene index for an organism.[8] A gene index is a non-redundant collection of sequences, in which all sequences derived from the same gene transcript are grouped together. They provide a foundation to be annotated, and are an essential component in several analyses, including cross-species comparative analyses, and determination of unique sub-sequences—useful for SAGE, and for creating custom microarray chips. In addition, a non-redundant cDNA collection is useful in the creation of a cDNA-based microarray probe set.

The technology relies upon the sequencing of cDNA libraries, essentially partial DNA copies of mRNA transcripts. These cDNA fragments are typically sequenced on an ABI sequencer, resulting in a set of binary trace-files (chromatographs). The program phred[236] is used to extract the nucleotide sequence and per-base quality values. The phred quality values (a.k.a. phred values) are calculated as $q = -10 \times \log_{10} p$, where p is the estimated error probability for a given base. Thus, a phred value of 10 correlates with an error probability of 10%, a phred value of 20 with an error probability of 1%, and so on.

Expressed sequence tags (ESTs) are sequences derived from cDNA libraries, and have several common features, illustrated in Figure 1. These include a region of the vector sequence, and a cloning restriction site in front of the cloned sequence. In the case of a short cDNA insert, vector sequence may also be seen at the far end of the EST read. Because we utilize oligo-dT primed, directionally cloned cDNAs, our 3' ESTs typically begin with a polyT stretch, which correlates with the reverse complement of the polyadenylation (polyA) tail from the 3' end of mRNA transcripts. Similarly, the reverse complement of a polyadenylation signal is expected 11–30 bases down-stream from the detected tail.[144] To determine tissue of origin in a pooled library environment, a synthetic oligo tag is

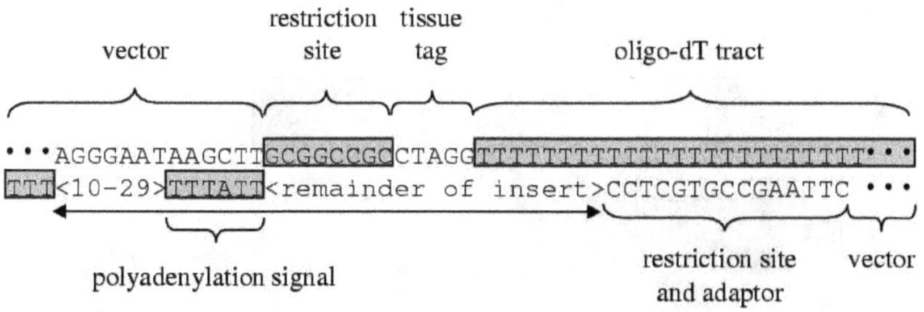

```
                    restriction  tissue
        vector         site       tag          oligo-dT tract

··· AGGGAATAAGCTTGCGGCCGCCTAGGTTTTTTTTTTTTTTTTTTTTTTTTTTTT···
TTT<10-29>TTTATT<remainder of insert>CCTCGTGCCGAATTC ···

        polyadenylation signal                  restriction site    vector
                                                 and adaptor
```

Fig. 1. Typical features found in ESTs derived from oligo-dT-primed, directionally cloned cDNA libraries.

inserted between the restriction site and oligo-dT during the creation of the cDNA libraries.

To maintain high rates of efficiency during the discovery process, we used the complementary approaches of cDNA library normalization and subtraction.[90] Both normalization and subtraction are used to reduce the prevalence of "undesired" clones from a cDNA library. In the context of gene discovery, undesired clones are either those we have seen before or are likely to see many times. Normalization addresses this issue by reducing the prevalence of the most abundant clones. Subtraction is a more focused procedure, reducing—ideally removing— the prevalence of a targeted set of clones.

2. Gene Discovery Pipeline Overview

We have developed an EST sequence processing and annotation pipeline, shown in Figure 2, to process 3' and 5' sequence data derived from cDNA clones. This pipeline supports all types of cDNA libraries—single-tissue or pooled, normalized or subtracted—from any species. The pooling of multiple tissues, all drawn from the same species, results in a more complex library and often enhances the effectiveness of serial subtraction, thereby improving the efficiency of the gene discovery process. However, the information regarding the tissue source of each clone is biologically interesting and useful. To collect this information from pooled multiple-tissue libraries, we use a method that "tags" each clone with an oligonucleotide sequence so that subsequent to generation of sequence data, the original tissue source can be determined.[279] While not directly useful as a definitive means for expression analysis, knowing the tissue an EST is derived from is a valuable resource. It should be emphasized that due to the subtraction processing, not all tissues in a pooled setting will reveal the presence of an EST for each

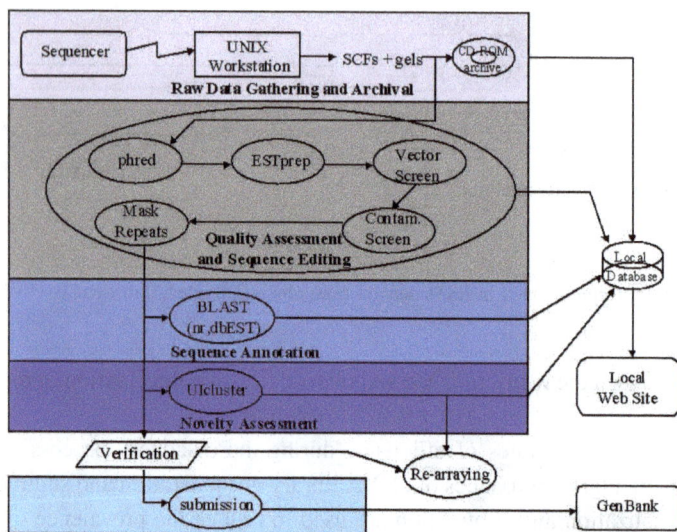

Fig. 2.   Overview of the EST-based gene discovery pipeline.

mRNA message expressed by that tissue. Once a clone from a transcript is used in a subtraction, the prevalence of clones from that transcript will be greatly reduced.

The pipeline shown in Figure 2 can be viewed as consisting of five major components:

(1) Raw Data Gathering and Archival
(2) Quality Assessment and Sequence Editing (Feature Identification)
(3) Sequence Annotation
(4) Novelty Assessment
(5) Deposition or Submission to Local and Public Databases

The raw data gathering and archiving steps must assure efficient and secure communication between the data gathering nodes—nominally ABI sequencers—and the networked system of computers dedicated to the subsequent processing steps. This represents a challenge for a number of reasons, both technical and logistical. The quality assessment and sequence editing steps must assure that all sequence data processed by the remainder of the pipeline are of sufficient quality to be reliable. Both sequence integrity in terms of confidence of base calls, as well as confidence that the sequence is oriented properly and free of contamination, are supported. In the sequence annotation step, a number of sequence characteristics must be identified, labeled and stored for entry into a local database as well as submission to public databases. Well-known repetitive motifs are identified, and an initial BLAST examination against known nucleotide and amino acid databases is performed. Novelty assessment is based on a sequence similarity metric, and serves multiple purposes in this pipeline. Initially, all new sequences are clustered into the existing local "UniGene" set to calculate the current rate of novelty of the library being sequenced. As the final step in the pipeline, ESTs that are of sufficient quality, and are not contaminated are submitted, with annotation, to public databases such as dbEST and our local databases.

3. Pipeline Component Details

In this section, the software components of the large-scale gene discovery pipeline shown in Figure 2 are described. The pipeline is described from a functional perspective. Then each phase of the pipeline is described in detail. First, we describe the set of algorithms and software tools developed to initially process all sequences to confirm orientation, detect tissue tags, search for polyadenylation sequence and signal, and trim for overall sequence quality. Second, we describe a precise, high-performance tool for forming clusters of sequences which are likely to have been derived from the same gene or gene family. Finally, our method for verifying clone identity, and the informatics required to support it, is described.

The sequence-processing pipeline can be broken into the five broad phases listed in Section 2. In the Data Gathering and Archiving phase the "raw" chromatograph (SCF) files are collected from the ABI sequencers. The SCF files may be directly available (*e.g.*, from ABI 3700 sequencers), or obtained through tracking and extraction from a gel image (*e.g.*, from an ABI 377 sequencer). The naming of clones and sequences follows a standardized nomenclature, described in detail in Subsection 3.1. The sequence names are imported directly into the ABI sequencing software from sample sheets automatically generated via a web-based interface.

In the Quality Assessment and Feature Editing phase, phred[236] is used to obtain the base-calls and per-base quality values directly from the SCF files. These sequence and quality files are then processed by ESTprep.[751] The ESTprep application identifies several categories of features, including the region of high-quality sequence, commonly found in EST sequences. A complete description of the processing by ESTprep is covered in Subsection 3.2.

After processing with ESTprep, the sequences and their associated quality files are trimmed according to the identified regions of high-quality sequences. The trimmed sequence is then assessed for contaminating or repetitive sequences using RepeatMasker (`http://ftp.genome.washington.edu/RM/RepeatMasker.html`). Contaminating sequences fall into three categories—mitochondrial, bacterial (*E. coli*), and vector sequences. If significant amounts of contamination are found, the sequence is removed from the pipeline. Likewise, if a hit to the vector sequence is found only at the 3'-most end of the sequence, the matching portion is removed rather than discarding the whole sequence. Such EST sequences are derived from short cDNA clones, where the EST sequence "reads through" the cDNA insert and into the vector beyond. RepeatMasker is also used to mask repetitive and low complexity elements within the sequences. Here, the set of species-specific repetitive sequences from RepBase[404] is used as the set of sequences that should be masked. The process of masking takes bases identified as repetitive or low complexity, and replaces them with N's.

Sequences that reach this part of the sequence processing pipeline are of sufficient quality, length, not contaminated, and have had any repeats masked. They enter the Annotation phase. These sequences are blasted[24] against the non-redundant nucleotide database from GenBank and the appropriate species' sequences from dbEST,[87] as an initial annotation.

The masked sequences are also used in the Novelty Assessment phase, which utilizes the UIcluster program,[836] and is fully described in Subsection 3.4. The process of clustering the EST sequences is useful both for creating a non-redundant set of sequences and in assessing a library's gene discovery rate. The discovery rate of a library provides critical feedback to the library creation group for determining when normalization and subtraction processes should be performed.

The final phase is that of Sequence Submission. All sequences that have sufficient quality and are not contaminated are promptly submitted to the dbEST division of GenBank according to the "bulk email" format. This information includes the trimmed sequence with annotation describing some of the identified features—detection of polyA tail and polyA signal, any repetitive elements, library of origin, and tissue of origin. A complete snapshot of all the data from the sequence processing pipeline is loaded into our local database, and is available from `http://genome.uiowa.edu`. Our local database also stores comprehensive annotation for all locally generated ESTs, including expression, clustering and mapping information.

3.1. *Raw Data Gathering and Archival*

The first step in this stage is the generation of sample sheets to be loaded into the ABI sequencing software. This is an important first step in ensuring that the sequences are assigned names that conform to our naming structure described below. The sample sheets also store data on the technician, PCR block, and plate type—96- or 384-well—for each sequencing run. After the chromatograph files (SCFs) have been generated, they are transferred to a central drop-point. The drop-point is implemented on a Linux workstation which is exporting disk partitions to both Macintosh and Windows using netatalk (http://netatalk.org) and Samba (http://samba.org) respectively.

Next, the chromatograph files are copied to the appropriate project directory onto a local RAID disk array. Each project has a separate directory hierarchy divided into (minimally) 3' and 5' sequence directories. Within the direction-specific directories a plate of sequences is stored within a directory hierarchy consisting of run date, machine name, and plate name. This allows us to track progress by date, and to identify systematic problems with a specific sequencer or capillary. The ABI-derived chromatograph files and the gel image, if run on a 377, are archived onto CD or DVD-ROM for long-term storage.

Because the EST processing pipeline works on large numbers of cDNA libraries, a clone naming convention was designed to quickly identify what library a given sequence or clone was derived from. The naming format is essentially a set of eight values separated by dashes. The first value denotes the originating institution. For all of our libraries, the value "UI" is used, indicating that those clones came from the University of Iowa. The second and third values denote the project and library codes. The plate, row, and column that specify a well are the fourth, fifth, and sixth values. Finally, the seventh and eighth values are for the replication number and replicating institution. These are important for tracking sets of clones that have been distributed, such as re-arrayed sets sent to Research Genetics. The replication number of "0" is reserved for the original "master" plates. So the clone name of "UI-R-A0-ad-e-04-0-UI" refers to a clone from the UI rat project (project code "R") library "A0". The clone is located at column "e", row 4 of plate "ad". The final two data points tell us that the clone is from the original master plate arrayed at the University of Iowa. Further information on our clone naming protocol can be found at http://ratEST.eng.uiowa.edu/localdocs/naming.html.

3.2. Sequence Assessment and Sequence Editing

As mentioned earlier, this stage both verifies the quality of the EST sequences, and identifies common features. The quality is assessed based upon the per-base values assigned by phred and detection of contaminated ESTs. The features to be identified are dependent on the EST end sequenced (3' or 5') and on the specifics of the cloning procedure.

Phred[236] is used to extract the initial base calls and per-base quality values from the "raw" SCFs. New chromatograph files are also generated from the ABI-based files. This ensures that the base-calls embedded within the SCF are consistent with those determined by phred. An additional benefit is that the SCF files generated by phred are also significantly smaller than those generated by the ABI sequencing software.

Next, the ESTprep[750] program is used to perform an initial quality assessment, and to identify features common to ESTs. The fundamental procedure underlying ESTprep is to first identify candidate sites representing the expected cloning site. These sites are then validated by searching for the expected vector sequence adjacent to the predicted restriction site. Based upon the identified location of the cloning site, other features may be identified if applicable, including library tag, polyA tail, and polyA signal. All features to be detected are configurable in the set of features to be identified, specifics of the features and in the allowable number and types of errors. For clarity, a diagram of these features is shown in Figure 3.

The following description of ESTprep is provided using a set of default parameters to provide context. Most of these parameters are configurable through modification of a configuration file shown in Figure 4. The first stage within ESTprep is an initial quality assessment. During this stage the sequence quality is verified, and any low-quality stretches of sequence at the beginning are removed to avoid detection of spurious features. If the average sequence quality over the first 200 bases is less than 20, the sequence is removed from the pipeline. Similarly, a leading stretch of sequence with fewer than 8 out of 20 bases greater than 20 will be removed if present.

Next, the program attempts to identify the restriction site used during the cloning process. The strategy used during the restriction site identification is to first identify high-quality restriction site candidates and to validate those candidates by looking for the expected vector sequence adjacent to the site. The quality of the candidate restriction sites is gauged by the number of errors, or edit distance,[309] away from the "correct" restriction site.

As mentioned above, the first feature identified is the restriction site. For our pipeline, we use an eight base recognition site. If the actual recognition site used

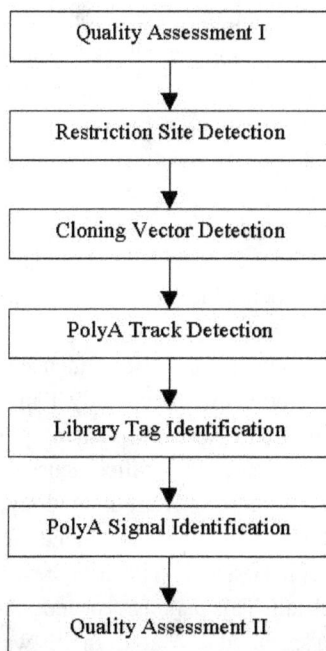

Fig. 3. ESTprep processing stages.

in cloning is not eight bases long, a "synthetic" restriction site is constructed using the last eight bases of the vector sequence (including the restriction site) prior to the cloned cDNA insert. In Figure 1, this sequence is the recognition sequence of the NotI restriction enzyme (GCGGCCGC). When the recognition site is less than eight bases, a synthetic restriction site is constructed. The adjacent portion from the vector/polylinker are prepended to make the synthetic restriction site. The presence of vector sequence leading up to the restriction site is used to confirm that this site corresponds to the bona fide cloning vector site. This is necessary because of errors inherent in the sequencing process. Sequence errors may occur in the restriction site, making it unrecognizable, and alternatively they may alter another region such that it then resembles the restriction site.

Feedback based upon empirical assessment by the sequencing group was used in refining the quality assessment parameters. ESTprep allows rejection of sequences based upon average sequence quality, number of phred 20 bases, and polyA tail length. The current defaults used in our EST sequence quality assessment are shown in Figure 4 and described in detail below.

For 5' EST sequences, detection of a restriction site with leading vector, and the assessment of quality is a complete analysis. However, for 3' ESTs several other features are assessed—polyA tail, polyA signal, and library tag. The polyA tail and polyA signal are representative of the polyA tail and signal in the mRNAs from which the cDNA library was derived. Because of the procedures used in library construction the polyA/T stretches in the cDNA library are tightly controlled. Thus, rather than a detected polyA tract of a few hundred nucleotides, the polyA tract of mRNAs from well constructed cDNA libraries averages approximately 26 nt in length. The expected length is a user-specified parameter in the prep.params file.

After the restriction site has been located, a sequence is parsed downstream to find the first nucleotide of the polyA tail - typically T in 3' ESTs. The algorithm for detecting the starting and ending positions of the polyA tail is surprisingly complex, involving a number of carefully crafted heuristics. A brief summary of this algorithm is provided here. Starting from the ending position of the putative library tag in the input sequence, a floating-point values array of T-density is constructed. This density array is parsed from a position beyond the maximum length (user-specified) of the polyA tail. This parsing proceeds backwards until a point at which the density exceeds a specified threshold. In this way, the right-most extent of the region richest in 'Ts' is located. If the T-rich region does not end with a T, then the ending point is recursively retracted by one, providing a more accurate identification of the polyA tail extent. Finally, if the polyA tail identified is not of sufficient length, the detection process is repeated, beginning one base right of the original starting base. Once the right-most extent of the polyA tail has been located, the process is resumed to the left to attempt to identify the left extend of the polyA tail. A lower density threshold (94% of original) is used so as not to truncate longer polyA tails, while mis-identifying smaller portions of polyA stretch immediately adjacent to the restriction site. If a valid-length of polyA tail can still not be identified, the sequence will not have a library tag associated with it.

If a polyA tail is identified, the search begins for a corresponding polyadenylation signal. Specification of custom sets of polyadenylation signals is supported. Typically, the two canonical polyadenylation signals (AAUAAA, AUUAAA) would be provided by the user, and a second set may be specified for other alternative signals. The default configuration is to accept the pair of canonical signals, and in addition, the alternative signals identified in Beaudoing et al.[72] Polyadenylation signals must be found within 11–30 nucleotides from the right-most end of an identified polyA tail in 3' ESTs.[144]

The final feature identified by ESTprep is the library tag. This is typically a

```
Sequence_Direction: FORWARD
Echo_Sequence: 2
Quality-Params: 10 8 20 100
Restriction-Site-Tag: 200 2 1 2 GCGGCCGC
Cloning-Vector: 6 GCCAAGCTAAAATAACCTCACTAAAGGGAATAAGCTT
LibTags: 1 1 1
CAGCC
rat-eye
polyAtail: T 10 .95 20 18 18 .65
polyAsignal: 11 30 2 0 0 0 TTTATT TTTAAT
polyAsignal-alt: 14 TTTACT TTTATA TTTATG TTTATC TTTAGT TTTAAA
TTTCTT CTTTTT TGTTTT AGCCCC TATATT TGTATT TCTATT TTCATT
```

Fig. 4.   An example parameter file for ESTprep.

stretch of 4–10 nucleotides inserted during library construction to aid in identifying the tissue of origin for a given EST sequence. For an in-depth description of the processes used in the library tag construction and detection, see Gavin *et al.*[279] As shown in Figure 1, the library tag lies between the restriction site and the polyA tail. Without identification of both the restriction site and polyA tail, identification of the library tag is not attempted. Library tag identification is performed by constructing a substring of sequence between the restriction site and the polyA tail. This sequence is then used in a local sequence alignment versus all possible library tags, as specified in the prep.params configuration file; see Figure 4.

The final stage in ESTprep processing is a second quality assessment. While fully user-configurable in prep.params, typically, a 20 base-pair sliding window is applied looking for the first region which contains 8 bases with phred quality values less than 10. The sequence trim location for the right-most end of the sequence is then the start of the first window meeting these criteria. Typically, if the resulting trimmed sequence would be less than 100 base-pairs long, the sequence is rejected from the pipeline, as dbEST does not accept sequences less that 100 bp in length. If the average quality of the trimmed sequence is less than 20 or if the fraction of bases with quality greater than 20 is too low (e.g., less than 53%) the sequence is similarly rejected from the pipeline.

The output of ESTprep is a summary file listing each feature identified. Consider a sequence file named "foo". The resulting summary file would be named "foo.prpsmry" (see Figure 6). A more verbose, and human readable descriptive transcript of the resulting analysis is also printed to STDOUT, which is typically redirected to a file for later analysis, or for posting to a web page for project monitoring.

The format of ESTprep's parameter file is shown in Figure 4. The Se-

```
>UI-R-CV0-bro-h-12-0-UI.s1 608 0 608 ABI
TGACGGCCAGTGCCAAGCTAAAATTAACCCTCACTAAAGGGAATAAGCTT
GCGGCCGCCAGCCTTTTTTTTTTTTTTTTTTTTTTTTTTTAACAATATTTGT
ACCGTTTTTATTTGTAAAAATAACCATCTGAATGCATGTCCATCGTATGT
TACAGGTAAGTACTTCATTGCATATTAGAGCACTCAGTAGTTGGGAAAGT
ATTAACCTGTGCTTGGGAAGATTCACACTGGTCCAAAGTCCTTCACTGAA
CCCAATGCCATTTTTCCTCATTTTTTACACTCAGGACGTTTACCAAAGTC
ACTCAACTCCAATCTACATCTTAAAATTACAGACGAAAAAATCCCACTGA
AATCATCACATAATAGTTTATGCTGTTACAAACACGTTTTACAAAATGTT
ACACTGGCATAAGATGTGTATCACCGTGCTTTGCAAGGATATATTGCACA
ATGCTGAAGCGTGGTCTGGAGACAAAACTGCTAAACAAAAGATTCTCCCC
TCGAGTCCTCAGATCACCGGGTAAGAAACACACAGAATTTCATCTTACAA
ACACGCTCAGATTGTCACTATCTTAAAACGCTGTCCTCCTATAATACTGA
CCTATGGG
```

Fig. 5.   An example 3' EST sequence.

```
CLONENAME: seq/UI-R-CV0-bro-h-12-0-UI.s1.seq
RESTRSITEFOUND: TRUE
VECTORFOUND: TRUE
VECTORLENGTH: 33
POLYATAILFOUND: TRUE
POLYATAILLENGTH: 25
LIBTAGFOUND: TRUE
LIBTAG: CAGCC rat-eye
POLYASIGNALFOUND: TRUE
POLYASIGNAL: TTTATT
TRIMLOC: 71 451
GOODQUALITY: 25.36
QUAL_FILTERS: 25.36 39.84
STATUS: GO
```

Fig. 6.   A Prep summary file resulting from running ESTprep on the sequence in Figure 5 using the params list in Figure 4.

quence_Direction field is used to denote which direction the EST sequence is expected to be in. Valid possible values are FORWARD and REVERSE. When processing FORWARD sequences, 3' relevant features are identified, including polyadenylation tail and signal, and library tag. The Echo_Sequence field controls the verbosity of output from ESTprep. Next are the parameters used in quality assessment. These parameters are specified by four integer numbers following the Quality-Params field. The first three values (10 8 20, above) specify the parameters used in the quality-based trimming procedure. These values specify that the left trim-site will be at the first position where there are 8 positions within a window of 20 that are less than phred quality of 10. The fourth value specifies the minimum number of bases that must remain after trimming. The example in Figure 4 specifies that a minimum of 100 nucleotides are required after trimming for the sequence to satisfy the quality criteria.

The Restriction-Site-Tag field is used to specify a recognition sequence identifying the restriction recognition site. In cases where the actual recognition site is less than eight bases long, the right-most eight base sequence is used, which may include a few bases of the cloning vector. Additional upstream vector sequence is provided with the Cloning-Vector field. Identification of the expected vector sequence upstream of an identified restriction site is used to validate the restriction site.

The LibTags field is used to define the set of expected library tags. The definition is divided onto three separate lines. The first line specifies the number of tags in the set, along with specification of allowable error properties. In the example above, the values "1 1 1" denote the fact that this library contains a single valid library tag, and that the detection scheme may allow up to one substitution or one insertion/deletion error when identifying the library tag. The second and third lines are used to specify the valid library tags and their correlated source tissues.

The polyAtail field is used in identifying the polyadenylation tail from 3' ESTs. Seven parameters are specified. The first is the character for which the tail will be identified. In the case of typical 3' ESTs this is T, the complement of A. The next four values specify limiting values on a subsequence in order for it to be identified as a polyA tail. The first of these (10) specifies a minimum polyA tail length of 10. The polyA tail is then extended from an original exact match of 10 T's. The next two parameters require that in the event that the expected tail sequence does not begin with a series of 10 consecutive T's, that the detected tail be 95% T over the first 20 bases. The 95% threshold also defines the criteria for determining the end of the polyA tail. As the tail is extended the relative percent T is calculated. The end of the tail is determined as the last position prior to the percent $T < 95\%$ that is a T. The fifth parameter specifies the maximum number of bases from the identified tail sequence that should be retained in the trimmed sequence. The final two values (18 and .65) are used to identify a probable polyA tail when the more stringent 95% parameter fails to identify exact tail boundaries. This is often the result of regions of low-quality sequence, or internal priming events.

The polyAsignal field is used to specify the set of potential polyA signals to be identified, and parameters to limit the search. The first two values limit the distance within which a polyA signal may be detected from an identified polyA tail. The values of 11 and 30 are used in the example above, as published in

Chen *et al.*[144] Therefore, a valid polyA signal must begin from 11 to 30 bases from the identified end of a polyA tail. The third value specifies the number of polyA signals in the set to be searched, in this case two. The next three numbers specify the allowable errors—missense, insertion, and deletion. By default these are set to zero when searching for a specific set of alternative polyA signals. The remaining words are the canonical polyA signal candidates.

The polyAsignal-alt field is similar to the polyAsignal field, but used to differentiate between canonical polyA signals (*e.g.*, TTTATT and TTTAAT) and alternative polyA signals. The first value specifies the number of alternative signals that may be identified. The remaining values specify the valid alternative polyA signal candidates. No errors are allowed during the identification of alternative polyA signals.

The contamination detection and repeat masking phase of the EST processing pipeline utilizes the RepeatMasker package as the core sequence similarity utility. The benefit of using RepeatMasker over an alternative alignment algorithm is its sensitivity in detecting distantly related sequences. This is especially important when masking repetitive elements, which may have diverged significantly from the canonical form. The cost of using RepeatMasker is measured in processing time. RepeatMasker utilizes cross_match to perform its sequence alignments which is an efficient implementation of the Smith-Waterman-Gotoh algorithm.[300, 780]

The computational overhead inherent in using RepeatMasker is only significant when screening versus a large database of sequences, such as the bacterial genome. In this case, a different alignment algorithm could have been selected for the different screening needs (bacterial, mitochondrial, vector, repeats). The identification of repetitive elements requires a sensitive alignment algorithm. In contrast, assessing for contaminants could be performed with a less sensitive algorithm, such as BLAST. However, the decision to use a single package simplified the design of the programs that execute and parse the resulting outputs.

RepBase[404] is used as the baseline database of repetitive elements to be masked for a given species. Any hit to a repetitive element is masked, meaning that the responsible sub-sequence is replaced base-for-base with a string of N's.

When assessing for bacterial or mitochondrial contamination, a sequence is considered contaminated if more than 85% of the sequence matches the bacterial or mitochondrial database. Assessing for vector contamination is slightly more complex. Two classes of contamination from vector sequences are possible. The first is complete vector contamination, *i.e.*, there was no insert. For this case, a criteria of 85% is used, just as in the assessment of bacterial and mitochondrial contamination. The other potential vector contaminant is that of a sequence derived from a short cDNA insert. In this case, only the end of the EST sequence is expected to match vector sequence. The remaining sequence is valid cDNA sequence. Because the amount of vector sequence may be limited, a low detection threshold is used. To limit the number of false-positive matches, an additional requirement is applied, requiring the detected vector sequence to extend to within 10 bases of the end of the trimmed sequence. When screening for vector contamination, it is important to utilize the sequence of the specific vector used in the cDNA library. This ensures that weak hits at the end of a short cDNA insert are detected.

Sequences detected with complete contamination are removed from the pipeline. For sequences with detected trailing vector sequence, the subsequence matching the vector is trimmed. Similar trimming is also performed on the quality file to maintain synchronicity with the trim file.

A final quality assessment utilizes a sample-sequence based verification. This is performed by rearraying eight clones from each plate of clones arrayed for sequencing. These sets of eight clones are arrayed into a new plate, referred to as a verification plates. These verification sequences are then compared to the original sequences. In the event that the two sequences (original and verification) do not match, the clone is marked.

The verification procedure provides valuable feedback on the consistency of clone-sequence correlation. As an example, the verification data has been used to identify a problematic plate that was re-sequenced. The verification data is also useful in prioritizing candidates for inclusion in non-redundant sets. Clones with positive verification results are preferentially selected, while those with negative verification results are selected against.

3.3. *Annotation*

An initial annotation of the ESTs is a useful set of data for investigators using the ESTs. All uncontaminated, high-quality ESTs, as determined from the previous phase, are blasted against the non-redundant nucleotide and amino acid databases from the National Center for Biotechnology Information (NCBI). When available, a database of species-specific ESTs is also blasted against to provide initial cross-references into UniGene clusters.

3.4. *Novelty Assessment*

The cDNA library normalization and subtraction techniques rely upon efficient novelty assessment to maximize discovery. A typical method used to assess novelty in cDNA libraries is based upon clustering the ESTs. The process of clustering partitions a set of input sequences based upon sequence similarity. This aids in assessment of library novelty and in the definition of non-redundant sets. The UIcluster clustering program[836] is used for our clustering analyses. This program can be run on a single CPU or in parallel, utilizing the MPI toolkit.[258] The fundamental strategy used in UIcluster is to add new sequences to an existing set of clusters one at a time. Each sequence is compared to a representative element from every cluster. If the sequence has significant similarity to a cluster's representative element, it is added to that cluster. In the event that the sequence has a minimal similarity to more than one representative sequence, it is added to the cluster it is most similar to, with stored annotation on which other clusters it matched. Masked EST sequences are used as the input to the clustering process. This restricts sequences from being grouped together based upon repetitive or low-complexity sequence.

To build the most accurate clustering possible, the available full-length mRNA sequences for the given species are added in to the clustering. Results from our Rat Gene Discovery and Mapping Project (T. Scheetz, in preparation; http:// ratEST.uiowa.edu) indicate that incorporation of mRNA sequences merge clusters that would otherwise be disjoint for a significant fraction of the mRNAs. Multiple clusters may arise from a single transcript due to cDNA library artifact such as internal priming or restriction sites, or from alternative processing such as alternative polyadenylation or alternative splicing.

The default criteria used to determine minimal sequence similarity is an alignment of 38 out of 40 (95%) bases. Both misread and insertion/deletion errors are allowed. The minimal alignment is extended to the maximal amount allowed by sequence homology. Sequences may match in either orientation, *i.e.*, forward or reverse complemented. The resulting alignment is saved in the output clustering file, an example of which is shown in Figure 7.

The example shows a cluster of three sequences. The representative element (primary) is denoted with the "@P" tag, followed by the name of the representative element. This line is followed by the nucleotide sequence of the representative element. Other cluster elements (secondaries) are denoted with the "@S" tag. The @S tag is followed by annotation specifying the extent of match between the secondary and the representative element. The first two numbers are integers representing the start of alignment in the primary and secondary respectively. The third number is the length of the alignment, and the fourth is the percent identity of the alignment. The final tag denotes the direction of the match. This can be one of FORWARD, REVCOMP, or ORPHAN. The ORPHAN tag is used when a sequence no longer matches the current primary. Orphans can only occur when the *repick* option is used, causing re-evaluation of the primary element during cluster-

```
@P: UI-R-DO1-cml-n-03-0-UI.sl 0
TTTTTTTTTTTTTTTTTTTGGTCAGGAAATTTTATTTGAACATTCTAAAGCAAGAATGCTTCAGATGTTAC
TTAAATGTCCCAGACAGGATTAACAAAATTAAATGTTTCTAAATTACAAATTTAGCTCCAGTAGGAGTTT
CATAAAAGAAGAAAACAACCCCCTCCCAAAAGAAGTATGACACACACATTCTGAAGAAACCCCAATGTTT
CATGCAATGGTAGGCAAGATGTAGAAGGCCACCCAAACCCATCTGTTTCTACACCAGTCATCACCCCGAA
GAGTCCTCCAGTCAATCTGTACATCCAAATGCATCCGGGAACCTACACCTACAAGACATTATTAATGTTA
TATACATTTATTGCCCCCCTTGGTTTTTTTAATAATTTCTTATGTAAAGCCTTCATTGAAACCCAAAAAA
AAAAAAAAAAAGGATGTAAGACTAACTTGGGGGTAGGGAGGGGAAGATAATCACTTTAGACATTCAGTTAA
AATGTAAATTATCTAAATCTCCAAATGTTTAATAAAAACAAGCATCTTCTCCATTTAACACCTTGCTTGT
TAACTGTACAGTAAATTGTATTATAGAGAGTACATCTCTATTTTCATACTGTATCTTCTTTGGATGGAAT
TGAGAAAGCTGGTTAATTTTAAGATAAATAAATGAGATTGATCCAACTAAGATTAAGATGACAGCAGATA
TATTCCATGCAGAATTTAATAGTTTTTAATTTGT

@S: UI-R-DO1-cmm-n-05-0-UI.sl 18 18 704 100.000000 FORWARD
TTTTTTTTTTTTTTTTTTTGGTCAGGAAATTTTATTTGAACATTCTAAAGCAAGAATGCTTCAGATGTTAC
TTAAATGTCCCAGACAGGATTAACAAAATTAAATGTTTCTAAATTACAAATTTAGCTCCAGTAGGAGTTT
CATAAAAGAAGAAAACAACCCCCTCCCAAAAGAAGTATGACACACACATTCTGAAGAAACCCCAATGTTT
CATGCAATGGTAGGCAAGATGTAGAAGGCCACCCAAACCCATCTGTTTCTACACCAGTCATCACCCCGAA
GAGTCCTCCAGTCAATCTGTACATCCAAATGCATCCGGGAACCTACACCTACAAGACATTATTAATGTTA
TATACATTTATTGCCCCCCTTGGTTTTTTTAATAATTTCTTATGTAAAGCCTTCATTGAAACCCAAAAAA
AAAAAAAAAAAGGATGTAAGACTAACTTGGGGGTAGGGAGGGGAAGATAATCACTTTAGACATTCAGTTAA
AATGTAAATTATCTAAATCTCCAAATGTTTAATAAAAACAAGCATCTTCTCCATTTAACACTTGCTTGT
TAACTGTACAGTAAATTGTATTATAGAGAGTACATCTCTATTTTCATACTGTATCTTCTTTGGATGGAAT
TGAGAAAGCTGGTTAATTTTAAGATAAATAAATGAGATTGATCCAACTAAGATTAAGATGACAGCAGATA
TATTCCATGCAGAATTTAATAG

@S: RPLAHO1TF 19 18 417 99.520384 FORWARD
TTTTTTTTTTTTTTTTTTTGGTCAGGAAATTTTATTTGAACATTCTAAAGCAAGAATGCTTCAGATGTTACT
TAAATGTCCCAGACAGGATTAACAAAATTAAATGTTTCTAAATTACAAATTTAGCTCCAGTAGGAGTTTC
ATAAAAGAAGAAAACAACCCCCTCCCAAAAGAAGTATGACACACACATTCTGAAGAAACCCCAATGTTTC
ATGCAATGGTAGGCAAGATGTAGAAGGCCACCCAAACCCATCTGTTTCTACACCAGTCATCACCCCGAAG
AGTCCTCCAGTCAATCTGTACATCCAAATGCATCCGGGAACCTACACCTACAAGACATTATTAATGTTAT
ATACATTTATTGCCCCCCTTGGATTTTTTAATAATTTCTTATGTAAAGCCTTCATTGAAACCCAAAAAAT
AAAAAAAGGATGTAAGACTAACTT
```

Fig. 7.   Excerpt from a UIcluster output.

ing. With this option, the longest constituent sequence is selected as the primary element.

To support efficient processing of 100,000's of ESTs several features have been integrated into UIcluster to accelerate the clustering process. Most significantly, incoming sequences are only compared to the primaries of each cluster, and a hashing scheme is used to efficiently perform an initial assessment of similarity. The hashing strategy computes a numeric value for each ξ-base subsequence, as shown in Equation 1. In this equation, K represents the size of the symbol alphabet—for nucleotide sequences, K is equal to 4. The variable Φ_i represents the ith letter in the subsequence. Thus each potential subsequence generates a unique hash value. Subsequences including a non-standard base call—*i.e.*, not A, C, G, or T—are not hashed.

$$H = \sum_{i=0}^{\xi-1} K^i \times \Phi_i \qquad (1)$$

For the default ξ of 8, this yields 4^8 possible values based on the four-character DNA alphabet {A, C, G or T}. The presence of any other symbol results in an invalid hash. The hash values for each cluster representative are maintained in a global data structure, the Global Hash Table. Hash values are also created for each incoming sequence, and a complete sequence similarity assessment is only performed if at least one hash value is shared between the incoming sequence and a cluster representative. This filtering strategy greatly reduces the time to compute a clustering of EST sequences, but with a significant increase in the amount of memory required to store the Global Hash Table.

The hash size is a run-time configurable option. The larger the hash that is used, the more efficient the comparison becomes. There are two practical limiting factors on the size of hash that can be used. The first is architecture specific. On a 32-bit architecture it is inefficient to use a hash size of more than 16 bases, since $2^{32} = 4^{16}$. The second limitation is determined by the similarity required to cluster two sequences together. The less strict the alignment criteria is, the shorter the maximum viable hash size. This reflects the increased impact of a single-base error has on the hashes. Take as an example the default clustering criteria of 38 out of 40bp. If a hash size of 16 is used two errors could cause all 16 base hashes within a 40 bp region to not match their error-free counterparts. Equation 2 provides a guide for determining the maximum viable ξ for a given clustering criterion, expressed in terms of N identical bases in a subsequence of length M.

$$\xi = \left\lfloor \frac{M}{M - N + 1} \right\rfloor \qquad (2)$$

Testing for cluster membership is evaluated first on clusters with the largest number of similar hash elements.

3.5. Submission to Local and Public Databases

Sequence, clone, and clustering data are deposited in a local database. The schema for this data is shown in Figure 8. The local database infrastructure also supports mapping information (not described). The sequence and clone records contain information on the sequence-specific features and annotation. The cluster information maintains the cluster assignments, local and UniGene, for each sequence. Additional cluster-specific information summarizing the features of the constituent sequences of each cluster is maintained in a separate table. Often, during the initial loading of data, certain information may not be available. The NCBI accession and GI numbers, and UniGene cluster membership are chief among these, as they are determined by remote sites (NCBI). To maintain a comprehensive data set, programs are available to update those data as the information becomes available.

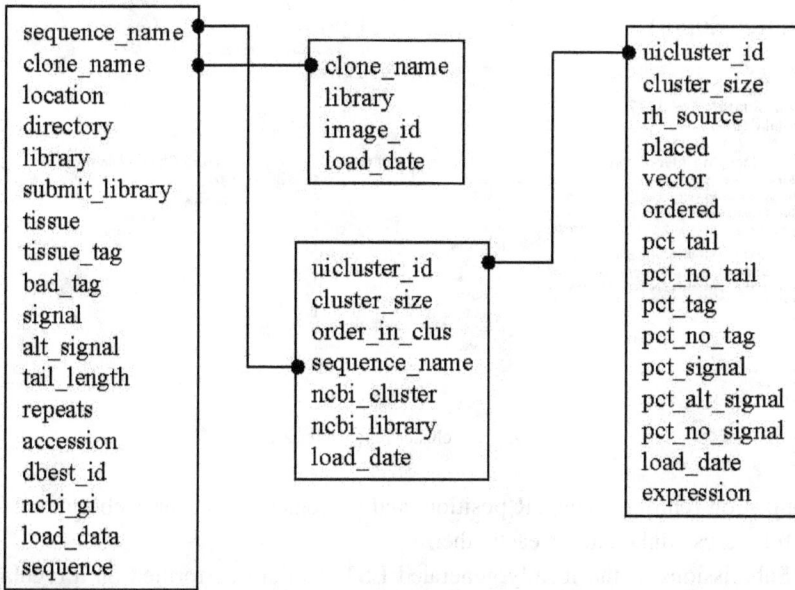

Fig. 8.   EST-specifc database schema.

Several Perl-based web interfaces have been implemented to allow access to the stored information via the Internet. Specifically, a search tool is available, as well as report generation interfaces for clone-, sequence-, and cluster-specific information. Examples of these interfaces are provided in the following figures. Figure 9 shows the top of the Cluster Report interface. A summary of the cluster is provided at the top of the report (shown), with a summary of data for each constituent sequence listed below the summary (not shown). This summary includes mapping status and assignment (if applicable), a summary of expression, a list of libraries the sequences were seen in, and a link for each constituent sequence to it's NCBI cluster. The remainder of the cluster report provides a summary of information for each constituent sequence. This includes links to the Sequence Report (described below) which presents a complete set of sequence-related information.

Figure 10 shows a screen capture of the Sequence Report. This interface provides access to all of the sequence feature information that is obtained throughout the sequence processing pipeline. The sequence information is presented in three discrete sections. First, the clone-related information: the clone name and IMAGE ID,[485] if applicable. Next, a summary of the mapping data is presented if applicable. The mapping data includes primers, radiation hybrid retention vector, and

Cluster Report

UIowa Rat EST Project

Home Page Database Search Library/Tissue Search Page Mapping Information Page

UIowa Cluster: Rn.UI.777
20 Member(s)

Mapping Information		Library Distribution		NCBI Cluster Distribution	
Number of Placed:	0	TIGR:	5	NA:	9
Number of Consensus Vector:	0	UI-R-BJ0:	2	Rn.6360:	11
Number of Primer Ordered:	1	UI-R-BJ1:	2	Thiosulfate	
Mapping Source:	MCW	UI-R-CS0:	1	sulphurtransferase	
		UI-R-CS0s:	1	(rhodanese)	
Expression Information		UI-R-CW0:	2		
Kidney(x4), NA(x2), Atrium-16.5(x2), Ileum(x2),		UI-R-DB0:	1		
rat-heart-pool(x2), rat-aorta-pool(x2), Bladder(x1),		UI-R-DV1:	1		
cartilage(x1), Brain(x1), Ventricle-16.5(x1), Eye(x1)		UI-R-E1:	1		
		UI-R-EA0:	2		
		UI-R-V0:	1		

Fig. 9.   The cluster report interface.

localization (chromosome, cR position, and placement bin). The web-interfaces are fully cross-linked to the each other.

Submissions of the locally generated EST data are performed on a regular basis to NCBI's dbEST.[87] In addition to the trimmed-for-quality EST sequence, additional information is submitted: (i) presence of polyA tail and polyA signal, including differentiating between canonical and alternative signal; (ii) information pertaining to the tissue from which the EST was derived, as determined by the identified library tag; (iii) how the cDNA library was constructed; and (iv) if any regions of the EST are similar to known repetitive elements. The file format used conforms to the specified "bulk" format described at the NCBI (http://www.ncbi.nlm.nih.gov/dbEST/how_to_submit.html) but is not submitted via email due to size constraints. Instead, submissions are made via FTP to a server at the NCBI.

4. Discussion

Because the processing requirements are data parallel, the number of ESTs that can be processed in a single day is easily expanded by adding additional processing nodes. We use PBS/OpenPBS[342] to distribute the processing jobs onto a cluster of workstations. To provide an insight into the relative computational resources required by the various components, a breakdown of the runtime on one thousand sequences is given Figure 11. All times were generated using a single-CPU system with 512MB of available RAM.

To date, we have processed approximately 500,000 EST sequences. This has required significant amounts of both computational and storage resources. Given

Sequence Report

Home Page Database Search Library/Tissue Search Page Mapping Information Page

`UI-R-A1-dw-f-02-0-UI.s1    3' EST`

Clone Name:	UI-R-A1-dw-f-02-0-UI
Clone IMAGE ID:	1771070
Mapping Information:	
Status:	**Mapped**

Left primer: GGTCAGGTATGAAGGAAGAGGG
Right primer: ACAGCATGACTCCTCCTAAGGG
Consensus vector:
20001000210100010002002022001001202102000010010012000011001210000000020000002000100100121200000020011

Chromosome: 4
Map location: 246.36 cR
Bin: D4GOT28 - D4RAT21

Map Source:	UIowa
UIowa UniGene Cluster:	Rn.UI.779
Tissue Library:	UI-R-A1 (UIowa), Lib.40 (NCBI)
NCBI UniGene:	Rn.6383
NCBI Entrez:	AA901246 (Accession #)
NCBI dbEST Entry:	4233747 (GI)
Tissue Source:	Lung
Lib Tag:	TTCCA
Poly A Signal:	TTTATT
Poly A Tail Length:	25
Sequence Length:	350

```
>UI-R-A1-dw-f-02-0-UI.s1    3' EST
TTTTTTTTTTTTTTTTTAAATGGGAAGTATATCAATTTCACTTTATTAGC
CTAATGCAGGAATATAAGGAAAAGGGGATTGAGGTCAGGTATGAAGGAAG
AGGGTTAAAACTTGTGACAGTAAGGATCCTTCAAAACTCGGATGCAGTCC
TCAGTCCTCAGTCACTGCAACGTCACCTCTGCCCACACAAACAGCCACCT
CCGGTTCCTGTCTCCACTTCTGAGCCCAGTAACTCCTCAATCCACGCCCA
GTTACAGTGTGATGAGGTCTACCAGGTTGCCCTTAGGAGGAGTCATGCTG
TCAGGAAAGAAATTCACTAAGGTGTCAAAGAGCTGAGTTCTCTGAGCATA
```

Fig. 10.   The sequence report interface.

the estimates in Figure 11, the complete set of 500,000 sequences would have required 6250 CPU hours to process. In addition, the amount of storage resources required is also prodigious. The chromatograph (SCF) files generated by the ABI software are approximately 190KB each. This represents 95GB of storage. To minimize the storage requirements of the SCF files, we generate a new SCF from the ABI SCF using phred. The benefit of storing the phred SCFs is the significant reduction in average SCF size from 190KB to 40KB. This reduces the amount of storage necessary from 95GB to 20GB.

Although the chromatograph files are the single largest file for any sequence,

Phred	2–3 minutes
ESTprep	2–3 minutes
Masking	1 hour
Annotation	5 minutes
Clustering	5–10 minute
Total	1.25 CPU hours

Fig. 11.   Estimated processing time for one sequencing gel.

the remaining files also require a significant amount of storage. These files include several versions of the sequence (original, trimmed, and masked) and quality files (original and trimmed), as well as several files of intermediate annotation (output from RepeatMasker, ESTprep, and BLAST). In all, these files average approximately 100KB to store per sequence.

REFERENCES

1. Practical data integration in biopharmaceutical R&D: Strategies and technologies. White paper, 3rd Millennium Inc., 125 Cambridge Park Drive, Cambridge, MA 02140, 2002.
2. S. Abiteboul, P. Buneman, and D. Suciu. *Data on the Web: From Relations to Semistructured Data and XML.* Morgan Kaufmann, 1999.
3. S. Abiteboul, R. Hull, and V. Vianu. *Foundations of Databases.* Addison-Wesley, 1995.
4. J. P. Abrahams, M. van den Berg, E. van Batenburg, and C. Pleij. Prediction of RNA secondary structure, including pseudoknotting, by computer simulation. *Nucleic Acids Research*, 18:3035–3044, 1990.
5. F. Achard, G. Vaysseix, and E. Barillot. XML, bioinformatics and data integration. *Bioinformatics*, 17(2):115–125, 2001.
6. J. Adachi and M. Hasegawa. *MOLPHY Version 2.3: Programs for Molecular Phylogenetics Based on Maximum Likelihood.* Number 28 in Computer Science Monographs. Institute of Statistical Mathematics, Tokyo, 1996.
7. M. D. Adams, S. E. Celniker, R. A. Holt, C. A. Evans, J. D. Gocayne, P. G. Amanatides, S. E. Scherer, et al. The genome sequence of *Drosophila melanogaster. Science*, 287(5461):2185–2195, 2000.
8. M. D. Adams, J. M. Kelley, J. D. Gocayne, M. Dubnick, M. H. Polymeropoulos, H. Xiao, C. R. Merril, A. Wu, B. Olde, R. F. Moreno, A. R. Kerlavage, W. R. Mc-Combie, and Venter J.C. Complementary DNA sequencing: Expressed sequence tags and the human genome project. *Science*, 252:1651–1656, 1991.
9. P. Agarwal and V. Bafna. The ribosome scanning model for translation initiation: Implications for gene prediction and full-length cDNA detection. In *Proceedings of 6th International Conference on Intelligent Systems for Molecular Biology*, pages 2–7, 1998.
10. R. Agrawal, T. Imielinski, and A. Swami. Mining association rules between sets of items in large databases. In *Proceedings of ACM-SIGMOD International Conference on Management of Data*, pages 207–216, 1993.
11. R. Agrawal, H. Mannila, R. Srikant, H. Toivonen, and A. I. Verkamo. Fast discovery of association rules. In *Advances in knowledge discovery and data mining*, pages 1–34. MIT Press, 1996.
12. R. Agrawal and S. Srikant. Fast algorithms for mining association rules. In *Proceedings of 20th International Conference on Very Large Data Bases*, pages 487–499, 1994.

13. B. Aguado and R. D. Campbell. Characterization of a human lysophosphatidic acid acyltransferase that is encoded by a gene located in the class III region of the human major histocompatibility complex. *Journal of Biological Chemistry*, 273:4096–4105, 1998.

14. Y. Akiyama and M. Kanehisa. NeuroFold: An RNA secondary structure prediction system using a Hopfi eld neural network. In *Proceedings of 3rd Genome Informatics Workshop*, 1992. In Japanese.

15. T. Akutsu. Dynamic programming algorithm for RNA secondary structure prediction with pseudoknots. *Discrete Applied Mathematics*, pages 45–62, 2000.

16. T. Akutsu, S. Miyano, and S. Kuhara. Algorithms for inferring qualitative models of biological networks. In *Proceedings of Pacifi c Symposium on Biocomputing 2000*, pages 293–304, 2000.

17. T. Akutsu and S. Miyano. Identifi cation of genetic networks from a small number of gene expression patterns under the boolean network model. In *Proceedings of Pacifi c Symposium on Biocomputing'99*, pages 17–28, 1999.

18. B. Alberts, A. Johnson, J. Lewis, M. Raff, K. Roberts, and P. Walter. *Molecular Biology of the Cell*. Garland Publishing, 4th edition, 2002.

19. N. N. Alexandrov and A. A. Mironov. Application of a new method of pattern recognition in DNA sequence analysis: A study of *E. coli* promoters. *Nucleic Acids Research*, 18:1847–1852, 1990.

20. A. A. Alizadeh, M. B. Eisen, R. E. Davis, C. Ma, I. S. Lossos, A. Rosenwald, J. C. Boldrick, H. Sabet, T. Tran, X. Yu, et al. Distinct types of diffuse large B-cell lymphoma identifi ed by gene expression profi ling. *Nature*, 403:503–511, 2000.

21. U. Alon, N. Barkai, D. A. Notterman, K. Gish, S.Y.D. Mack, and J. Levine. Broad patterns of gene expression revealed by clustering analysis of tumor colon tissues probed by oligonucleotide arrays. *Proc. Natl. Acad. Sci. USA*, 96:6745–6750, 1999.

22. P. Aloy and R. B. Russell. Interrogating protein interaction networks through structural biology. *Proc. Natl. Acad. Sci. USA*, 99(9):5896–5901, 2002.

23. S. F. Altschul, W. Gish, W. Miller, E. W. Myers, and D. J. Lipman. Basic local alignment search tool. *Journal of Molecular Biology*, 215:403–410, 1990.

24. S. F. Altschul, T. L. Madden, A. A. Schaffer, J. Zhang, Z. Zhang, W. Miller, and D. J. Lipman. Gapped BLAST and PSI-BLAST: A new generation of protein database search programs. *Nucleic Acids Research*, 25(17):3389–3402, 1997.

25. M. A. Andrade, S. I. O'Donoghue, and B. Rost. Adaptation of protein surfaces to subcellular location. *Journal of Molecular Biology*, 276:517–525, 1998.

26. M. A. Andrade and C. Sander. Bioinformatics: From genome data to biological knowledge. *Current Opinion in Biotechnology*, 8:675–683, 1997.

27. Angiosperm Phylogeny Group. An ordinal classifi cation for the families of owering plants. *Annals of the Missouri Botanical Garden*, 85:531–553, 1998.

28. D. R. Appling. Genetic approaches to the study of protein-protein interactions. *Methods*, 19(2):338–349, 1999.

29. R. Apweiler, T. K. Attwood, A. Bairoch, A. Bateman, E. Birney, M. Biswas, P. Bucher, L. Cerutti, F. Corpet, M. D. R. Croning, R. Durbin, L. Falquet, W. Fleischmann, J. Gouzy, H. Hermjakob, N. Hulo, I. Jonassen, D. Kahn, A. Kanapin, Y. Karavidopoulou, R. Lopez, B. Marx, N. J. Mulder, T. M Oinn, M. Pagni, F. Servant, C. J. A. Sigrist, and E. M. Zdobnov. The InterPro database, an integrated documen-

tation resource for protein families, domains, and functional sites. *Nucleic Acids Research*, 29:37–40, 2001.

30. R. Apweiler, T. K. Attwood, A. Bairoch, A. Bateman, E. Birney, M. Biswas, P. Bucher, L. Cerutti, L. Corpet, M. D. Croning, R. Durbin, L. Falquet, W. Fleischman, J. Gouzy, H. Hermjakob, N. Hulo, I. Jonassen, D. Kahn, A. Kanapin, Y. Karavidopoulou, R. Lopez, B. Marx, N. J. Mulder, T. M. Oinn, M. Pagni, F. Servant, C. J. Sirgist, and E. M. Zdobnov. InterPro—an integrated documentation resource for protein families, domains, and functional sites. *Bioinformatics*, 16:1145–1150, 2000.

31. Arabidopsis Genome Initiative. Analysis of the genome sequence of the owering plant *Arabidopsis thaliana*. *Nature*, 408:796–815, 2000.

32. E. Ardini, R. Agresti, E. Tagliabue, M. Greco, P. Aiello, L. T. Yang, S. Menard, and J. Sap. Expression of protein tyrosine phosphatase alpha (RPTPalpha) in human breast cancer correlates with low tumor grade, and inhibits tumor cell growth *in vitro* and *in vivo*. *Oncogene*, 19(43):4979–4987, 2000.

33. S. A. Armstrong, J. E. Staunton, L. B. Silverman, R. Pieters, M. L. den Boer, M. D. Minden, S. E. Sallan, E. S. Lander, T. R. Golub, and S. J. Korsmeyer. MLL translocations specify a distinct gene expression profi le that distinguishes a unique leukemia. *Nature Genetics*, 30:41–47, 2002.

34. G. E. Arnold, A. K. Dunker, S. L. Johns, and R. J. Douthart. Use of conditional probabilities for determining relationships between amino acid sequence and protein secondary structure. *Proteins*, 12(4):382–399, 1992.

35. M. Ashburner et al. Gene ontology: Tool for the unifi cation of biology. *Nature Genetics*, 25(1):25–29, 2000.

36. W. R. Atchley and W. M. Fitch. A natural classifi cation of the basic helix-loop-helix class of transcription factors. *Proc. Natl. Acad, Sci. USA*, 94:5172–5176, 1997.

37. T. K. Attwood, M. J. Blythe, D. R. Flower, A. Gaulton, J. E. Mabey, N. Maudling, L. McGregor, A. L. Mitchell, G. Moulton, K. Paine, and P. Scordis. PRINTS and PRINTS-S shed light on protein ancestry. *Nucleic Acids Research*, 30:239–241, 2002.

38. S. Audic and J.-M. Claverie. Detection of eukaryotic promoters using Markov transition matrices. *computer & Chemistry*, 21(4):223–227, 1997.

39. T. B. Bailey and C. Elkan. Fitting a mixture model by expectation maximization to discover motifs. *Intelligent Systems for Molecular Biology*, 2:28–36, 1994.

40. A. Bairoch. The ENZYME database in 2000. *Nucleic Acids Research*, 28:304–305, 2000.

41. A. Bairoch and R. Apweiler. The SWISS-PROT protein sequence database and its supplement TrEMBL in 2000. *Nucleic Acids Research*, 28(1):45–48, 2000.

42. I. V. Bajić. *Digital Signal Processing Techniques in the Analysis of DNA/RNA and Protein Sequences*. Bsceng thesis, University of Natal, Durban, South Africa, 1998.

43. V. B. Bajić and I. V. Bajić. *ANN in DNA regulatory region recognitions: The case of promoters*. International Joint Conference on Neural Networks, Washington, DC, tutorial, cd edition, July 1999.

44. V. B. Bajić and I. V. Bajić. Neural network system for promoter recognition. In *Future Directions for Intelligent Systems and Information Science*, chapter 14, pages 288–305. Physica-Verlag, 2000.

45. V. B. Bajić, I. V. Bajić, and W. Hide. A new method of spectral analysis of DNA/RNA and protein sequences. In *Proceedings of 1st International Conference on Bioinformatics of Genome Regulation and Structure*, volume 1, pages 120–123, 1998.

46. V. B. Bajić and S. H. Seah. Dragon Gene Start Finder: An advanced system for fi nding approximate locations of the start of gene transcriptional units. *Genome Research*, 13:1923–1929, 2003.

47. V. B. Bajić and S. H. Seah. Dragon Gene Start Finder identifi es approximate locations of the 5' ends of genes. *Nucleic Acids Research*, 31(13):3560–3563, 2003.

48. V. B. Bajić, S. H. Seah, A. Chong, S. P. T. Krishnan, J. L. Y. Koh, and V. Brusic. Computer model for recognition of functional transcription start sites in polymerase II promoters of vertebrates. *Journal of Molecular Graphics & Modeling*, 21(5):323–332, 2003.

49. V. B. Bajić, A. Chong, S. H. Seah, and V. Brusic. Intelligent system for vertebrate promoter recognition. *IEEE Intelligent Systems*, 17(4):64–70, 2002.

50. V. B. Bajić. Comparing the success of different prediction software in sequence analysis: A review. *Briefi ngs in Bioinformatics*, 1(3):214–228, 2000.

51. V. B. Bajić, S. H. Seah, A. Chong, G. Zhang, J. L. Y. Koh, and V. Brusic. Dragon Promoter Finder: Recognition of vertebrate RNA polymerase II promoters. *Bioinformatics*, 18(1):198–199, 2002.

52. P. G. Baker and A. Brass. Recent development in biological sequence databases. *Current Opinion in Biotechnology*, 9:54–58, 1998.

53. P. Baldi. Computing with arrays of bell shaped and sigmoid functions. In *Proceedings of IEEE Neural Information Processing Systems*, volume 3, pages 728–734, 1990.

54. P. Baldi, S. Brunak, Y. Chauvin, and A. Krogh. Hidden Markov models for human genes: Periodic patterns in exon sequences. In *Theoretical and Computational Methods in Genome Research*, pages 15–32, 1997.

55. P. Baldi and Y. Chauvin. Hidden Markov models of G-protein-coupled receptor family. *Journal of Computational Biology*, 1:311–335, 1994.

56. P. Baldi and S. Brunak. *Bioinformatics: The Machine Learning Approach*. MIT Press, 1999.

57. P. Baldi and S. Brunak. *Bioinformatics: Adaptive Computation and Machine learning*. MIT Press, 2nd edition, 2001.

58. P. Baldi and A. D. Long. A Bayesian framework for the analysis of microarray expression data: regularized t-test and statistical inferences of gene changes. *Bioinformatics*, 17(6):509–519, 2001.

59. J. D. Banfi eld and A. E. Raftery. Model-based Gaussian and non-Gaussian clustering. *Biometrics*, 49:803–822, 1993.

60. R. E. Banks, M. J. Dunn, D. F. Hochstrasser, J. C. Sanchez, W. Blackstock, D. J. Pappin, and P. J. Selby. Proteomics: New perspectives, new biomedical opportunities. *Lancet*, 356:1749–1756, 2000.

61. H. Bannai, Y. Tamada, O. Maruyama, K. Nakai, and S. Miyano. Extensive feature detection of N-terminal protein sorting signals. *Bioinformatics*, 18(2):298–305, 2002.

62. W. C. Barker, F. Pfeiffer, and D. G. George. Superfamily classifi cation in PIR international protein sequence database. *Methods in Enzymology*, 266:59–71, 1996.

63. O. A. Bashkirov, E. M. Braverman, and I. B. Muchnik. Theoretical foundations of the potential function method in pattern recognition learning. *Automation and Remote Control*, 25:629–631, 1964.

64. A. Bateman, E. Birney, L. Cerruti, R. Durbin, L. Etwiller, S. R. Eddy, S. Griffiths-Jones, K. L. Howe, M. Marshall, and E. L. L. Sonnhammer. The Pfam protein families database. *Nucleic Acids Research*, 30(1):276–280, 2002.

65. A. Bateman, E. Birney, R. Durbin, S.R. Eddy, K. L. Howe, and E. L. L. Sonnhammer. The Pfam protein families database. *Nucleic Acids Research*, 28:263–266, 2000.

66. A. Bateman, E. Birney, R. Durbin, S. R. Eddy, R. D. Finn, and E. L. L. Sonnhammer. Pfam 3.1: 1313 multiple alignments and profile HMMs match the majority of proteins. *Nucleic Acids Research*, 27(1):260–262, 1999.

67. A. Bauer and B. Kuster. Affinity purification-mass spectrometry: Powerful tools for the characterization of protein complexes. *European Journal of Biochemistry*, 270(4):570–578, 2003.

68. S. M. Baxter and J. S. Fetrow. Sequence- and structure-based protein function prediction from genomic information. *Current Opinion in Drug Discovery & Development*, 4:291–295, 2001.

69. R. J. Bayardo, R. Agrawal, and D. Gunopulos. Constraint-based rule mining in large dense databases. In *Proceedings of 15th International Conference on Data Engineering*, pages 188–197, 1999.

70. R. J. Bayardo. Efficiently mining long patterns from databases. In *Proceedings of ACM SIGMOD International Conference on Management of Data*, pages 85–93, 1998.

71. C. R. Beals, C. M. Sheridan, C. W. Turck, P. Gardner, and G. R. Crabtree. Nuclear export of nf-atc enhanced by glycogen synthase kinase-3. *Science*, 275:1930–1933, 1997.

72. E. Beaudoing, S. Freier, J. R. Wyatt, J.-M. Claverie, and D. Gautheret. Patterns of variant polyadenylation signal usage in human genes. *Genome Research*, 10:1001–1010, 2000.

73. A. Ben-Dor, R. Shamir, and Z. Yakhini. Clustering gene expression patterns. *Journal of Computational Biology*, 6:281–297, 1999.

74. A. Bender and J. R. Pringle. Use of a screen for synthetic lethal and multicopy suppressee mutants to identify two new genes involved in morphogenesis in *Saccharomyces cerevisiae*. *Molecular and Cellular Biology*, 11(3):1295–1305, 1991.

75. C. Benoist, K. O'Hare, R. Breathnach, and P. Chambon. The ovalbumin gene—sequence of putative control regions. *Nucleic Acids Research*, 8:127–142, 1980.

76. D. A. Benson, I. Karsch-Mizrachi, D. J. Lipman, J. Ostell, B. A. Rapp, and D. L. Wheeler. GenBank. *Nucleic Acids Research*, 28:15–18, 2000.

77. D. A. Benson, I. Karsch-Mizrachi, D. J. Lipman, J. Ostell, B. A. Rapp, and D. L. Wheeler. Genbank. *Nucleic Acids Research*, 30:17–20, 2002.

78. D. Benton. Recent changes in the GenBank on-line services. *Nucleic Acids Research*, 18(6):1517–1520, 1990.

79. C. E. Bessey. The phylogenetic taxonomy of owering plants. *Annals of the Missouri Botanical Garden*, 2:109–164, 1915.

80. J. C. Bezdek. *Fuzzy Mathematics in Pattern Classification*. Cornell University, Ithaca, N. Y., 1973.

81. A. P. Bird, M. H. Taggart, R. D. Nicholas, and D. R. Higgs. Non-methylated CpG-rich islands at the human alpha-globin locus: Implications for evolution of the alpha-globin pseudogene. *EMBO Journal*, 6:999–1004, 1986.

82. C. M. Bishop. *Neural Networks for Pattern Recognition*. Clarendon Press, 1995.

83. M. Bittner, P. Meltzer, Y. Chen, Y. Jiang, E. Seftor, M. Hendrix, M. Radmacher, R. Simon, et al. Molecular classification of cutaneous malignant melanoma by gene expression profiling. *Nature*, 406:536–540, 2000.

84. C. L. Blake and C. J. Marz. UCI machine learning repository, 1998. See `http://www.ics.uci.edu/~mlearn/MLRepository.html`.

85. T. Blumenthal. Gene clusters and polycistronic transcription in eukaryotes. *Bioessays*, 20(6):480–487, 1998.

86. B. Boeckmann, A. Bairoch, R. Apweiler, M.-C. Blatter, A. Estreicher, E. Gasteiger, M. J. Martin, K. Michoud, C. O'Donovan, I. Phan, S. Pilbout, and M. Schneider. The SWISS-PROT protein knowledgebase and its supplement TrEMBL in 2003. *Nucleic Acids Research*, 31(1):365–370, 2003.

87. M. S. Boguski, T. M. J. Lowe, and C. M. Tolstoshev. dbEST—database for 'expressed sequence tags". *Nature Genetics*, 4:332–333, 1993.

88. M. S. Boguski and C. M. Tolstoshev. Gene discovery in dbEST. *Science*, 265:1993–1994, 1994.

89. M. V. Boland and R. F. Murphy. A neural network classifier capable of recognizing the patterns of all major subcellular structures in uorescence microscope images of HeLa cells. *Bioinformatics*, 17(12):1213–1223, 2001.

90. M. F. Bonaldo, G. Lennon, and M. B. Soares. Normalization and substraction: Two approaches to facilitate gene discovery. *Genome Research*, 6:791–806, 1996.

91. P. Bork and E. Koonin. Ready for a motif submission? a proposed checklist. *Trends in Biochemical Sciences*, 20:104, 1995.

92. P. Bork and E. V. Koonin. Predicting functions from protein sequences—where are the bottlenecks? *Nature Genetics*, 18:313–318, 1998.

93. P. Bork, T. Dandekar, Y. Diaz-Lazcoz, F. Eisenhaber, M. Huynen, and Y. Yuan. Predicting function: From genes to genomes and back. *Journal of Molecular Biology*, 283:707–725, 1998.

94. M. Borodovsky and J. D. McIninch. GENEMARK: Parallel gene recognition for both DNA strands. *Computers and Chemistry*, 17(2):123–133, 1993.

95. M. Borodovsky, J. D. McIninch, E.V. Koonin, K.E. Rudd, C. Medigue, and A. Danchin. Detection of new genes in a bacterial genome using Markov models for three gene classes. *Nucleic Acids Research*, 23:3554–3562, 1995.

96. N. K. Bose and P. Liang. *Neural Network Fundamentals with Graphs, Algorithms, and Applications*. McGraw-Hill, 1996.

97. H. Bourlard and N. Morgan. Continuous speech recognition by connectionist statistical methods. *IEEE Transactions on Neural Networks*, 4(6):893–909, 1993.

98. P. S. Bradley, U. M. Fayyad, and C. A. Reina. Scaling EM (expectation-maximization) clustering to large databases. Technical Report MSR-TR-98-35, Microsoft Research, November 1998.

99. R. K. Brayton, S. W. Director, G. D. Hachtel, and L. Vidigal. A new algorithm for statistical circuit design based on quasi-Newton methods and function splitting. *IEEE Transactions on Circuits and Systems*, CAS-26:784–794, 1979.

100. P. Brazhnik, A. de la Fuente, and P. Mendes. Gene networks: How to put the function in genomics. *Trends in Biotechnology*, 20:467–472, 2002.

101. L. Breiman, L. Friedman, R. Olshen, and C. Stone. *Classifi cation and Regression Trees*. Wadsworth and Brooks, 1984.

102. B. Brejová, D. Brown, and T. Vinař. Optimal spaced seeds for hidden Markov models, with application to homologous coding regions. In *Proceedings of 14th Annual Symposium on Combinatorial Pattern Matching*, pages 42–54, 2003.

103. B. Brejová, D. Brown, and T. Vinař. Vector seeds: An extension to spaced seeds allows substantial improvements in sensitivity and specifi city. In *Proceedings of 3rd Annual Workshop on Algorithms in Bioinformatics*, pages 39–54, 2003.

104. S. E. Brenner. Errors in genome annotation. *Trends in Genetics*, 15(4):132–133, 1999.

105. S. Brin, R. Motwani, and C. Silverstein. Beyond market basket: Generalizing association rules to dependence rules. *Data Mining and Knowledge Discovery*, 2:39–68, 1998.

106. S. Brin, R. Motwani, J. Ullman, and S. Tsur. Dynamic itemset counting and implication rules for market basket data. In *Proceedings of ACM-SIGMOD International Conference on Management of Data*, pages 255–264, 1997.

107. R. J. Brooker. *Genetics: Analysis and Principles*. Addison-Wesley, 1999.

108. B. R. Brooks, R. E. Bruccokeri, B. D. Olafson, D. J. States, S. Swaminathan, and M. Karplus. CHARMM: A program for macromolecular energy, minimization, and dynamics calculations. *Journal of Computational Chemistry*, 4:187–217, 1983.

109. D. Broomhead and D. Lowe. Multivariable function interpolation and adaptive networks. *Complex Systems*, 2:321–355, 1988.

110. M. Brown and C. Wilson. RNA pseudoknot modeling using intersections of stochastic context free grammars with applications to database search. In *Proceedings of Pacifi c Symposium on Biocomputing*, pages 109–125, 1996.

111. M. P. Brown, W. N. Grundy, D. Lin, N. Cristianini, C. W. Sugnet, T. S. Furey, M. Ares Jr, and D. Haussler. Knowledge-based analysis of microarray gene expression data by using support vector machines. *Proc. Natl. Acad. Sci. USA*, 97(1):262–267, 2000.

112. N. P. Brown, C. Leroy, and C. Sander. MView: A web-compatible database search for multiple alignment viewer. *Bioinformatics*, 14:380–381, 1998.

113. R. K. Brummitt, editor. *Vascular Plant Families and Genera*. Royal Botanical Gardens, Kew, England, 1992.

114. S. J. Brunak, J. Engelbrecht, and S. Knudsen. Prediction of human mRNA donor and acceptor sites from the DNA sequence. *Journal of Molecular Biology*, 220:49–65, 1991.

115. P. Bucher and E. N. Trifonov. Compilation and analysis of eukaryotic Pol II promoter sequences. *Nucleic Acids Research*, 14:10009–10026, 1986.

116. P. Bucher. Weight matrix description of four eukaryotic RNA polymerase II promoter elements derived from 502 unrelated promoter sequences. *Journal of Molecular Biology*, 212:563–578, 1990.

117. J. Buhler. Effi cient large-scale sequence comparison by locality-sensitive hashing. *Bioinformatics*, 17:419–428, 2001.

118. J. Buhler, U. Keich, and Y. Sun. Designing seeds for similarity search in genomic DNA. In *Proceedings of 7th Annual International Conference on Computational Biology*, pages 67–75, 2003.

119. J. Buhler and M. Tompa. Finding motifs using random projections. In *Proceedings of 5th Annual International Conference on Computational Biology*, pages 69–76, 2001.

120. F. R. Burden and D. A. Winkler. A quantitative structure-activity relationships model for the acute toxicity of substituted benzenes to *tetrahymena pyriformis* using Bayesian-regularised neural networks. *Chemical Research in Toxicology*, 13:436–440, 2000.

121. C. Burge. *Identification of genes in human genomic DNA*. PhD thesis, Stanford University, Stanford, CA, 1997.

122. C. Burge and S. Karlin. Finding the genes in genomic DNA. *Current Opinion on Structural Biology*, 8:346–354, 1998.

123. C. Burge and S. Karlin. Prediction of complete gene structures in human genomic DNA. *Journal of Molecular Biology*, 268:78–94, 1997.

124. C. J. C. Burges. A tutorial on support vector machines for pattern recognition. *Data Mining and Knowledge Discovery*, 2(2):121–167, 1998.

125. S. K. Burley and R. G. Roeder. Biochemistry and structural biology of transcription factor IID (TFIID). *Annual Review of Biochemistry*, 65:769–799, 1996.

126. D. M. Burns, V. Horn, J. Paluh, and C. Yanofsky. Evolution of the tryptophan synthetase of fungi: Analysis of experimentally fused *Escherichia coli* tryptophan synthetase alpha and beta chains. *Journal of Biological Chemistry*, 265(4):2060–2069, 1990.

127. M. Burset and R. Guigo. Evaluation of gene structure prediction programs. *Genomics*, 34:353–367, 1996.

128. C. elegans Sequencing Consortium. Genome sequence of the nematode *C. elegans*: A platform for investigating biology. *Science*, 282(5396):2012–2018, 1998.

129. A. Caccone and J. R. Powell. DNA divergence among hominoids. *Evolution*, 43:925–942, 1989.

130. Y.-D. Cai and K.-C. Chou. Nearest neighbour algorithm for predicting protein subcellular location by combining functional domain composition and pseudo-amino acid composition. *Biochemical and Biophysical Research Communications*, 305:407–411, 2003.

131. Y.-D. Cai, X.-J. Liu, X. Xu, and K.-C. Chou. Support vector machines for prediction of protein subcellular location. *Molecular Cell Biology Research Communications*, 4:230–233, 2001.

132. C. R. Cantor. Orchestrating the human genome project. *Science*, 248:49–51, 1990.

133. M. Caria. *Measurement Analysis: An Introduction to the Statistical Analysis of Laboratory Data in Physics, Chemistry, and the Life Sciences*. Imperial College Press, 2000.

134. R. Casadio, P. Fariselli, G. Finocchiaro, and P. L. Martelli. Fishing new proteins in the twilight zone of genomes: The test case of outer membrane proteins in *escherichia coli* k12, *escherichia coli* o157:h7, and other gram-negative bacteria. *Protein Science*, 12:1158–1168, 2003.

135. M. Caudill. *Neural Networks Primer*. Miller Freeman Publications, 1989.

136. J. Cedano, J. A. Pérez-Ponsa, and E. Querol. Relation between amino acid composition and cellular location of proteins. *Journal of Molecular Biology*, 266(3):594–600, 1997.

137. S. M. Chabregas, D. D. Luche, and L. P. Farias. Dual targeting properties of the N-terminal signal sequence of *arabidopsis thaliana* thi1 protein to mitochondria and chloroplasts. *Plant Molecular Biology*, 46:639–650, 2001.

138. D. M. Chao and R. A. Young. Activation without a vital ingredient. *Nature*, 383:119–120, 1996.

139. M. W. Chase, D. E. Soltis, R. G. Olmstead, D. Morgan, D. H. Les, B. D. Mishler, M. R. Duvall, R. A. Price, H. G. Hills, Y.-L. Qui, K. A. Kron, J. H. Rettig, E. Conti, J. D. Palmer, J. R. Manhart, K. J. Sytsma, H. J. Michaels, H. J. Kress, W. J. Karol, K. G. Clark, M. Hedrén, B. S. Gaut, R. K. Jansen, K.-J. Kim, C. F. Wimpee, J. F. Smith, G. R. Furnier, S. H. Strauss, Q.-Y. Xiang, G. M. Plunkett, P. S. Soltis, S. M. Swensen, S. E. Williams, P. A. Gadek, C. J. Quinn, L. E. Eguiarte, E. Golenberg, G. H. Learns, S. W. Graham, S. C. H. Barrett, S. Dayanandan, and V. Albert. Phylogenetics of seed plants: An analysis of nucleotide sequences from the plastid gene rbcL. *Annals of the Missouri Botanical Garden*, 80:528–580, 1993.

140. Y. Chauvin. A back-propagation algorithm with optimal use of hidden units. *Advances in Neural Information Processing Systems*, 1:519–526, 1989.

141. Y. Chauvin and D. Rumelhart. *Backpropagation: Theory, Architectures, and Applications*. Lawrence Erlbaum, 1995.

142. M. Chee, R. Yang, E. Hubbell, A. Berno, X. C. Huang, D. Stern, J. Winkler, D. J. Lockhart, M. S. Morris, and S. P. Fodor. Accessing genetic information with high-density DNA arrays. *Science*, 274(5287):610–614, 1996.

143. C. H. Chen, editor. *Fuzzy Logic and Neural Network Handbook*. McGraw-Hill, 1996.

144. F. Chen, C. C. MacDonald, and J. Wilusz. Cleavage site determinants in the mammalian polyadenylation signal. *Nucleic Acids Research*, 23:2614–2620, 1995.

145. I.-M. A. Chen and Victor M. Markowitz. An overview of the object-protocol model (OPM) and OPM data management tools. *Information Systems*, 20(5):393–418, 1995.

146. J. Chen, N.-H. Chua, D. Strauss, and L. Wong. Extracting Kozak consensus sequence using Kleisli. In *Proceedings of 1st International Conference on Bioinformatics of Genome Regulation and Structure*, pages 218–223, 1998.

147. J. Chen, D. Strauss, and L. Wong. Using Kleisli to bring out features in BLASTP results. In *Genome Informatics 1998*, pages 102–111, 1998.

148. J. Chen, L. Zhang, and L. Wong. A protein patent query system powered by Kleisli. In *Proceedings of ACM SIGMOD International Conference on Management of Data*, pages 593–595, 1998.

149. P. P. S. Chen. The entity-relationship model: Toward a unified view of data. *ACM Transaction on Database Systems*, 1(1):9–36, 1976.

150. Q. Chen, G. Hertz, and G. D. Stormo. PromFD 1.0: A computer program that predicts eukaryotic pol II promoters using strings and IMD matrices. *Computer Applications in the Biosciences*, 13:29–35, 1997.

151. Q. Chen, G. Z. Hertz, and G. D. Stormo. MATRIX SERACH 1.0: A computer program that scans DNA sequences for transcriptional elements using a database of weight matrices. *Computer Applications in the Biosciences*, 13:29–35, 1995.

152. S. Chen, C. F. N. Cowan, and P. M. Grant. Orthogonal least squares learning algorithm for radial basis function network. *IEEE Transactions on Neural Networks*, 2(2):302–309, 1991.

153. T. Chen, H. L. He, and G. M. Church. Modeling gene expression with differential equations. In *Proceedings of Pacific Symposium on Biocomputing'99*, pages 29–40, 1999.

154. V. Cherkassky and F. Mulier. *Learning from Data*. John Wiley & Sons, 1998.

155. D. Cheung, J. Han, V. Ng, A. Fu, and Y. Fu. A fast distributed algorithm for mining association rules. In *Proceedings of 4th International Conference on Parallel and Distributed Information Systems*, pages 31–44, 1996.

156. G. Chiaromonte, V. B. Yap, and W. Miller. Scoring pairwise genomic sequence alignments. In *Proceedings of Pacific Symposium on Biocomputing*, pages 115–126, 2002.

157. R. J. Cho, M. J. Campbell, E. A. Winzeler, L. Steinmetz, A. Conway, L. Wodicka, T. G. Wolfsberg, A. E. Gabrielian, D. Landsman, D. J. Lockhart, and R. W. Davis. A genome-wide transcriptional analysis of the mitotic cell cycle. *Molecular Cell*, 2:65–73, 1998.

158. K. P. Choi and L. Zhang. Sensitive analysis and efficient method for identifying optimal spaced seeds. *Journal of Computer and System Sciences*, 2003. To appear.

159. K.-C. Chou. Prediction of protein cellular attributes using pseudo-amino acid composition. *PROTEINS: Structure, Function and Genetics*, 44:60, 2001.

160. K.-C. Chou. Using subsite coupling to predict signal peptides. *Protein Engineering*, 14(2):75–79, 2001.

161. K.-C. Chou and Y.-D. Cai. Using functional domain composition and support vector machines for prediction of protein subcellular location. *Journal of Biological Chemistry*, 48:45765–45769, 2002.

162. S. Y. Chung and L. Wong. Kleisli, a new tool for data integration in biology. *Trends in Biotechnology*, 17(9):351–355, 1999.

163. P. Cirri, P. Chiarugi, G. Camici, G. Manao, G. Raugei, G. Cappugi, and G. Ramponi. The role of Cys12, Cys17, and Arg18 in the catalytic mechanism of low-m(r) cytosolic phosphotyrosine protein phosphatase phosphatase. *European Journal of Biochemistry*, 214:647–657, 1993.

164. M. G. Claros, S. Brunak, and G. von Heijne. Prediction of N-terminal protein sorting signals. *Current Opinion in Structural Biology*, 7:394–398, 1997.

165. M. G. Claros and P. Vincens. Computational method to predict mitochondrially imported proteins and their targeting sequences. *European Journal of Biochemistry*, 241:779–786, 1996.

166. J. Claverie. Computational methods for the identification of genes in vertebrate genomic sequences. *Human Molecular Genetics*, 6(10):1735–1744, 1997.

167. J. Claverie, O. Poirot, and F. Lopez. The difficulty of identifying genes in anonymous vertebrate sequences. *Computer & Chemistry*, 21(4):203–214, 1997.

168. J. Claverie and I. Sauvaget. Assessing the biological significance of primary structure consensus patterns using sequence databanks: I. Heat-shock and glucocorticoid control elements in eukaryotic promoters. *Computer Applications in the Biosciences*, 1:95–104, 1985.

169. D. Cochrane, C. Webster, G. Masih, and J. McCafferty. Identification of natural ligands for SH2 domains from a phage display cDNA library. *Journal of Molecular Biology*, 297(1):89–97, 2000.

170. E. F. Codd. A relational model for large shared data bank. *Communications of the ACM*, 13(6):377–387, 1970.

171. F. Collins and D. Galas. A new five-year plan for the U.S. Human Genome Project. *Science*, 262:43–46, 1993.

172. R. R. Copley, T. Doerks, I. Letunic, and P. Bork. Protein domain analysis in the era of complete genomes. *FEBS Letters*, 513:129–134, 2002.

173. J. Corden, B. Wasylyk, A. Buchwalder, P. Sassone-Corsi, C. Kedinger, and P. Chambon. Promoter sequence of eukaryotic protein-coding genes. *Science*, 209:1406–1414, 1980.

174. F. Corpet and B. Michot. RNAlign program: Alignment of RNA sequences using both primary and secondary structure. *Computer Applications in the Biosciences*, 7:347–352, 1994.

175. F. Corpet, F. Servant, J. Gouzy, and D. Kahn. ProDom and ProDom-CG: Tools for protein domain analysis and whole-genome comparisons. *Nucleic Acids Research*, 28:267–269, 2000.

176. C. Creighton and S. Hanash. Mining gene expression databases for association rules. *Bioinformatics*, 19(1):79–86, 2003.

177. F. H. Crick, L. Barnett, S. Brenner, and R. J. Watts-Tobin. General nature of the genetic code for proteins. *Nauchni trudove na Visshiia meditsinski institut, Sofia*, 192:1227–1232, 1961.

178. N. Cristianini and J. Shawe-Taylor. *An Introduction to Support Vector Machines.* Cambridge University Press, 2000.

179. A. Cronquist. *An Integrated System of Classification of Flowering Plants.* Columbia University Press, 1981.

180. S. H. Cross and A. P. Bird. CpG islands and genes. *Current Opinion in Genetics and Development*, 5:309–314, 1995.

181. S. H. Cross, V. H. Clark, and A. P. Bird. Isolation of CpG islands from large genomics clones. *Nucleic Acids Research*, 27:2099–2107, 1999.

182. M. Cserz'o, F. Eisenhaber, B. Eisenhaber, and I. Simon. On filtering false positive transmembrane protein predictions. *Protein Engineering*, 15(9):745–752, 2002.

183. E. Dam, K. Pleij, and D. Draper. Structural and functional aspects of RNA pseudoknots. *Biochemistry*, 31(47):11665–11676, 1992.

184. T. Dandekar, B. Snel, M. Huynen, and P. Bork. Conservation of gene order: A fingerprint of proteins that physically interact. *Trends in Biochemical Sciences*, 23(9):324–328, 1998.

185. R. Das, J. Junker, D. Greenbaum, and M. B. Gerstein. Global perspectives on proteins: Comparing genomes in terms of folds, pathways and beyond. *Pharmacogenomics Journal*, 1:115–125, 2001.

186. S. Datta. Exploring relationships in gene expressions: A partial least squares approach. *Gene Expression*, 9:257–264, 2001.

187. S. Datta and S. Datta. Comparisons and validation of statistical clustering techniques for microarray gene expression data. *Bioinformatics*, 19(4):459–466, 2003.

188. S. B. Davidson, C. Overton, and P. Buneman. Challenges in integrating biological data sources. *Journal of Computational Biology*, 2(4):557–572, 1995.

189. S. Davidson, C. Overton, V. Tannen, and L. Wong. BioKleisli: A digital library for biomedical researchers. *International Journal of Digital Libraries*, 1(1):36–53, 1997.

190. R. V. Davuluri, I. Grosse, and M. Q. Zhang. Computational identification of promoters and first exons in the human genome. *Nature Genetics*, 29(4):412–417, 2001.

191. M. O. Dayhoff. *Atlas of Protein Sequence and Structure, volumes 1–5, supplements 1–3*. National Biomedical Research Foundation, Washington, DC., 1965-1978.

192. M. O. Dayhoff, R. M. Schwartz, and B. C. Orcutt. A model of evolutionary change in proteins. *Atlas of Protein Sequence and Structure*, 5(Suppl. 3):345–352, 1978.

193. M. O. Dayhoff. The origin and evolution of protein superfamilies. *Federation Proceedings*, 35:2132–2138, 1976.

194. A. L. Delcher, S. Kasif, R. D. Fleischmann, J. Peterson, O. White, and S. L. Salzberg. Alignment of whole genomes. *Nucleic Acids Research*, 27:2369–2376, 1999.

195. G. Deleage, C. Combet, C. Blanchet, and C. Geourjon. ANTHEPROT: An integrated protein sequence analysis software with client/server capabilities. *Computers in Biology and Medicine*, 31:259–267, 2001.

196. B. Demeler and G. W. Zhou. Neural network optimization for *E.coli* promoter prediction. *Nucleic Acids Research*, 19:1593–1599, 1991.

197. A. P. Dempster, N. M. Laird, and D. B. Rubin. Maximum likelihood from incomplete data via the EM algorithm. *Journal of the Royal Statistical Society, Series B*, 39:1–38, 1977.

198. J. Demsar, B. Zupan, M.W. Kattan, J.R. Beck, and I. Bratko. Naive Bayesian-based nomogram for prediction of prostate cancer recurrence. *Studies in Health Technolnology Informatics*, 68:436–441, 1999.

199. M. Deng, S. Mehta, F. Sun, and T. Chen. Inferring domain-domain interactions from protein-protein interactions. *Genome Research*, 12(10):1540–1548, 2002.

200. J. M. Denu and J. E. Dixon. Protein tyrosine phosphatases: mechanisms of catalysis and regulation. *Current Opinion in Chemistry & Biology*, 2(5):633–641, 1998.

201. J. M. Denu, J. A. Stuckey, M. A. Saper, and J. E. Dixon. Form and function in protein dephosphorylation. *Cell*, 87:361–364, 1996.

202. S. J. DeRose. XQuery: A unified syntax for linking and querying general XML documents. In *Position Papers of QL'98—The Query Languages Workshop*, 1998.

203. D. Devos and A. Valencia. Intrinsic errors in genome annotation. *Trends in Genetics*, 17:429–431, 2001.

204. L. Devroye, L. Gyorfi, and G. Lugosi. *A Probabilistic Theory of Pattern Recognition*. Springer-Verlag, 1996.

205. P. D'haeseleer, X. Wen, S. Fuhrman, and R. Somogyi. Mining the gene expression matrix: Inferring gene relationships from large scale gene expression data. In *Information Processing in Cells and Tissues*, pages 203–212. Plenum, 1998.

206. K. Dolinski et al. Saccharomyces genome database, 2003. See http://www.yeastgenome.org.

207. G. Dong and J. Li. Efficient mining of emerging patterns: Discovering trends and differences. In *Proceedings of 5th ACM SIGKDD International Conference on Knowledge Discovery & Data Mining*, pages 15–18, 1999.

208. G. Dong, J. Li, and X. Zhang. Discovering jumping emerging patterns and experiments on real datasets. In *Proceedings of 9th International Database Conference on Heterogeneous and Internet Databases*, pages 15–17, 1999.

209. G. Dong, X. Zhang, L. Wong, and J. Li. CAEP: Classification by aggregating emerging patterns. In *Proceedings of 2nd International Conference on Discovery Science*, pages 30–42, 1999.

210. S. Dong and D. B. Searls. Gene structure prediction by linguistic methods. *Genomics*, 162:705–708, 1994.

211. T. A. Down and T. J. Hubbard. Computational detection and location of transcription start sites in mammalian genomic DNA. *Genome Research*, 12(3):458–461, 2002.

212. A. Drawid, R. Jansen, and M. Gerstein. Genome-wide analysis relating expression level with protein subcellular localization. *Trends in Genetics*, 16:426–430, 2000.

213. A. Drawid and M. Gerstein. A Bayesian system integrating expression data with sequence patterns for localizing proteins: Comprehensive application to the yeast genome. *Journal of Molecular Biology*, 301:1059–1075, 2000.

214. R. O. Duda and P. E. Hart. *Pattern Classification and Scene Analysis*. Wiley, 1973.

215. R. Durbin, S. R. Eddy, A. Krogh, and G. Mitchison, editors. *Biological Sequence Analysis: Probabilistics Models of Proteins and Nucleic Acids*. Cambridge University Press, 1998.

216. W. S. Dynan and R. Tjian. Control of eukaryotic messenger RNA synthesis by sequence-specific DNA-binding proteins. *Nature*, 316:774–778, 1985.

217. B. A. Eckman, J. S. Aaronson, J. A. Borkowski, W. J. Bailey, K. O. Elliston, A. R. Williamson, and R. A. Blevins. The Merck Gene Index browser: An extensible data integration system for gene finding, gene characterization, and EST data mining. *Bioinformatics*, 14(1):2–13, 1998.

218. A. Economou. Bacterial secretome: The assembly manual and operating instructions. *Molecular Membrane Biology*, 19(3):159–169, 2002.

219. S. R. Eddy. Hidden Markov models. *Current Opinion in Structural Biology*, 6:361–365, 1996.

220. S. R. Eddy and R. Durbin. RNA sequence analysis using covariance models. *Nucleic Acids Research*, 22:2079–2088, 1994.

221. S. R. Eddy, G. Mitchison, and R. Durbin. Maximum discrimination hidden Markov models of sequence consensus. *Journal of Computational Biology*, 2(9–23), 1995.

222. M. Edman, T. Jarhede, M. Sjöström, and A. Wieslander. Different sequence patterns in signal peptides from mycoplasmas, other gram-positive bacteria, and escherichia coli: A multivariate data analysis. *PROTEINS: Structure, Function, and Genetics*, 35:195–205, 1999.

223. A. Efstratiadis, J. W. Posakony, T. Maniatis, R. M. Lawn, C. O'Connell, R. A. Spritz, J. K. DeRiel, B. G. Forget, S. M. Weissman, J. L. Slightom, A. E. Blechl, O. Smithies, F. E. Baralle, C. C. Shoulders, and N. J. Proudfoot. The structure and evolution of the human beta-globin gene family. *Cell*, 21:653–668, 1980.

224. C. Ehresmann, F. Baudin, M. Mougel, P. Romby, J. P. Ebel, and B. Ehresmann. Probing the structure of RNAs in solution. *Nucleic Acids Research*, 15:9109–9112, 1987.

225. M. B. Eisen, P. T. Spellman, P. O. Brown, and D. Bostein. Cluster analysis and display of genome-wide expression patterns. *Proc. Nat. Acad. Sci. USA*, 95:14863–14868, 1998.

226. F. Eisenhaber, B. Eisenhaber, W. Kubina, S. Maurer-Stroh, G. Neuberger, G. Schneider, and M. Wildpaner. Prediction of lipid posttranslational modifications and localization signals from protein sequences: Big-Π, NMT and PTS1. *Nucleic Acids Research*, 31(13):3631–3634, 2003.

227. F. Eisenhaber and P. Bork. Wanted: Subcellular localization of proteins based on sequence. *Trends in Cell Biology*, 8:169–170, 1998.

228. F. Eisenhaber and P. Bork. Evaluation of human-readable annotation in biomolecular sequence databases with biological rule libraries. *Bioinformatics*, 15:528–535, 1999.

229. J. L. Elman. Finding structure in time. *Cognitive Science*, 14:179–211, 1990.

230. O. Emanuelsson, A. Elofsson, G. von Heijne, and S. Cristobal. In silico prediction of the peroxisomal proteome in fungi, plants, and animals. *Journal of Molecular Biology*, 330(2):443–456, 2003.

231. O. Emanuelsson, H. Nielsen, and G. von Heijne. ChloroP, a neural network-based method for predicting chloroplast transit peptides and their cleavage sites. *Protein Science*, 8(5):978–984, 1999.

232. O. Emanuelsson. Predicting protein subcellular localisation from amino acid sequence information. *Briefings in Bioinformatics*, 3(4):361–376, 2002.

233. A. J. Enright and C. A. Ouzounis. Functional associations of proteins in entire genomes by means of exhaustive detection of gene fusions. *Genome Biology*, 2(9):RESEARCH0034, 2001.

234. A. J. Enright, I. Iliopoulos, N. C. Kyrpides, and C. A. Ouzounis. Protein interaction maps for complete genomes based on gene fusion events. *Nature*, 402:86–90, 1999.

235. T. Etzold and P. Argos. SRS: Information retrieval system for molecular biology data banks. *Methods in Enzymology*, 266:114–128, 1996.

236. B. Ewing, L. Hillier, M. C. Wendl, and P. Green. Base-calling of automated sequencer traces using phred. I. Accuracy assessment. *Genome Research*, 8(3):175–178, 1998.

237. S. Faisst and S. Meyer. Compilation of vertebrate encoded transcription factors. *Nucleic Acids Research*, 20:3–16, 1992.

238. L. Falquet, M. Pagni, P. Bucher, N. Hulo, C. J. A. Sigrist, K. Hofmann, and A. Bairoch. The PROSITE database, its status in 2002. *Nucleic Acids Research*, 30(1):235–238, 2002.

239. J. B. Fant, C. D. Preston, and J. A. Barrett. Isozyme evidence of the parental origin and possible fertility of the hybrid *Potamogeton* x *uitans Roth*. *Plant Systematics and Evolution*, 229(1/2):45–57, 2001.

240. R. Farber, A. Lapedes, and K. Sirotkin. Determination of eukaryotic protein coding regions using neural networks and information theory. *Journal of Molecular Biology*, 226:471–479, 1992.

241. L. Fausett. *Fundamentals of Neural Networks*. Prentice-Hall, 1994.

242. U. Fayyad and K. Irani. Multi-interval discretization of continuous-valued attributes for classification learning. In *Proceedings of 13th International Joint Conference on Artificial Intelligence*, pages 1022–1029, 1993.

243. B. Fazi, M. J. Cope, A. Douangamath, et al. Unusual binding properties of the SH3 domain of the yeast actin-binding protein Abp1: Structural and functional analysis. *Journal of Biological Chemistry*, 277(7):5290–5298, 2002.

244. Z.-P. Feng. An overview on predicting the subcellular location of a protein. *In Silico Biology*, 2(3):291–303, 2002.

245. Z.-P. Feng and C.-T. Zhang. Prediction of the subcellular location of prokaryotic proteins based on the hydrophobicity index of amino acids. *International Journal of Biological Macromolecules*, 28:255–261, 2001.

246. M. F. Fernandez, A. Morishima, and D. Suciu. Efficient evaluation of XML middleware queries. In *Proceedings of ACM SIGMOD International Conference on Management of Data*, pages 103–114, 2001.

247. J. W. Fickett. Finding genes by computer: The state of the art. *Trends in Genetics*, 12(8):316–320, 1996.

248. J. W. Fickett and R. Guigo. Computational gene identification. In *Internet for Molecular Biologist*, pages 73–100. Horizon Scientific Press, 1996.

249. J. W. Fickett and A. G. Hatzigeorgiou. Eukaryotic promoter recognition. *Genome Research*, 7(9):861–878, 1997.

250. S. Fields and O. Song. A novel genetic system to detect protein-protein interactions. *Nature*, 340(6230):245–246, 1989.

251. R. A. Fisher. The use of multiple measurements in taxonomic problems. *Annals of Eugenics*, 7:179–188, 1936.

252. D. Florescu and D. Kossmann. Storing and querying XML data using RDBMS. *Data Engineering Bulletin*, 22(3):27–34, 1999.

253. S. P. Fodor, J. L. Read, M. C. Pirrung, L. Stryer, A. T. Lu, and D. Solas. Light-directed spatially addressable parallel chemical synthesis. *Science*, 251:767–773, 1991.

254. D. B. Fogel. *Evolutionary Computation*. IEEE Press, 2nd edition, 2000.

255. P. L. Forey, C. J. Humphries, I. J. Kitching, R. W. Scotland, D. J. Siebert, and D. M. Williams. *Cladistics: A Practical Course in Systematics*. Clarendon Press, 1992.

256. V. Di Francesco, J. Granier, and P.J. Munson. Protein topology recognition from secondary structure sequences—applications of the hidden Markov models to the alpha class proteins. *Journal of Molecular Biology*, 267:446–463, 1997.

257. S. Frank, A. Lustig, T. Schulthess, J. Engel, and R. A. Kammerer. A distinct seven-residue trigger sequence is indispensable for proper coiled-coil formation of the human macrophage scavenger receptor oligomerization domain. *Journal of Biological Chemistry*, 275:11672–11677, 2000.

258. H. Franke, P. Hochschild, P. Pattnaik, and M. Snir. An efficient implementation of MPI on IBM SP1. In *Proceedings of 23rd Annual International Conference on Parallel Processing*, volume 3, pages 197–201, 1994.

259. K. Frech, P. Dietze, and T. Werner. ConsInspector 3.0: New library and enhanced functionality. *Computer Applications in the Biosciences*, 13:109–110, 1997.

260. K. Frech, G. Herrmann, and T. Werner. Computer-assisted prediction, classification, and delimitation of protein binding sites in nucleic acids. *Nucleic Acids Research*, 21:1655–1664, 1993.

261. K. Frech, K. Quandt, and T. Werner. Muscle actin genes: A first step towards computational classification of tissue specific promoters. *In Silico Biology*, 1:29–38, 1998.

262. K. Frech and T. Werner. Specific modelling of regulatory units in DNA sequences. In *Proceedings of Pacific Symposium on Biocomputing*, pages 151–162, 1997.

263. N. Friedman, M. Linial, I. Nachman, and D. Pe'er. Using Bayesian networks to analyse expression data. *Journal of Computational Biology*, 7:601–620, 2000.

264. T.-T. Friess, N. Cristianini, and C. Campbell. The kernel adatron algorithm: A fast and simple learning procedure for support vector machines. In *Proceedings of 15th International Conference on Machine Learning*, 1998.

265. D. Frishman, A. Mironov, and M. Gelfand. Starts of bacterial genes: Estimating the reliability of computer predictions. *Gene*, 234:257–265, 1999.

266. K. J. Fryxell. The coevolution of gene family trees. *Trends in Genetics*, 12(9):364–369, 1996.

267. Y. Fujiwara, M. Asogawa, and K. Nakai. Prediction of mitochondrial targeting signals using hidden Markov models. In *Proceedings of 8th International Workshop on Genome Informatics*, pages 53–60, 1997.

268. Y. Fukunishi and Y. Hayashizaki. Amino-acid translation program for full-length cDNA sequences with frame-shift errors. *Physiological Genomics*, 5:81–87, 2001.

269. T. S. Furey, N. Cristianini, N. Duffy, D. W. Bednarski, M. Schummer, and D. Haussler. Support vector machine classification and validation of cancer tissue samples using microarray expression data. *Bioinformatics*, 16(10):906–914, 2000.

270. J. S. Garavelli, Z. Hou, N. Pattabiraman, and R. M. Stephens. The RESID database of protein structure modifications and the NRL-3D sequence-structure database. *Nucleic Acids Research*, 29:199–201, 2001.

271. M. Gardiner-Garden and M. Frommer. CpG islands in vertebrate genomes. *Journal of Molecular Biology*, 196:261–282, 1987.

272. J. L. Gardy, C. Spencer, K. Wang, M. Ester, G. E. Tusnády, I. Simon, S. Hua, K. deFays, C. Lambert, K. Nakai, and F. S. L. Brinkman. PSORT-B: Improving protein subcellular localization for gram-negative bacteria. *Nucleic Acids Research*, 31(13):3613–3617, 2003.

273. M. Garey and D. Johnson. *Computers and Intractability: A Guide to the Theory of $\mathcal{NP}$-completeness*. W. H. Freeman, 1979.

274. R. B. Gary and G. D. Stormo. Graph-theoretic approach to RNA modeling using comparative data. *Intelligent Systems for Molecular Biology*, 3:75–80, 1995.

275. A. P. Gasch, M. Huang, S. Metzner, D. Botstein, S. J. Elledge, and P. O. Brown. Genomic expression responses to DNA-damaging agents and the regulatory role of the yeast ATR homolog Mec1p. *Molecular Biology of the Cell*, 12:2987–3003, 2001.

276. A. P. Gasch, P. T. Spellman, C. M. Kao, O. Carmel-Harel, M. B. Eisen, G. Storz, D. Botstein, and P. O. Brown. Genomic expression programs in the response of yeast cells to environmental changes. *Molecular Biology of the Cell*, 11:4241–4257, 2000.

277. A. Gasch and M. B. Eisen. Exploring the conditional coregulation of yeast gene expression through fuzzy k-means clustering. *Genome Biology*, 3(11):research0059.1–0059.22, 2002.

278. A. C. Gavin, M. Bosche, R. Krause, et al. Functional organization of the yeast proteome by systematic analysis of protein complexes. *Nature*, 415(6868):141–147, 2002.

279. A. J. Gavin, T. E. Scheetz, C. A. Roberts, B. O'Leary, T. A. Braun, V. C. Sheffield, M. B. Soares, J. P. Robinson, and T. L. Casavant. Pooled library tissue tags for est-based gene discovery. *Bioinformatics*, 18(9):1162–1166, 2002.

280. H. Ge, Z. Liu, G. Church, and M. Vidal. Correlation between transcriptome and interactome mapping data from *Saccharomyces cerevisiae*. *Nature Genetics*, 29(4):482–486, 2001.

281. J. Gehrke, V. Ganti, R. Ramakrishnan, and W. Y. Loh. BOAT—optimistic decision tree construction. In *Proceedings of ACM-SIGMOD International Conference on Management of Data*, pages 169–180, 1999.
282. M. S. Gelfand. Prediction of function in DNA sequence analysis. *Journal of Computational Biology*, 2(1):87–115, 1995.
283. M. S. Gelfand, A. A. Mironov, and P. A. Pezner. Gene recognition via spliced sequence alignment. *Proc. Natl. Acad. Sci. USA*, 93:9061–9066, 1996.
284. F. W. Gembicki. *Vector optimization for Control with Performance and Parameter Sensitivity Indices*. PhD thesis, Case Western Reserve University, Cleveland, Ohio, 1974.
285. Gene Ontology Consortium. Creating the gene ontology resource: Design and implementation. *Genome Research*, 11:1425–1433, 2001.
286. H.-H. Gerdes and C. Kaether. Green uorescent protein: Applications in cell biology. *FEBS Letters*, 389:44–47, 1996.
287. D. Ghosh. Status of the transcription factors database. *Nucleic Acids Research*, 21:2091–2093, 1993.
288. S. Ghosh and P. P. Majumder. Mapping a quantitative trait locus via the EM algorithm and Bayesian classification. *Genetic Epidemiology*, 19(2):97–126, 2000.
289. G. J. Gibson and C. F. N. Cowan. On the decision regions of multilayer perceptrons. *Proceedings of the IEEE*, 78(10):1590–1594, 1990.
290. W. Gish and D. J. States. Identification of protein coding regions by database similarity search. *Nature Genetics*, 3(3):266–272, 1993.
291. C. A. Goble, R. Stevens, G. Ng, S. Bechhofer, N. M. Paton, P. G. Baker, M. Peim, and A. Brass. Transparent access to multiple bioinformatics information sources. *IBM Systems Journal*, 40:532–552, 2001.
292. A. Goffeau, R. Aert, M. L. Agostini-Carbone, A. Ahmed, M. Aigle, L. Alberghina, K. Albermann, et al. The yeast genome directory. *Nature*, 387(6632 Supplement):5, 1997.
293. A. Goffeau, B. G. Barrell, H. Bussey, R. W. Davis, B. Dujon, H. Feldmann, F. Galibert, J. D. Hoheisel, C. Jacq, M. Johnston, E. J. Louis, H. W. Mewes, Y. Murakami, P. Philippsen, H. Tettelin, and S. G. Oliver. Life with 6000 genes. *Science*, 274(5287):546–567, 1996.
294. C. S. Goh, A. A. Bogan, M. Joachimiak, et al. Co-evolution of proteins with their interaction partners. *Journal of Molecular Biology*, 299(2):283–293, 2000.
295. D. S. Goldfarb, J. Gariepy, G. Schoolnik, and R. D. Kornberg. Synthetic peptides as nuclear localization signals. *Nature*, 322:641–644, 1986.
296. E. Golemis. *Protein-Protein Interactions: A Molecular Cloning Manual*. Cold Spring Harbor Laboratory Press, 2002.
297. T. R. Golub, D. K. Slonim, P. Tamayo, C. Huard, M. Gaasenbeek, J. P. Misirov, H. Coller, M. L. Loh, J. R. Downing, M. A. Caligiuri, C. D. Bloomfield, and E. S. Lander. Molecular classification of cancer: Class discovery and class prediction by gene expression monitoring. *Science*, 286(15):531–537, 1999.
298. S. M. Gomez and A. Rzhetsky. Towards the prediction of complete protein–protein interaction networks. In *Proceedings of Pacific Symposium on Biocomputing*, pages 413–424, 2002.

299. D. Gorlich. Nuclear protein import. *Current Opinion in Cell Biology*, 9(3):412–419, 1997.

300. O. Gotoh. An improved algorithm for matching biological sequences. *Journal of Molecular Biology*, 162:705–708, 1982.

301. A. Grigoriev. A relationship between gene expression and protein interactions on the proteome scale: Analysis of the bacteriophage T7 and the yeast *Saccharomyces cerevisiae*. *Nucleic Acids Research*, 29(17):3513–3519, 2001.

302. U. Grob and K. Stuber. Recognition of ill-defined signals in nucleic acid sequences. *Computer Applications in the Biosciences*, 4:79–88, 1988.

303. S. Guha, R. Rastogi, and K. Shim. CURE: an efficient clustering algorithm for large databases. In *Proceedings of ACM-SIGMOD International Conference on Management of Data*, pages 73–84, 1998.

304. R. Guigo. Computational gene identification: An open problem. *Computers & Chemistry*, 21(4):215–222, 1997.

305. R. Guigo. Assembling genes from predicted exons in linear time with dynamic programming. *Journal of Computational Biology*, 5:681–702, 1998.

306. R. Guigo, S. Knudsen, N. Drake, and T. Smith. Prediction of gene structure. *Journal of Molecular Biology*, 226:141–157, 1992.

307. A. P. Gultyaev, F. H. D. van Batenburg, and C. W. A. Pleij. The computer simulation of RNA folding pathways using a genetic algorithm. *Journal of Molecular Biology*, 250:37–51, 1995.

308. M. Gunduz, M. Ouchida, K. Fukushima, H. Hanafusa, T. Etani, S. Nishioka, K. Nishizaki, and K. Shimizu. Genomic structure of the human ING1 gene and tumor-specific mutations detected in head and neck squamous cell carcinomas. *Cancer Research*, 60:3143–3146, 2000.

309. D. Gusfield. *Algorithms on Strings, Trees, and Sequences*. Cambridge University Press, 1997.

310. S. P. Gygi, Y. Rochon, B. R. Franza, and R. Aebersold. Correlation between protein and mRNA abundance in yeast. *Molecular and Cellular Biology*, 19(3):1720–1730, 1999.

311. L. M. Haas, P. M. Schwarz, P. Kodali, E. Kotlar, J. E. Rice, and W. C. Swope. DiscoveryLink: A system for integrated access to life sciences data sources. *IBM Systems Journal*, 40(2):489–511, 2001.

312. G. Habeler, K. Natter, G. G. Thallinger, M. E. Crawford, S. D. Kohlwein, and Z. Trajanoski. Ypl.db: The yeast protein localization database. *Nucleic Acids Research*, 30:80–83, 2002.

313. D. H. Haft, B. J. Loftus, D. L. Richardson, F. Yang, J. A. Eisen, I. T. Paulsen, and O. White. TIGRFAMS: A protein family resource for the functional identification of proteins. *Nucleic Acids Research*, 29:41–43, 2001.

314. S. Hahn, S. Buratowski, P. A. Sharp, and L. Guarente. Yeast TATA-binding protein TFIID binds to TATA elements with both consensus and nonconsensus sequences. *Proc. Natl. Acad. Sci. USA*, 86:5718–5722, 1989.

315. J. M. Hall, M. K. Lee, B. Newman, et al. Linkage of early-onset familial breast cancer to chromosome 17q21. *Science*, 250(4988):1684–1689, 1990.

316. M. A. Hall. *Correlation-based feature selection machine learning*. PhD thesis, Department of Computer Science, University of Waikato, New Zealand, 1998.

317. R. Hamming. *Coding and Information Theory*. Prentice Hall, 1982.
318. A. Hamosh, A. F. Scott, J. Amberger, C Bocchini, V. Valle, and V. A. McKusick. Online Mendelian Inheritance in Man (OMIM), a knowledgebase of human genes and genetic disorders. *Nucleic Acids Research*, 30(1):52–55, 2002.
319. J. Han and Y. Fu. Discovery of multiple-level association rules from large databases. In *Proceedings of 21st International Conference on Very Large Data Bases*, pages 420–431, 1995.
320. J. Han, J. Pei, and Y. Yin. Mining frequent patterns without candidates generation. In *Proceedings of ACM-SIGMOD International Conference on Management of Data*, pages 1–12, 2000.
321. S. Hannenhalli and S. Levy. Promoter prediction in the human genome. *Bioinformatics*, 17:S90–S96, 2001.
322. R. Harr, M. Haggstrom, and P. Gustafsson. Search algorithm for pattern match analysis of nucleic acid sequences. *Nucleic Acids Research*, 11:2943–2957, 1983.
323. S. Harroch, G. C. Furtado, W. Brueck, J. Rosenbluth, J. Lafaille, M. Chao, J. D. Buxbaum, and J. Schlessinger. A critical role for the protein tyrosine phosphatase receptor type Z in functional recovery from demyelinating lesions. *Nature Genetics*, 32(3):411–414, 2002.
324. J. A. Hartigan and M. A. Wong. A k-means clustering algorithm. *Applied Statistics*, 28:100–108, 1979.
325. B. Hassibi and D. G. Stork. Second order derivatives for network pruning: Optimal brain surgeon. *Advances in Neural Information Processing Systems*, 5:164–172, 1993.
326. T. Hastie and W. Stuetzle. Principal curves. *Journal of the American Statistical Association*, 84:502–516, 1989.
327. T. Hastie and R. Tibshirani. Discriminant adaptive nearest neighbor classification. *IEEE Transactions on Pattern Analysis and Machine Intelligence*, 18(6):607–616, 1996.
328. T. Hastie, R. Tibshirani, and J. Friedman. *The Elements of Statistical Learning: Data Mining, Inference, and Prediction*. Springer-Verlag, 2001.
329. A. Hatzigeorgiou, T. Harrer, N. Mache, and M. Reczko. The gene sequence analysis system DIANA. In *Bioinformatics: From Nucleic Acids and Proteins to Cell Metabolism*, pages 19–28. Wiley-VCH Verlag, 1995.
330. A. Hatzigeorgiou, N. Mache, and M. Reczko. Functional site recognition of the DNA sequence by artificial neural networks. In *Proceedings of IEEE International Joint Symposia on Intelligence and Systems*, pages 12–17, 1996.
331. A. Hatzigeorgiou, N. Mache, J. Wieland, M. Reczko, and A. Zell. Erkennung von promotoren und kodierenden bereichen in eukaryontischen genomischen sequenzen mit neuronalen netzen. In *Proceedings of Bioinformatik 94*, pages 70–74, 1994.
332. A. Hatzigeorgiou and M. Reczko. Recognition of protein coding regions and reading frames in DNA using neural networks. In *Proceedings of World Congress on Neural Networks*, volume 3, pages 136–138, 1995.
333. A. Hatzigeorgiou and M. Reczko. Gene identification with neural networks. In *Proceedings of Symposium on Control, Optimization, and Supervision*, volume 1, pages 140–143, 1996.

334. A. G. Hatzigeorgiou. *Mathematical models for feature recognition in nucleotide sequences*. Dr. rer. nat. dissertation, University of Jena, Germany, 2000.

335. A. G. Hatzigeorgiou. Translation initiation start prediction in human cDNAs with high accuracy. *Bioinformatics*, 18(2):343–350, 2002.

336. W. Hayes and M. Borodovsky. How to interpret anonymous genome? Machine learning approach to gene identification. *Genome Research*, 8:1154–1171, 1998.

337. T. R. Hazbun and S. Fields. Networking proteins in yeast. *Proc. Natl. Acad. Sci. USA*, 98(8):4277–4278, 2001.

338. R. Hecht-Nielsen. *Neurocomputing*. Addison-Wesley, 1990.

339. D. Heckerman. Bayesian networks for knowledge discovery. In *Advances in Knowledge Discovery and Data Mining*, pages 273–305, 1996. MIT Press.

340. C. Helbing, C. Veilette, K. Riabowol, R. N. Johnston, and I. Garkavetsev. A novel candidate tumor suppressor, ING1, is involved in the regulation of apoptosis. *Cancer Research*, 57:1255–1258, 1997.

341. J. Henderson, S. Salzberg, and K. Fasman. Finding genes in human DNA with a hidden markov model. *Journal of Computational Biology*, 4(2):119–126, 1997.

342. R. Henderson and D. Tweten. Portable batch system: External reference specification. Technical report, NASA Ames Research Center, 1996.

343. J. G. Henikoff, E. A. Greene, S. Pietrovski, and L. Henikoff. Increased coverage of protein families with the Blocks database servers. *Nucleic Acids Research*, 28:228–230, 2000.

344. S. Henikoff. Scores for sequence searches and alignments. *Current Opinion in Structural Biology*, 6:353–360, 1996.

345. S. Henikoff, E. A. Greene, S. Pietrokovski, P. Bork, T. K. Attwood, and L. Hood. Gene families: The taxonomy of protein paralogs and chimeras. *Science*, 278:609–614, 1997.

346. S. Henikoff and J. G. Henikoff. Amino acid substitution matrices from protein blocks. *Proc. Natl. Acad. Sci. USA*, 89(22):10915–10919, 1992.

347. S. Henikoff and J.G. Henikoff. Automated assembly of protein blocks for database searching. *Nucleic Acids Research*, 19:6565–6572, 1991.

348. D. Hennessy, B. Buchanan, D. Subramanian, P. A. Wilkosz, and J. M. Rosenberg. Statistical methods for the objective design of screening procedures for macromolecular crystallization. *Acta Crystallogr. D Biol. Crystallogr.*, 56(7):817–827, 2000.

349. W. Hennig. *Phylogenetic Systematics*. University of Illinois Press, Urbana, Illinois, 1966.

350. J. Hertz, A. Krogh, and R. G. Palmer. *Introduction to the Theory of Neural Computation*. Adison-Wesley, 1991.

351. G. R. Hicks and N. V. Raikhel. Protein import into the nucleus: an integrated view. *Annual Review of Cell and Developmental Biology*, 11:155–188, 1995.

352. D. M. Hillis and J. J. Bull. An empirical test of bootstrapping as a method for assessing confidence in phylogenetic analysis. *Systematic Biology*, 42:182–192, 1993.

353. T. Hirokawa, B.-C. Seah, and S. Mitaku. Sosui: Classification and secondary structure prediction system for membrane proteins. *Bioinformatics*, 14(4):378–379, 1998.

354. D. S. Hirschberg. Algorithms for the longest common subsequence problems. *Journal of the ACM*, 24(4):664–675, 1977.
355. L. Hirschman, J. C. Park, J. Tsujii, L. Wong, and C. H. Wu. Accomplishments and challenges in literature data mining for biology. *Bioinformatics*, 18:1553–1561, 2002.
356. J. D. Hirst and M. J. Sternberg. Prediction of structural and functional features of protein and nucleic acid sequences by artificial neural networks. *Biochemistry*, 31:7211–7218, 1992.
357. H. Hishigaki, K. Nakai, T. Ono, A. Tanigami, and T. Takagi. Assessment of prediction accuracy of protein function from protein-protein interaction data. *Yeast*, 18(6):523–521, 2001.
358. Y. Ho, A. Gruhler, A. Heilbut, et al. Systematic identification of protein complexes in *saccharomyces cerevisiae* by mass spectrometry. *Nature*, 415(6868):180–183, 2002.
359. D. S. Hochbaum. Approximating covering and packing problems. In D. S. Hochbaum, editor, *Approximation Algorithms for NP-hard Problems*, chapter 3, pages 94–143. PWS, 1997.
360. K. M. Hoffmann, N. K. Tonks, and D. Barford. The crystal structure of domain 1 of receptor protein-tyrosine phosphatase mu. *Journal of Biological Chemistry*, 272(44):27505–27508, 1997.
361. K. Hofmann, P. Bucher, L. Falquet, and A. Bairoch. The PROSITE database, its status in 1999. *Nucleic Acids Research*, 27(1):215–219, 1999.
362. M. C. Honeyman, V. Brusic, N. Stone, and L. C. Harrison. Neural network-based prediction of candidate T-cell epitopes. *Nature Biotechnology*, 16(10):966–969, 1998.
363. S. B. Hoot, S. Magallón, and P. R. Crane. Phylogeny of basal eudicots based on three molecular data sets: atpB, *rbcL*, and 18S nuclear ribosomal DNA sequences. *Annals of the Missouri Botanical Garden*, 86:1–32, 1999.
364. S. Horai, K. Hayasaka, R. Kondo, K. Tsuganes, and N. Takahata. Recent African origin of modern humans revealed by complete sequences of hominoid mitochondrial DNAs. *Proc. Natl. Acad. Sci. USA*, 92:532–536, 1995.
365. P. B. Horton and M. Kanehisa. An assessment of neural network and statistical approaches for prediction of *E.coli* promoter sites. *Nucleic Acids Research*, 20:4331–4338, 1992.
366. P. Horton. *String Algorithms and Machine Learning Applications for Computational Biology*. PhD thesis, University of California-Berkeley, Berkeley, CA, 1997.
367. P. Horton and K. Nakai. A probabilitistic classification system for predicting the cellular localization sites of proteins. *Intelligent Systems for Molecular Biology*, 4:109–115, 1996.
368. P. Horton and K. Nakai. Better prediction of protein cellular localization sites with the k nearest neighbours classifier. *Intelligent Systems for Molecular Biology*, 5:147–152, 1997.
369. S. Hua and Z. Sun. Support vector machine approach for protein subcellular localization prediction. *Bioinformatics*, 17(8):721–728, 2001.
370. H. Huang, C. Xiao, and C. H. Wu. ProClass protein family database. *Nucleic Acids Research*, 28:273–276, 2000.
371. X. Huang, M. D. Adams, H. Zhou, and A. R. Kerlavage. A tool for analyzing and annotating genomic sequences. *Genomics*, 46:37–45, 1997.

372. T. Hubbard et al. The Ensembl genome database project. *Nucleic Acids Research*, 30(1):38–41, 2002.

373. T. R. Hughes, M. J. Marton, A. R. Jones, C. J. Roberts, R. Stoughton, C. D. Armour, H. A. Bennett, E. Coffey, H. Dai, Y. D. He, M. J. Kidd, A. M. King, et al. Functional discovery via a compendium of expression profiles. *Cell*, 102:109–126, 2000.

374. W.-K. Huh, J. V. Falvo, L. G. Gerke, A. S. Carroll, R. W. Howson, J. S. Weissman, and E. K. O'Shea. Global analysis of protein localization in budding yeast. *Nature*, 425:686–691, 2003.

375. A. D. Huitema, R. A. Mathot, M. M. Tibben, J. H. Schellens, S. Rodenhuis, and J. H. Beijnen. Validation of techniques for the prediction of carboplatin exposure: application of Bayesian methods. *Clinical Pharmacology & Therapeutics*, 67(6):621–630, 2000.

376. B. Hussain and M. R. Kabuka. A novel feature recognition neural network and its application to character recognition. *IEEE Transactions on Pattern Analysis and Machine Intelligence*, 16:98–106, 1994.

377. G. B. Hutchinson. The prediction of vertebrate promoter regions using different hexamer frequency analysis. *Computer Applications in the Biosciences*, 12:391–398, 1996.

378. G. B. Hutchinson and M. R. Hayden. The prediction of exons through an analysis of spliceable open reading frames. *Nucleic Acids Research*, 20:3453–3462, 1992.

379. J. Hutchinson. *The Families of Flowering Plants*. Clarendon Press, 2nd edition, 1959.

380. T. Ideker, O. Ozier, B. Schwikowski, and A. F. Siegel. Discovering regulatory and signalling circuits in molecular interaction networks. *Bioinformatics*, 18(Suppl. 1):S233–S240, 2002.

381. S. Ieong, M. Y. Kao, T. W. Lam, W. K. Sung, and S. M. Yiu. Predicting RNA secondary structures with arbitrary pseudoknots by maximizing the number of stacking pairs. In *Proceedings of 2nd IEEE International Symposium on Bioinformatics and Bioengineering*, pages 183–190, 2001.

382. P. Indyk and R. Motwani. Approximate nearest neighbors: Towards removing the curse of dimensionality. In *Proceedings of 30th Annual ACM Symposium on Theory of Computing*, pages 604–613, 1998.

383. I. P. Ioshikhes and M. Q. Zhang. Large-scale human promoter mapping using CpG islands. *Nature Genetics*, 26(1):61–63, 2000.

384. C. Iseli, C. V. Jongeneel, and P. Bucher. ESTScan: A program for detecting, evaluating, and reconstructing potential coding regions in EST sequences. *Intelligent Systems for Molecular Biology*, 7:138–148, 1999.

385. M. Ishikawa. A structural learning algorithm with forgetting of link weights. Technical Report TR-90-7, Electrotechnical Laboratories, Tsukuba City, Japan, 1990.

386. T. Ito, T. Chiba, R. Ozawa, M. Yoshida, M. Hattori, and Y. Sakaki. A comprehensive two-hybrid analysis to explore the yeast protein interactome. *Proc. Natl. Acad. Sci. USA*, 98(8):4569–4574, 2001.

387. L. M. Iyer, L. Aravind, P. Bork, K. Hofmann, A. R. Mushegian, I. B. Zhulin, and E. V. Koonin. Quod erat demonstrandum? the mystery of experimental validation of apparently erroneous computational analyses of protein sequences. *Genome Biology*, 2:research0051, 2001.

388. T. Jaakkola, M. Diekhans, and D. Haussler. A discriminative framework for detecting remote protein homologies. *Journal of Computational Biology*, 7((1-2)):95–114, 2000.

389. B. Jagla and J. Schuchhardt. Adaptive encoding neural networks for the recognition of human signal peptide cleavage sites. *Bioinformatics*, 16(3):245–250, 2000.

390. R. Jansen, D. Greenbaum, and M. Gerstein. Relating whole genome expression data with protein-protein interactions. *Genome Research*, 12(1):37–46, 2002.

391. J. W. Jarvik and C. A. Telmer. Epitope tagging. *Annual Review of Genetics*, 32:601–618, 1998.

392. R. Javahery, A. Khachi, K. Lo, B. Zenzie-Gregory, and S. T. Smale. DNA sequence requirements for transcriptional initiator activity in mammalian cells. *Molecular and Cellular Biology*, 14:116–127, 1994.

393. F. V. Jensen. *An Introduction to Bayesian Networks*. Springer-Verlag, 1996.

394. F. V. Jensen. *Bayesian Networks and Decision Graphs*. Springer, 2001.

395. C. Ji, R. R. Snapp, and D. Psaltis. Generalizing smoothness constraints from discrete samples. *Neural Computation*, 2(2):188–197, 1990.

396. Z. Jia, D. Barford, A. J. Flint, and N. K. Tonks. Structural basis for phosphotyrosine peptide recognition by protein tyrosine phosphatase 1B. *Science*, 268:1754–1758, 1995.

397. J. Jiang and H. J. Jacob. EbEST: An automatic tool using expressed sequence tags to delineate gene structure. *Genome Research*, 8(3):268–275, 1998.

398. G. H. John. *Enhancements to the Data Mining Process*. PhD thesis, Stanford University, 1997.

399. I. T. Jolliffe. *Principal Component Analysis*. Springer Verlag, Berlin, 1986.

400. D. T. Jones, W. R. Taylor, and J. M. Thornton. The rapid generation of mutation data matrices from protein sequences. *Computer Applications in the Biosciences*, 8:275–282, 1992.

401. J. Jones, J. K. Field, and J. M. Risk. A comparative guide to gene prediction tools for the bioinformatics amateur. *International Journal of Oncology*, 20:697–705, 2002.

402. N. C. Jones, P. W. J. Rigby, and E. B. Ziff. Trans-acting protein factors and the regulation of eukaryotic transcription: lessons from studies on DNA tumor viruses. *Genes & Development*, 2:267–281, 1988.

403. C. P. Joshi, H. Zhou, X. Huang, and V. L. Chiang. Context sequences of translation initiation codon in plants. *Plant Molecular Biology*, 35:993–1001, 1997.

404. J. Jurka. RepBase Update: A database and an electronic journal of repetitive elements. *Trends in Genetics*, 9:418–420, 2000.

405. G. V. Kaas. An exploratory technique for investigating large quantities of categorical data. *Applied Statistics*, 29:119–127, 1980.

406. K. Kaiser and M. Meisterernst. The human general co-factors. *Trends in Biochemical Sciences*, 21:342–345, 1996.

407. R. A. Kammerer, T. Schulthess, R. Landwehr, A. Lustig, J. Engel, U. Aebi, and M. O. Steinmetz. An autonomous folding unit mediates the assembly of two-stranded coiled coils. *Proc. Natl. Acad. Sci. USA*, 95:13419–13424, 1998.

408. Z. Kan, E. C. Rouchka, W. R. Gish, and D. J. States. Gene structure prediction and alternative splicing analysis using genomically aligned ESTs. *Genome Research*, 115:889–900, 2001.

409. M. D. Kane, T. A. Jatkoe, C. R. Stumpf, J. Lu, J. D. Thomas, and S. J. Madore. Assessment of the sensitivity and specificity of oligonucleotide (50mer) microarrays. *Nucleic Acids Research*, 28:4552–4557, 2000.

410. M. Kanehisa, S. Goto, S. Kawashima, and A. Nakaya. The KEGG database at GenomeNet. *Nucleic Acids Research*, 30(1):42–46, 2002.

411. N. Kaplan, A. Vaaknin, and M. Linial. PANDORA: Keyword-based analysis of protein sets by integration of annotation sources. *Nucleic Acids Research*, 31:5617–5626, 2003.

412. N. B. Karayiannis. Reformulated radial basis neural networks trained by gradient descent. *IEEE Transactions on Neural Networks*, 10(3):657–671, 1999.

413. S. Karlin and S. F. Altschul. Applications and statistics for multiple high-scoring segments in molecular sequences. *Proc. Natl. Acad. Sci. USA*, 90:5873–5877, 1993.

414. S. Karlin and S. F. Altschul. Methods for assessing the statistical significance of molecular sequence features by using general scoring schemes. *Proc. Natl. Acad. Sci. USA*, 87:2264–2268, 1990.

415. E. D. Karnin. A simple procedure for pruning back-propagation trained neural networks. *IEEE Transactions on Neural Networks*, 1(2):239–242, 1990.

416. P. D. Karp. Database links are a foundation for interoperability. *Trends in Biotechnology*, 14:273–279, 1996.

417. P. D. Karp, M. Krummenacker, S. Paley, and J. Wagg. Integrated pathway-genome databases and their role in drug discovery. *Trends in Biotechnology*, 17:275–281, 1999.

418. G. Karypis, E.-H. Han, and V. Kumar. CHAMELEON: A hierarchical clustering algorithm using dynamic modeling. *IEEE COMPUTER*, 32(8):68–75, 1999.

419. J. Kasanov, G. Pirozzi, A. J. Uveges, and B. K. Kay. Characterizing class I WW domains defines key specificity determinants and generates mutant domains with novel specificities. *Chemistry & Biology*, 8(3):231–241, 2001.

420. S. Kasif and A. L. Delcher. Modeling biological data and structure with probabilistic networks. In *Computational Methods in Molecular Biology*, pages 335–352. Elsevier, 1998.

421. L. Kaufman and P. J. Rousseeuw. *Finding Groups in Data: An Introduction to Cluster Analysis*. Wiley, 1990.

422. J. Kaufmann and T. S. Smale. Direct recognition of initiator elements by a component of the transcription factor IID complex. *Genes & Development*, 8:821–829, 1994.

423. J. Kaufmann, C. P. Verrijzer, J. Shao, and S. T. Smale. CIF, an essential cofactor for TFIID-dependent initiator function. *Genes & Development*, 10:873–886, 1996.

424. H. Kawaji, C. Sch'onbach, Y. Matsuo, J. Kawai, Y. Okazaki, Y. Hayashizaki, and H. Matsuda. Exploration of novel motifs derived from mouse cDNA sequences. *Genome Research*, 12:367–378, 2002.

425. U. Keich, M. Li, B. Ma, and J. Tromp. On spaced seeds of similarity search. *Discrete Applied Mathematics*, 2003. To appear.

426. P. Kemmeren, N. L. van Berkum, J. Vilo, et al. Protein interaction verification and functional annotation by integrated analysis of genome-scale data. *Molecular Cell*, 9(5):1133–1143, 2002.

427. W. J. Kent. BLAT—the BLAST-like alignment tool. *Genome Research*, 12(4):656–664, 2002.

428. W. J. Kent and A. M. Zahler. Conservation, regulation, synteny, and introns in a large-scale C. briggsae-C. elegans genomic alignment. *Genome Research*, 10(8):1115–1125, 2000.

429. B. Kerem, J. M. Rommens, J. A. Buchanan, et al. Identification of the cystic fibrosis gene: Genetic analysis. *Science*, 245(4922):1073–1080, 1989.

430. J. Khan, J. S. Wei, M. Ringner, L. H. Saal, M. Ladanyi, F. Westerman, F. Berthold, M. Schwab, C. R. Antonescu, C. Peterson, and P. S. Meltzer. Classification and diagnostic prediction of cancers using gene expression profiling and artificial neural networks. *Nature Medicine*, 7(6):673–679, 2001.

431. S. H. Kim, G. J. Suddath, G. J. Quigley, A. McPherson, J. L. Sussman, A. H. J. Wang, N. C. Seeman, and A. Rich. Three-dimensional tertiary structure of yeast phenylalanine transfer RNA. *Science*, 185:435–439, 1974.

432. R. D. King, M. Saqi, R. Sayle, and M. J. Sternberg. DSC: Public domain protein secondary structure predication. *Computer Applications in the Biosciences*, 13(4):473–474, 1997.

433. H. Kishino, T. Miyata, and M. Hasegawa. Maximum likelihood inference of protein phylogeny and the origin of chloroplasts. *Journal of Molecular Evolution*, 17:368–376, 1990.

434. D. Kisman, B. Ma, and M. Li. tPatternHunter: Gapped, fast, and translated homology search. Manuscript, Bioinformatics Solutions Inc., 2003.

435. J. Kleffe, K. Hermann, W. Vahrson, B. Wittig, and V. Brendel. GeneGenerator—a exible algorithm for gene prediction and its application to maize sequences. *Bioinformatics*, 14(3):232–243, 1998.

436. M. Klemettinen, H. Mannila, P. Ronkainen, H. Toivonen, and A. Verkamo. Finding interesting rules from large sets of discovered association rules. In *Proceedings of 3rd International Conference on Information and Knowledge Management*, pages 401–408, 1994.

437. P. S. Klosterman, M. Tamura, S. R. Holbrook, and S. E. Brenner. SCOR: a structural classification of RNA database. *Nucleic Acids Research*, 30(1):392–394, 2002.

438. A. Klug and J. W. Schwabe. Protein motifs 5: Zinc fingers. *FASEB Journal*, 9(8):597–604, 1995.

439. S. Knudsen. Promoter2.0: For the recognition of Pol II promoter sequences. *Bioinformatics*, 15(5):356–361, 1999.

440. T. Kohonen. Self-organized formation of topologically correct feature maps. *Biological Cybernetics*, 43:59–69, 1982.

441. T. Kohonen. Learning vector quantization for pattern recognition. Technical Report TKK-F-A601, Helsinki University of Technology, 1986.

442. T. Kohonen. *Self-Organization and Associative Memory*. Springer-Verlag, 2nd edition, 1989.

443. T. Kohonen. Improved versions of learning vector quantizations. In *Proceedings of International Joint Conference on Neural Networks*, volume I, pages 545–550, 1990.

444. T. Kohonen. The self-organizing map. *Proceedings of the IEEE*, 78(9):1464–1480, 1990.

445. T. Kohonen. *Self-Organizing Maps*. Springer-Verlag, 2nd edition, 1997.

446. Y. V. Kondrakhin, A. E. Kel, N. A. Kolchanov, A. G. Romashchenko, and L. Milanesi. Eukaryotic promoter recognition by binding sites for transcription factors. *Computer Applications in the Biosciences*, 11:477–488, 1995.

447. D. A. Konings and R. R. Gutell. A comparison of thermodynamic foldings with comparatively derived structures of 16S and 16S-like rRNAs. *RNA*, 1(6):559–574, 1995.

448. R. D. Kornberg. RNA polymerase II transcription control. *Trends in Biochemical Sciences*, 21:325–326, 1996.

449. Z. Kou, W. W. Cohen, and R. F. Murphy. Extracting information from text and image proteomics. In *Proceedings of 3rd ACM SIGKDD Workshop on Data Mining in Bioinformatics*, pages 2–9, 2003.

450. M. Kozak. An analysis of vertebrate mRNA sequences: Intimations of translational control. *The Journal of Cell Biology*, 115:887–903, 1991.

451. M. Kozak. Initiation of translation in prokaryotes and eukaryotes. *Gene*, 234:187–208, 1999.

452. M. Kozak. An analysis of 5'-noncoding sequences from 699 vertebrate messenger RNAs. *Nucleic Acids Research*, 15:8125–8148, 1987.

453. A. Krogh. Gene fi nding: Putting the parts together. In *Guide to Human Genome Computing*, chapter 11, pages 261–274. Academic Press, 2nd edition, 1998.

454. A. Krogh. Two methods for improving performance of an HMM and their application for gene fi nindg. *Intelligent Systems for Molecular Biology*, 5:179–186, 1997.

455. A. Krogh. An introduction to hidden Markov models for biological sequences. In *Computational Methods in Molecular Biology*, pages 45–62. Elsevier, 1998.

456. A. Krogh, M. Brown, I. S. Mian, K. Sjolander, and D. Haussler. Hidden Markov models in computational biology: Applications to protein modeling. *Journal of Molecular Biology*, 235:1501–1531, 1994.

457. A. Krogh, B. Larsson, G. von Heijne, and E. L. L. Sonnhammer. Predicting transmembrane protein topology with a hidden Markov model: Application to complete genomes. *Journal of Molecular Biology*, 305:567–580, 2001.

458. N. X. Krueger, M. Streuli, and H. Saito. Structural diversity and evolution of human receptor-like protein tyrosine phosphatases. *EMBO Journal*, 9(10):3241–3252, 1990.

459. J. K. Kruschke. Creating local and distributed bottlenecks in hidden layers of back-propagation networks. In *Proceedings of 1988 Connectionist Models Summer School*, pages 120–126, 1988.

460. J. K. Kruschke. Improving generalization in back-propagation networks with distributed bottlenecks. In *Proceedings of International Joint Conference on Neural Networks*, volume I, pages 443–447, 1989.

461. M. Kukar, I. Kononenko, and T. Silvester. Machine learning in prognosis of the femoral neck fracture recovery. *Artifi cial Intelligence in Medicine*, 8(5):431–451, 1996.

462. A. Kumar and M. Snyder. Protein complexes take the bait. *Nature*, 415(6868):123–124, 2002.

463. A. Kumar, S. Agarwal, J. A. Heyman, S. Matson, M. Heidtman, S. Piccirillo, L. Umansky, A. Drawid, R. Jansen, Y. Liu, K.-H. Cheung, P. Miller, M. Gerstein, G. S. Roeder, and M. Snyder. Subcellular localization of the yeast proteome. *Genes & Development*, 16:707–719, 2002.

464. S. Kumar, K. Tamura, I. B. Jakobsen, and M. Nei. Mega: Molecular evolutionary genetics analysis, version 2.0, 2000. Published by authors. Pennsylvania State University, University Park; and Arizona State University, Tempe.

465. S. Y. Kung. *Digital Neural Networks*. Prentice-Hall, 1993.

466. J. Kyte and R. F. Doolittle. A simple method for displaying the hydropathic character of a protein. *journal of Molecular Biology*, 157:105–132, 1982.

467. E. S. Lander, L. M. Linton, B. Birren, C. Nusbaum, M. C. Zody, J. Baldwin, K. Devon, K. Dewar, et al. Initial sequencing and analysis of the human genome. *Nature*, 409(6822):861–921, 2001.

468. W. H. Landschulz, P. F. Johnson, and S. L. McKnight. The leucine zipper: A hypothetical structure common to a new class of DNA binding proteins. *Science*, 240:1759–1764, 1988.

469. W. H. Landschulz, P. F. Johnson, and S. L. McKnight. The DNA binding domain of the rat liver nuclear protein C/EBP is bipartite. *Science*, 243:1681–1688, 1989.

470. K. J. Lang and A. H. Waibel. A time-delay neural network architecture for isolated word recognition. *Neural Networks*, 3:23–43, 1990.

471. P. Langley, W. Iba, and K. Thompson. An analysis of Bayesian classifi er. In *Proceedings of 10th National Conference on Artifi cial Intelligence*, pages 223–228. AAAI Press, 1992.

472. P. Langley. *Elements of Machine Learning*. Morgan Kaufmann, 1996.

473. A. S. Lapedes, C. Barnes, C. Burks, R. M. Farber, and K. M. Sirotkin. Applications of neural networks and other machine learning algorithms to DNA sequence analysis. In *Computers and DNA*, pages 157–182. Addison-Wesley, 1989.

474. F. Larsen, G. Gundersen, R. Lopez, and H. Prydz. CpG islands as gene markers in the human genome. *Genomics*, 13:1095–1107, 1992.

475. R. A. Laskowski. PDBsum: Summaries and analyses of PDB structures. *Nucleic Acids Research*, 29:221–222, 2001.

476. D. S. Latchman. *Eukaryotic Transcription Factors*. Academic Press, 1991.

477. R. P. Laura, A. S. Witt, H. A. Held, et al. The Erbin PDZ domain binds with high affi nity and specifi city to the carboxyl termini of delta-catenin and ARVCF. *Journal of Biological Chemistry*, 277(15):12906–12914, 2002.

478. S. L. Lauritzen. The EM algorithm for graphical association models with missing data. *Computational Statistics and Data Analysis*, 19:191–201, 1995.

479. Y. Le Cun, J. S. Denker, and S. A. Solla. Optimal brain damage. *Advances in Neural Information Processing Systems*, II:598–605, 1990.

480. P. Leder and M. Nirenberg. RNA codewords and protein synthesis II: Nucleotide sequence of a valine RNA codeword. *Proc. Natl. Acad. Sci. USA*, 52:420–427, 1964.

481. S. D. Lee, D. W. Cheung, and B. Kao. Is sampling useful in data mining? A case in the maintenance of discovered association rules. *Data Mining and Knowledge Discovery*, 2:233–262, 1998.

482. S.-J. Lee and H.-L. Tsai. Pattern fusion in feature recognition neural networks for handwritten character recognition. *IEEE Transactions on Systems, Man, and Cybernetics*, 28(4):612–625, 1998.

483. P. Legrain, J. Wojcik, and J. M. Gauthier. Protein–protein interaction maps: A lead towards cellular functions. *Trends in Genetics*, 17(6):346–352, 2001.

484. L.-G. Lei, S.-S. Zhang, and Z.-Y. Yu. The karyotype and the evolution of *Gymnotheca*. *Acta Botanica Boreali-Occidentalis Sinica*, 11(6):41–46, 1991.

485. G. G. Lennon, C. Auffray, M. Polymeropoulos, and M. B. Soares. The I.M.A.G.E. Consortium: An integrated molecular analysis of genomes and their expression. *Genomics*, 33:151–152, 1996.

486. H. L. Levin. *Ancient Invertebrates and Their Living Relatives*. Prentice Hall, 1999.

487. M. Levitt and M. Gerstein. A unified statistical framework for sequence comparison and structural comparison. *Proc. Natl. Acad. Sci. USA*, 95:5913–5920, 1998.

488. S. Levy, L. Compagnoni, E. W. Myers, and G. D. Stormo. Xlandscape: The graphical display of word frequencies in sequences. *Bioinformatics*, 14:74–80, 1998.

489. S. Lewis, M. Ashburner, and M. G. Reese. Annotating eukaryote genomes. *Current Opinion in Structural Biology*, 10:349–354, 2000.

490. F. Li and G. D. Stormo. Selection of optimal DNA oligos for gene expression arrays. *Bioinformatics*, 17:1067–1076, 2001.

491. J. Li, G. Dong, and K. Ramamohanarao. DeEPs: Instance-based classification using emerging patterns. In *Proceedings of 4th European Conference on Principles and Practice of Knowledge Discovery in Databases*, pages 191–200, 2000.

492. J. Li, H. Liu, J. R. Downing, A. E.-J. Yeoh, and L. Wong. Simple rules underlying gene expression profiles of more than six subtypes of acute lymphoblastic leukemia (ALL) patients. *Bioinformatics*, 19:71–78, 2003.

493. J. Li, H. Liu, and L. Wong. A comparative study on feature selection and classification methods using a large set of gene expression profiles. In *Proceedings of 13th International Conference on Genome Informatics*, pages 51–60, 2002.

494. J. Li, S.-K. Ng, and L. Wong. Bioinformatics adventures in database research. In *LNCS 2572: Proceedings of 9th International Conference on Database Theory*, pages 31–46, 2003.

495. J. Li, K. R., and G. Dong. The space of jumping emerging patterns and its incremental maintenance algorithms. In *Proceedings of 17th International Conference on Machine Learning*, pages 551–558, 2000.

496. J. Li and L. Wong. Corrigendum: Identifying good diagnostic genes or genes groups from gene expression data by using the concept of emerging patterns. *Bioinformatics*, 18:1407–1408, 2002.

497. J. Li and L. Wong. Geography of differences between two classes of data. In *Proceedings 6th European Conference on Principles of Data Mining and Knowledge Discovery*, pages 325–337, 2002.

498. J. Li and L. Wong. Identifying good diagnostic genes or genes groups from gene expression data by using the concept of emerging patterns. *Bioinformatics*, 18:725–734, 2002.

499. J. Li and L. Wong. Solving the fragmentation problem of decision trees by discovering boundary emerging patterns. In *Proceedings of IEEE International Conference on Data Mining*, pages 653–656, 2002.

500. M. Li, B. Ma, D. Kisman, and J. Tromp. PatternHunter II: Highly sensitive and fast homology search. *Journal of Bioinformatics and Computational Biology*, 2003. To appear.

501. W.-H. Li. *Molecular Evolution*. Sinauer Associates, 1997.

502. H.-X. Liang. Karyomorphology of *Gymnotheca* and phylogeny of four genera in Saururaceae. *Acta Botanica Yunnanica*, 13(3):303–307, 1991.
503. H.-X. Liang. Study on the pollen morphology of Saururaceae. *Acta Botanica Yunnaninca*, 14(4):401–404, 1992.
504. H.-X. Liang. On the systematic significance of oral organogenesis in Saururaceae. *Acta Phytotaxonomica Sinica*, 32(5):425–432, 1994.
505. H.-X. Liang. On the evolution and distribution in Saururaceae. *Acta Botanica Yunnanica*, 17(3):255–267, 1995.
506. H.-X. Liang and S. C. Tucker. Comparative study of the oral vasculature in Saururaceae. *American Journal of Botany*, 77:607–623, 1990.
507. H.-X. Liang and S. C. Tucker. Floral ontogeny of *Zippelia Begoniaefolia* and its familial affinity: Saururaceae or Piperaceae? *American Journal of Botany*, 82(5):681–689, 1995.
508. D.-I. Lin and Z. M. Kedem. Pincer-search: A new algorithm for discovering the maximum frequent set. In *Proceedings of 6th International Conference on Extending Database Technology*, pages 105–119, 1998.
509. K. Lin, A. E. Ting, J. Wang, and L. Wong. Hunting TPR domains using Kleisli. In *Proceedings of 9th International Workshop on Genome Informatics*, pages 173–182, 1998.
510. M. Linial. How incorrect annotation evolved—the case of short ORFs. *Trends in Biotechnology*, 21:298–300, 2003.
511. M. Linial and G. Yona. Methodologies for target selection in structural genomics. *Progress in Biophysics & Molecular Biology*, 73:297–320, 2000.
512. D. Lipman and W. Pearson. Rapid and sensitive protein similarity searches. *Science*, 227:1435–1441, 1985.
513. R. P. Lippmann. An introduction to computing with neural nets. *IEEE ASSP Magazine*, pages 4–22, 1987.
514. H. Liu and R. Setiono. Chi2: Feature selection and discretization of numeric attributes. In *Proceedings of IEEE 7th International Conference on Tools with Artificial Intelligence*, pages 338–391, 1995.
515. H. Liu and L. Wong. Data mining tools for biological sequences. *Journal of Bioinformatics and Computational Biology*, 1(1):139–168, 2003.
516. M. D. Lledó, P. O. Karis, M. B. Crespo, M. F. Fay, and M. W. Chase. Phylogenetic position and taxonomic status of the genus *Aegialitis* and subfamilies Staticoideae and Plumbaginoideae (Plumbaginaceae): Evidence from plastid DNA sequences and morphology. *Plant Systematics and Evolution*, 229:107–124, 2001.
517. L. Lo Conte, S. E. Brenner, T. J. P. Hubbard, C. Chothia, and A. G. Murzin. SCOP database in 2002: Refinements accommodate structural genomics. *Nucleic Acids Research*, 30:264–267, 2002.
518. D. J. Lockhart, H. Dong, M.C. Byrne, M.T. Follettie, M.V. Gallo, M.S. Chee, M. Mittmann, C. Wang, M. Kobayashi, H. Horton, and E.L. Brown. Expression monitoring by hybridization to high-density oligonucleotide arrays. *Nature Biotechnology*, 14(13):1675–1680, 1996.
519. W. Y. Loh and Y. S. Shih. Split selection methods for classification trees. *Statistica Sinica*, 7:815–840, 1997.

520. W. Y. Loh and N. Vanichsetakul. Tree-structured classification via generalized discriminant analysis. *Journal of American Statistical Association*, 83:715–728, 1988.

521. T. A. Longacre, M. H. Chung, D. N. Jensen, and M. R. Hendrickson. Proposed criteria for the diagnosis of well-differentiated endometrial carcinoma. A diagnostic test for myoinvasion. *American Journal of Surgical Pathology*, 19(4):371–406, 1995.

522. A. Loria and T. Pan. Domain structure of the ribozyme from eubacterial ribonuclease P. *RNA*, 2(6):551–563, 1996.

523. I. S. Lossos, R. Breuer, O. Intrator, and A. Lossos. Cerebrospinal uid lactate dehydrogenase isoenzyme analysis for the diagnosis of central nervous system involvement in hematooncologic patients. *Cancer*, 88(7):1599–1604, 2000.

524. B. G. Louis and M. C. Ganoza. Signals determining translational start-site recognition in eukaryotes and their role in prediction of genetic reading frames. *Molecular Biology Reports*, 13:103–115, 1988.

525. T. M. Lowe and S. R. Eddy. tRNAscan-SE: A program for improved detection of transfer RNA genes in genomic sequence. *Nucleic Acids Research*, 25(5):955–964, 1997.

526. T. M. Lowe and S. R. Eddy. A computational screen for methylation guide snoRNAs in yeast. *Science*, 283:1168–1171, 1999.

527. L. Lu, H. Lu, and J. Skolnick. MULTIPROSPECTOR: An algorithm for the prediction of protein-protein interactions by multimeric threading. *Proteins*, 49(3):350–364, 2002.

528. A. V. Lukashin, V. V. Anshelevich, B. R. Amirikyan, A. I. Gragerov, and M. D. Frank-Kamenetskii. Neural network models for promoter recognition. *Journal of Biomolecular Structure & Dynamics*, 6:1123–1133, 1989.

529. A. V. Lukashin and M. Borodovsky. GeneMark.hmm: New solutions for gene finding. *Nucleic Acids Research*, 26(4):1107–1115, 1998.

530. R. B. Lyngso and C. N. S. Pedersen. RNA pseudoknot prediction in energy-based models. *Journal of Computational Biology*, 7(3-4):409–427, 2000.

531. R. B. Lyngso, M. Zuker, and C. N. S. Pedersen. Fast evaluation of internal loops in RNA secondary structure prediction. *Bioinformatics*, 15(6):440–445, 1999.

532. T. J. Lyons, A. P. Gasch, L. A. Gaither, D. Botstein, P. O. Brown, and D. J. Eide. Genome-wide characterization of the Zap1p zinc-responsive regulon in yeast. *Proc. Natl. Acad. Sci. USA*, 97:7957–7962, 2000.

533. B. Ma, J. Tromp, and M. Li. PatternHunter: Faster and more sensitive homology search. *Bioinformatics*, 18(3):440–445, 2002.

534. G. MacBeath and S. L. Schreiber. Printing proteins as microarrays for high-throughput function determination. *Science*, 289(5485):1760–1763, 2000.

535. N. Mache and P. Levi. Detection of eukaryotic POL II promoters with multi-state time-delay neural network. In *IMISE Report No. 1: Proceedings of German Conference on Bioinformatics*, Leipzip, 1996. Institut fuer Medizinische Informatik, Statistik und Epidemilogie.

536. R. Mack and M. Hehenberger. Text-based knowledge discovery: Search and mining of life-sciences documents. *Drug Discovery Today*, 7(11 Supplement):S89–S98, 2002.

537. J. MacQueen. Some methods for classification and analysis of multivariate observations. *Proceedings of 5th Berkeley Symposium on Mathematical Statistics and Probability*, 1:281–297, 1967.

538. H. Mangalam. The Bio* toolkits—a brief overview. *Briefings in Bioinformatics*, 3(3):296–302, 2002.

539. O. L. Mangasarian, W. Nick Street, and W. H. Wolberg. Breast cancer diagnosis and prognosis via linear programming. *Operations Research*, 43(4):570–577, 1995.

540. H. Mannila, H. Toivonen, and A. I. Verkamo. Efficient algorithms for discovering association rules. In *Proceedings of AAAI'94 Workshop on Knowledge Discovery in Databases (KDD'94)*, pages 181–192, 1994.

541. E. M. Marcotte, M. Pellegrini, H.-L. Ng, D. W. Rice, T. O. Yeates, and D. Eisenberg. Detecting protein function and protein-protein interactions from genome sequences. *Science*, 285:751–753, 1999.

542. E. M. Marcotte, M. Pellegrini, M. J. Thompson, T. O. Yeates, and D. Eisenberg. A combined algorithm for genome-wide prediction of protein function. *Nature*, 402:83–86, 1999.

543. E. M. Marcotte, I. Xenarios, and D. Eisenberg. Mining literature for protein-protein interactions. *Bioinformatics*, 17(4):359–363, 2001.

544. E. Marshall and E. Pennisi. NIH launches the final push to sequence the genome. *Science*, 272(5259):188–189, 1996.

545. M. M. Martínez-Ortega and E. Rico. Seed morphology and its systematics significance in some *Veronica* species (Scrophulariaceae) mainly from the Western Mediterranean. *Plant Systematics and Evolution*, 228:15–32, 2001.

546. C. Mathe, M. F. Sagot, T. Schiex, and P. Rouze. Current methods of gene prediction, their strengths and weaknesses. *Nucleic Acids Research*, 30(19):4103–4117, 2002.

547. C. K. Mathews and K. E. Van Holde. *Biochemistry*. Benjamin Cummings, 2nd edition, 1996.

548. D. M. Mathews, J. Sabina, M. Zuker, and D. H. Turner. Expanded sequence dependence of thermodynamic parameters improves prediction of RNA secondary structure. *Journal of Molecular Biology*, 288(5):911–940, 1999.

549. S. Matis, Y. Xu, M. Shah, X. Guan, J. R. Einstein, R. Mural, and E. Uberbacher. Detection of RNA polymerase II promoters and polyadenylation sites in human DNA sequence. *Computers & Chemistry*, 20:135–140, 1996.

550. H. Matsuda. Detection of conserved domains in protein sequences using a maximum-density subgraph algorithm. *IEICE Transactions on Fundamentals of Electronics, Communications, and Computer Sciences*, E83-A:713–721, 2000.

551. H. Matsuda, T. Ishihara, and A. Hashimoto. Classifying molecular sequences using a linkage graph with their pairwise similarities. *Theoretical Computer Science*, 210:305–325, 1999.

552. L. R. Matthews, P. Vaglio, J. Reboul, et al. Identification of potential interaction networks using sequence-based searches for conserved protein-protein interactions or 'interologs". *Genome Research*, 11(12):2120–2126, 2001.

553. V. Matys, E. Fricke, R. Geffers, E. Gossling, M. Haubrock, R. Hehl, K. Hornischer, D. Karas, A. E. Kel, O. V. Kel-Margoulis, D. U. Kloos, S. Land, B. Lewicki-Potapov, H. Michael, R. Munch, et al. TRANSFAC: Transcriptional regulation, from patterns to profiles. *Nucleic Acids Research*, 31(1):374–378, 2003.

554. L. A. McCue, K. A. McDonough, and C. E. Lawrence. Functional classification of cnmp-binding proteins and nucleotide cyclases with implications for novel regulatory pathways in mycobacterium tuberculosis. *Genome Research*, 10(2):204–219, 2000.

555. P. McGarvey, H. Huang, W. C. Barker, B. C. Orcutt, and C. H. Wu. The PIR website: New resource for bioinformatics. *Bioinformatics*, 16:290–291, 2000.

556. D. J. McGeoch. On the predictive recognition of signal peptide sequences. *Virus Research*, 3:271–286, 1985.

557. S. McKnight and R. Tjian. Transcriptional selectivity of viral genes in mammalian cells. *Cell*, 46:795–805, 1986.

558. M. Mehta, R. Agrawal, and J. Rissanen. SLIQ: A fast scalable classifier for data mining. In *Proceedings of International Conference on Extending Database Technology*, pages 18–32, 1996.

559. K. Melén, A. Krogh, and G. von Heijne. Reliability measures for membrane protein topology prediction algorithms. *Journal of Molecular Biology*, 327:735–744, 2003.

560. S.-W. Meng, Z.-D. Chen, D.-Z. Li, and H.-X. Liang. Phylogeny of Saururaceae inferred from matR sequence data. *Acta Botanica Sinica*, 43(6):653–656, 2001.

561. S.-W. Meng and H.-X. Liang. Comparative embryology on Saururaceae. *Acta Botanica Yunnanica*, 19(1):67–74, 1997.

562. K. M. L. Menne, H. Hermjakob, and R. Apweiler. A comparison of signal sequence prediction methods using a test set of signal peptides. *Bioinformatics*, 16:741–742, 2000.

563. H. W. Mewes, D. Frishman, C. Gruber abd B. Geier, D. Haase, A. Kaps, K. Lemcke, G. Mannhaupt, F. Pfeiffer, C. Schuller, S. Stocker, and B. Weil. MIPS: A database for genomes and protein sequences. *Nucleic Acids Research*, 28:37–40, 2000.

564. H. W. Mewes, D. Frishman, U. Guldener, et al. MIPS: A database for genomes and protein sequences. *Nucleic Acids Research*, 30(1):31–34, 2002.

565. L. Milanesi, M. Muselli, and P. Arrigo. Hamming-clustering method for signal prediction in 5' and 3' regions of eukaryotic genes. *Computer Applications in the Biosciences*, 12:399–404, 1996.

566. L. Milanesi and I. Rogozin. Prediction of human gene structure. In *Guide to Human Genome Computing*, pages 215–259. Academic Press, 2nd edition, 1998.

567. R. Milo, S. Shen-Orr, S. Itzkovitz, N. Kashtan, D. Chklovskii, and U. Alon. Network motifs: Simple building blocks of complex networks. *Science*, 298:824–827, 2002.

568. J. M. Mingot, E. A. Espeso, E. Díez, and M. A. Penalva. Ambient ph signaling regulates nuclear localization of the *Aspergillus nidulans* pacc transcription factor. *Molecular and Cellular Biology*, 21(5):1688–1699, 2001.

569. M. Minsky and S. Papert. *Perceptrons*. MIT Press, 1969.

570. P. J. Mitchell and R. Tjian. Transcriptional regulation in mammalian cells by sequence-specific DNA binding proteins. *Science*, 245:371–245, 1998.

571. T. M. Mitchell. *Machine Learning*. McGraw-Hill, 1997.

572. K. Miyahara and F. Yoda. *Printed Japanese Character Recognition Based on Multiple Modified LVQ Neural Network*. IEEE Press, 1996.

573. B. Modrek, A. Resch, C. Grasso, and C. Lee. Genome-wide detection of alternative splicing in expressed sequences of human genes. *Nucleic Acids Research*, 29:2850–2859, 2001.

574. K. V. K. Mohan and C. D. Atreya. Novel organelle-targeting signals in viral proteins. *Bioinformatics*, 19:10–13, 2003.

575. S. M'oller, M. D. R. Croning, and R. Apweiler. Evaluation of methods for the prediction of membrane spanning regions. *Bioinformatics*, 17(7):646–653, 2001.

576. J. Moody and C. Darken. Learning with localized receptive fields. In *Proceedings of 1988 Connectionist Models Summer School*, pages 133–143, 1988.

577. J. Moody and C. Darken. Fast learning in networks of locally-tuned processing units. *Neural Computation*, 1(2):281–294, 1989.

578. D. P. Mortlock, P. Sateesh, and J. W. Innis. Evolution of N-terminal sequences of the vertebrate HOXA13 protein. *Mammalian Genome*, 11:151–158, 2000.

579. R. Mott. Maximum likelihood estimation of the statistical distribution of Smith-Waterman local sequence similarity scores. *Bulletin of Mathematical Biology*, 54:59–75, 1992.

580. J. Moult, K. Fidelis, A. Zemla, and T. Hubbard. Critical assessment of methods of protein structure prediction (CASP): Round IV. *Proteins*, Suppl. 5:2–7, 2001.

581. M. C. Mozer and P. Smolensky. Skeletonization: A technique for trimming the fat from a network via relevance assessment. *Advances in Neural Information Processing*, 1:107–115, 1989.

582. N. J. Mulder, R. Apweiler, T. K. Attwood, A. Bairoch, D. Barrell, A. Bateman, D. Binns, M. Biswas, P. Bradley, P. Bork, P. Bucher, R. R. Copley, E. Courcelle, U. Das, R. Durbin, L. Falquet, W. Fleischmann, S. Griffiths-Jones, D. Haft, N. Harte, N. Hulo, D. Kahn, A. Kanapin, M. Krestyaninova, R. Lopez, I. Letunic, D. Lonsdale, V. Silventoinen, S. E. Orchard, M. Pagni, D. Peyruc, C. P. Ponting, J. D. Selengut, F. Servant, C. J. A. Sigrist, R. Vaughan, and E. M. Zdobnov. The interpro database, 2003 brings increased coverage and new features. *Nucleic Acids Research*, 31:315–318, 2003.

583. A. Muller, R. M. MacCallum, and M. J. Sternberg. Benchmarking PSI-BLAST in genome annotation. *Journal of Molecular Biology*, 293:1257–1271, 1999.

584. M. E. Mulligan and W. R. McClure. Analysis of the occurrence of promoter-sites in DNA. *Nucleic Acids Research*, 14:109–126, 1986.

585. K. Mullis, F. Faloona, S. Scharf, et al. Specific enzymatic amplification of DNA in vitro: The polymerase chain reaction. *Cold Spring Harbour Symposium on Quantitative Biology*, 51(Pt. 1):263–273, 1986.

586. K. B. Mullis. The unusual origin of the polymerase chain reaction. *Scientific American*, 262(4):56–61, 64–5, 1990.

587. K. Murakami and T. Takagi. Gene recognition by combination of several gene-finding programs. *Bioinformatics*, 14(8):665–675, 1998.

588. V. Murino. Structured neural networks for pattern recognition. *IEEE Transactions on Systems, Man, and Cybernetics—Part B: Cybernetics*, 28(4):553–561, 1998.

589. R. F. Murphy, M. V. Boland, and M. Velliste. Towards a systematics for protein subcellular location: Quantitative description of protein localization patterns and automated analysis of uorescence microscope images. *Intelligent Systems for Molecular Biology*, 8:251–259, 2000.

590. J. Murvai, K. Vlahovicek, E. Barta, and S. Pongor. The sbase protein domain library, release 8.0: A collection of annotated protein sequence segments. *Nucleic Acids Research*, 29(1):58–60, 2001.

591. M. T. Musavi, W. Ahmed, K. H. Chan, K. B. Faris, and D. M. Hummels. On the training of radial basis function classifiers. *Neural Networks*, 5(4):595–603, 1992.

592. T. Nagashima, D. Silva, L. Socha, N. Petrovsky, H. Suzuki, R. Saito, T. Kasukawa, I. Kurochkin, A. Konagaya, and C. Sch'onbach. Inferring higher functional information for RIKEN mouse full-length cDNA clones with FACTS. *Genome Research*, 13(6):1520–1533, 2003.

593. R. Nair, P. Carter, and B. Rost. NLSdb: Database of nuclear localization signals. *Nucleic Acids Research*, 31:397–399, 2003.

594. R. Nair and B. Rost. Inferring sub-cellular localization through automated lexical analysis. *Bioinformatics*, 18(Suppl. 1):S78–S86, 2002.

595. R. Nair and B. Rost. Sequence conserved for subcellular localization. *Protein Science*, 11:2836–2847, 2002.

596. R. Nair and B. Rost. LOC3D: Annotate sub-cellular localization for protein structures. *Nucleic Acids Research*, 31:3337–3340, 2003.

597. K. Nakai and P. Horton. PSORT: A program for detecting sorting signals in proteins and predicting their subcellular localization. *Trends in Biochemical Sciences*, 24:34–36, 1999.

598. K. Nakai. Refinement of the prediction methods of signal peptides for the genome analyses of *Saccharomyces cerevisiae* and *Bacillus subtilis*. In *Proceedings of 7th International Workshop on Genome Informatics*, pages 72–81, 1996.

599. K. Nakai. Protein sorting signals and prediction of subcellular localization. *Advances in Protein Chemistry*, 54:277–344, 2000.

600. K. Nakai. Prediction of *in vivo* fates of proteins in the era of genomics and proteomics. *Journal of Structural Biology*, 134:103–116, 2001.

601. K. Nakai and M. Kanehisa. Expert system for predicting protein localization sites in gram-negative bacteria. *PROTEINS: Structure, Function, and Genetics*, 11:95–110, 1991.

602. K. Nakai and M. Kanehisa. A knowledge base for predicting protein localization sites in eukaryotic cells. *Genomics*, 14:897–911, 1992.

603. M. Nakao. Improved accuracy of PSORTII with feature selection and DANN, November 2003. Private Communication.

604. H. Nakashima and K. Nishikawa. Discrimination of intracellular and extracellular proteins using amino acid composition and residue-pair frequences. *Journal of Molecular Biology*, 238:54–61, 1994.

605. K. Nakata, M. Kanehisa, and J. V. Maizel. Discriminant analysis of promoter regions in *Escherichia coli* sequences. *Computer Applications in the Biosciences*, 4:367–371, 1988.

606. Growth of genbank, 2002. Available at www.ncbi.nlm.nih.gov/Genbank/genbankstats.html.

607. S. B. Needleman and C. D. Wunsch. A general method applicable to the search for similarities in the amino acid sequence of two proteins. *Journal of Molecular Biology*, 48:444–453, 1970.

608. S.-K. Ng and M. Wong. Toward routine automatic pathway discovery from on-line scientific text abstracts. *Genome Informatics*, 10:104–112, 1999.

609. S. K. Ng, Z. Zhang, and S. H. Tan. Integrative approach for computationally inferring protein domain interactions. *Bioinformatics*, 19(8):923–929, 2003.

610. W. Nickel. The mystery of nonclassical protein secretion: A current view on cargo proteins and potential export routes. *European Journal of Biochemistry*, 270(10):2109–2119, 2003.

611. H. Nielsen, J. Engelbrecht, S. Brunak, and G. von Heijne. Identification of prokaryotic and eukaryotic signal peptides and prediction of their cleavage sites. *Protein Engineering*, 10:1–6, 1997.

612. H. Nielsen and A. Krogh. Prediction of signal peptides and signal anchors by a hidden Markov model. *Intelligent Systems for Molecular Biology*, 6:122–130, 1998.

613. H. Nielsen. Hot papers in bioinformatics, interview by Eugene Russo. *The Scientist*, 13(13):8, 1999.

614. H. Nielsen. Machine learning approaches for the prediction of signal peptides and other protein sorting signals. *Protein Engineering*, 12(1):3–9, 1999.

615. J. Nilsson, S. Stahl, J. Lundeberg, et al. Affinity fusion strategies for detection, purification, and immobilization of recombinant proteins. *Protein Expression and Purification*, 11(1):1–16, 1997.

616. M. W. Nirenberg and J. H. Matthaei. The dependence of cell-free protein synthesis in *E. coli* upon naturally occurring or synthetic polyribonucleotides. *Proc. Natl. Acad. Sci. USA*, 47:1588–1602, 1961.

617. K. Nishikawa and T. Ooi. Correlation of the amino acid composition of a protein to its structural and biological characters. *Journal of Biochemistry*, 91(5):1821–1824, 1982.

618. T. Nishikawa, T. Ota, and T. Isogai. Prediction whether a human cDNA sequence contains initiation codon by combining statistical information and similarity with protein sequences. *Bioinformatics*, 16:960–967, 2000.

619. K. C. Nixon. Winclada (beta) ver. 0.9.9, 1999. Published by author. Ithaca, NY.

620. N. K. Gray and M. Wickens. Control of translation initiation in animals. *Annual Review of Cells & Developmental Biology*, 14:399–458, 1998.

621. C. D. Novina and A. L. Roy. Core promoters and transcriptional control. *Trends in Genetics*, 9:351–355, 1996.

622. S. J. Nowlan and G. E. Hinton. Simplifying neural networks by soft weight-sharing. *Neural Computation*, 4(4):473–493, 1992.

623. R. Nussinov and A. B. Jacobson. Fast algorithm for predicting the secondary structure of single-stranded RNA. *Proc. Natl. Acad. Sci. USA*, 77(11):6309–6313, 1980.

624. R. Nussinov, J. Owens, and J. V. Maizel. Sequence signals in eukaryotic upstream regions. *Biochimica et Biophysica Acta*, 866:109–119, 1986.

625. N. Ogawa, J. DeRisi, and P. O. Brown. New components of a system for phosphate accumulation and polyphosphate metabolism in *Saccharomyces cerevisiae* revealed by genomic expression analysis. *Molecular Biology of the Cell*, 11:4309–4321, 2000.

626. U. Ohler, S. Harbeck, H. Niemann, E. Noth, and M. G. Reese. Interpolated Markov chains for eukaryotic promoter recognition. *Bioinformatics*, 15(5):362–369, 1999.

627. U. Ohler, G. C. Liao, H. Niemann, and G. M. Rubin. Computational analysis of core promoters in the Drosophila genome. *Genome Biology*, 3(12):RESEARCH0087, 2002.

628. U. Ohler, H. Niemman, G.-C. Liao, and G. M. Rubin. Joint modeling of DNA sequence and physical properties to improve eukaryotic promoter recognition. *Bioinformatics*, 17(Suppl 1):S199–S206, 2001.

629. U. Ohler and M. G. Reese. Detection of eukaryotic promoter regions using stochastic language models. In *Molekulare Bioinformatik*, pages 89–100. Shaker, 1998.

630. U. Ohler, G. Stemmer, S. Harbeck, and H. Niemann. Stochastic segment models of eukaryotic promoter regions. In *Proceedings of Pacifi c Symposium on Biocomputing*, pages 380–391, 2000.

631. H. Okada. Karyomorphology and relationship in some genera of Saururaceae and Piperaceae. *Botanical Magazine Tokyo*, 99:289–299, 1986.

632. T. Okamoto, T. Suzuki, and N. Yamamoto. Microarray fabrication with covalent attachment of DNA using bubble jet technology. *Nature Biotechnology*, 18(4):438–441, 2000.

633. S. Oliver. Guilt-by-association goes global. *Nature*, 403:601–603, 2000.

634. A. B. Olshen and A. N. Jain. Deriving quantitative conclusions from microarray expression data. *Bioinformatics*, 18(7):961–970, 2002.

635. M. V. Olson. The human genome project. *Proc. Natl. Acad. Sci. USA*, 90:4338–4344, 1993.

636. E. M. O'Neill, A. Kaffman, E. R. Jolly, and E. K. O'Shea. Regulation of pho4 nuclear localization by the pho80-pho85 cyclin-cdk complex. *Science*, 271:209–212, 1996.

637. M. C. O'Neill. Training back-propagation neural networks to defi ne and detect DNA-binding sites. *Nucleic Acids Research*, 19(2):313–318, 1991.

638. M. C. O'Neill. Consensus methods for fi nding and ranking DNA binding sites: Application to *Escherichia coli* promoters. *Journal of Molecular Biology*, 207:301–310, 1989.

639. E. K. O'Shea, R. Rutkowski, and P. S. Kim. Evidence that the leucine zipper is a coiled coil. *Science*, 243:538–542, 1989.

640. A. O'Shea-Greenfi eld and S. T. Smale. Roles of TATA and initiator elements in determining the start site location and direction of RNA polymerase II transcription. *Journal of Biological Chemistry*, 267(2):1391–1402, 1992.

641. R. Overbeek, M. Fonstein, M. D'Souza, et al. The use of gene clusters to infer functional coupling. *Proc. Natl. Acad. Sci. USA*, 96(6):2896–2901, 1999.

642. R. Overbeek, N. Larsen, G. D. Pusch, M. D'Souza, E. Selkov Jr., N. Kyrpides, M. Fonstein, N. Maltsev, and E. Selkov. WIT: Integrated system for high-throughput genome sequence analysis and metabolic reconstruction. *Nucleic Acids Research*, 28:123–125, 2000.

643. T. Oyama, K. Kitano, K. Satou, and T. Ito. Extraction of knowledge on protein-protein interaction by association rule discovery. *Bioinformatics*, 18(5):705–714, 2002.

644. S. Pages, A. Belaich, J. P. Belaich, et al. Species-specifi city of the cohesin-dockerin interaction between *clostridium thermocellum* and *clostridium cellulolyticum*: Prediction of specifi city determinants of the dockerin domain. *Proteins*, 29(4):517–527, 1997.

645. C. Papanicolaou, M. Gouy, and J. Ninio. An energy model that predicts the correct folding of both the tRNA and 5S RNA molecules. *Nucleic Acids Research*, 12:31–44, 1984.

646. J. Park, M. Chen, and P. Yu. An effective hash-based algorithm for mining association rules. In *Proceedings of ACM-SIGMOD International Conference on Management of Data*, pages 175–186, 1995.

647. J. Park, S. A. Teichmann, T. Hubbard, and C. Chothia. Intermediate sequences increase the detection of homology between sequences. *Journal of Molecular Biology*, 273:349–354, 1997.

648. K.-J. Park and M. Kanehisa. Prediction of protein subcellular locations by support vector machines using compositions of amino acids and amino acid pairs. *Bioinformatics*, 19(13):1656–1663, 2003.

649. T. Park, S.-G. Yi, S. Lee, S. Y. Lee, D.-H. Yoo, J.-I. Ahn, and Y.-S. Lee. Statistical tests for identifying differentially expressed genes in time-course microarray experiments. *Bioinformatics*, 19(6):694–703, 2003.

650. F. Pazos and A. Valencia. Similarity of phylogenetic trees as indicator of protein-protein interaction. *Protein Engineering*, 14(9):609–614, 2001.

651. F. M. G. Pearl, N. Martin, J. E. Bray, D. W. A. Buchan, A. P. Harrison, D. Lee, G. A. Reeves, A. J. Shepherd, I. Sillitoe, A. E. Todd, J. M. Thornton, and C. A. Orengo. A rapid classification protocol for the CATH domain database to support structural genomics. *Nucleic Acids Research*, 29:223–227, 2001.

652. J. Pearl. *Causality*. Cambridge University Press, 2000.

653. D. A. Pearlman, D. A. Case, J. C. Caldwell, G. L. Seibel, C. Singh, P. Weiner, and P. A. Kollman. *AMBER 4.0*. University of California, San Francisco, 1991.

654. W. R. Pearson. Searching protein sequence libraries: Comparison of the sensitivity and selectivity of the Smith-Waterman and FASTA algorithms. *Genomics*, 11:635–650, 1991.

655. W. R. Pearson and D. J. Lipman. Improved tools for biological sequence comparison. *Proc. Natl. Acad. Sci. USA*, 85:2444–2448, 1988.

656. A. G. Pedersen, P. Baldi, Y. Chauvin, and S. Brunak. The biology of eukaryotic promoter prediction—a review. *Computers & Chemistry*, 23:191–207, 1999.

657. A. G. Pedersen, P. Baldi, S. Brunak, and Y. Chauvin. Characterization of prokaryotic and eukaryotic promoters using hidden Markov models. *Intelligent Systems for Molecular Biology*, 4:182–191, 1996.

658. A. Gorm Pedersen and H. Nielsen. Neural network prediction of translation initiation sites in eukaryotes: Perspectives for EST and genome analysis. *Intelligent Systems for Molecular Biology*, 5:226–233, 1997.

659. H. R. B. Pelham. The retention signal for soluble proteins of the endoplasmic reticulum. *Trends in Biochemical Sciences*, 15:482–486, 1990.

660. M. Pellegrini, E. M. Marcotte, M. J. Thompson, D. Eisenberg, and T. O. Yeates. Assigning protein functions by comparative genome analysis: Protein phylogenetic profiles. *Proc. Natl. Acad. Sci. USA*, 96:4285–4288, 1999.

661. F. E. Penotti. Human DNA TATA boxes and transcription initiation sites. *Journal of Molecular Biology*, 213:37–52, 1990.

662. R. C. Perier, V. Praz, T. Junier, C. Bonnard, and P. Bucher. The eukaryotic promoter database (EPD). *Nucleic Acids Research*, 28:302–303, 2000.

663. C. M. Perou, T. Sorlie, M. B. Eisen, M. van de Rijn, S. S. Jeffrey, C. A. Rees, J. R. Pollack, D. T. Ross, H. Johnsen, L. A. Akslen, et al. Molecular portraits of human breast tumours. *Nature*, 406:747–752, 2000.

664. A. E. Pertiz, R. Kierzek, N. Sugimoto, and D. H. Turner. Thermodynamic study of internal loops in oligoribonucleotides: Symmetric loops are more stable than asymmetric loops. *Biochemistry*, 30:6428–5436, 1991.

665. G. Pesole, S. Liuni, G. Grillo, F. Licculli, A. Larizza, W. Makalowski, and C. Sac-
     cone. UTRdb and UTRsite: Specialized databases of sequences and functional el-
     ements of 5' and 3' untranslated regions of eukaryotic mRNAs. *Nucleic Acids Re-
     search*, 28:193–196, 2000.

666. J. D. Peterson, L. A. Umayam, T. Dickinson, E. K. Hickey, and O. White. The com-
     prehensive microbial resource. *Nucleic Acids Research*, 29:123–125, 2001.

667. J. Platt. A resource-allocating network for function interpolation. *Neural Computa-
     tion*, 3(2):213–225, 1991.

668. J. Platt. Fast training of support vector machines using sequential minimal optimiza-
     tion. In *Advances in Kernel Methods—Support Vector Learning*. MIT Press, 1998.

669. D. C. Plaut, S. J. Nowlan, and G. E. Hinton. Experiments on learning by back propa-
     gation. Technical Report CMU-CS-86-126, Carnegie-Mellon University, Pittsburgh,
     PA 15213, 1986.

670. T. Poggio and F. Girosi. Regularization algorithms for learning that are equivalent to
     multilayer networks. *Science*, 247:978–982, 1990.

671. R. M. Polhill, P. H. Raven, and C. H. Stirton. Evolution and systematics of the Legu-
     minosae. In *Advances in Legume Systematics: Part I.*, pages 1–26. Royal Botanic
     Gardens, Kew, England, 1981.

672. M. H. Polymeropoulos, J. J. Higgins, L. I. Golbe, et al. Mapping of a gene for parkin-
     son's disease to chromosome 4q21-q23. *Science*, 274(5290):1197–1199, 1996.

673. L. Ponger, L. Duret, and Mouchiroud. Determination of CpG islands: Expression in
     early embryo and isochore structure. *Genome Research*, 11:1854–1860, 2001.

674. L. Ponger and D. Mouchiroud. CpGProD: Identifying CpG islands associated with
     transcription start sites in large genomic mammalian sequences. *Bioinformatics*,
     18:631–633, 2002.

675. E. Portugaly, I. Kifer, and M. Linial. Selecting targets for structural determination by
     navigating in a graph of protein families. *Bioinformatics*, 18:899–907, 2002.

676. M. J. D. Powell. Radial basis functions for multivariable interpolation: A review. In
     Algorithms for Approximation. Clarendon Press, 1987.

677. V. Praz, R. Perier, C. Bonnard, and P. Bucher. The eukaryotic promoter database,
     EPD: New entry types and links to gene expression data. *Nucleic Acids Research*,
     30(1):322–324, 2002.

678. D. S. Prestridge. SIGNAL SCAN: A computer program that scans DNA sequences
     for eukaryotic transcriptional elements. *Computer Applications in the Biosciences*,
     7:203–206, 1991.

679. D. S. Prestridge. Predicting Pol II promoter sequences using transcription factor bind-
     ing sites. *Journal of Molecular Biology*, 249:923–932, 1995.

680. D. S. Prestridge. SIGNAL SCAN 4.0: Additional databases and sequence formats.
     Computer Applications in the Biosciences, 12:157–160, 1996.

681. D. S. Prestridge. Computer software for eukaryotic promoter analysis: Review. *Meth-
     ods in Molecular Biology*, 130:265–295, 2000.

682. D. S. Prestridge and C. Burks. The density of transcriptional elements in promoter
     and non-promoter sequences. *Human Molecular Genetics*, 2:1449–1453, 1993.

683. D. S. Prestridge and G. Stormo. SIGNAL SCAN 3.0: New database and program
     features. *Computer Applications in the Biosciences*, 9:113–115, 1993.

684. K.D. Pruitt and D.R. Maglott. RefSeq and LocusLink: NCBI gene-centered resources. *Nucleic Acids Research*, 29:137–140, 2001.

685. M. Ptashne and A. Gann. Transcriptional activation by recruitment. *Nature*, 386:567–577, 1997.

686. C. H. Pui and W. E. Evans. Acute lymphoblastic leukemia. *New England Journal of Medicine*, 339:605–615, 1998.

687. N. Qian and T.J. Sejnowski. Predicting the secondary structure of globular proteins using neural network models. *Journal of Molecular Biology*, 202:865–884, 1988.

688. Y. L. Qiu, J. Lee, F. Bernasconi-Quadroni, D. E. Soltis, P. S. Soltis, M. Zanis, E. A. Zimmer, Z. Chen, V. Savolainen, and M. W. Chase. The earliest angiosperms: evidence from mitochondrial, plastid, and nuclear genomes. *Nature*, 402:404–407, 1999.

689. K. Quandt, K. Frech, H. Karas, E. Wingender, and T. Werner. MatInd and MatInspector: New fast and versatile tools for detection of consensus matches in nucleotide sequence data. *Nucleic Acids Research*, 23:4878–4884, 1995.

690. K. Quandt, K. Grote, and T. Werner. GenomeInspector: A new approach to detect correlation patterns of elements on genomic sequences. *Computer Applications in the Biosciences*, 12:405–413, 1996.

691. K. Quandt, K. Grote, and T. Werner. GenomeInspector: Basic software tools for analysis of spatial correlations between genomic structures within megabase sequences. *Genomics*, 33:301–304, 1996.

692. J. R. Quinlan. Induction of decision trees. *Machine Learning*, 1:81–106, 1986.

693. J. R. Quinlan. *C4.5: Program for Machine Learning*. Morgan Kaufmann, 1993.

694. G. Raddatz, M. Dehio, T.F. Meyer, and C. Dehio. PrimeArray: genome-scale primer design for DNA-microarray construction. *Bioinformatics*, 17:98–99, 2001.

695. J. C. Rain, L. Selig, H. De Reuse, et al. The protein-protein interaction map of *Helicobacter pylori*. *Nature*, 409(6817):211–215, 2001.

696. T. H. Rainer, P. K. Lam, E. M. Wong, and R. A. Cocks. Derivation of a prediction rule for post-traumatic acute lung injury. *Resuscitation*, 42(3):187–196, 1999.

697. S. Rampone. Recognition of splice junctions on DNA sequences by BRAIN learning algorithm. *Bioinformatics*, 14(8):676–684, 1998.

698. G. Ramsay. DNA chips: State-of-the art. *Nature Biotechnology*, 16:40–44, 1998.

699. R. Rastogi and K. Shim. Public: A decision tree classifi er that integrates building and pruning. In *Proceedings of 24th International Conference on Very Large Data Bases*, pages 404–415, 1998.

700. S. Raudys. How good are support vector machines? *Neural Network*, 13(1):17–19, 2000.

701. R. Reed. Pruning algorithms—a survey. *IEEE Transactions on Neural Networks*, 4(5):740–747, 1993.

702. M. G. Reese. *Erkennung von Promotoren in pro- und eukaryontischen DNA-Sequenzen durch K̈unstliche Neuronale Netze*. Diploma work, University of Heidelberg, Germany, 1994.

703. M. G. Reese. Application of a time-delay neural network to promoter annotation in the *Drosophila melanogaster* genome. *Computers & Chemistry*, 26(1):51–56, 2001.

704. M. G. Reese and F. H. Eeckman. Novel neural network prediction system for human promoters and splice sites. In *Proceedings of Workshop on Gene-Finding and Gene Structure Prediction*, 1995.

705. M. G. Reese and F. H. Eeckman. Time-delay neural networks for eukaryotic promoter prediction, 1999. Unpublished.

706. M. G. Reese and F.H. Eeckman. Novel neural network algorithms for improved eukaryotic promoter site recognition. *Genome Science and Technology*, 1(1):45, 1995.

707. M. G. Reese, N. L. Harris, and F. H. Eeckman. Large scale sequencing specifi c neural networks for promoter and splice site recognition. In *Proceedings of Pacifi c Symposium on Biocomputing*, 1996.

708. M. G. Reese, G. Hartzell, N. l. Harris, U. Ohler, J. F. Abril, and S. E. Lewis. Genome annotation assessment in Drosophila melanogaster. *Genome Research*, 10:483–501, 2000.

709. A. Rehm, P. Stern, H. L. Ploegh, and D. Tortorella. Signal peptide cleavage of a type I membrane protein, hcmv us11, is dependent on its membrane anchor. *EMBO Journal*, 20(7):1573–1583, 2001.

710. A. Reinhardt and T. Hubbard. Using neural networks for prediction of the subcellular location of proteins. *Nucleic Acids Research*, 26(9):2230–2236, 1998.

711. J. Rice. *Mathematical Statistics and Data Analysis*. Wadsworth, 1988.

712. M. D. Richard and R. P. Lippmann. Neural networks classifi ers estimate Bayesian *a posteriori* probabilities. *Neural Computation*, 3:461–483, 1991.

713. S. K. Riis and A. Krogh. Improving prediction of protein secondary structure using structured neural networks and multiple sequence alignments. *Journal of Computational Biology*, 3:163–183, 1996.

714. RIKEN Genome Exploration Research Group Phase II Team and FANTOM Consortium. Functional annotation of a full-length mouse cDNA collection. *Nature*, 409:685–690, 2001.

715. E. Rivas and S. Eddy. A dynamic programming algorithm for RNA structure prediction including pseudoknots. *Journal of Molecular Biology*, 285:2053–2068, 1999.

716. J. Robie, J. Lapp, and D. Schach. XML Query Language (XQL). In *Position Papers of QL'98—The Query Languages Workshop*, 1998.

717. R. G. Roeder. The role of general initiation factors in transcription by RNA polymerase II. *Trends in Biochemical Sciences*, 21:327–335, 1996.

718. I. B. Rogozin, A. V. Kochetov, F. A. Kondrashov, E. V. Koonin, and L. Milanesi. Presence of ATG triplets in 5' untranslated regions of eukaryotic cDNAs correlates with a 'weak' context of the start codon. *Bioinformatics*, 17(10):890–900, 2001.

719. D. A. Rosenblueth, D. Thieffry, A. M. Huerta, H. Salgado, and J. Collado-Vides. Syntactic recognition of regulatory regions in *Escherichia coli*. *Computer Applications in the Biosciences*, 12(5):415–422, 1996.

720. C. M. Ross, J. B. Kaplan, M. E. Winkler, and B. P. Nichols. An evolutionary comparison of *Acinetobacter calcoaceticus* trpF with trpF genes of several organisms. *Molecular Biology and Evolution*, 7(1):74–81, 1990.

721. P. Ross-Macdonald, P. S. R. Coelho, T. Roemer, S. Agarwal, A. Kumar, R. Jansen, K.-H. Cheung, A. Sheehan, D. Symoniatis, L. Umansky, M. Heidtman, F. K. Nelson, H. Iwasaki, K. Hager, M. Gerstein, P. Miller, G. S. Roeder, and M. Snyder. Large-scale

analysis of the yeast genome by transposon tagging and gene disruption. *Nature*, 402:413–418, 1999.
722. B. Rost. PHD: Predicting one-dimensional protein structure by profile based neural networks. *Methods in Enzymology*, 266:525–539, 1996.
723. B. Rost. Did evolution leap to create the protein universe? *Current Opinion in Structural Biology*, 12:409–416, 2002.
724. B. Rost, P. Fariselli, and R.Casadio. Topology prediction for helical transmembrane proteins at 86% accuracy. *Protein Science*, 5:1704–1718, 1996.
725. B. Rost and C. Sander. Combining evolutionary information and neural networks to predict protein secondary structure. *Proteins*, 19:55–72, 1994.
726. A. Roy, S. Govil, and R. Miranda. Algorithm to generate radial basis function (RBF)-like nets for classification problems. *Neural Networks*, 8(2):179–201, 1995.
727. M. A. Roytberg, T. V. Astahova, and M. S. Gelfand. Combinatorial approaches to gene recognition. *Computer & Chemistry*, 21(4):229–235, 1997.
728. D. E. Rumelhart, G. E. Hinton, and R. J. Wiliams. Learning internal representations by error propagation. In *Parallel Distributed Processing*, volume 1, pages 318–362. M.I.T. Press, 1986.
729. D. E. Rumelhart, G. E. Hinton, and R. J. Williams. Learning representations by back-propagating errors. *Nature*, 323:533–536, 1986.
730. D. E. Rumelhart, J .L. McClelland, and the PDP Research Group, editors. *Parallel Distributed Processing*, volume 1 & 2. MIT Press, 1986.
731. S. Russell, J. Binder, D. Koller, and K. Kanazawa. Local learning in probabilistic networks with hidden variables. In *Proceedings of 14th Joint International Conference on Artificial Intelligence, volume 2*, pages 1146–1152, 1995.
732. R. Rymon. Search through systematic set enumeration. In *Proceedings of 3rd International Conference on Principles of Knowledge Representation and Reasoning*, pages 539–550, 1992.
733. A. Saito, T. Furukawa, S. Fukushige, S. Koyama, M. Hoshi, Y. Hayashi, and A. J. Horii. p24/ING1-ALT1 and p47/ING1-ALT2, distinct alternative transcripts of p33/ING1. *Human Genetics*, 45:177–181, 2000.
734. R. Saito, H. Suzuki, and Y. Hayashiazaki. Interaction generality, a measurement to assess the reliability of a protein-protein interaction. *Nucleic Acids Research*, 30:1163–1168, 2002.
735. R. Saito, H. Suzuki, and Y. Hayashizaki. Construction of reliable protein-protein interaction networks with a new interaction generality measure. *Bioinformatics*, 19(6):756–763, 2003.
736. H. Sakoe and S. Chiba. Dynamic programming algorithm optimization for spoken word recognition. *IEEE Transactions on Acoustics, Speech and Signal Processing*, 26:43–49, 1987.
737. A. A. Salamov, T. Nishikawa, and M. A. Swindells. Assessing protein coding region integrity in cDNA sequencing projects. *Bioinformatics*, 14:384–390, 1998.
738. S. L. Salzberg. Decision trees and Markov chains for gene finding. In *Computational Methods in Molecular Biology*, pages 187–206. Elsevier, 1998.
739. S. L. Salzberg, A. L. Delcher, K. H. Fasman, and J. Henderson. A decision tree system for finding genes in DNA. *Journal of Computational Biology*, 5(4):667–680, 1998.

740. S. L. Salzberg. On comparing classifiers: Pitfalls to avoid and a recommended approach. *Data Mining and Knowledge Discovery*, 1(3):317–328, 1997.

741. S. L. Salzberg, A. L. Delcher, S. Kasif, and O. White. Microbial gene identification using interpolated Markov models. *Nucleic Acids Research*, 26(2):544–548, 1998.

742. R. Sandy. *Statistics for Business and Economics*. McGrawHill, 1989.

743. D. Sankoff. Simultaneous solution of RNA folding, alignment and protosequence problems. *SIAM Journal on Applied Mathematics*, 45(5):810–825, 1985.

744. V. M. Sarich and A. C. Wilson. Immunological time scale for hominid evolution. *Science*, 158:1200–1203, 1967.

745. O. Sasson, N. Linial, and M. Linial. The metric space of proteins-comparative study of clustering algorithms. *Bioinformatics*, 18(Supplement):S14–S21, 2002.

746. O. Sasson, A. Vaaknin, H. Fleischer, E. Portugaly, Y. Bilu, N. Linial, and M. Linial. ProtoNet: Hierarchical classification of the protein space. *Nucleic Acids Research*, 31:348–352, 2003.

747. K. Satou, G. Shibayama, T. Ono, Y. Yamamura, E. Furuichi, S. Kuhara, and T. Takagi. Finding association rules on heterogeneous genome data. In *Proceedings of Pacific Symposium on Biocomputing*, pages 397–480, 1997.

748. A. Savasere, E. Omiecinski, and S. Navathe. An efficient algorithm for mining association rules in large databases. In *Proceedings of 21st International Conference on Very Large Data Bases*, pages 432–443, 1995.

749. C. Scharfe, P. Zaccaria, K. Hoertnagel, M. Jaksch, T. Klopstock, R. Lilland H. Prokisch, K.-D. Gerbitz, H. W. Mewes, and T. Meitinger. MITOP: Database for mitochondria-related proteins, genes and diseases. *Nucleic Acids Research*, 27:153–155, 1999.

750. T. E. Scheetz, N. Trivedi, C. A. Roberts, T. Kucaba, B. Berger, N. L. Robinson, C. L. Birkett, A .J. Gavin, B. O'Leary, T. A. Braun, M. F. Bonaldo, J. P. Robinson, V. C. Sheffield, M. B. Soares, and T. L. Casavant. ESTprep: Preprocessing cDNA sequence reads. *Bioinformatics*, 19(11):1318–1324, 2003.

751. T. E Scheetz and J. J. Laffin and B. Berger and S. Mackerly and S. A. Baumes, R. Brown II, S. Chang, J. Coco, J. Conklin, K. Crouch, M. Donohue, G. Doonan, C. Estes, M. Eyestone, K. Fishler, J. Gardiner, L. Guo, B. Johnson, C. Keppel, R. Kreger, M. Lebeck, R. Marcelino, V. Miljkovich, M. Perdue, L. Qui, J. Rehmann, R.S. Reiter, B. Rhoads, K. Schaefer, C. Smith, I. Sunjevaric, K. Trout, N. Wu, C. L. Birkett, J. Bischof, B. Gackle, A. Gavin, B. Mokrzycki, C. Moressi, B. OLeary, K. Pedretti, C. Roberts, M. Smith, D. Tack, N. Trivedi, T. Kucaba, T. Freeman, J. Lin, M.F. Bonaldo, T. L. Casavant, V. C. Sheffield, M. B. Soares. High-throughput gene discovery in the rat. *Genome Research*, in press.

752. M. Schena, D. Shalon, R. W. Davis, and P. O. Brown. Quantitative monitoring of gene expression patterns with a complementary DNA microarray. *Science*, 270:467–470, 1995.

753. M. Scherf, A. Klingenhoff, and T. Werner. Highly specific localization of promoter regions in large genomic sequences by PromoterInspector: A novel context analysis approach. *Journal of Molecular Biology*, 297:599–606, 2000.

754. K. Schittowski. NLQPL: A FORTRAN-subroutine solving constrained nonlinear programming problems. *Annals of Operations Research*, 5:485–500, 1985.

755. G. Schneider, S. Sj öling, E. Wallin, P. Wrede, E. Glaser, and G. von Heijne. Feature-extraction from endopeptidase cleavage sites in mitochondrial targeting peptides. *PROTEINS*, 30:49–60, 1998.

756. C. Sch önbach, P. Kowalski-Saunders, and V. Brusic. Data warehousing in molecular biology. *Briefi ngs in Bioinformatics*, 1:190–198, 2000.

757. B. Scholkopf and A. J. Smola. *Learning with Kernels*. MIT Press, 2002.

758. G. D. Schuler, J. A. Epstein, H. Ohkawa, and J. A. Kans. Entrez: Molecular biology database and retrieval system. *Methods in Enzymology*, 266:141–162, 1996.

759. B. Schwikowski, P. Uetz, and S. Fields. A network of protein-protein interactions in yeast. *Nature Biotechnology*, 18:1257–1261, 2000.

760. B. Schwikowski, P. Uetz, and S. Fields. A network of protein-protein interactions in yeast. *Nature Biotechnology*, 18:1257–1261, 2000.

761. J. A. Scott, E. L. Palmer, and A. J. Fischman. How well can radiologists using neural network software diagnose pulmonary embolism? *American Journal of Roentgenology*, 175(2):399–405, 2000.

762. B. E. Segee and M. J. Carter. Fault tolerance of pruned multilayer networks. In *Proceedings of International Joint Conference on Neural Networks*, volume II, pages 447–452, 1991.

763. H. P. Selker, J. L. Griffi th, S. Patil, W. J. Long, and R. B. D'Agostino. A comparison of performance of mathematical predictive methods for medical diagnosis: identifying acute cardiac ischemia among emergency department patients. *Journal of Investigative Medicine*, 43(5):468–476, 1995.

764. J. Selletin and B. Mitschang. Data-intensive intra- & internet applications—Experiences using Java and CORBA in the World Wide Web. In *Proceedings of 14th IEEE International Conference on Data Engineering*, pages 302–311, 1998.

765. F. Servant, C. Bru, S. Carrere, et al. ProDom: Automated clustering of homologous domains. *Briefi ngs in Bioinformatics*, 3(3):246–251, 2002.

766. D. Sha and V. B. Bajić. Adaptive on-line ANN learning algorithm and application to identifi cation of non-linear systems. *Informatica: An International Journal of Computing and Informatics*, 23(4):251–259, 1999.

767. D. Sha and V. B. Bajić. On-line adaptive learning rate BP algorithm for MLP and application to an identifi cation problem. *Journal of Applied Computer Science*, 7(2):67–82, 1999.

768. D. Sha and V. B. Bajić. On-line hybrid learning algorithm for MLP in identifi cation problems. *Computers & Electrical Engineering, An International Journal*, 28(6):587–598, 2002.

769. J. Shafer, R. Agrawal, and M. Mehta. SPRINT: A scalable parallel classifi er for data mining. In *Proceedings of 22nd International Conference on Very Large Data Bases*, pages 544–555, 1996.

770. D. Shalon, S. J. Smith, and P. O. Brown. A DNA microarray system for analyzing complex DNA samples using two-color uorescent probe hybridization. *Genome Research*, 6(7):639–645, 1996.

771. J. Sietsma and R. J. F. Dow. Neural net pruning—why and how. In *Proceedings of IEEE International Conference on Neural Networks*, volume I, pages 325–333, 1988.

772. J. Sietsma and R. J. F. Dow. Creating artifi cial neural networks that generalize. *Neural Networks*, 4(1):67–69, 1991.

773. V. L. Singer, C. R. Wobbe, and K. Struhl. A wide variety of DNA sequences can functionally replace yeast TATA element for transcriptional activation. *Genes & Development*, 4:636–645, 1990.

774. S. T. Smale. Generality of a functional initiator consensus sequence. *Gene*, 182:13–22, 1997.

775. S. T. Smale. Transcription initiation from TATA-less promoters within eukaryotic protein coding genes. *Biochimica et Biophysica Acta*, 1351:73–88, 1997.

776. S. T. Smale and D. Baltimore. The initiator as a transcriptional control element. *Cell*, 57:103–111, 1989.

777. I. Small, H. Wintz, K. Akashi, and H. Mireau. Two birds with one stone: Genes that encode products targeted to two or more compartments. *Plant Molecular Biology*, 38:265–277, 1998.

778. G. P. Smith. Filamentous fusion phage: Novel expression vectors that display cloned antigens on the virion surface. *Science*, 228(4705):1315–1317, 1985.

779. H. O. Smith and K. W. Wilcox. A restriction enzyme from *Hemophilus in uenzae* I: Purification and general properties. *Journal of Molecular Biology*, 51(2):379–391, 1970.

780. T. F. Smith and M. S. Waterman. Identification of common molecular subsequences. *Journal of Molecular Biology*, 147:195–197, 1981.

781. E. E. Snyder and G. D. Stormo. Identification of coding regions in genomic DNA sequences: An application of dynamic programming and neural networks. *Nucleic Acids Research*, 21:607–613, 1993.

782. E. E. Snyder and G. D. Stormo. Identification of protein coding regions in genomic DNA. *Journal of Molecular Biology*, 248:1–18, 1995.

783. L. A. Soinov, M. A. Krestyaninova, and A. Brazma. Towards reconstruction of gene networks from expression data by supervised learning. *Genome Biology*, 4(1):R6.1–9, 2003.

784. V. Solovyev and A. Salamov. The Gene-Finder computer tools for analysis of human and model organisms genome sequences. *Intelligent Systems for Molecular Biology*, 5:294–302, 1997.

785. V. V. Solovyev, A. A. Salamov, and C. B. Lawrence. Predicting internal exons by oligonucleotide composition and discriminant analysis of spliceable open reading frames. *Nucleic Acids Research*, 22:5156–5163, 1994.

786. D. E. Soltis, P. S. Soltis, M. W. Chase, M. E. Mort, D. C. Albach, M. Zanis, V. Savolainen, W. H. Hahn, S. B. Hoot, M. F. Fay, M. Axtell, S. M. Swensen, L. M. Prince, W. J. Kress, K. C. Nixon, and J. S. Farris. Angiosperm phylogeny inferred from 18S rDNA, rbcL, and atpB sequences. *Botanical Journal of the Linnean Society*, 133(4):381–461, 2000.

787. D. E. Soltis, P. S. Soltis, D. L. Nickrent, L. A. Johnson, W. J. Hahn, S. B. Hoot, J. A. Sweere, R. K. Kuzoff, K. A. Kron, M. W. Chase, S. M. Swensen, E. A. Zimmer, S. M. Chaw, L. J. Gillespie, W. J. Kress, and K. J. Sytsma. Angiosperm phylogeny inferred from 18S ribosomal DNA sequences. *Annals of the Missouri Botanical Garden*, 84:1–49, 1997.

788. D. F. Specht. Probabilistic neural networks for classification, mapping or associative memory. In *Proceedings of IEEE International Conference on Neural Networks*, volume 1, pages 525–532, 1988.

789. D. F. Specht. Probabilistic neural networks. *Neural Networks*, 3:109–118, 1990.

790. D. F. Specht. A general regression neural network. *IEEE Transactions on Neural Networks*, 2:568–576, 1991.

791. D. F. Specht. Probabilistic neural networks and general regression neural networks. In *Fuzzy Logic and Neural Network Handbook*, chapter 3, McGraw-Hill, 1996.

792. P. T. Spellman, G. Sherlock, M. Q. Zhang, V. R. Iyer, K. Anders, M. B. Eisen, P. O. Brown, D. Botstein, and B. Futcher. Comprehensive identification of cell cycle-regulated genes of the yeast saccharomyces cerevisiae by microarray hybridization. *Molecular Biology of the Cell*, 9(12):3273–3297, 1998.

793. E. Sprinzak, S. Sattath, and H. Margalit. How reliable are experimental protein-protein interaction data? *Journal of Molecular Biology*, 327(5):919–923, 2003.

794. R. Srikant and R. Agrawal. Mining generalized association rules. In *Proceedings of 21st International Conference on Very Large Data Bases*, pages 407–419, 1995.

795. R. Staden. Computer methods to locate signals in nucleic acid sequences. *Nucleic Acids Research*, 12:505–519, 1984.

796. R. Staden. Methods to define and locate patterns of motifs in sequences. *Computer Applications in the Biosciences*, 4:53–60, 1988.

797. J. E. Stajich, D. Block, K. Boulez, S. E. Brenner, S. A. Chervitz, C. Dagdigian, G. Fuellen, J. G. Gilbert, I. Korf, H. Lapp, H. Lehvaslaiho, C. Matsalla, C. J. Mungall, B. I. Osborne, M. R. Pocock, P. Schattner, M. Senger, L. D. Stein, E. Stupka, M. D. Wilkinson, and E. Birney. The Bioperl toolkit: Perl modules for the life sciences. *Genome Research*, 12(10):1611–1618, 2002.

798. B. J. Stapley, L. A. Kelley, and M. J. E. Sternberg. Predicting the subcellular location of proteins from text using support vector machines. In *Proceedings of Pacific Symposium on Biocomputing*, pages 374–385, 2002.

799. L. A. Stargell and K. Struhl. Mechanisms of transcriptional activation in vivo: Two steps forward. *Trends in Genetics*, 8:311–315, 1996.

800. E. W. Steeg. Neural networks, adaptive optimization, and RNA secondary structure prediction. In *Artificial Intelligence and Molecular Biology*, pages 121–160, 1993.

801. G. D. Stormo. DNA binding sites: Representation and discovery. *Bioinformatics*, 16(1):16–23, 2000.

802. G. D. Stormo. Gene finding approaches for eukaryotes. *Genome Research*, 10:394–397, 2000.

803. G. D. Stormo. Computer methods for analyzing sequence recognition of nucleic acids. *Annual Review of Biophysics and Biophysical Chemstry*, 17:241–63, 1988.

804. G. D. Stormo, T. D. Schneider, and L. M. Gold. Characterization of translational initiation sites in *E. coli*. *Nucleic Acids Research*, 10(9):2971–2996, 1982.

805. G. D. Stormo, T. D. Schneider, L. M. Gold, and A. Ehrenfeucht. Use of the 'perceptron' algorithm to distinguish translational initiation sites in E. coli. *Nucleic Acids Research*, 10:2997–3010, 1982.

806. M. Streuli, N. X. Krueger, T. Thai, M. Tang, and H. Saito. Distinct functional roles of the two intracellular phosphatase-like domains of the receptor-linked protein tyrosine phosphatases LCA and LAR. *EMBO Journal*, 9:2399–2407, 1990.

807. M. Struyvé, M. Moons, and J. Tommassen. Carboxyl-terminal phenylalanine is essential for the correct assembly of a bacterial outer membrane protein. *Journal of Molecular Biology*, 218:141–148, 1991.

808. C. M. Stultz, R. Nambudripad, R. H. Lathrop, and J. V. White. Predicting protein
     structure with probabilistic models. In *Protein Structural Biology in Biomedical Re-
     search*, pages 447–506, 1997.

809. M. Suwa. Fraction of proteins annotatable by sequence similarity, November 2003.
     Private Communication.

810. D. L. Swofford. *PAUP: Phylogenetic Analysis Using Parsimony, Ver. 4.OB8*. Sinauer
     Associates, 2001.

811. S.-H. Sze and P. A. Pevzner. Las Vegas algorithms for gene recognition: Suboptimal
     and error-tolerant spliced alignment. *Journal of Computational Biology*, 4(3):297–
     310, 1997.

812. S.-H. Sze, M. Roytberg, M. Gelfand, A. Mironov, T. Astakhova, and P. Pevzner. Al-
     gorithms and software for support of gene identification experiments. *Bioinformatics*,
     14(1):14–19, 1998.

813. J. E. Tabaska, R. B. Cary, H. N. Gabow, and G. D. Stormo. An RNA folding method
     capable of identifying pseudoknots and base triples. *Bioinformatics*, 14(8):691–699,
     1998.

814. T. Takagi and M. Sugeno. Derivation of fuzzy control rules from human operator's
     control actions. In *Proceedings of IFAC Symposium on Fuzzy Information, Knowl-
     edge Representation, and Decision Analysis*, pages 55–60, 1983.

815. A. Takhtajan. *Diversity and Classification of Flowering Plants*. Columbia University
     Press, 1997.

816. P. Tamayo, D. Slonim, J. Mesirov, Q. Zhu, S. Kitareewan, E. Dmitrovsky, E. S. Lan-
     der, and T. R. Golub. Interpreting patterns of gene expression with self-organizing
     maps: Methods and application to hematopoietic differentiation. *Proc. Natl. Acad.
     Sci. USA*, 96:2907–2912, 1999.

817. C. K. Tang and D. E. Draper. Unusual mRNA pseudoknot structure is recognized by
     a protein translational repressor. *Cell*, 57:531–536, 1989.

818. R. L. Tatusov, A. R. Mushegian, P. Bork, N. R. Brown, W. S. Hayes, M. Borodovski,
     K. E. Rudd, and E. V. Koonin. Metabolism and evolution of *haemophilus in uenzae*
     deduced from a whole genome comparison with *escherichia coli*. *Current Biology*,
     6:279–291, 1996.

819. R. L. Tatusov, D. A. Natale, I. V. Garkavtsev, T. A. Tatusova, U. T. Shankavaram,
     B. S. Rao, B. Kiryutin, M. Y. Galperin, N. D. Fedorova, and E. V. Koonin. The COG
     database: New developments in phylogenetic classification of proteins from complete
     genomes. *Nucleic Acids Research*, 29:22–28, 2001.

820. T. A. Tatusova and T. L. Madden. BLAST 2 Sequences—a new tool for comparing
     protein and nucleotide sequences. *FEMS Microbiology Letters*, 174:247–250, 1999.

821. P. Taubert. Leguminosae. In *Die Naturlichen P anzenfamilien*. W. Engelmann, 1894.

822. N. R. Temkin, J. Holubkov, J. E. Machamer, H. R. Winn, and S. S. Dikmen. Clas-
     sification and regression trees (CART) for prediction of function at 1 year following
     head trauma. *Journal of Neurosurgery*, 82(5):764–771, 1995.

823. M. Thattai and A. van Oudenaarden. Intrinsic noise in gene regulatory networks.
     Proc. Natl. Acad. Sci. USA, 98:8614–8619, 2001.

824. G. Thimm and E. Fiesler. Evaluating pruning methods. In *Proceedings of Interna-
     tional Symposium on Artificial Neural Networks*, volume A2, pages 20–25, 1995.

825. J. D. Thompson, T. J. Gibson, F. Plewniak, F. Jeanmougin, and D. G. Higgins. The CLUSTAL-X windows interface: Flexible strategies for multiple sequence alignment aided by quality analysis tools. *Nucleic Acids Research*, 25(24):4876–4882, 1997.

826. J. D. Thompson, D. G. Higgins, and T. J. Gibson. CLUSTAL W: Improving the sensitivity of progressive multiple sequence alignment through sequence weighting, position-specifi c gap penalties, and weight matrix choice. *Nucleic Acids Research*, 22:4673–4680, 1994.

827. R. Tibshirani, T. Hastie, B. Narasimhan, and G. Chu. Diagnosis of multiple cancer types by shrunken centroids of gene expression. *Proc. Natl. Acad. Sci. USA*, 99(10):6567–6572, 2002.

828. B. Tinland, Z. Koukolikova-Nicola, M. N. Hall, and B. Hohn. The t-DNA-linked vird2 protein contains two distinct functional nuclear localization signals. *Proc. Natl. Acad. Sci. USA*, 89:7442–7446, 1992.

829. I. Tinoco, P. N. Borer, B. Dengler, M. D. Levine, O. C. Uhlenbeck, D. M. Crothers, and J. Gralla. Improved estimation of secondary structure in ribonucleic acids. *Nature New Biology*, 246:40–41, 1973.

830. I. Tinoco, O. C. Uhlenbeck, and M. D. Levine. Estimation of secondary structure in ribonucleic acids. *Nature*, 230:362–367, 1971.

831. S. Tiwari, S. Ramachandran, A. Bhattacharya, S. Bhattacharya, and R. Ramaswamy. Prediction of probable genes by Fourier analysis of genomic sequences. *Computer Applications in the Biosciences*, 13(3):263–270, 1997.

832. H. Toivonen. Sampling large databases for association rules. In *Proceedings of 22rd International Conference on Very Large Data Bases*, pages 134–145, 1996.

833. A. H. Tong, B. Drees, G. Nardelli, et al. A combined experimental and computational strategy to defi ne protein interaction networks for peptide recognition modules. *Science*, 295(5553):321–324, 2002.

834. N. K. Tonks and B. G. Neel. From form to function: Signaling by protein tyrosine phosphatases. *Cell*, 87(3):361–364, 1996.

835. E. N. Trifonov. Inferring context of regulatory sequence elements. *Computer Applications in the Biosciences*, 12:423–429, 1996.

836. N. Trivedi, J. Bischof, S. Davis, K. Pedretti, T. E. Scheetz, T. A. Braun, C. A. Roberts, N. L. Robinson, V. C. Sheffi eld, M. B. Soares, and T. L. Casavant. Parallel creation of non-redundant gene indices from partial mRNA transcripts. *Future Generation Computer Systems*, 18(6):863–870, 2002.

837. O. Troyanskaya, M. Cantor, G. Sherlock, P. Brown, T. Hastie, R. Tibshirani, D. Botstein, and R. B. Altman. Missing value estimation methods for DNA microarrays. *Bioinformatics*, 17:520–525, 2001.

838. A. Tsuji and I. Tamai. Organic anion transporters. *Pharm. Biotechnol.*, 12:471–491, 1999.

839. S. C. Tucker. Floral development in *Saururus cernuus*: 1. Floral initiation and stamen development. *American Journal of Botany*, 62(3):289–301, 1975.

840. S. C. Tucker. In orescence and ower development in the Piperaceae: I. *Peperomia. American Journal of Botany*, 67(5):686–702, 1980.

841. S. C. Tucker. In orescence and oral development in *Houttuynia cordata* (Saururaceae). *American Journal of Botany*, 68(8):1017–1032, 1981.

842. S. C. Tucker. In orescence and ower development in the Piperaceae: II. In orescence development of *Piper*. *American Journal of Botany*, 69(5):743–752, 1982.

843. S. C. Tucker. In orescence and ower development in the Piperaceae: III. In orescence development of *Piper*. *American Journal of Botany*, 69(9):1389–1401, 1982.

844. S. C. Tucker. Initiation and development of in orescence and ower in *Anemopsis californica* (Saururaceae). *American Journal of Botany*, 72(1):20–31, 1985.

845. S. C. Tucker and A. W. Douglas. Floral structure, development, and relationships of paleoherbs: *Saruma, Cabomba, Lactoris*, and selected Piperales. In D. W. Taylor and L. J. Hickey, editors, *Flowering Plants: Origin, Evolution, and Phylogeny*. Chapman & Hall Press, 1996.

846. S. C. Tucker, A. W. Douglas, and H.-X. Liang. Utility of ontogenetic and conventional characters in determining phylogenetic relationships of Saururaceae and Piperaceae (Piperales). *Systematic Botany*, 18(4):414–441, 1993.

847. E. P. Turton, D. J. Scott, M. Delbridge, S. Snowden, and R. C. Kester. Ruptured abdominal aortic aneurysm: A novel method of outcome prediction using neural network technology. *European Journal of Vascular and Endovascular Surgery*, 19(2):184–189, 2000.

848. E. C. Uberbacher and R. J. Mural. Locating protein-coding regions in human DNA sequences by a multiple sensor-neural network approach. *Proc. Natl. Acad. Sci. USA*, 88:11261–11265, 1991.

849. P. Uetz, L. Giot, G. Cagney, T. A. Mansfi eld, R. S. Judson, et al. A comprehensive analysis of protein-protein interactions in Saccharomyces cerevisiae. *Nature*, 403(6770):623–627, 2000.

850. J. D. Ullman. *Principles of Database and Knowledgebase Systems I*. Computer Science Press, 1989.

851. A. Ureta-Vidal, L. Ettwiller, and E. Birney. Comparative genomics: Genome-wide analysis in metazoan eukaryotes. *Nature Reviews Genetics*, 4(4):251–262, 2003.

852. P. E. Utgoff. An incremental ID3. In *Proceedings of 5th International Conference on Machine Learning*, pages 107–120, 1988.

853. A. Valencia and F. Pazos. Computational methods for the prediction of protein interactions. *Current Opinion in Structural Biology*, 12(3):368–373, 2002.

854. K. H. van Wely, J. Swaving, R. Freudl, and A. J. Driessen. Translocation of proteins across the cell envelope of gram-positive bacteria. *FEMS Microbiology Reviews*, 25(4):437–454, 2001.

855. V. N. Vapnik. *The Nature of Statistical Learning Theory*. Springer, 1995.

856. V. N. Vapnik. *Statistical Learning Theory*. Wiley-Interscience, 1998.

857. V. Veljković, I. Ćosić, B. Dimitrijević, and Lalović. Is it possible to analyse DNA and protein sequences by the methodsof digital signal processing? *IEEE Transactions on Biomedical Engineering*, 32(5):337–341, 1985.

858. V. Veljković and I. Slavić. Simple general-model pseudopotential. *Physical Review Letters*, 29(5):105–107, 1972.

859. J. C. Venter, M. D. Adams, E. W. Myers, P. W. Li, R. J. Mural, G. G. Sutton, H. O. Smith, et al. The sequence of the human genome. *Science*, 291(5507):1304–1351, 2001.

860. C. P. Verrijzer and R. Tjian. TAFs mediate transcriptional activation and promoter selectivity. *Trends in Biochemical Sciences*, 21:338–342, 1996.

861. G. von Heijne. Patterns of amino acids near signal-sequence cleavage sites. *European Journal of Biochemistry*, 133:17–21, 1983.

862. G. von Heijne. A new method for predicting signal sequence cleavage sites. *Nucleic Acids Research*, 14:4683–4690, 1986.

863. G. von Heijne. *Sequence Analysis in Molecular Biology: Treasure Trove or Trivial Pursuit*. Academic Press, 1987.

864. C. von Mering, R. Krause, B. Snel, et al. Comparative assessment of large-scale data sets of protein-protein interactions. *Nature*, 417(6887):399–403, 2002.

865. J. L. Vriesema, H. G. van der Poel, F. M. Debruyne, J. A. Schalken, L. P. Kok, and M. E. Boon. Neural network-based digitized cell image diagnosis of bladder wash cytology. *Diagnostic Cytopathology*, 23(3):171–179, 2000.

866. A. Waibel, T. Hanazawa, G. Hinton, K. Shikano, and K. J. Lang. Phoneme recognition using time-delay neural networks. *IEEE Transactions on Acoustics, Speech, and Signal Processing*, 37(3):328–339, 1989.

867. A. Waibel, H. Sawai, and K. Shikano. Modularity and scaling in large phonemic neural networks. *IEEE Transactions on Acoustics, Speech, and Signal Processing*, 37(12):1888–1898, 1989.

868. A. J. Walhout, R. Sordella, X. Lu, et al. Protein interaction mapping in *c. elegans* using proteins involved in vulval development. *Science*, 287(5450):116–122, 2000.

869. M. Walker, V. Pavlovic, and S. Kasif. A comparative genomic method for computational identification of prokaryotic translation initiation sites. *Nucleic Acids Research*, 30(14):3181–3191, 2002.

870. L. Wang. Multi-associative neural networks and their applications to learning and retrieving complex spatio-temporal sequences. *IEEE Transactions on Systems, Man, and Cybernetics—Part B: Cybernetics*, 29(1):73–82, 1999.

871. W. Q. Wang, J. P. Sun, and Z. Y. Zhang. An overview of the protein tyrosine phosphatase superfamily. *Current Topics in Medicinal Chemistry*, 3(7):739–748, 2003.

872. B. Wasylyk. Transcription elements and factors of RNA polymerase B promoters of higher eukaryotes. *Critical Reviews Biochemistry*, 23:77–120, 1988.

873. J. D. Watson. The human genome project: Past, present, and future. *Science*, 248:44–49, 1990.

874. J. D. Watson and R. M. Cook-Deegan. Origins of the human genome project. *FASEB Journal*, 5:8–11, 1991.

875. J. D. Watson and F. H. C. Crick. Molecular structure of nucleic acids: A structure for deoxyribose nucleic acid. *Nature*, 171(4356):737–738, 1953.

876. A. S. Weigend, D. E. Rumelhart, and B. A. Huberman. Back-propagation, weight-elimination, and time series prediction. In D. Touretzky, J. Elman, T. Sejnowski, and G. Hinton, editors, *Proceedings of Connectionist Models Summer School*, pages 105–116, 1990.

877. A. S. Weigend, D. E. Rumelhart, and B. A. Hubermann. Generalization by weight-elimination with application to forecasting. *Advances in Neural Information Processing*, 3:875–882, 1991.

878. R. O. J. Weinzierl. *Mechanism of Gene Expression*. Imperial College Press, 1999.

879. T. Werner. Models for prediction and recognition of eukaryotic promoters. *Mammalian Genome*, 10:168–175, 1999.

880. J. Westbrook, Z. Feng, S. Jain, T. N. Bhat, N. Thanki, V. Ravichandran, G. L. Gilliland, W. Bluhm, H. Weissig, D. S. Greer, P. E. Bourne, and H. E. Berman. The Protein Data Bank: Unifying the archive. *Nucleic Acids Research*, 30(1):245–248, 2002.

881. J. Weston, F. Perez-Cruz, O. Bousquet, O. Chapelle, A. Elisseeff, and B. Scholkopf. Feature selection and tranduction for prediction of molecular bioactivity for drug design. *Bioinformatics*, 19(6):764–771, 2003.

882. D. L. Wheeler, D. M. Church, A. E. Lash, D. D. Leipe, T. L. Madden, J. U. Pontius, G. D. Schuler, L. M. Schriml, T. A. Tatusova, L. Wagner, and B. A. Rapp. Database resources of the national center for biotechnology information: 2002 update. *Nucleic Acids Research*, 30:13–16, 2002.

883. D. Whitley and C. Bogart. The evolution of connectivity: Pruning neural networks using genetic algorithms. In *Proceedings of International Joint Conference on Neural Networks*, volume I, pages 134–137, 1990.

884. B. Widrow and M. A. Lehr. 30 years of adaptive neural networks: Perceptron, madaline, and backpropagation. *Proceedings of the IEEE*, 78(9):1415–1442, 1990.

885. B. Widrow, R. G. Winter, and R. A. Baxter. Layered neural nets for pattern recognition. *IEEE Transactions on Acoustics, Speech, and Signal Processing*, 36:1109–1118, 1988.

886. S. R. Wiley, R. J. Kraus, and J. E. Mertz. Functional binding of TATA box binding component of transcription factor TFIID to the −30 region of TATA-less promoters. *Proc. Natl. Acad. Sci. USA*, 89:5814–5818, 1992.

887. E. Wingender. Transcription regulating proteins and their recognition sequences. *Critical Reviews in Eukaryotic Gene Expression*, 1:11–48, 1990.

888. I. H. Witten and E. Frank. *Data Mining: Practical Machine Learning Tools and Techniques with Java Implementation*. Morgan Kaufmann, 2000.

889. C. R. Wobbe and K. Struhl. Yeast and human TATA-binding proteins have nearly identical DNA sequence requirement for transcription in vitro. *Molecular and Cellular Biology*, 10:3859–3867, 1990.

890. L. Wodicka, H. Dong, M. Mittmann, M.-H. Ho, and D. J. Lockhart. Genome-wide expression monitoring in *saccharomyces cerevisiae*. *Nature Biotechnology*, 15:1359–1367, 1997.

891. C. R. Woese, O. Kandler, and M. L. Wheelis. Towards a natural system of organisms: Proposal for the domains Archaea, Bacteria, and Eucarya. *Proc. Natl. Acad. Sci. USA*, 87:4576–4579, 1990.

892. J. Wojcik and V. Sch¨achter. Protein-protein interaction map inference using interacting domain profile pairs. *Bioinformatics*, 17(Supplement 1):S296–S305, 2001.

893. S. L. Wolfe. *Introduction to Cell and Molecular Biology*. Wadsworth, 1995.

894. L. Wong. Kleisli, a functional query system. *Journal of Functional Programming*, 10(1):19–56, 2000.

895. L. Wong. Kleisli, its exchange format, supporting tools, and an application in protein interaction extraction. In *Proceedings of IEEE International Symposium on Bio-Informatics and Biomedical Engineering*, pages 21–28, 2000.

896. L. Wong. PIES, a protein interaction extraction system. In *Proceedings of Pacific Symposium on Biocomputing*, pages 520–531, 2001.

897. L. Wong. Technologies for integrating biological data. *Briefi ngs in Bioinformatics*, 3(4):389–404, 2002.

898. V. Wood, R. Gwilliam, M.A. Rajandream, M. Lyne, R. Lyne, A. Stewart, J. Sgouros, N. Peat, et al. The genome sequence of *Schizosaccharomyces pombe*. *Nature*, 415(6874):871–880, 2002.

899. R. Wooster, S. L. Neuhausen, J. Mangion, et al. Localization of a breast cancer susceptibility gene, brca2, to chromosome 13q12-13. *Science*, 265(5181):2088–2090, 1994.

900. J. Wootton and S. Federhen. Statistics of local complexity in amino acid sequences and sequence databases. *Computers and Chemistry*, 17:149–163, 1993.

901. C. T. Workman and G. D. Stormo. ANN-Spec: A method for discovering transcription factor binding sites with improved specifi city. In *Proceedings of Pacifi c Symposium on Biocomputing*, pages 112–123, 2000.

902. C. H. Wu, S. Zhao, and H. L. Chen. A protein class database organized with PROSITE protein groups and PIR superfamilies. *Journal of Computational Biology*, 3:547–562, 1996.

903. C. H. Wu, H. Huang, L. Arminski, J. Castro-Alvear, Y. Chen, Z. Hu, R. S. Ledley, K. C. Lewis, H.-W. Mewes, B. C. Orcutt, B. E. Suzek, A. Tsugita, C. R. Vinayaka, L.-S. Yeh, J. Zhang, and W. C. Barker. The Protein Information Resource: An integrated public resource of functional annotation of proteins. *Nucleic Acids Research*, 30(1):35–37, 2002.

904. C. H. Wu, C. Xiao, Z. Hou, Z. Huang, and W. C. Barker. iProClass: An integrated, comprehensive, and annotated protein classifi cation database. *Nucleic Acids Research*, 29:52–54, 2001.

905. K. C. Wu, J. T. Bryan, M. I. Morasso, S. I. Jang, J. H. Lee, J. M. Yang, L. N. Marekov, D. A. Parry, and P. M. Steinert. Coiled-coil trigger motifs in the 1B and 2B rod domain segments are required for the stability of keratin intermediate fi laments. *Molecular Biology of the Cell*, 11:3539–3558, 2000.

906. Z.-Y. Wu. An outline of phytogeography. *The Society of Botanists in Yunnan Province*, 1:44–45, 1984.

907. Z.-Y. Wu, Y.-C. Tang, A.-M. Lu, and Z.-D. Chen. On primary subdivisions of the magnoliophyta—towards a new scheme for an eight-class system of classifi cation. *Acta Phytotaxonomica Sinica*, 36:385–402, 1998.

908. Z.-Y. Wu and W.-C. Wang. A preliminary study on tropical and subtropical ora in Yunnan I. *Acta Phytotaxonomica Sinica*, 6(2):183–254, 1957.

909. Z.-Y. Wu and W.-C. Wang. Some corrections on the paper "A preliminary study on tropical and subtropical ora in Yunnan I". *Acta Phytotaxonomica Sinica*, 7(2):193–196, 1958.

910. I. Xenarios, L. Salwinski, X. J. Duan, P. Higney, S. Kim, and D. Eisenberg. DIP, the Database of Interacting Proteins: A research tool for studying cellular networks of protein interactions. *Nucleic Acids Research*, 30(1):303–305, 2002.

911. D. Xu, G. Li, L. Wu, J. Zhou, and Y. Xu. PRIMEGENS: robust and effi cient design of gene-specifi c probes for microarray analysis. *Bioinformatics*, 18(11):1432–1437, 2002.

912. Y. Xu, R. J. Mural, and E. C. Uberbacher. Inferring gene structures in genomic sequences using pattern recognition and expressed sequence tags. *Intelligent Systems for Molecular Biology*, 5:344–353, 1997.

913. Y. Xu and E. C. Uberbacher. Automated gene identification in large-scale genomic sequences. *Journal of Computational Biology*, 4(3):325–338, 1997.

914. Y. Xu and E. C. Uberbacher. Computational gene prediction using neural networks and similarity search. In *Computational Methods in Molecular Biology*, chapter 7, pages 109–128. Elsevier, 1998.

915. T. Yada, M. Ishikawa, H. Tanaka, and K. Asai. Extraction of hidden Markov model representations of signal patterns in DNA sequences. In *Proceedings of Pacific Symposium on Biocomputing*, pages 686–696, 1996.

916. F. Yang, L. G. Moss, and G. N. Phillips. The molecular structure of green uorescent protein. *Nature Biotechnology*, 14(10):1246, 1996.

917. R. J. Yarger, G. Reese, and T. King. *MySQL & mSQL*. O'Reilly, 1999.

918. E.-J. Yeoh, M. E. Ross, S. A. Shurtleff, W. K. William, D. Patel, R. Mahfouz, F. G. Behm, S. C. Raimondi, M. V. Reilling, A. Patel, C. Cheng, D. Campana, D. Wilkins, X. Zhou, J. Li, H. Liu, C.-H. Pui, W. E. Evans, C. Naeve, L. Wong, and J. R. Downing. Classification, subtype discovery, and prediction of outcome in pediatric acute lymphoblastic leukemia by gene expression profiling. *Cancer Cell*, 1:133–143, 2002.

919. G. Yona, N. Linial, and M. Linial. ProtoMap: Automatic classification of protein sequences and hierarchy of protein families. *Nucleic Acids Research*, 28:49–55, 2000.

920. M. Yoshida, K. Fukuda, and T. Takagi. PNAD-CSS: A workbench for constructing a protein name abbreviation dictionary. *Bioinformatics*, 16:169–175, 2000.

921. D. Yu, S. Chatterjee, G. Sheikholeslami, and A. Zhang. Efficiently detecting arbitrary shaped clusters in very large datasets with high dimensions. Technical Report 98-08, Department of Computer Science and Engineering, State University of New York, Buffalo, New York, November 1998.

922. Z. Yuan. Prediction of protein subcellular locations using Markov chain models. *FEBS Letters*, 451:23–26, 1999.

923. Z. Yuan and R. D. Teasdale. Prediction of golgi type II membrane proteins based on their transmembrane domains. *Bioinformatics*, 18:1109–1115, 2002.

924. B. P. Yuhas, Jr. M. H. Goldstein, T. J. Sejnowski, and R. E. Jenkins. Neural network model of sensory integration for improved vowel recognition. *Proceedings of the IEEE*, 78(10):1658–1668, 1990.

925. M. J. Zaki, S. Parthasarathy, M. Ogihara, and W. Li. New algorithms for fast discovery of association rules. In *Proceedings of 3rd International Conference on Knowledge Discovery and Data Mining*, pages 283–286, 1997.

926. L. Zawel and D. Reinberg. Initiation of transcription by RNA polymerase II: A multistep process. *Progress in Nucleic Acid Research and Molecular Biology*, 44:67–108, 1993.

927. E. M. Zdobnov and R. Apweiler. InterProScan—an integration platform for the signature-recognition methods in InterPro. *Bioinformatics*, 17:847–848, 2001.

928. F. Zeng, R. Yap, and L. Wong. Using feature generation and feature selection for accurate prediction of translation initiation sites. In *Proceedings of 13th International Conference on Genome Informatics*, pages 192–200, 2002.

929. M. Zeremski, J. E. Hill, S. S. Kwek, I. A. Grigorian, K. V. Gurova, I. V. Garkevt-sev, L. Diatchenko, E. V. koonin, and A. V. Gudkov. Structure and regulation of the mouse ING1 gene. Three alternative transcripts encode two PHD finger proteins that have opposite effects on p53 function. *Journal of Biological Chemistry*, 274:32172–32181, 1999.

930. M. Q. Zhang. Computational prediction of eukaryotic protein-coding genes. *Nature Reviews Genetics*, 3(9):698–709, 2002.

931. M. Q. Zhang. Identification of human gene core promoter in silico. *Genome Research*, 8:319–326, 1998.

932. R. Zhang, G. Evans, F. J. Rotella, E. M. Westbrook, D. Beno, E. Huberman, A. Joachimiak, and F. R. Collart. Characteristics and crystal structure of bacterial inosine-5'-monophosphate dehydrogenase. *Biochemistry*, 38:4691–4700, 1999.

933. T. Zhang. Association rules. In *Proceedings of 4th Pacific-Asia Conference on Knowledge Discovery and Data Mining*, pages 245–256, 2000.

934. T. Zhang, R. Ramakrishnan, and M. Livny. BIRCH: an efficient data clustering method for very large databases. In *Proceedings of ACM-SIGMOD International Conference on Management of Data*, pages 103–114, Montreal, Canada, June 1996.

935. X. Zhang, G. Dong, and K. Ramamohanarao. Exploring constraints to efficiently mine emerging patterns from large high-dimensional datasets. In *Proceedings of 6th ACM SIGKDD International Conference on Knowledge Discovery & Data Mining*, pages 310–314, 2000.

936. Z. Y. Zhang. Protein-tyrosine phosphatases: biological function, structural character-istics, and mechanism of catalysis. *Critical Reviews in Biochemistry and Molecular Biology*, 33(1):1–52, 1998.

937. G. Zhou and J. Si. Subset-based training and pruning of sigmoid neural networks. *Neural Networks*, 12(1):80–89, 1999.

938. X. Zhou, M. Cahoon, P. Rosa, and L. Hedstrom. Expression, purification, and char-acterization of inosine 5'-monophosphate dehydrogenase from *Borrelia burgdorferi*. *Journal of Biological Chemistry*, 272:21977–21981, 1997.

939. H. Zhu, M. Bilgin, R. Bangham, et al. Global analysis of protein activities using proteome chips. *Science*, 293(5537):2101–2105, 2001.

940. A. Zien, G. Raatsch, S. Mika, B. Schoelkopf, C. Lemmem, A. Smola, T. Lengauer, and K.R. Mueller. Engineering support vector machine kernels that recognize transla-tion initiation sites. In *Proceedings of German Conference on Bioinformatics*, pages 37–43, 1999.

941. A. Zien, G. Raatsch, S. Mika, B. Schoelkopf, T. Lengauer, and K.R. Mueller. En-gineering support vector machine kernels that recognize translation initiation sites. *Bioinformatics*, 16(9):799–807, 2000.

942. M. Zilversmit, P. O'Grady, and R. Desalle. Shallow gemomics, phylogenetics, and evolution in the family Drosophilidae. In *Proceedings of Pacific Symposium on Bio-computing*, pages 512–523, 2002.

943. M. Zuker. On finding all suboptimal foldings of an RNA molecule. *Science*, 244:48–52, 1989.

944. M. Zuker and P. Stiegler. Optimal computer folding of large RNA sequences us-ing thermodynamics and auxiliary information. *Nucleic Acids Research*, 9:133–148, 1981.

LIST OF CONTRIBUTORS

Ivan V. Bajić
Dept of Electrical & Computer Engineering
University of Miami
1251 Memorial Drive, Coral Gables, FL 33146-0640, USA
Email: ivan_bajic@ieee.org

Vladimir B. Bajić
Institute for Infocomm Research
21 Heng Mui Keng Terrace, Singapore 119613
Email: bajicv@i2r.a-star.edu.sg

Winona C. Barker
Georgetown University Medical Center
National Biomedical Research Foundation
Washington, DC 20007, USA
Email: wb8@georgetown.edu

Daniel Brown
School of Computer Science
University of Waterloo
200 University Ave W., Waterloo, ON N2L 3G1, Canada
Email: browndg@monod.uwaterloo.ca

511

Thomas L. Casavant
Center for Bioinformatics & Computational Biology
University of Iowa
5017 Seamans Center, Iowa City, IA 52242, USA
Email: tomc@eng.uiowa.edu

Paul Horton
Computational Biology Research Center
National Institute of Advanced Industrial Science and Technology
Aomi Frontier Building 17F, 2-43 Aomi
Koutou-ku, Tokyo 135-0064, Japan
Email: horton-p@aist.go.jp

Noam Kaplan
Dept of Biological Chemistry
Alexander Silberman Institute of Life Sciences
Hebrew University of Jerusalem
Givat-Ram, Jerusalem 91904, Israel
Email: kaplann@cc.huji.ac.il

Prasanna Kolatkar
Genome Institute of Singapore
60 Biopolis Street, Genome #02-01, Singapore 138672
Email: kolatkarp@gis.a-star.edu.sg

Jinyan Li
Institute for Infocomm Research
21 Heng Mui Keng Terrace, Singapore 119613
Email: jinyan@i2r.a-star.edu.sg

Ming Li
School of Computer Science
University of Waterloo
200 University Ave W., Waterloo, ON N2L 3G1, Canada
Email: mli@pythagoras.math.uwaterloo.ca

Xinzhong Li
Dept of Computing
Imperial College
180 Queens Gate, London SW7 2AZ, UK
Email: xinzhong@doc.ic.ac.uk

Kui Lin
College of Life Sciences
Beijing Normal University
No. 19, Xinjiekouwai Street, Beijing 100875, P. R. China
Email: linkui@bnu.edu.cn

Michal Linial
Dept of Biological Chemistry
Alexander Silberman Institute of Life Sciences
Hebrew University of Jerusalem
Givat-Ram, Jerusalem 91904, Israel
Email: michall@cc.huji.ac.il

Huiqing Liu
Institute for Infocomm Research
21 Heng Mui Keng Terrace, Singapore 119613
Email: huiqing@i2r.a-star.edu.sg

Jianhua Liu
Genome Institute of Singapore
60 Biopolis Street, Genome #02-01, Singapore 138672
Email: liujh@gis.a-star.edu.sg

Bin Ma
Dept of Computer Science
University of Western Ontario
London, ON N6A 5B7, Canada
Email: bma@uwo.ca

Hideo Matsuda
Dept of Bioinformatic Engineering
Graduate School of Information Science and Technology
Osaka University
1-3 Machikaneyama-cho, Toyonaka, Osaka 560-8531, Japan
Email: matsuda@ist.osaka-u.ac.jp

Shao-Wu Meng
Institute for Infocomm Research
21 Heng Mui Keng Terrace, Singapore 119613
Email: swmeng@i2r.a-star.edu.sg

Lance Miller
Genome Institute of Singapore
60 Biopolis Street, Genome #02-01, Singapore 138672
Email: millerl@gis.a-star.edu.sg

Yuri Mukai
Computational Biology Research Center
National Institute of Advanced Industrial Science and Technology
Aomi Frontier Building 17F, 2-43 Aomi
Koutou-ku, Tokyo 135-0064, Japan
Email: yuri-mukai@aist.go.jp

Kenta Nakai
Human Genome Center
Institute of Medical Science
University of Tokyo
4-6-1 Shirokanedai, Minato-ku, Tokyo 108-8639, Japan
Email: knakai@ims.u-tokyo.ac.jp

See-Kiong Ng
Institute for Infocomm Research
21 Heng Mui Keng Terrace, Singapore 119613
Email: skng@i2r.a-star.edu.sg

Ori Sasson
School of Computer Science and Engineering
Hebrew University of Jerusalem
Givat-Ram, Jerusalem 91904, Israel
Email: ori@cs.huji.ac.il

Todd E. Scheetz
Center for Bioinformatics & Computational Biology
University of Iowa
5017 Seamans Center, Iowa City, IA 52242, USA
Email: tscheetz@eng.uiowa.edu

Christian Schönbach
Biomedical Knowledge Discovery Team
Bioinformatics Group
RIKEN Genomic Sciences Center
Yokohama 230-0045, Japan
Email: schoen@gsc.riken.jp

Wing-Kin Sung
School of Computing
National University of Singapore
3 Science Drive 2, Singapore 117543
Email: ksung@comp.nus.edu.sg

Soon-Heng Tan
Institute for Infocomm Research
21 Heng Mui Keng Terrace, Singapore 119613
Email: soonheng@i2r.a-star.edu.sg

Anthony Tung
School of Computing
National University of Singapore
3 Science Drive 2, Singapore 117543
Email: atung@comp.nus.edu.sg

Haiyan Wang
Max-Planck Institute for Molecular Genetics
Ihnestrasse 73, 14195 Berlin, Germany
Email: whyinsa@yahoo.com

Limsoon Wong
Institute for Infocomm Research
21 Heng Mui Keng Terrace, Singapore 119613
Email: limsoon@i2r.a-star.edu.sg

Cathy H. Wu
Georgetown University Medical Center
National Biomedical Research Foundation
Washington, DC 20007, USA
Email: wuc@georgetown.edu

Roland Yap
School of Computing
National University of Singapore
3 Science Drive 2, Singapore 117543
Email: ryap@comp.nus.edu.sg

www.ingramcontent.com/pod-product-compliance
Lightning Source LLC
Chambersburg PA
CBHW052116230326
41598CB00079B/3707